Collision Theory for Atoms and Molecules

NATO ASI Series

Advanced Science Institutes Series

A series presenting the results of activities sponsored by the NATO Science Committee, which aims at the dissemination of advanced scientific and technological knowledge, with a view to strengthening links between scientific communities.

The series is published by an international board of publishers in conjunction with the NATO Scientific Affairs Division

A	**Life Sciences**	Plenum Publishing Corporation
B	**Physics**	New York and London
C	**Mathematical and Physical Sciences**	Kluwer Academic Publishers Dordrecht, Boston, and London
D	**Behavioral and Social Sciences**	
E	**Applied Sciences**	
F	**Computer and Systems Sciences**	Springer-Verlag
G	**Ecological Sciences**	Berlin, Heidelberg, New York, London,
H	**Cell Biology**	Paris, and Tokyo

Recent Volumes in this Series

Volume 189—Band Structure Engineering in Semiconductor Microstructures
edited by R. A. Abram and M. Jaros

Volume 190—Squeezed and Nonclassical Light
edited by P. Tombesi and E. R. Pike

Volume 191—Surface and Interface Characterization by Electron Optical Methods
edited by A. Howie and U. Valdrè

Volume 192—Noise and Nonlinear Phenomena in Nuclear Systems
edited by J. L. Muñoz-Cobo and F. C. Difilippo

Volume 193—The Liquid State and Its Electrical Properties
edited by E. E. Kunhardt, L. G. Christophorou, and L. H. Luessen

Volume 194—Optical Switching in Low-Dimensional Systems
edited by H. Haug and L. Bányai

Volume 195—Metallization and Metal-Semiconductor Interfaces
edited by I. P. Batra

Volume 196—Collision Theory for Atoms and Molecules
edited by F. A. Gianturco

Series B: Physics

Collision Theory for Atoms and Molecules

Edited by

F. A. Gianturco

The University of Rome
Cittá Universitaria
Rome, Italy

Plenum Press
New York and London
Published in cooperation with NATO Scientific Affairs Division

Proceedings of a NATO Advanced Study Institute on
Collision Theory for Atoms and Molecules,
held September 14-26, 1987,
in Cortona, Italy

Library of Congress Cataloging in Publication Data

NATO Advanced Study Institute on Collision Theory for Atoms and Molecules (1987: Cortona, Italy)
Collision theory for atoms and molecules / edited by F. A. Gianturco.
p. cm.—(NATO ASI series. Series B, Physics; v. 196)
"Proceedings of a NATO Advanced Study Institute on Collision Theory for Atoms and Molecules, held September 14-26, 1987, in Cortona, Italy"—T.p. verso.
Bibliography: p.
Includes index.
ISBN 0-306-43202-1
1. Collisions (Nuclear physics)—Congresses. 2. Electron-molecule scattering—Congresses. 3. Photons—Scattering—Congresses. I. Gianturco, Francesco A., date. II. Title. III. Series.
QC794.6.C6N39 1987 89-30388
539.7′54—dc19 CIP

A Division of Plenum Publishing Corporation
233 Spring Street, New York, N.Y. 10013

Printed in the United States of America

PREFACE

The NATO-Advanced Study Institute on "Collision Theory for Atoms and Molecules" was made possible by the main sponsorship and the generous financial support of the NATO Scientific Affairs Division in Brussels, Belgium. Special thanks are therefore due to the late Dr. Mario Di Lullo and to Dr. Craig Sinclair, of this Division, who repeatedly advised us and kept us aware of administrative requirements. The Institute was also assisted by the financial aid from the Scientific Committees for Chemistry and Physics of the Italian National Research Council (CNR).

The search and selection of a suitable location, one which participants would easily reach from any of Italy's main airports, was ably aided by the Personnel of the Scuola Normale Superiore of Pisa and made possible by its Directorship. Our thanks therefore go to its present director, Prof. L. Radicati, and to its past director, Prof. E. Vesentini who first agreed to our use of their main building in Pisa and of their palatial facilities at the "Palazzone" in Cortona.

The local administration in the town of Cortona was once more very generous with their time and their share of concerts and exhibits that were made available to us during the weeks of the Advanced Study Institute. It is therefore a pleasure for us to thank Mr. Favilli and his collaborators. The hospitality of the whole staff at the Residence "Neumann" and at the "Palazzone" itself was also warm and attentive to the needs of a rather large group of hungry and thirsty cosmopolitan scientists. We remember with gratitude all of them.

My special and warm thanks also go to my organizers in loco, Dr. G. Stefani and Dr. M. Venanzi who never failed to willingly accept all the unexpected and numerous chores created by the entire correspondence and reservation procedures for many months before the ASI. Dr. Venanzi and Ms Filomena Reveruzzi must also be thanked for their ubiquitous presence during the long opening hours of our Institute Office at the "Palazzone", and for their generous disposition in helping all participants as much as possible.

Finally, last but not least, I am most grateful to Tuscany, its beautiful landscape and its fine weather, which encouraged everyone to relax, thus allowing the surroundings to work their magic for stimulating and inspiring the participants.

The preparation of this book was aided in the last stages by the professional competence and cooperation of Ms Patrizia Michetti.

Franco A. Gianturco
ASI Director
Dept. of Chemistry
University of Rome, Italy

CONTENTS

NON-ADIABATIC SCATTERING PROCESSES

ADIABATIC PROCESSES IN MOLECULAR SCATTERING

INTERNATIONAL SCIENTIFIC COMMITTEE

Prof. P.G. Burke
Dept. of Applied Mathematics
and Theoretical Physics
The Queen's University
Belfast, Northhern Ireland (U.K.)

Prof. F.A. Gianturco
Dept. of Chemistry
University of Rome
Città Universitaria
Rome (I)

Prof. C.J. Joachain
Physique Théorique
Université Libre de Bruxelles, Bruxelles
and
Institut de Physique Corpisculaire
Université de Louvain, Louvain-La-Neuve
Belgium

INTRODUCTION

The present volume is a more formal and organized presentation of a series of lectures given by various scientists during a Nato Advanced Study Institute on the same subject which was held in Cortona, Italy, during the month of September 1987. The aim of the collection is not simply to present the latest scientific results in a limited area of atomic and molecular physics but rather to provide a broad audience of researchers, those either approaching the field for the first time or those wanting to enlarge their expertise in it, with individual and self-consistent treatments for a wide variety of phenomena which entail collisional events between atoms, molecules and electrons, photons or ions. Therefore, while it appears as a result of the ASI, it also hopes to extend its relevance and validity even further as a useful guide for graduate schools and research groups.

It is a well known fact that all energy processes are eventually attributable to reactions which occur on a molecular, atomic or nuclear scale. Although we have been able to exploit such processes even if our understanding of them has been incomplete, the usual empirical approach cannot be used as a basis for the current quest of better and more efficient, economic and self-sufficient energy sources. It has therefore become clear that pure and applied science has much to offer in the way of rational approaches to the various aspects of energy storing, exchanging and dissipation processes that occur in atomic and molecular gases, either neutral or ionized. The role that atomic and molecular science is likely to play in gas-phase solar energy conversion, just to cite one example, will include supplying basic understanding of several competing photoexcitation processes and of collision processes that involve excited states and energy-transfer encounters. The interaction of solar photons with all the constituents of the earth's atmosphere, including pollutants, is also within the purview of atomic and molecular science. Moreover, the whole area of gaseous electronics, i.e. the study of gaseous systems whose electrical behaviour can be varied by external means, requires a detailed knowledge at a microscopic level of several collisional processes that have a very direct bearing on the macroscopic features of power switches, fluorescent lamps and chemical transformation in the presence of electrical currents.

The various Parts and Chapters of the present volume therefore try to present under the same roof a broad variety of microscopic collisional processes that are often treated in isolation, or that apparently have very little to do with each other. There is a unifing element, however, and this is offered by the theoretical tools, of either classical mechanics or quantum mechanics, and by the specific ways in which the structural properties of the various atomic and molecular systems are probed by the collisional events which are being studied.

Thus, the three different Parts of the volume provide a broad grouping of the topics discussed; moreover, each one focuses on a specific aspect of the scattering process. Part I deals with the dynamical theories in which photons and electrons are used as projectiles against atoms and molecules, the latter targets providing the additional structural features that diversify the outcoming types of cross sections. Thus, molecular photoionisation, electronic excitation and internal energy storage or dissipation are the results of a unified theoretical treatment prosented by P.G. Burke in the first Chapter of Part I, while the same array of dynamical observables and an even more detailed discussion of the seemingly simpler processes with atomic targets are discussed in the following Chapter by C.J. Joachain. The presence of an intense external field obviously modifies the scattering and complicates the theoretical picture, as is shown by the discussion on atomic systems carried out by M. Gavrila in Chapter III.

The use and implementation of theoretical models is greatly conditioned by the availability of reliable and detailed experiments, this being particularly true for the strongly quantal conditions under which photons and electrons exchange energy and momentum with atomic and molecular systems. Thus, Chapters IV, V and VI discuss three different classes of experiments that have greatly helped, in recent years, to elucidate the nature of the specific forces at play and of the structural features of the target systems. G. Stefani (Chpt. IV) describes the role of coincidence measurements of secondary electrons, under different conditions in probing excited atomic states, while J. Kessler (Chpt. V) uses polarized electrons to get from the measurements additional features of the dynamics that would not be oberved otherwise. The complications of additional degrees of freedom in the target molecules are clearly shown by the experiments described by M. Tronc (Chpt. VI). The theoretical treatment of photodissociation in triatomic systems is discussed in detail in Chapter VII by R. Schinke.

Part II fo the volume addresses primarily the new set of theoretical complications presented by the existence of several electronic states that are brought into being by the crossed perturbations between either two atomic partners or a molecule and an atom as collisional partners. The dynamical event is in turn the prime cause for these new states to be coupled together, and therefore one often needs to decide, depending on the relative collision energies or the specific cross sections involved, whether or not classical treatments are acceptable. B.H. Bransden analyses the case of two interacting atoms in the first Chapter of this Part, while in Chapter II V. Sidis examines the different frameworks under which molecular systems could be more efficiently analysed. Here too, experimental evidence is very usefully discussed vis à vis previous theoretical treatments and Chapter III thus summarizes a broad class of proton-molecule experiments where the interplay between adiabatic and non-adiabatic effects can be clearly traced along a series of different molecular systems (presented by M. Noll and P. Toennies in Capter IV). The following experimental section deals with another class of non-adiabatic collisions which involves metastable atoms and where the dynamical processes are directly responsible for coupling together the various electronic states that are available for each specific system (B. Brunetti and F. Vecchiocattivi in Chpt. V).

The third Part of the volume deals with collisional processes with molecular partners where the prime theoretical aim is to provide some simple and yet reliable rule for discussing the enormous manifold of internal levels involved in energy exchange, deposition and storage processes of even the simplest molecular gases. Chapter I of this Part (by F.A. Gianturco) discusses mainly rotationally inelastic events in atom-molecule collisions where only neutral partners are concerned. The attempts at producing general laws and scaling procedures are reviewed in detail and the most currently used methods to reduce computational efforts are analysed and compared.

Finally, the last Chapter (by W. Reinhardt) discusses a new way in which important results of classical mechanics and specific relations among classical trajectories can be cast into different forms thereby aiding our understanding of the behaviour of quantal systems. Such an approach can often unrevel, through the recognition of specific patterns among trajectories, the chaotic motion that can set in under specific, and hopefully recognizable, dynamical conditions.

In conclusion, the various theoretical models all deal essentially with the same question: how to obtain, from a detailed analysis of relevant microscopic encounters and elementary coupling forces, the behaviour and the macroscopic features of increasingly more complicated gaseous mixtures. Some of the answers provided here apply only to very simple situations or to highly specific systems, but the authors usually offer possible extending procedures, together with a reliable compass, to lead us on from where we stand now.

Hopefully, that will be the subject of a future Volume, after the next ASI once again assesses the progress of collision theory in atomic and molecular science.

Franco A. Gianturco, Editor
Department of Chemistry
Città Universitaria
Rome, Italy

Rome, November 1988

PART I

SCATTERING PROCESSES WITH PHOTONS AND ELECTRONS AS PROJECTILES

ELECTRON AND PHOTON COLLISIONS WITH MOLECULES

P. G. Burke

Department of Applied Mathematics & Theoretical Physics
The Queen's University of Belfast
Belfast, N. Ireland, BT7 1NN

ABSTRACT

These lectures review the theory of electron and photon collisions with molecules. After an introduction which discusses the role of these collisions in various applications and provides references to other review articles and monographs, a section is included which summarises the basic processes which can occur when an electron or a photon collides with a molecule. The next section presents the theory of electron-molecule collisions from two viewpoints. Firstly when a laboratory frame representation is used and secondly when a molecular frame representation is adopted. In the latter case the adiabatic nuclei approximation is introduced and its limitations are discussed. In the next section the theory of molcular photoionisation is presented and the close link with electron molecule collisions emphasised. The next section contains a discussion of computational methods. Particular attention is paid to approximations which have been made in the representation of the exchange and polarisation potentials. A short review is then given of methods used in solving the coupled integrodifferential equations which are satisfied by the wave function of the colliding or ejected electrons. In the next section we consider methods for including the nuclear motion which go beyond the adiabatic nuclei approximation. Finally in the last section we consider some further illustrative results of electron molecule collisions and molecular photoionisation calculations and experiments.

INTRODUCTION

The study of the interaction of electrons and photons with molecules is a field which is of fundamental interest as a branch of collision theory and is of importance in many applications where photons, electrons and molecules are involved (eg plasma physics, laser physics, atmospheric and interstellar science, isotope separation, MHD power generation, electrical discharges, radiation chemistry and radiophysics). The subject has seen an explosion of interest both experimentally and theoretically in recent years. This has been stimulated on the experimental side by the increasing availability of electron beams with millivolt energy resolution, by synchrotron radiation sources and by intense tunable lasers and on the theoretical side by the development of new and deeper understanding of the basic physical processes and by the introduction of new computational methods which enable accurate predictions to be made using the current generation of computers.

This intense research activity has resulted in the publication of many recent review articles, monographs and proceedings of meetings, workshops and symposia. These include reviews of the theory of electron molecule collisions by Itikawa [1], Burke [2], Lane [3], Norcross and Collins [4], Thompson [5], Gianturco and Jain [6], and Morrison [7], while a comprehensive review of electron molecule scattering experiments has been written by Trajmar et al [8]. In addition recent books covering the field of electron and photon collisions with molecules include those edited by Rescigno et al [9], Gianturco and Stefani [10], McGlynn et al [11], Shimamura and Takayanagi [12], and Pitchford et al [13].

In these lectures we will concentrate on recent developments of this subject since the lectures published by the author [14] in the proceedings of previous NATO ASI on atomic and molecular collision theory. The main developments since then have been the introduction of new methods and the refinement of existing methods for studying low energy electron molecule collisions, the increasing power and availability of computers enabling, for example, accurate low energy electronic excitation cross sections to be calculated for the first time, the increasing understanding of the role of resonances and threshold phenomena both in electron scattering and in photoionisation and the increasing sophistication of experiments providing much more stringent tests of the theoretical models. Finally, while the present lectures are self contained the above references should be consulted for further discussion of the theory and for a more detailed presentation of the relevant experiments.

SUMMARY OF THE BASIC PROCESSES

The basic processes which occur in electron and photon molecule collisions include:

Rotational and vibrational excitation

$$e^- + AB(vj) \rightarrow e^- + AB(v'j') \tag{1}$$

where here and in the following equations we refer for notational convenience to a diatomic molecule although the processes apply equally well to polyatomic molecules.

Electronic excitation (with or without rotational and vibrational excitation).

$$e^- + AB(vj) \rightarrow e^- + AB^*(v'j') \tag{2}$$

Dissociative attachment.

$$e^- + AB \rightarrow A + B^- \tag{3}$$

Dissociative recombination (in the case of positive molecular ions)

$$e^- + AB^+ \rightarrow A + B \tag{4}$$

Radiative recombination

$$e^- + AB^+ \rightarrow AB + h\nu \tag{5}$$

Dielectronic recombination

$$e^- + AB^+ \rightarrow AB^{**} \rightarrow AB^* + h\nu \tag{6}$$

where AB^{**} is an intermediate resonance state where two electrons are in excited electronic orbitals.

Ionisation

$$e^- + AB \rightarrow AB^+ + 2e^- \tag{7}$$

Dissociation

$$e^- + AB \rightarrow A + B + e^- \tag{8}$$

Dissociative ionisation

$$e^- + AB \rightarrow A + B^+ + 2e^- \tag{9}$$

Photoionisation (with or without resolved rotational and vibrational excitation)

$$h\nu + AB(vj) \rightarrow AB^+(v'j') + e^- \tag{10}$$

Photodissociation

$$h\nu + AB \rightarrow A + B^* \tag{11}$$

Dissociative photoionisation

$$h\nu + AB \rightarrow A + B^+ + e^- \quad . \tag{12}$$

We see that these processes are far more varied than those which occur in electron and photon collisions with atoms because of the possibility of exciting nuclear degrees of freedom as well as electronic degrees of freedom. In addition, the multicentred and non-spherical nature of the electon molecule interaction considerably complicates the solution of the collision problem even if we neglect the nuclear motion by freezing the nuclei during the collision (which is the basic assumption made in the fixed-nuclei approximation discussed later in these lectures). We will also see that resonances play a much more important role in molecular collisions than in atomic collisions. For example, non-resonant vibrational excitation is very small since the colliding electron is typically only in the vicinity of the molecule $\sim 10^{-16}$ seconds which is too short a time compared with vibration times $\sim 10^{-14}$ seconds to transfer energy and momentum to the nuclear vibrational motion with high probability. Hence, vibrational excitation only proceeds with high probability in resonance regions or for very low energy electrons when the time which the electron spends in the vicinity of the molecule is increased substantially. This is illustrated in figure 1 which shows the vibrational excitation cross section for $e^- - N_2$ scattering. This figure shows the experimental results of Ehrhardt and Willman [16] compared with boomerang model calculations of Dubé and Herzenberg [15]. It is seen that the vibrational excitation cross section is negligible for energies which are removed from the $^2\Pi_g$ resonance at ~ 2.3 eV. This resonance has a width ~ 0.3 eV giving a lifetime of $\geq 10^{-14}$ seconds. Resonances also control both the dissociative attachment, dissociative recombination and dielectronic recombination processes defined by eqs. (3), (4) and (6). For example, in the scattering of electrons with a few eV by H_2 we have

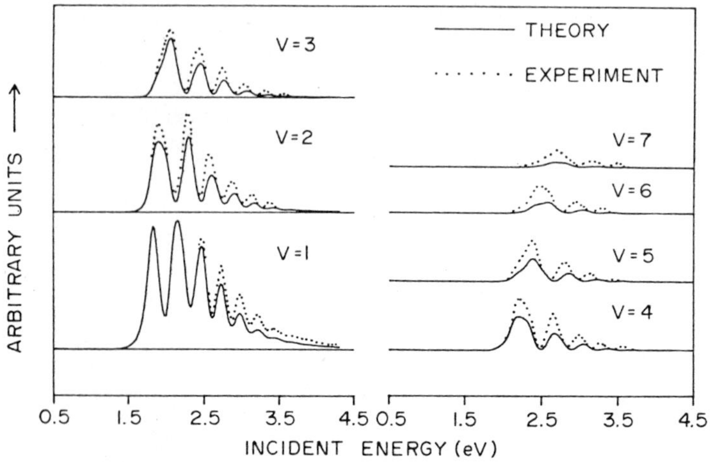

Fig. 1. Vibrational excitation of N_2 by electron impact in the neighbourhood of the $^2\Pi_g$ resonance (figure 3 from Dubé and Herzenberg [15]).

$$e^- + H_2(X\ ^1\Sigma_g^+\ v = 0) \rightarrow H_2^-(^2\Sigma_u^+) \begin{cases} \rightarrow H_2(X\ ^1\Sigma_g^+,\ v') + e^- & (13a) \\ \rightarrow H + H^- & (13b) \end{cases}$$

The intermediate resonance state H_2^- can decay either by leaving the molecule in its ground state or in a vibrationally excited state or by dissociating to H + H^-. The role of resonances in electron molecule scattering was comprehensively reviewed by Schulz [17] some years ago while a recent review of their role in molecular photoionisation has been written by Dehmer et al [18].

Also, we note that the long-range interaction between an electron and a molecule is more complicated than that between an electron and an atom. For a linear molecule, the potential has the asymptotic form

$$V(\underline{r},\underline{\hat{R}}) \underset{r\to\infty}{\sim} -\frac{\mu}{r^2}P_1(\cos\theta) - \frac{Q}{r^3}P_2(\cos\theta) - \frac{\alpha_o}{2r^4} - \frac{\alpha_2}{2r^4}P_2(\cos\theta) \qquad (14)$$

where $\underline{\hat{R}}$ is a unit vector along the inter-nuclear axis, $\underline{r}$ is the coordinate of the colliding electron referred to the centre of gravity of the molecule and $\cos\theta = \underline{\hat{r}}.\underline{\hat{R}}$. For polar molecules, the dipole moment μ is non-zero and the corresponding term in the potential dominates the rotational excitation cross section as well as giving rise to angular distributions which are strongly peaked in the forward direction. For molecules with large dipole moments, the dipole term in the potential also supports bound or virtual states which leads to associated threshold phenomena in the cross sections. This is illustrated in figure 2 which shows vibrational excitation cross sections for HF measured by Rohr and Linder [19]. Close to threshold these cross sections exhibited pronounced peaks which are associated with virtual states near each of the vibrational excitation thresholds. Similar peaks have been seen for other polar molecules such as HCl and HBr as well as for some non-polar molecules such as SF_6. The latter example shows that the long-range potential is not an essential ingredient of such peaks at least for polyatomic molecules. Finally, we note that the long-range nature of the quadrupole Q and polarisation α_o and α_2 terms in the potential are also important in rotational excitation and in low-energy collisions particularly for non-polar molecules.

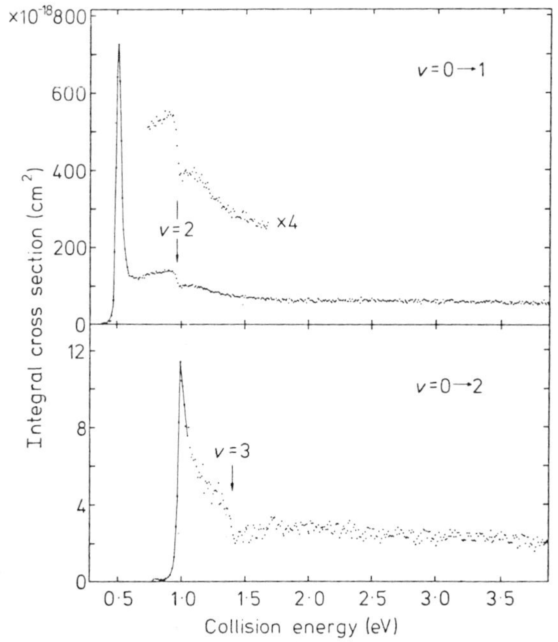

Fig 2. Integral excitation functions $\sigma_v(E)$ for v = 0 → 1 and v = 0 → 2 in e^-- HF collisions (figure 5 from Rohr and Linder [19]).

THEORY OF ELECTRON-MOLECULE COLLISIONS

Laboratory-frame representation

We now turn to the derivation of the basic equation describing electron molecule collisions. The Schrödinger equation describing the electron molecule system is

$$(H_m + T + V)\Psi = E\Psi \tag{15}$$

where H_m is the Hamiltonian for the target molecule, T is the kinetic energy of relative motion of the electron-molecule system

$$T = -\frac{1}{2\mu}\nabla^2, \qquad \mu = \frac{M}{1+M} \approx 1 \tag{16}$$

where M is the molecular mass and where we use atomic units with $m_e = \hbar = e = 1$, and V is the electron-molecule interaction potential

$$V = \sum_j \frac{1}{|\underline{r} - \underline{r}_j|} - \sum_i \frac{Z_i}{|\underline{r}-\underline{R}_i|} \equiv V(\underline{R},\underline{r}_m,\underline{r}). \tag{17}$$

Here $\underline{R}$ stands symbolically for the positions $\underline{R}_i$ of all the nuclei in the molecule, $\underline{r}_m$ for the set of coordinates $\underline{r}_j$ of the electrons in the molecule and $\underline{r}$ for the coordinate of the incident electron. The total energy of the system in the reference frame where the centre-of-mass of the whole system is at rest is E.

We expand the total wave function in the form

$$\Psi = \mathcal{A} \sum_{\alpha} \Phi_{\alpha} (\underline{R}, \underline{r}_m) F_{\alpha} (\underline{r}) + \sum_{\beta} \chi_{\beta} (\underline{R}, \underline{r}_m, \underline{r}) \alpha_{\beta} \tag{18}$$

where the first expansion is a sum over target eigenstates and pseudo-states Φ_{α} (representing target polarisation effects) satisfying

$$\langle\Phi_{\alpha} | H_m | \Phi_{\alpha'}\rangle = E_{\alpha} \delta_{\alpha\alpha'} , \tag{19}$$

while the second expansion is a sum over quadratically integrable functions χ_{β} representing short-range correlation effects. The suffix α represents the rotational-vibrational state as well as the electronic state of the target. The operator $\mathcal{A}$ antisymmetrises the first expansion with respect to interchange of the space and spin coordinates of any pair of electrons where we note that for notational simplicity we have suppressed the spin coordinates of the electrons in eq. (18). In practise for molecules of interest in these lectures where relativistic effects can be neglected the spin of the target can be coupled to the spin of the scattered electron to form an eigenstate of the total spin and its z component which are conserved in the collision.

We now substitute eq. (18) into eq. (15) and project onto the functions Φ_{α} and χ_{β}. This yields coupled equations for the functions $F_{\alpha}(\underline{r})$ and the coefficients α_{β}. Eliminating the coefficients α_{β} then yields the coupled integrodifferential equations

$$(\nabla^2 + k_{\alpha}^2)F_{\alpha}(\underline{r}) = 2\sum_{\beta} [V_{\alpha\beta}(\underline{r})F_{\beta}(\underline{r})+\int(W_{\alpha\beta}(\underline{r},\underline{r}')+K_{\alpha\beta}(\underline{r},\underline{r}'))F_{\beta}(\underline{r}')d\underline{r}'] \tag{20}$$

where

$$k_{\alpha}^2 = 2(E - E_{\alpha}) \tag{21}$$

and where the direct potential

$$V_{\alpha\beta}(\underline{r}) = \int \int \Phi_{\alpha}^* (\underline{R},\underline{r}_m) V(\underline{R},\underline{r}_m,\underline{r}) \Phi_{\beta}(\underline{R},\underline{r}_m) d\underline{R}d\underline{r}_m . \tag{22}$$

The exchange potential $W_{\alpha\beta}(\underline{r},\underline{r}')$, arising from the antisymmetrisation terms, and the correlation potential (or optical potential) $K_{\alpha\beta}(\underline{r},\underline{r}')$ arising from the quadratically integrable functions in eq. (18) will not be written down explicitly here but will be considered in more detail later.

It follows immediately from eq. (22) that the direct potential has the asymptotic form already defined by eq. (14). The dipole and quadrupole terms arise from the first two terms in a multipole expansion of $V(\underline{R},\underline{r}_m,\underline{r})$ while the polarisation terms arise in second-order from virtual transitions from the target state of interest to the other states in expansion (18) coupled to it by the dipole operater. We will see later that the polarisation potential is the most difficult part of the interaction to obtain an accurate representation for.

In contrast to the direct potential, the exchange and correlation potentials are short range, vanishing exponentially as a rule at large distances. Both of these can now be accurately calculated for diatomic molecules and for simple polyatomic molecules.

The scattering amplitudes and cross sections are obtained by considering the asymptotic form of the total wave function. The asymptotic form corresponding to an electron incident upon the target molecule in an initial state Φ_{α} is

$$\Psi_\alpha \underset{r\to\infty}{\longrightarrow} \Phi_\alpha e^{ik_\alpha z} + \sum_\beta \Phi_\beta f_{\beta\alpha}(\theta\ \phi) \frac{e^{ik_\beta r}}{r} \tag{23}$$

where $f_{\beta\alpha}(\theta\ \phi)$ is the scattering amplitude corresponding to the scattered electron moving after the collision in direction $(\theta\ \phi)$ leaving the molecule in the final state Φ_β. We note that dissociating states of the molecule are implicitly allowed for in eq. (23) by the inclusion of the appropriate continuum vibrational states of H_m in expansions (18) and (23). The differential cross section for a transition from an initial state Φ_α to a final state Φ_β, with a corresponding transition in the spin magnetic quantum number of the scattered electron, is then given by

$$\frac{d\sigma_{\beta\alpha}}{d\Omega} = \frac{k_\beta}{k_\alpha} |f_{\beta\alpha}(\theta\ \phi)|^2 \tag{24}$$

in units of a_o^2/steradian and the total cross section is obtained by integrating over all scattering angles.

In order to reduce eq. (24) to a form which is suitable for low-energy collision calculations we expand the functions $F_\alpha(\underline{r})$ in terms of angular functions $X^{p\mu}_{h\ell}(\hat{r})$ which transform according to the appropriate irreducible representation (IR) of the molecular point group [6,20]

$$F_\alpha(\underline{r}) = \sum_{h\ell} r^{-1} f^\alpha_{h\ell}(r)\ X^{p_\alpha \mu_\alpha}_{h\ell}(\hat{r}) \tag{25}$$

where p_α denotes the particular IR, μ_α distinguishes each component of the basis if the IR has dimension greater than one and h distinguishes different bases of the same IR that correspond to the same angular momentum ℓ. These angular functions can be expanded in terms of real or complex spherical harmonics

$$X^{p\mu}_{h\ell}(\hat{\underline{r}}) = \sum_m b^{p\mu}_{h\ell m}\ Y_{\ell m}(\hat{r}) \tag{26}$$

where the $b^{p\mu}_{h\ell m}$ corresponds to a unitary transformation between the bases.

We then obtain a set of coupled ordinary integrodifferential equations for the reduced radial functions $f^\alpha_{h\ell}(r)$. We will discuss the corresponding equations in the molecular frame of reference in the next section and will consider methods for their solution in a later section.

For fast collisions the interaction between the electron and the molecule lasts for a short time and the first Born approximation then usually provides an accurate description of the collision. The scattering amplitude in the first Born approximation is given by

$$f^{Born}_{\beta\alpha}(\theta\ \phi) = -\frac{1}{2\pi} \int\int\int e^{i\underline{K}\cdot\underline{r}}\ \Phi^*_\beta(\underline{R},\underline{r}_m)\ V(\underline{R},\underline{r}_m,\underline{r}) \times \Phi_\alpha(\underline{R},\underline{r}_m)\ d\underline{R}\ d\underline{r}_m\ d\underline{r} \tag{27}$$

when the momentum transfer $\underline{K} = \underline{k}_\alpha - \underline{k}_\beta$. Massey [21] introduced the Born approximation to describe rotational excitation cross sections for rotating dipoles where it can be used at all energies because the potential is dominated by the long-range r^{-2} dipole term. The Born approximation also gives accurate results at all energies for high incident electron angular momenta. It has consequently been widely used to augment or to correct calculations which use more accurate approaches for low partial waves. As an example, the multipole-extracted adiabatic nuclei (MEAN) approximation of Norcross and Padial [22] often uses the Born approximation to correct the scattering amplitude calculation in the adiabatic nuclei approximation in the case of polar molecules.

Molecular-Frame Representation: The Adiabatic Nuclei Approximation

While the theory developed in the previous section is completely general and has been used to describe low energy electron collisions with light diatomic molecules such as H_2 and N_2 a major difficulty arises because of the large number of channels which are coupled after the partial wave decomposition of the scattered electron wave function has been carried out. Thus, for a linear closed shell molecule the orbital angular momentum ℓ of the scattered electron and the rotational angular momentum $\underline{j}$ of the molecule are coupled to give an eigenstate of the total angular momentum $\underline{J}$ and its z component M_J which are conserved in the collision as discussed by Arthurs and Dalgarno [23]. Thus we define

$$\mathcal{Y}^{JM_J}_{\ell j}(\hat{\underline{r}},\hat{\underline{R}}) = \sum_{m_\ell m_j} (\ell\, m_\ell\, j\, m_j | JM_J)\, Y_{\ell m_\ell}(\hat{\underline{r}})\, Y_{jm_j}(\hat{\underline{R}}). \qquad (28)$$

It follows that for each J,M_J and parity $\pi = (-1)^{\ell+j}$ combination we have to solve coupled radial equations corresponding to the quantum numbers $v\ell j$ (where v is the vibrational quantum number) for each electronic state considered. This becomes prohibitively expensive for heavier molecules where many v, ℓ and j values must be included in the expansion to give convergence.

These difficulties can to a large extent be overcome by adopting an approximation similar to that used in molecule bound-state calculations. In this case the Born-Oppenheimer separation of the electronic and nuclear motion is adopted where the electronic motion is first determined in the fixed field of the nuclei. The molecular rotational and vibrational motion is then included in a second step of the calculation. This approximation owes its validity to the large ratio of the nuclear mass to the electronic mass and can be adopted in electron molecule collision processes when the time of the collision is much shorter than the molecular rotation and/or vibration time. The approximation is thus expected to be valid when the scattered electron energy is not close to threshold or when the energy is not close to a narrow resonance but, as we shall see, it can be extended to apply to these situations as well.

This approximation, called the adiabatic nuclei or fixed-nuclei approximation, is not new but was first used to describe the scattering of electrons by homonuclear diatomic molecules by Stier [24] and Fisk [25] in the 1930's. However, in the last few years it has gained acceptance as the basis of computational methods which are yielding the most accurate low-energy cross sections for complex molecules.

In order to formulate this approximation it is convenient to adopt a frame of reference which is rigidly attached to the molecule as illustrated in figure 3 for a diatomic molecule. In this figure A and B are the two nuclei, G is their centre of gravity which is taken as the origin of

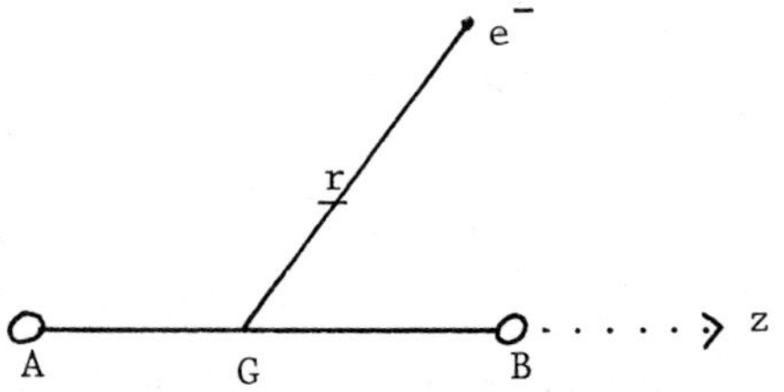

Fig. 3. Molecular-Frame for a Diatomic Molecule

coordinates and the z axis is chosen to lie along the internuclear axis. This molecular frame of reference can be related to the laboratory frame of reference by a unitary transformation defined by the equation

$$Y_{\ell m}(\underline{\hat{r}}') = \sum_{m'} D^{\ell *}_{mm'}(\alpha\ \beta\ \gamma)\ Y_{\ell m'}(\underline{\hat{r}}) \ . \tag{29}$$

Here the primed coordinates refer to the laboratory frame and the unprimed coordinates to the molcular frame, while the molecular frame is orientated in a direction defined by the Euler angles $(\alpha\ \beta\ \gamma)$ in the laboratory frame. Also, we have adopted the notation of Rose [26] for the Wigner rotation matrices $D^{\ell}_{mm'}(\alpha\ \beta\ \gamma)$.

In contrast to eq. (15) the fixed-nuclei approximation starts from the solution of the Schrödinger equation

$$(H_{e\ell} + T + V)\psi = E\psi \tag{30}$$

where $H_{e\ell}$ is the electronic part of the Hamiltonian for the target molecule assuming that the target nuclei have fixed coordinates $\underline{R}$. It follows that $H_{e\ell}$ is related to H_m used in eq. (15) by the equation

$$H_m = H_{e\ell} + T_{\underline{R}} \tag{31}$$

where $T_{\underline{R}}$ is the kinetic energy of relative motion of the nuclei. The remaining quantities T and V in eq. (30) have the same meaning as in eq. (15).

The solution of eq. (30) proceeds in an analogous way to the solution of eq. (15) except that now the nuclear coordinates only appear parametrically in eq. (30) and we have to resolve the equations for each combination of coordinates $\underline{R}$ of interest. We expand the wave function ψ in the form

$$\psi = \mathcal{A} \sum_i \phi_i(\underline{R},\underline{r}_m)\ G_i(\underline{r}) + \sum_j \xi_j(\underline{R},\underline{r}_m,\underline{r})\ b_j \tag{32}$$

where the first expansion is a sum over target electronic eigenstates and pseudo-states ϕ_i satisfying

$$\langle\phi_i|H_{e\ell}|\phi_{i'}\rangle = \varepsilon_i\ \delta_{ii'} \tag{33}$$

and the second expansion is a sum over quadratically integrable functions representing short-range correlation effects. The suffix i now represents only the electronic state of the target for fixed nuclear coordinates $\underline{R}$. As before, we now substitute eq. (32) into eq. (30) and project onto the functions ϕ_i and ξ_j. This yields coupled equations for the functions $G_i(\underline{r})$ and the coefficients b_j. Eliminating the coefficients b_j then yields the coupled integrodifferential equations

$$(\nabla^2 + k_i^2)\ G_i(\underline{r}) = 2 \sum_j [V^{FN}_{ij}(\underline{r})\ G_j(\underline{r}) + \int (W^{FN}_{ij}(\underline{r},\underline{r}') + K^{FN}_{ij}(\underline{r},\underline{r}'))G_j(\underline{r}')d\underline{r}'] \tag{34}$$

where

$$k_i^2 = 2(E - \varepsilon_i) \tag{35}$$

and where the fixed-nuclei direct potential is defined by

$$V^{FN}_{ij}(\underline{r}) = \int \phi_i^*(\underline{R},\underline{r}_m)\ V(\underline{R},\underline{r}_m,\underline{r})\ \phi_j(\underline{R},\ \underline{r}_m)\ d\underline{r}_m \tag{36}$$

which depends parametrically on the nuclear coordinate $\underline{R}$. In a similar way the exchange potential $W^{FN}_{ij}(\underline{r},\underline{r}')$ and the correlation potential $K^{FN}_{ij}(\underline{r},\underline{r}')$ depend parametrically on $\underline{R}$.

The final step in order to reduce eqs. (34) to a form which are suitable for numerical evaluation is to separate out the angular variables by expanding $G_i(\underline{r})$ in terms of the appropriate symmetry adapted angular functions $X^{p\mu}_{h\ell}(\underline{r})$ defined in eq. (26). We write

$$G_i(\underline{r}) = \Sigma_{h\ell}\ r^{-1}\ g_{ih\ell}\ (r)\ X^{p\mu}_{h\ell}\ (\hat{\underline{r}}) \tag{37}$$

and similar expansions for the potential matrices, which leads immediately to coupled radial integrodifferential equations of the form

$$\left(\frac{d^2}{dr^2} - \frac{\ell(\ell+1)}{r^2} + k_i^2\right)g_s\ (r) = 2\ \Sigma_{s'}\ [V^{FN}_{ss'}\ (r)\ g_{s'}\ (r)$$

$$+ \int_0^\infty (W^{FN}_{ss'}(r,r') + K^{FN}_{ss'}\ (r,r'))g_{s'}(r')dr'] \tag{38}$$

where for notational convenience we have combined the electronic index i and the components $h\ell$ of the IR of the molecular point group into one index s. We now look for solutions of eqs. (38) satisfying the boundary conditions

$$g_{ss'}\ (0) = 0$$

$$g_{ss'}\ (r) \underset{r\to\infty}{\sim} k_i^{-\frac{1}{2}}\ (\sin\theta_s\ \delta_{ss'} + \cos\theta_s\ K_{ss'}) \tag{39}$$

where $\theta_s = k_i r - \frac{1}{2}\ell\pi$ and the second index s' on $g_{ss'}$ denotes the linearly independent solutions of eqs. (38). This equation defines the real symmetric K-matrix $K_{ss'}$ from which the S-matrix can be obtained from the matrix equation

$$\underline{S} = \frac{\underline{1} + i\ \underline{K}}{\underline{1} - i\ \underline{K}}\ . \tag{40}$$

The scattering amplitude in the molecular frame of reference is then given by

$$f_{ii'}(\hat{\underline{k}}.\hat{\underline{r}}) = \sum_{\substack{h\ell h'\ell'\\ p\mu}} \frac{2\pi}{i\ k_i}\ X^{p\mu}_{h\ell}\ (\hat{\underline{k}})\ X^{p\mu}_{h'\ell'}(\hat{\underline{r}})\ i^{\ell-\ell'}(S^{p\mu}_{ih\ell,i'h'\ell'}$$

$$- \delta_{ii'}\ \delta_{hh'}\ \delta_{\ell\ell'}) \tag{41}$$

where $\hat{\underline{k}}$ and $\hat{\underline{r}}$ are the initial and final directions of the scattered electron. The corresponding expression for the scattering amplitude in the laboratory frame of reference can be immediately obtained using transformation (29). The resultant scattering amplitude has been written down by Gianturco and Jain [6].

So far we have shown how the scattering amplitude can be calculated at each fixed configuration of nuclear coordinates $\underline{R}$. Rotational and vibrational excitation scattering amplitudes can be extracted from this fixed-nuclei scattering amplitude using an approach originally proposed by Chase [27] in a study of neutron scattering by nuclei. In the case of diatomic molecules the relevant scattering amplitude can be written as

$$f_{ivjm\ i'v'j'm'}\ (\hat{\underline{k}}.\hat{\underline{r}}) = \langle R^{-1}\ \chi_{iv}(R)\cdot Y_{jm}\ (\hat{\underline{R}})|f_{ii'}(\hat{\underline{k}}.\hat{\underline{r}};\underline{R})|$$

$$\times\ R^{-1}\ \chi_{i'v'}(R)\ Y_{j'm'}(\hat{R})\rangle \tag{42}$$

where $f_{ii'}(\hat{k}.\hat{r};R)$ is the scattering amplitude defined by eq. (41), where we have explicitly denoted its parametric dependence on the nuclear coordinates $\underline{R}$, and $\chi_{iv}(R)$ and $Y_{jm}(\hat{R})$ are the molecular vibrational and rotational eigenfunctions. This approach was first used by Altshular [28] in a study of low energy scattering of electrons by polar molecules. As we have already mentioned the method is valid provided that the collision time is short compared with the vibration and/or rotation times. This approximation is now widely used in such situations.

However, it is well known that this adiabatic nuclei (AN) approximation breaks down when long-range interactions play a dominant role in the collision as is the case for polar molecules. The calculation in the AN approximation then leads to divergent total cross sections. Nevertheless the angular momentum of the scattered electron $\underline{\ell}$ still couples to the internuclear coordinates $\underline{R}$ when the electron is close to the molecule, although it no longer does so in the asymptotic region. In this case Chang and Fano [29] suggested that an accurate cross section can be obtained by making a frame transformation from the AN approximation, which is still used in an internal region, to the laboratory frame which is used in an external region. An accurate result for the cross section can also be obtained by treating the low partial waves in the AN approximation and the high partial waves using the laboratory frame, which is essentially an angular momentum frame transformation approach.

The AN approximation also breaks down close to rotational and vibrational excitation thresholds giving cross sections which are non-zero at threshold. Chang and Temkin [31] suggested that AN cross sections should be multiplied by the ratio of the final to the initial wave numbers k_f/k_i forcing the cross section to go correctly to zero at threshold. However this approach does not give accurate absolute cross sections in this limit. A qualitative study of the error in the AN approximation close to threshold was made by Morrison et al [32] who carried out laboratory frame vibrational close coupling calculations based on eq. (20) and AN calculations using identical model potentials. Their results are presented in figure 4 for $e^- - H_2$ collisions. This figure shows that the AN approximation gives serious errors close to threshold for vibrational excitation but calamitous errors close to threshold for ro-vibrational excitation. Methods for correcting the AN approximation close to threshold without solving the full laboratory frame equation with vibrational close coupling have been discussed recently by Morrison [7] however there is still much work to be done to obtain a definitive solution.

We conclude this section by illustrating the other situation in which the simple AN approximation fails: viz in the neighbourhood of narrow resonances. We show in figure 5 the total cross section for $e^- - N_2$ scattering calculated using an AN R-matrix theory by Burke et al [33] compared with the experimental measurements of Kennerly [34]. We see that the experimental results show structure in the neighbourhood of the 2.3 eV $^2\Pi_g$ resonance due to the fact that the resonance life-time and the nuclear vibration time are the same order of magnitude giving rise to vibrational effects in the resonance decay. This structure is absent in the AN theory. On the other hand, both experiment and theory are devoid of structure in the neighbourhood of the 25 eV $^2\Sigma_u$ resonance where the resonance life-time is much shorter than the vibration time. The AN approximation in this case gives accurate results. The AN approximation has been modified to enable narrow resonance structures to be treated and some of these modifications will be considered later in these lectures.

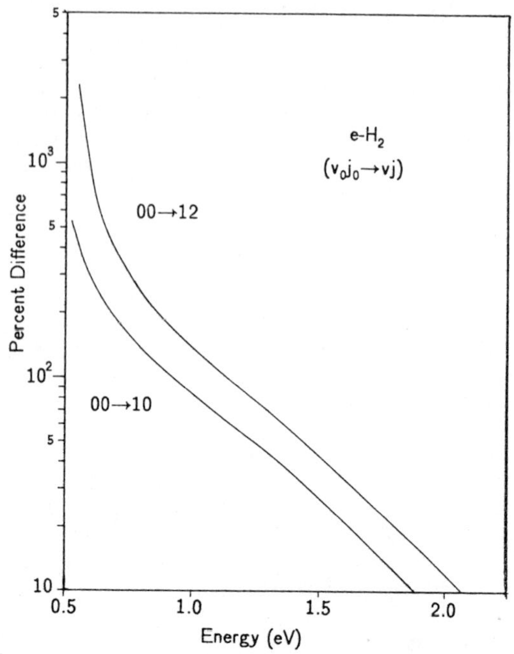

Fig. 4. Percentage deviation of the AN $e^- - H_2$ integrated cross section for the (00→10) and (00→12) excitation for the LFCC values (figure 1 from Morrison et al [32])

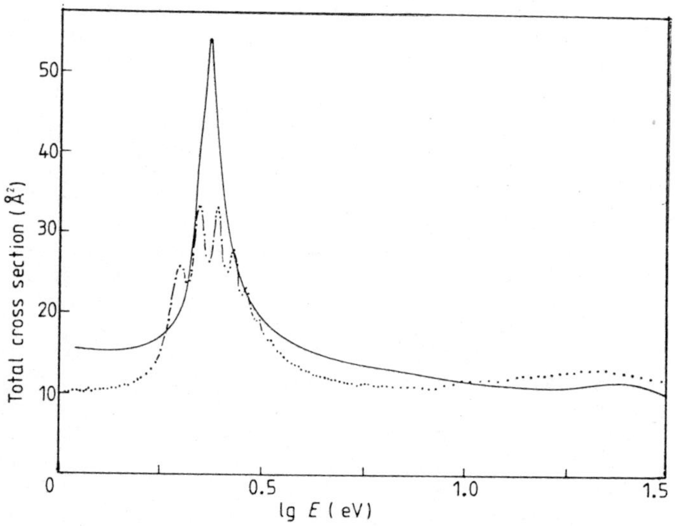

Fig. 5. The total cross section for $e^- - N_2$ scattering between 1 and 31.6 eV in Å^2. Full curve, AN R-matrix calculation. Dots, experiment of Kennerly [34] (fig. 3 from Burke et al [33]).

THEORY OF MOLECULAR PHOTOIONISATION

In this section we will briefly review the theory of molecular photoionisation describing process (10). We note that processes (11) and (12) can be described by a straightforward extension of this theory, either by allowing the final electronic state to be bound or by allowing the final molecular ion to dissociate after photoionisation. We note that the final state in eq. (10) corresponds to the collison of an electron with a molecular ion. Hence we expect that the theory of electron molecule collisions developed in the previous section will be central to describing photoionisation.

In accordance with the fixed-nuclei approximation we assume that the molecule has a fixed orientation in space during the photoionisation process. We introduce a laboratory frame of reference in which the z' axis is defined by the polarisation direction (assumed linear) of the incident photon as illustrated in figure 6. We assume, as we did for electron molecule collisions in the previous section, that the molecular frame in which the electron molecule calculation is carried out is orientated in a direction defined by the Euler angles ($\alpha\ \beta\ \gamma$). (see eq. (29)).

The photoionisation cross section leaving the molecular ion in the final electronic state denoted by j with the photoelectron ejected in direction $\underline{k}'$ is given in the dipole length approximation by

$$\frac{d\sigma_j^L}{d\Omega'} = 4\pi^2\alpha a_o^2\ \omega\ |\langle\psi_{jE}^{(-)}\ (\hat{\underline{k}}')|\hat{\underline{\varepsilon}}.\underline{D}^L|\psi_o\rangle|^2 \qquad (43)$$

and in the dipole velocity approximation by

$$\frac{d\sigma_j^V}{d\Omega'} = \frac{4\pi^2\alpha a_o^2}{\omega}\ |\langle\psi_{jE}^{(-)}\ (\hat{\underline{k}}')|\hat{\underline{\varepsilon}}.\underline{D}^V|\psi_o\rangle|^2 \qquad (44)$$

where α is the fine structure constant, a_o is the Bohr radius of the hydrogen atom, ω is the incident photon energy in atomic units and $\underline{D}^L$ and $\underline{D}^V$ are the dipole-length and dipole-velocity operators which in the former case has the spherical components.

$$D_\mu^L = \left(\frac{4\pi}{3}\right)^{\frac{1}{2}} \sum_i r_i\ Y_{1\mu}\ (\hat{\underline{r}}_i') \qquad (45)$$

where the summation is over all the electronic coordinates of the target molecule. Light which is linearly polarised along the z' axis corresponds to $\mu = 0$ while circular polarisation with helicity $\pm$ 1 along this axis corresponds to $\mu = \pm 1$. Finally, ψ_o and $\psi_{jE}^{(-)}$ are the initial bound state and final scattering state of the molecule satisfying the normalisation conditions

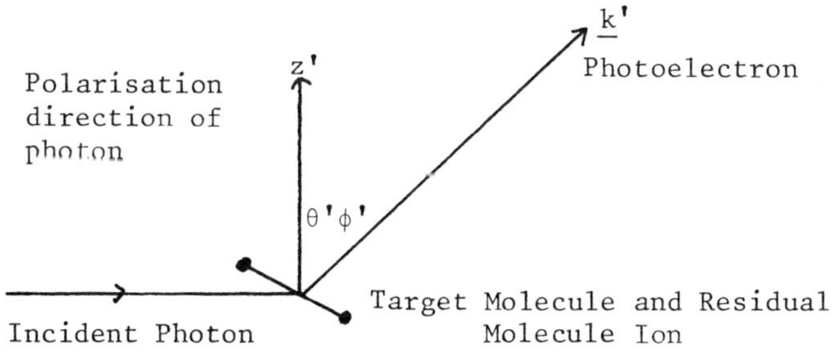

Fig. 6. Frame for Molecular Photoionisation Calculations

$$<\psi_o|\psi_o> = 1 \tag{46}$$

and

$$<\psi^{(-)}_{jE}|\psi^{(-)}_{j'E'}> = \delta_{jj'}\,\delta(E-E') \tag{47}$$

and where the boundary condition satisfied by $\psi^{(-)}_{jE}$ is given by

$$\psi^{(-)}_{jE}(\hat{\underline{k}}') = \psi^{inc}_{jE} + \psi^{ing}_{jE} \quad . \tag{48}$$

Here ψ^{inc}_{jE} corresponds to a Coulomb modified plane wave in direction $\hat{\underline{k}}'$ incident upon the residual molecular ion in the electronic state denoted by j and ψ^{ing}_{jE} corresponds to ingoing spherical waves in all open channels. In addition to eqs. (43) and (44) we can define an equivalent dipole acceleration form for the cross section. If exact wave functions are used for ψ_o and $\psi^{(-)}_{jE}$ all these forms give identical results. However in practise using approximate wave functions they are different. A measure of the accuracy of the calculation is then often taken to be the closeness of the length and velocity cross section although this is not a sufficient condition for the results to be accurate.

As we have already remarked $\psi^{(-)}_{jE}$ corresponds to an electron-molecular ion collision state. Hence it can be related to the wave function defined by eq. (32). We find that

$$\psi^{(-)}_{iE} = \sum_{\substack{h\ell h'\ell' \\ p\mu}} i^{\ell'+1}\, e^{-i\sigma_{\ell'}}\, X^{p\mu}_{h'\ell'}(\hat{\underline{k}})\, \psi^{p\mu}_{ih\ell} \tag{49}$$

where $\psi^{p\mu}_{ih\ell}$ is a solution of eqs. (37) and (38) defined in the molecular frame of reference satisfying the boundary conditions

$$\begin{aligned} g_{ss'}(0) &= 0 \\ g_{ss'}(r) &\underset{r\to\infty}{\sim} (2\pi k_i)^{-\frac{1}{2}}\,(e^{-i\theta_s} S^*_{ss'} - e^{i\theta_s}\delta_{ss'}) \end{aligned} \tag{50}$$

where for a residual molecular ion with charge z we have

$$\begin{aligned} \theta_s &= k_i r - \tfrac{1}{2}\ell\pi - \eta_i \ln 2k_i r + \sigma_s \\ \eta_i &= -z/k_i \\ \sigma_s &= \arg\Gamma(\ell + 1 + i\eta_i) \end{aligned} \tag{51}$$

As in eqs. (37) and (38) the index s represents (ihℓ). Also $\hat{\underline{k}}$ is the direction of the ejected photoelectron also referred to the molecular frame of reference.

Our final step in the evaluation of eqs. (43) and (44) is to transform the dipole length and velocity operators from the laboratory frame to the molecular frame. We find using eqs. (29) and (45) that

$$D^L_\mu = (\frac{4\pi}{3})^{\frac{1}{2}} \sum_i \sum_m r_i\, D^{1*}_{\mu m}(\alpha\ \beta\ \gamma)\, Y_{1m}(\hat{\underline{r}}_i) \tag{52}$$

The dipole length and velocity integrals can then be carried out in the molecular frame and the cross section transferred to the laboratory frame. We find after some algebra that after averaging over all orientations of the molecular axis

$$(\frac{d\sigma}{d\Omega'})_{AV} = \frac{\sigma_j}{4\pi}(1 + \beta_j\, P_2(\cos\theta')) \tag{53}$$

where σ_j is the total photoionisation cross section and β_j is the asymmetry parameter. Explicit expressions for these quantities have been given in the case of linear molcules by Burke [14].

The angular dependence of the photoionisation cross section given by eq. (53) was shown by Yang [35] to follow firstly from the fact that the transition operator is electric dipole, secondly that only one direction is defined by the photon polarisation direction in the case of randomly orientated molecules and finally that by parity conservation interference effects between even and odd parities giving rise to a $P_1(\cos\theta')$ cannot occur.

Finally we note that when the spin polarisation of the photoelectrons is observed as well as their angular distributions further parameters in addition to σ_j and β_j are required to completely describe the photoionisation process. The theory in this case has been developed by Cherepkov [36] who showed that for unorientated non-chiral molecules five independent parameters σ_j, β_j, A_j, γ_j and η_j are required where A_j, γ_j and η_j describe the spin polarization state. In the case of unorientated chiral molecules ten parameters are required to completely specify the final state [36].

COMPUTATIONAL METHODS

In this section we review recent computational methods for solving the coupled integrodifferential equations (38) and the calculation of the corresponding cross sections. First we consider the role of the direct exchange and polarisation potentials in the collision and we review approximate representations for these potentials which have been adopted. We then turn to methods for solving the coupled integrodifferential equations (38) including single-centre expansion methods and recent L^2 methods.

The Interaction Potentials

In order to calculate accurate cross sections at low electron impact energies it is necessary to include accurately the effect of the direct, exchange and correlation potentials. This is illustrated in figure 7 where we compare theory and experimental results for the total elastic $e^- - H_2$ cross section. Even at an energy of 1 Ryd, using only the static (S)

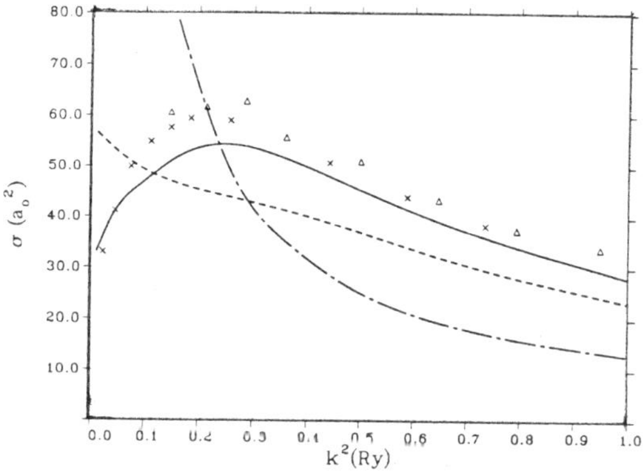

Fig.7. Comparison of theoretical and experimental total elastic e^--H_2 cross sections as a function of electron energy. Theoretical results (Linear algebraic equations method [37]): chain-static (S); dash-static exchange (SE); line-SE + polarisation (SEP).Experimental results: triangles-Hoffmann et al [38]; crosses - Dalba et al [39] (fig. 1 from Collins and Schneider [37]).

interaction (corresponding to the inclusion of only terms in the direct potential in eq. (38) which coupled to the ground electronic state) leads to errors of over thirty percent. The situation is improved by including electron exchange with respect to the ground state target. However, this static exchange (SE) calculation does not reproduce the maximum in the cross section near 4 eV. These features are only obtained by including polarisation-correlation effects through virtual excitations to the closed electronic channels. The full interaction is needed to properly represent the broad resonance in the $^2\Sigma_u$ state.

Local exchange potentials. In the case of diatomic molecules the exchange potential in eqs. (34) and (38) can be treated exactly by one of the methods to be discussed later in this section. However, for complex polyatomic molecules most work has so far been carried out using a simplified treatment based on a local approximation. In this way, the coupled integrodifferential equations are reduced to coupled differential equations which are much easier to solve.

One of the most successful and widely used methods of obtaining a local exchange potential is the Hara free-electron gas exchange (HFEGE) model (Hara [40]). In this approach the total wave function is assumed to be composed of plane waves which are antisymmetrised in accordance with the Pauli exclusion principle and the exchange energy obtained by summing all states up to the Fermi level. The resultant form of the potential is

$$V_{ex}^{HFEGE}(\underline{r}) = -\frac{2}{\pi} k_F(\underline{r}) \left(\tfrac{1}{2} + \frac{1-\eta^2}{4\eta} \ell n \left|\frac{1+\eta}{1-\eta}\right| \right) \tag{54}$$

where

$$k_F(\underline{r}) = (3\pi^2 \rho(\underline{r}))^{1/3}, \eta(\underline{r}) = (k^2 + 2I + k_F^2(\underline{r}))^{\frac{1}{2}}/k_F(\underline{r}), \tag{55}$$

I is the ionisation potential of the molecule and $\rho(\underline{r})$ is the molecular charge distribution. It can be seen that as $r \to \infty$ the numerater of η should be k and not $(k^2 + 2I)^{\frac{1}{2}}$ as in eq. (55). This has led to the introduction of the asymptotically adjusted form of the HFEGE potential where

$$\eta(\underline{r}) = (k^2 + k_F^2(\underline{r}))^{\frac{1}{2}}/k_F \ . \tag{56}$$

This is usually called the AAFEGE model. Alternatively, I can be made an adjustable parameter in the model.

While the HFEGE potential has enjoyed considerable success it does not ensure that the continuum orbital representing the scattered electron is orthogonal to the fully occupied bound orbitals of the same symmetry which is a requirement of the exact solution. This orthogonalisation can be achieved by adding additional inhomogeneous terms in the scattering equation of the form.

$$\sum_\alpha \lambda_\alpha \phi_\alpha(\underline{r}) \tag{57}$$

as suggested by Burke and Chandra [41]. The λ_α are Lagrange multipliers which are chosen to enforce orthogonality of the continuum orbital $G(\underline{r})$ to the appropriate bound orbitals $\phi_\alpha(r)$ so that

$$\int \phi_\alpha^*(\underline{r})\, G(\underline{r})\, dr = 0 \ . \tag{58}$$

Salvini and Thompson [42] and Jain and Thompson [43, 44, 45, 46] have shown that for electron scattering by polyatomic molecules the HFEGE model plus orthogonalisation gives a successful treatment of exchange.

Another approach for defining a local exchange potential is the semiclassical exchange (SCE) approximation. If one performs a Taylor expansion about $\underline{r}$, the point in space where the exchange integral is computed, one finds that [47, 48]

$$I = \int \phi^*(\underline{r}_1)|\underline{r} - \underline{r}_1|^{-1} G(\underline{r}_1)d\underline{r}_1 = \int |\underline{r}'|^{-1} \exp[(\nabla_\phi + \nabla_G).\underline{r}']\phi^*(\underline{r})G(\underline{r})d\underline{r}' \tag{59}$$

where $\underline{r}' = \underline{r}_1 - \underline{r}$ and the ∇ operators act on either the bound electron wave function ϕ or on the continuum function G depending on their subscripts. Using spherical polar coordinates we can solve the integral giving

$$I = -\frac{4\pi}{|\nabla_\phi + \nabla_G|^2} \phi^*(\underline{r})\ G(\underline{r}) \tag{60}$$

For high energy collisions, the bound functions are slowly varying compared with the continuum functions and hence we can disregard the ∇_ϕ operator compared with the ∇_G operator. We then find the following expansion for the semiclassical exchange (SCE) potential

$$V_{ex}^{SCE}(\underline{r}) = \tfrac{1}{2}(2k^2 - V_{st}(\underline{r})) - \tfrac{1}{2}\{[2k^2 - V_{st}(\underline{r})]^2 + 4\pi\rho(\underline{r})\}^{\frac{1}{2}} \tag{61}$$

where we have summed over all the bound orbitals. Hence the SCE potential is energy dependent and depends on the molecular charge distribution $\rho(\underline{r})$. The extension of this potential to lower collision energies can be carried out starting from eq. (60) and including the effect of the bound electron momenta on the scattered electron. This has been discussed recently by Gianturco and Scialla [48]. The SCE potential has been used for a number of studies of diatomic molecules as well as recently being employed for $e^- - CH_4$ scattering [49].

Separable exchange potentials. It has been pointed out by Rescigno and Orel [50] and by Schneider and Collins [51] that the large computational effort necessary to evaluate the exchange terms is a result of the non separable character of the exchange interaction rather than of its non-locality. The problem of including exchange may therefore be simplified by constructing a separable approximation to the exchange kernel in eq. (34). The exchange kernel $W_{ij}(\underline{r},\underline{r}')$ may be expanded as

$$W_{ij}(\underline{r},\underline{r}') = \sum_\alpha \chi_\alpha(\underline{r})\ \gamma_\alpha \chi_\alpha(\underline{r}') \tag{62}$$

where the $\{\chi_\alpha\}$ are an orthonormal basis of Gaussian or Slater functions giving a diagonal representation of the exchange potential with eigenvalues γ_α. These basis functions are then expanded in terms of the symmetry adapted angular functions $X_{h\ell}^{p\mu}(\underline{r})$ defined by eq. (26). It has been found in practise that a ten term separable representation of exchange in eq. (62) is sufficient to give $e^- -$ LiH static exchange eigenphase sums accurate to better than ten percent. Further, since the exchange potential in the form of eq. (62) can be treated non-iteratively in the solution of the integrodifferential equations, improvements of computational speed by factors of between three and ten can be obtained over methods which use direct numerical quadrature.

Polarisation potentials. In many early low-energy electron-molecule collision calculations, the polarisation potential was represented by its asymptotic form eq. (14) for some r > a and was cut-off at shorter distances by an empirical function. Thus we write

$$V_{pol}(\underline{r},\hat{\underline{R}}) = -\left(\frac{\alpha_o}{2r^4} + \frac{\alpha_2}{2r^4} P_2(\cos\theta)\right) f_c(r,r_c) . \tag{63}$$

Several forms of the cut-off function f_c have been considered [41,52] one of the most frequently used being

$$f_c(r,r_c) = 1 - \exp[-(r/r_c)^6] \tag{64}$$

where the cut-off parameter r_c is usually determined by fitting some "landmark" feature in the cross section such as the position of a resonance. For example in $e^- - N_2$ scattering the position of the $^2\Pi_g$ resonance at 2.3 eV has been used in the fit. This approach may be seriously called into question since frequently the cut-off function will be reflecting an attempt to compensate for important non-adiabatic or dynamical effects which have been neglected and which will be different in different symmetry states of the electron-molecule system. Further, it is found in many cases that the collision cross sections at low energies and in the vicinity of resonances can depend sensitively on the value of the parameter r_c again throwing doubt on the validity of this approach.

There have been a number of attempts to incorporate non-adiabatic effects by going beyond the above simple cut-off procedure [53]. One of the earliest approaches was suggested by Temkin [54] in the context of electron-atom scattering. In our case, this corresponds to replacing the target ground state wave function $\phi_o(\underline{R},\underline{r}_m)$ by a wave function which depends on the coordinates of the scattered electron $\phi_o(\underline{R},\underline{r}_m) + \phi_{POL}(\underline{R},\underline{r}_m,\underline{r})$. Here ϕ_{POL} represents the polarising effect of the scattered electron and is cut-off when the scattered electron lies within the target electrons. This polarised-orbital approach gives rise to a direct potential with the correct asymptotic form as well as an additional exchange potential. It has been applied to $e^- - H_2^+$ by Temkin and Vasavada [55] and to $e^- - H_2$ by Lane and Henry [56]. However it has not received much attention in recent years partly because it is difficult to apply and partly because it still involves a cut-off, albeit one which has more physical justification than f_c in eq. (63).

Another parameter-free polarisation potential has recently been introduced for atoms by O'Connell and Lane [57] and applied to diatomic molecules by Padial and Norcross [58] and to polyatomic molecules by Gianturco et al [59] with rather good results. The method is based on the free-electron-gas (FEG) model, whereby the correlation energy can be expressed in terms of the molecular charge density in the high and low density limits. However the correlation energy defined in this way does not have the correct asymptotic behaviour. Consequently the rather ad-hoc procedure is adopted of joining the asymptotic form of the potential to the short range FEG form where they cross on the boundary of the atom or molecule. It is found that this crossing radius is only weakly dependent on the target molecule [58] giving further credence to the method.

Non-adiabatic polarisation effects can also be represented by including polarised pseudo-states in expansion (32). This approach was first applied to $e^- - H$ scattering by Damburg and Karule [60] who showed that including a p-wave state with the radial form

$$P(r) = \left(\frac{32}{129}\right)^{\frac{1}{2}} (r^2 + \tfrac{1}{2} r^3)e^{-r} \tag{65}$$

represented the full polarisability of the 1s ground state. We note that including 2p H-atom eigenstates contributes 65.8% of the polarisability while including all the bound p-states still only includes 81.4% of the polarisability. More recently Burke and Mitchell [61] have shown how polarised pseudo-states can be calculated for a general atom. The static dipole polarizability of an atomic system in state ϕ_o is defined by the equation

$$\alpha_\mu = 2 \sum_k \int dk \frac{|\langle\phi_o|D^L_\mu|\phi_k\rangle|^2}{E_k - E_o} \tag{66}$$

where the summation and integration in this equation are taken over all states ϕ_k including the continuum which are coupled to state ϕ_o by the dipole length operator defined by eq. (45). The polarised pseudo-state ϕ_p is defined by the requirement that the summation and integration in eq. (66) are replaced by a single term

$$\alpha_\mu = \frac{2|\langle\phi_o|D^L_\mu|\phi_p\rangle|^2}{E_p - E_o} \tag{67}$$

where

$$\langle\phi_p|\phi_p\rangle = 1 \tag{68}$$

and where

$$\langle\phi_p|H|\phi_p\rangle = E_p \tag{69}$$

where H is the atomic Hamiltonian. We can obtain a variational approximation to ϕ_p by considering the functional

$$J = \langle\phi^t_p|H - E_o|\phi^t_p\rangle - 2\langle\phi^t_p|D^L_\mu|\phi_o\rangle \tag{70}$$

which is stationary for small variations of the trial function ϕ^t_p about the exact polarised pseudo-state ϕ_p.

In the molecular case,the polarised pseudo-state method has been applied to $e^- - H_2$ scattering by Schneider [62] and by Nesbet et al [63] and recently to $e^- - N_2$ scattering by Noble et al [64]. In the case of linear molecules there are two components of the polarisability α_{11} and $\alpha_\perp$ in directions along and perpendicular to the molecular axis. For H_2 or N_2 in their $^1\Sigma_g^+$ ground state the corresponding pseudo-states representing these two components of polarisability have symmetries $^1\Sigma_u$ and $^1\Pi_u$ respectively. It follows that the corresponding expansion (32) must include these two pseudo-states in addition to the ground electronic state. We show in figure 8 the total cross section for $e^- - N_2$ scattering including pseudo-states in the $^2\Sigma_g^+$ scattering state [64]. A series of calculations were carried out where the pseudo-states were expanded in a basis ξ_j in the form

$$\phi^t_p = \sum_j a_j \xi_j \tag{71}$$

and results are shown where four and ten terms are retained in this expansion. Without pseudo-states, the $^2\Sigma_g$ contribution to the cross section below the $^2\Pi_g$ resonance is far too large. When the pseudo-states are included this contribution decreases giving good agreement with experiment when ten terms are retained in expansion (71). The results appear to be converging as the number of terms in eq. (71) is increased but further work is necessary to confirm this convergence and to include target correlation effects which were omitted in this work which used the SCF N_2 target of Nesbet [66].

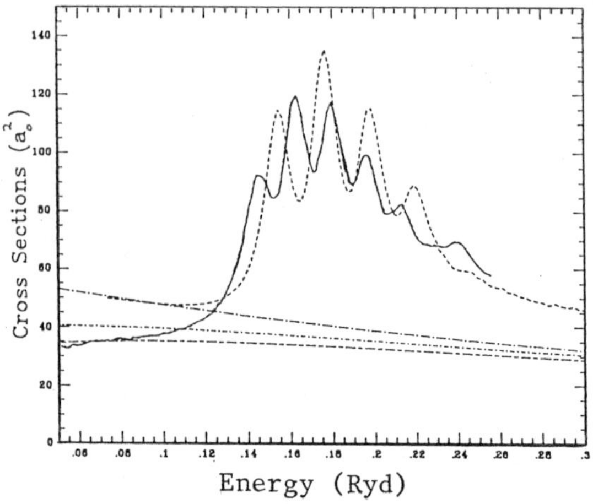

Fig. 8. The total cross section for e-N_2 scattering. —— , experiment of Kennerly [33], - - -, theory with no pseudo states, Gillan et al [65], —·—· , $^2\Sigma_g^+$ contribution with no pseudo-states; -·-·-, $^2\Sigma_g^+$ contribution with four configurations; —-—-, $^2\Sigma_g^+$ contribution with ten configurations from Gillan et al [64])

A number of other approaches for including the polarisation potential are based on the optical potential of many-body theory. It was shown by Bell and Squires [67] that this potential contains the full information about the electron-target interaction. Klonover and Kaldor [68-71] included the potential in second-order for e - H_2 scattering, correcting the T-matrix calculated omitting the long-range polarisation potential. More recently Ficocelli Varracchio and Lamanna [72] have revived this approach and considered a more sophisticated non-perturbative representation of the optical potential based on the Random Phase Approximation (RPA). Their results are in good agreement with other ab-initio calculations for e - H_2 scattering.

We also remark that an appropriate choice of the quadratically integrable basis functions in the second expansion in eq. (32) can also represent long-range polarisation effects. Schneider and Collins [73] accurately represented polarisation effects for e^- - N_2 scattering in this way using the linear algebraic equations method discussed later in these lectures.

In conclusion we observe that while an accurate representation of the polarisation potential in low energy electron-molecule collisions remains a difficult problem, recent work has shown that accurate results can now be obtained.

Solution of the Fixed-Nuclei Equations

We now turn to methods for solving the coupled integrodifferential equations (38) or equivalent equations which describe the collision of an electron with a molecule which is fixed in space. These methods include the most accurate ab-initio methods which are now available for calculating low energy electron molecule collision cross sections.

Single-centre Expansion Methods. This method is straightforward to apply and was amongst the earliest methods to be used in practice. It is still being used for electron collisions by many polyatomic molecules involving one heavy atom and several hydrogen atoms such as H_2O, H_2S, CH_4 and NH_3 [6]. However, it is computationally expensive since usually many partial waves need to be coupled to obtain converged cross sections.

The basic procedure is to expand the potentials and the wave function describing the colliding electron in a single-centre expansion about the centre-of-mass of the molecule. The equations can then be solved either directly using standard integration algorithms such as the de Vogelaire and Numerov algorithms (eg Burke and Sinfailam [74], Chandra [75], Raseev [76]) or converted to a set of coupled integral equations (eg Sams and Kouri [77], Henry et al [78], Morrison [79]). The numerical methods are simply modified to treat local or separable exchange potentials, however new iterative and non-iterative techniques were required to treat exchange exactly (eg Collins et al [80], Robb and Collins [81]).

When the colliding electron lie outside the charge distributions of the target states retained in expansion (32), the coupled integro-differential equations (38) reduce to coupled differential equations with long-range coupling potentials of the form r^{-n}. In this region, the single-centre expansion method converges fast and is used to augment methods which use more sophisticated approaches in the inner region such as the linear algebraic equations method and the R-matrix method discussed later in this section. Norcross [82] has reviewed progress in methods for the solution of the coupled equations in the asymptotic region and a fast program for solving these equations has been written by Noble and Nesbet [83].

Linear Algebraic Equations Method. The Linear Algebraic (LA) method is the first of three methods which we now discuss which are capable of yielding highly accurate ab-initio results, the other two being the R-matrix method and the Schwinger variational method. The LA method, which is similar in spirit to the approach used by Seaton [84] for electron atom collisions, was introduced for electron molecule collisions by Collins and Schneider [51,85]. It starts from the coupled integrodifferential equations (38) in integral equation form which we can write in matrix notation as

$$\underline{g}(r) = \underline{G}^1(r) - \int_0^\infty \underline{G}^0(r,r') \int_0^\infty \underline{U}(r',r'')\, \underline{g}(r'')\, dr'\, dr'' \qquad (72)$$

where the potential $\underline{U}$ includes the direct, exchange and correlation potentials in eq. (38) and where the Green's function

$$\underline{G}^0(r,r') = \begin{cases} \underline{G}^1(r)\, \underline{G}^2(r') & r < r' \\ \\ \underline{G}^2(r)\, \underline{G}^1(r') & r > r' \end{cases} \qquad (73)$$

with

$$\begin{aligned} \underline{G}^1(r) &= \underline{k}r\, j_{\underline{\ell}}(\underline{k}r) \underset{r\to\infty}{\sim} \sin(\underline{k}r - \tfrac{1}{2}\underline{\ell}\pi) \\ \underline{G}^2(r) &= -\, r\, n_{\underline{\ell}}(\underline{k}r) \underset{r\to\infty}{\sim} \underline{k}^{-1} \cos(\underline{k}r - \tfrac{1}{2}\underline{\ell}\pi)\ . \end{aligned} \qquad (74)$$

Both $\underline{G}^1$ and $\underline{G}^2$ are diagonal matrices in the space of the channels retained in eqs. (38) or (72) and $j_\ell(\underline{k}r)$ and $n_\ell(\underline{k}r)$ are the spherical Bessel and Neumann functions. The integrals in eq. (72) are now replaced by quadratures out to a radius $r = a$ beyond which the direct potential assumes its asymptotic form. Thus

$$\underline{g}(r_\alpha) = \underline{G}^1(r_\alpha) - \sum_\beta w_\beta \underline{G}^0(r_\alpha, r_\beta) \sum_\gamma w_\gamma\, \underline{U}(r_\beta, r_\gamma)\, \underline{g}(r_\gamma) \qquad (75)$$

where r_α, r_β, r_γ and w_β, w_γ are the quadrature points and weights respectively. Equation (75) together with the boundary conditions at r = a then reduce to a set of linear-algebraic equations for the unknown functions $\underline{g}$ at the quadrature points r_α which have the form

$$\underline{M}\,\underline{g} = \underline{a} \tag{76}$$

where the matrix $\underline{M}$ is of order n x m where n is the number of channels retained in eq. (75) and m is the number of quadrature points. These equations are solved by an interation-variation method (Schneider and Collins [86]) which avoids storage of the full matrix.

It is useful at this point to discuss in greater detail the form of the correlation potential contained in $\underline{U}$. Returning to our original expansion (32) we introduce, following Feshbach [87,88], the projection operators P and Q where P projects onto the target states retained in the first expansion and Q projects onto the quadratically integrable functions retained in the second expansion. Hence we can rewrite eq. (32) in the form

$$\psi = (P + Q)\psi \;. \tag{77}$$

We can make use of flexibility in the short-range part of the functions G_i in eq. (32) to choose the P and Q spaces to be orthogonal. Hence P and Q satisfy the equations

$$P^2 = P, \quad Q^2 = Q, \quad PQ = 0. \tag{78}$$

Substituting eq. (77) into the Schrödinger equation (30) and projecting onto the spaces P and Q then yields the equations

$$P(H - E)\ (P + Q)\psi \ = 0 \tag{79a}$$

$$Q(H - E)\ (P + Q)\psi \ = 0 \tag{79b}$$

where for convenience we have written

$$H = H_{e\ell} + T + V. \tag{80}$$

We can solve eq. (79b) for $Q\psi$ giving

$$Q\psi \ = -\ Q\ \frac{1}{Q(H-E)Q}\ Q\ HP\psi. \tag{81}$$

Substituting this result for $Q\psi$ into eq. (79a) then gives

$$P\ [H - PHQ\ \frac{1}{Q(H-E)Q}\ QHP - E]\ P\psi \ = 0 \tag{82}$$

which is equivalent to eqs. (34) and, after separating the angular variables, to eqs. (38). We write eq. (82) as

$$[PHP + \mathcal{V} - E]\ P\psi \ = 0 \tag{83}$$

where the correlation potential

$$\mathcal{V} = -\,PHQ\ \frac{1}{Q(H-E)Q}\ QHP \;. \tag{84}$$

This potential is sometimes called the "optical potential" although more usually this term is reserved for the full potential including the direct and exchange potentials which are contained in PHP.

In the LA method, Q space is represented by a square-integrable multicentre basis while P space is represented as we have seen by single-centre channel functions. In this way the slow convergence of single-centre expansions due to the nuclear singularities is removed. In addition, the integrals required in QHQ in eq. (84) are evaluated using standard multicentre quantum chemistry codes which, when combined with a separable representation of exchange, gives a very efficient procedure for evaluating the cross section. The method has been widely used for diatomic molecules and we will discuss some results obtained using it later in these lectures.

R-Matrix Method. The R-matrix method was first introduced by Wigner [89,90] and Wigner and Eisenbud [91] in an analysis of nuclear resonance reactions. It was developed as an ab-initio method for electron-atom collisions by Burke et al [92] and by Burke and Robb [93] and for electron-molecule collisions by Schneider [94,95], Burke et al [96] and Noble [97]. Recently the method has been used to describe collisions of electrons and positrons with many diatomic molecules.

The method proceeds by partitioning the interaction regions into two parts depending on whether the colliding electron is inside or outside of a sphere of radius r = a which is chosen to just envelope the charge distribution of the molecular states of interest. This is illustrated in figure 9. In the internal region the interaction is strong non-local and multicentred and the electron-molecule complex behaves in a similar way to a bound state. Hence molecular bound state methods and codes can be used in this region. In the external region the interaction is weak, local and the wave function can be accurately represented by a single-centre expansion.

In the internal region the wave function is expanded in a multicentre basis in the molecular frame of reference for fixed separation of the nuclei. We introduce the basis ψ_k in analogy with eqs. (32) and (37) by the equation

$$\psi_k = \mathcal{A} \sum_{sj} \Phi_s \, u_j(r) \, a_{sjk} + \sum_j \xi_j \, b_{jk} \tag{85}$$

where the channel functions Φ_s are obtained by combining the multicentre target states ϕ_i with the angular (and spin) functions for the scattered electron, the u_j are radial basis functions defined over the range $0 \le r \le a$, the ξ_j are the usual multicentre quadratically integrable functions which also vanish by r = a and the coefficients a_{sjk} and b_{jk} are obtained by diagonalising $H + L_b$ in the basis giving

$$\langle \psi_k | H + L_b | \psi_{k'} \rangle = E_k \, \delta_{kk'} \tag{86}$$

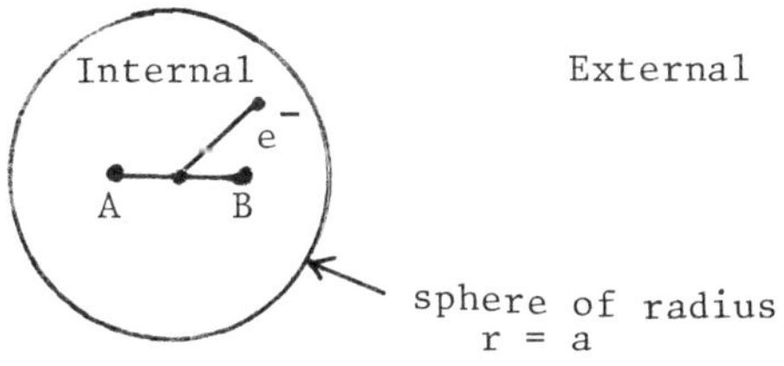

Fig. 9. Partitioning of space in the R-matrix method

where again H is defined by eq. (80). The integral in eq (86) is carried out over the internal region and L_b is a surface projection operator introduced by Bloch [98] which ensures that $H + L_b$ is Hermitian. It is defined by

$$L_b = \tfrac{1}{2} \sum_s | \Phi_s > \delta(r - a) \left(\frac{d}{dr} - \frac{b}{r}\right) < \Phi_s | \tag{87}$$

where b is an arbitrary parameter.

We now solve the Schrödinger equation (30) in the internal region by expanding the wave function in terms of this basis. To do this we rewrite eq. (30) as

$$(H + L_b - E)\psi = L_b \psi \tag{88}$$

which has the formal solution

$$\psi = (H + L_b - E)^{-1} L_b \psi . \tag{89}$$

The inverse operator in this equation can be expanded in terms of the basis ψ_k giving

$$|\psi > = \sum_k \frac{|\psi_k >< \psi_k| L_b |\psi >}{E_k - E} . \tag{90}$$

We project this equation onto the channel functions Φ_s and evaluate it on the boundary of the internal region. If as in eqs. (37) and (38) we introduce the channel function g_s by

$$g_s = < \Phi_s | \psi > \tag{91}$$

we obtain the matrix equation

$$\underline{g}(a) = \underline{R}. \left(r \frac{d\underline{g}}{dr} - b\underline{g}\right)_{r=a} \tag{92}$$

where the n x n dimensional R-matrix is defined by

$$R_{ss'} = \frac{1}{2a} \sum_k \frac{w_{sk} w_{s'k}}{E_k - E} \tag{93}$$

and where the surface amplitudes w_{sk} are given by

$$w_{sk} = \sum_j u_j(a)\, a_{sjk} . \tag{94}$$

The dimension n is the number of channels retained in eq. (85).

Eqs. (92) and (93) are the basic equations of the R-matrix method. They give the logarithmic derivative of the channel functions g_s on the boundary of the inner region at any energy E in terms of the surface amplitudes w_{sk} and eigenenergies E_k. These latter quantities are obtained from a single diagonalisation in eq. (86) of $H + L_b$ in the internal region. The matrix elements and diagonalisation required in eq. (86) can be computed using standard molecular structure codes. Once the logarithmic derivative of g_s is determined on the boundary, then the K-matrix and S-matrix can be determined by solving the coupled eqs. (38) in the external region (where they reduce to coupled ordinary differential equations with long-range coupling potentials behaving as r^{-n} using standard methods as discussed in the Section on Single-centre Expansion Methods.

In practise, care must be taken in the choice of the radial basis functions u_j in order that the summation in eq. (93) converges rapidly.

Recent work has shown that these functions are best chosen to be solutions of a model second-order differential equation satisfying homogeneous boundary conditions

$$u_j(0) = 0$$
$$\frac{a}{u_j} \left.\frac{du_j}{dr}\right|_{r=a} = b \tag{95}$$

which are Lagrange orthogonalised to the target molecular orbitals. In this case, the summation in eq. (93) must be augmented by a correction, suggested by Buttle [99], for the neglected high lying levels. The procedure for defining and calculating the basis functions u_j has been described in detail by Tennyson et al [100].

The Schwinger Variational Method. The Schwinger variational method is the third completely ab-initio approach which we will discuss. It has been developed and applied to multichannel electron molecule collision processes by McKoy and co-workers (eg Takatsuka and McKoy [101,102], Gibson et al [103], Huo et al [104]). The method has recently been used to obtain accurate cross sections for a variety of diatomic and polyatomic molecules.

The method starts from the usual Lippmann-Schwinger integral equation for the transition operator

$$T = V + V\, G_o^+\, T \tag{96}$$

where V is the total electron-molecule interaction potential defined by eq. (17) and G_o^+ is the outgoing-wave free particle Green's function associated with $E - H_{e\ell} - T$ given in eq. (30). The potential is first written in separable form in terms of a discrete basis set $\{|\alpha\rangle\}$ according to

$$V = \sum_{\alpha\beta} V|\alpha\rangle\, [V^{-1}]_{\alpha\beta}\, \langle\beta|V \tag{97}$$

where $[V^{-1}]_{\alpha\beta}$ signifies the inverse of the matrix whose elements are $\langle\alpha|V|\beta\rangle$. It is then straightforward to show (eg Lovelace [105], Adhikari and Sloan [106]) that the Schwinger variational functional for the T-matrix is given by

$$\langle \underline{k}_f|T|\underline{k}_i\rangle = \sum_{\alpha\beta} \langle \underline{k}_f|V|\alpha\rangle\, [D^{-1}]_{\alpha\beta}\, \langle\beta|V|\underline{k}_i\rangle \tag{98}$$

where $[D^{-1}]_{\alpha\beta}$ signifies the inverse of the matrix with elements

$$D_{\alpha\beta} = \langle\alpha|V - V\, G_o^+\, V|\beta\rangle \tag{99}$$

and where $|\underline{k}_i\rangle$ and $|\underline{k}_f\rangle$ are the plane wave states

$$|\underline{k}_i\rangle = (2\pi)^{-3/2}\, e^{i\underline{k}_i\cdot\underline{r}}\,, \qquad |\underline{k}_f\rangle = (2\pi)^{-3/2}\, e^{i\underline{k}_f\cdot\underline{r}} \tag{100}$$

An important feature of this result is that the discrete basis functions $\{|\alpha\rangle\}$ only appear in conjunction with the potential V, hence these functions need only have the range of V. Therefore, unlike the Kohn variational method, the Schwinger basis functions are not required to satisfy the asymptotic boundary conditions. The method involves the evaluation of two types of matrix element $\langle \underline{k}|V|\alpha\rangle$ and $\langle\alpha|V\, G_o^+ V|\beta\rangle$. The first can be evaluated in closed form when GTO's are used to represent the $\{|\alpha\rangle\}$ and the second can be evaluated by writing it in the form

$$\langle\alpha|V\,G_o^+V|\beta\rangle = \sum_{ij} \langle\alpha|V|i\rangle\ \langle i|G_o^+|j\rangle\ \langle j|V|\beta\rangle \tag{101}$$

where $\{|i\rangle\}$ and $\{|j\rangle\}$ are bases which need not be the same as $\{|\alpha\rangle\}$.

The method has been refined through an iterative prescription in which numerical oscillatory functions are used to augment the discrete basis. This gives faster convergence when long-range potentials are important in the collision. In addition other related variational prescriptions have been developed such as a $\tilde{C}$-functional suitable for electronically inelastic collisions [102,107].

Continuum Multiple-scattering Method. The continuum multiple scattering method (CMSM) was developed as a tool for studying electron molecule collisions and molecular photoionisation by Dill and Dehmer [108] and has been reviewed by Dehmer and Dill [109]. The approach has also been independently considered by Ziesche and John [110]. It is particularly useful for polyatomic molecules where the ab-initio methods just discussed are difficult to apply.

The method, first used in molecular bound state studies (eg Johnson [111]), involves the partitioning of space into three regions as illustrated in figure 10 for a diatomic molecule. In regions I the electron-molecule potential is represented by direct and local exchange terms; in region II the potential is approximated by a constant; while in region III a polarisation potential is added to the direct and local exchange potentials. Usually the potentials in regions I and III are spherically symmetric although this restriction has been relaxed.

The form assumed for the wave function representing the colliding electron in the three regions is as follows. In region I surrounding each nucleus

$$\psi_{I_i} = \sum_{\ell m} a_{\ell m}^{I_i}\, f_\ell^{I_i}\,(k\,r_i)\; Y_{\ell m}\,(\hat{\underline{r}}_i)\ . \tag{102}$$

In the region between the inner spheres and the outer sphere

$$\psi_{II} = \sum_{\ell m} a_{\ell m}^{II}\, j_\ell\,(k\,r_o)\; Y_{\ell m}\,(\hat{\underline{r}}_o) + \sum_i \sum_{\ell m} b_{\ell m}^{II_i}\, n_\ell(k\,r_i)\; Y_{\ell m}\,(\hat{\underline{r}}_i) \tag{103}$$

Finally, in the outer region

$$\psi_{III} = \sum_{\ell m} [a_{\ell m}^{III}\; f_\ell^{III}\,(k\,r_o) + b_{\ell m}^{III}\; g_\ell^{III}\,(k\,r_o)]\; Y_{\ell m}(\hat{\underline{r}}_o) \tag{104}$$

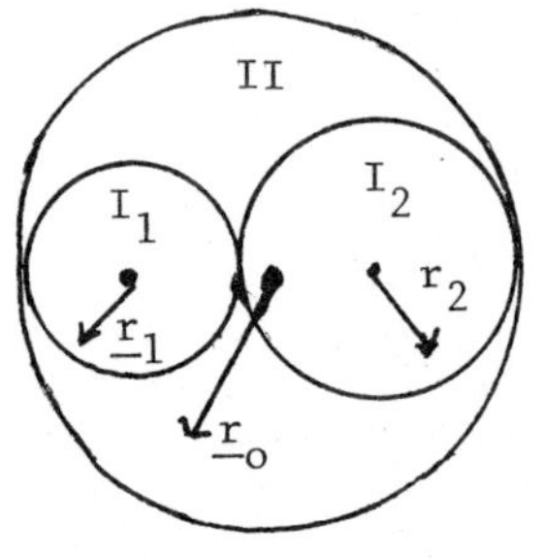

Fig. 10. Partitioning of space in the continuum multiple scattering method

In these equations f_ℓ and g_ℓ are the regular and irregular solutions in the given potentials in the appropriate regions, and j_ℓ and n_ℓ are the regular and irregular spherical Bessel functions. The coefficients $a_{\ell m}$ and $b_{\ell m}$ are obtained by matching the functions and derivatives on all the spherical boundaries. The K-matrix and cross section can then be obtained by matching to the asymptotic form defined by eqs. (32), (37) and (39).

Clearly, the CMSM method suffers from two main defects. Firstly the potential in region II is by no means constant. Secondly, the cut-off in the polarisation potential on the boundary of region III is a very ad-hoc procedure. Nevertheless, this approach has proved very useful in order to survey many molecules with a relatively modest amount of computational effort.

Other L^2 Methods. The R-matrix method and the Schwinger variational method can be regarded as members of a class of methods in which the wave function is expanded in terms of quadratically integrable functions which are only accurate over a limited region of space. This enables standard molecular structure codes, which have been developed over many years, to be used with little modification. A number of other methods which use quadratically integrable functions have enjoyed success in the study of electron molecule collisions and molecular photoionisation.

The Stiltjes-moment method, introduced by Langhoff [112,113] replaces the continuum states in the equations (43) and (44) for the photoionisation cross sections by a discrete basis defined by expanding the total wave function in the form

$$\psi_k = \sum_j \xi_j \, b_{jk} \tag{105}$$

where ξ_i are quadratically integrable functions and the coefficients b_{jk} are determined by diagonalising the total Hamiltonian

$$\langle \psi_k | H | \psi_{k'} \rangle = E_k \, \delta_{kk'} \quad . \tag{106}$$

These two equations can be contrasted with the corresponding equations (85) and (86) used in the R-matrix method. The method then proceeds by defining the oscillator strengths

$$\tilde{f}_i = \frac{2}{3} \tilde{\varepsilon}_i | \langle \psi_i | \quad D^L_{\mu=0} \quad | \psi_o \rangle |^2 \tag{107}$$

where $\tilde{\varepsilon}_i = E_i - E_o$. This equation provides a discrete representation of eq. (43).

In order to extract the continuum oscillator strength we introduce the spectral moments

$$S(-k) = \sum_{i=1}^{m} \tilde{f}_i \, \tilde{\varepsilon}_i^{-k} \qquad 0 \leq k \leq 2n - 1 \tag{108}$$

where m is the number of basis terms retained in eq. (105). If $m \gg n$ then eq. (108) provides an accurate representation of these moments. The principal representation of these moments is then constructed so that

$$S(-k) = \sum_{i=1}^{n} f_i \, \varepsilon_i^{-k} \qquad 0 \leq k \leq 2n - 1 \tag{109}$$

where there are 2n equations in the 2n unknowns ε_i and f_i, $i = 1,\ldots,n$. The ε_i and f_i provided a smoothing of the original $\tilde{\varepsilon}_i$ and $\tilde{f}_i$. The Stieltjes distribution is then defined by

$$f_S^n(\varepsilon) = 0 , \quad 0 < \varepsilon < \varepsilon_1$$
$$f_S^n(\varepsilon) = \sum_{i=1}^{j} f_i , \quad \varepsilon_j < \varepsilon < \varepsilon_{j+1} \tag{110}$$
$$f_S^n(\varepsilon) = \sum_{i=1}^{n} f_i = S(0), \quad \varepsilon_n < \varepsilon .$$

The corresponding Stieltjes densities, obtained from the slopes of the straight line segments connecting successive midpoints of the vertical portions of the Stieltjes distribution histogram defined by eqs. (10) is

$$\frac{df_S^n}{d\varepsilon} = 0 \quad 0 < \varepsilon < \varepsilon_1$$
$$\frac{df_S^n}{d\varepsilon} = \frac{\frac{1}{2}(f_{j+1}^n + f_j^n)}{\varepsilon_{j+1} - \varepsilon_j} , \quad \varepsilon_j < \varepsilon < \varepsilon_{j+1} \tag{111}$$
$$\frac{df_S^n}{d\varepsilon} = 0 , \quad \varepsilon_n < \varepsilon .$$

This can be shown to converge to the exact continuum oscillator strength distribution in the limit $n \to \infty$ and hence provides an approximation for the photoionisation oscillator strength distribution for finite n.

Alternative procedures have also been discussed in which one of the ε_i in eq. (109) is fixed at the energy of interest and the remaining ε_i and f_i obtained by solving the coupled equations.

We also mention here the stabilisation method introduced by Hazi and Taylor [114] which exploits the fact that the inner region of the wave function may be approximated up to a normalisation factor by quadratically integrable basis functions. If the Hamiltonian is diagonalised in a sufficiently large basis the eigenvalues approximating resonances become stable with respect to the size of the basis and exhibit a slope with respect to changes in the basis size proportional to the width of the resonance.

Other L^2 methods which have been particularly useful for obtaining resonance positions and widths, are the complex coordinate and complex scaling methods which have been reviewed by Ho [115].

Expansion in Spheroidal Coordinates. A basic difficulty with the single-centre expansion methods discussed in the Section on Single-centre Expansion Methods, is their slow convergence for all molecules except those containing hydrogen atoms. We have seen that multicentre expansion methods similar to those used in bound state calculations overcome this difficulty. An alternative approach in the case of diatomic molecules is to use two-centre spheroidal coordinates defined by

$$\xi_i = \frac{r_{iA} + r_{iB}}{R} , \quad \eta_i = \frac{r_{iA} - r_{iB}}{R} , \quad \phi \tag{112}$$

where r_{iA} and r_{iB} are the distances of the ith electron from nuclei A and B, R is the internuclear separation and ϕ is the azimuthal angle about the molecular axis. The wave function describing the scattered electron is then expended as

$$F(\underline{r}) = \sum_{\ell m} \frac{1}{R} \frac{f_{\ell m}(\xi)}{(\xi^2 - 1)^{\frac{1}{2}}} Y_{\ell m}(\eta, \phi) \tag{113}$$

and coupled integrodifferential equations are then derived for the $f_{\ell m}$.

This approach has been used by Hara [40,116] and Bell [117] for $e - H_2$ scattering and by Crees and Moores [118] for $e^- - N_2$ scattering. Recently it has been the basis of work on the H_2 photoionisation by Hara and collaborators [119,120,121].

Partial-differential-equations Method. An interesting new method which also avoids the slow convergence problems associated with the single-centre expansion methods has been suggested by Temkin and collaborators [122,123, 124]. In this case the partial integrodifferential equations (34) are solved directly. For a diatomic or linear molecule we can use the cylindrical symmetry of the molecule to reduce the equations to two-dimensions in the variables (r,θ) which are defined in figure 3. In this case the corresponding two dimensional Laplacian can be written as

$$\nabla^2 = \frac{\partial^2}{\partial r^2} + \frac{1}{r^2}\left[\frac{\partial^2}{\partial \theta^2} + \cot\theta\,\frac{\partial}{\partial\theta} - \frac{m^2}{\sin^2\theta}\right] \tag{114}$$

where m is the usual azimuthal quantum number which is conserved in this case. The non-local exchange potential terms can be treated non-iteratively due to their special form [124].

Recently, the method has been extended to allow for correlation effects in the target wave function and results obtained for $e - N_2$ collisions [125] including just the ground electronic state. In general, the effect of including target correlation is to decrease the phase shift and in the case of $^2\Sigma_g$ scattering to increase the cross section.

The extension of this method to include more than one electronic state or to non-linear molecules will be computationally very demanding and it remains to be seen how competitive it will be in such cases.

VIBRATIONAL EXCITATION AND DISSOCIATIVE ATTACHMENT

In this section we consider methods for treating the nuclear motion which go beyond the adiabatic nuclei approximation considered in the Section on the Molecular Frame Representation. Our emphasis will be on vibrational excitation and dissociative attachment

Hybrid Theory of Vibrational Excitation

One of the most straightforward ways of including non-adiabatic effects in vibrational excitation is to retain the vibrational term in the Hamiltonian but still treat the rotational motion adiabatically. Hence, instead of solving eq. (30), we now solve the equation

$$(H_{e\ell} + T_{vib} + T + V)\tilde{\psi} = E\,\tilde{\psi} \tag{115}$$

where T_{vib} is the kinetic energy operator for the nuclear vibrational motion and where other quantities have the same meaning as in eq. (30). Instead of using eq. (32) we expand the wave function $\tilde{\psi}$ in the form

$$\tilde{\psi} = \mathcal{A}\sum_{iv} \phi_i(\underline{R},\underline{r}_m)\chi_{iv}(\underline{R})\,H_{iv}(\underline{r}) + \sum_j \eta_j(\underline{R},\underline{r}_m,\underline{r})c_j \tag{116}$$

where we adopt a frame of reference in which the molcule has a fixed orientation and where $\chi_{iv}(\underline{R})$ are the molecular vibrational wave functions. In the case of a diatomic molecule, these wave functions depend only on the radial distance R. Substituting eq. (116) into eq. (115), projecting onto the functions $\phi_i \chi_{iv}$ and η_i and eliminating the coefficients c_j then yields coupled integrodifferential equations for the functions H_{iv} which are similar in form to eqs. (34). However the potentials now couple the target vibrational as well as electronic states. These equations can be solved by separating out the angular variables using an expansion analogous to eq. (37) giving a set of coupled radial integrodifferential equations for the radial parts of the function H_{iv}.

Equations of this type have been solved by Chandra and Temkin [126,127], Temkin [128] and Choi and Poe [129, 130]. However they are computationally very demanding since for each electronic and vibrational level an expansion is required in the angular momentum of the scattered electron which for molecules composed of heavy atoms is slowly convergent. Nevertheless, almost converged results were obtained for e - N_2 collisions by Chandra and Temkin who considered only the ground electronic state and adopted a local exchange potential and a polarisation potential with an adjustable parameter which was set to give the correct $^2\Pi_g$ resonance position.

Resonance Theory

We have already remarked that the adiabatic-nuclei approximation breaks down in the neighbourhood of narrow resonances, and it is just in these energy regions where experiment shows that the vibrational excitation cross sections are large. In this section we discuss methods in which vibrational excitation and dissociative attachment cross sections are treated from the outset as resonance phenomena. We will first consider an approach developed by Herzenberg and collaborators [15,131,132,133,134].

As an example of the role of resonances in vibrational excitation we show in figure 11 the potential energy curves and nuclear wave functions which arise in the boomerang model for e - N_2 scattering introduced by Herzenberg [133] and Birtwistle and Herzenberg [135]. The incident electron is capture into an intermediate resonance state of N_2^- as illustrated in the figure. The nuclei then move apart and are reflected at the outer turning point of the resonance state before the electron is emitted leaving the molecule in an excited vibrational state.

For each final value of the internuclear coordinates $\underline{R}$ we define a set of resonance states ψ_n^r as solutions of the Schrödinger equation

$$(H - W_n(\underline{R})\ \psi_n^r = 0 \tag{117}$$

subject to the outgoing wave Siegert [136] boundary conditions

$$\psi_n^r \underset{r\to\infty}{\sim} \sum_s \Phi_s \, e^{ik_{ns}r} a_{ns} \tag{118}$$

where the Φ_s are the target channel functions, where the k_{ns} are wave numbers for the colliding electron defined by

$$k_{ns}^2 = 2\ (W_n - \varepsilon_s) \tag{119}$$

and where the a_{ns} are normalisation constants. In these equations, we have adopted the notation given earlier by eqs. (35), (80) and (85).

If the total energy W_n lies below the first electronic threshold, then all channels are closed and hence the wave numbers k_{ns} are pure imaginary. The corresponding wave function defined by eq. (118) then decays exponentially asymptotically and corresponds to a bound state of the electron plus molecule system. On the other hand if the energy is such that some channels are open then the Siegert prescription ensures that

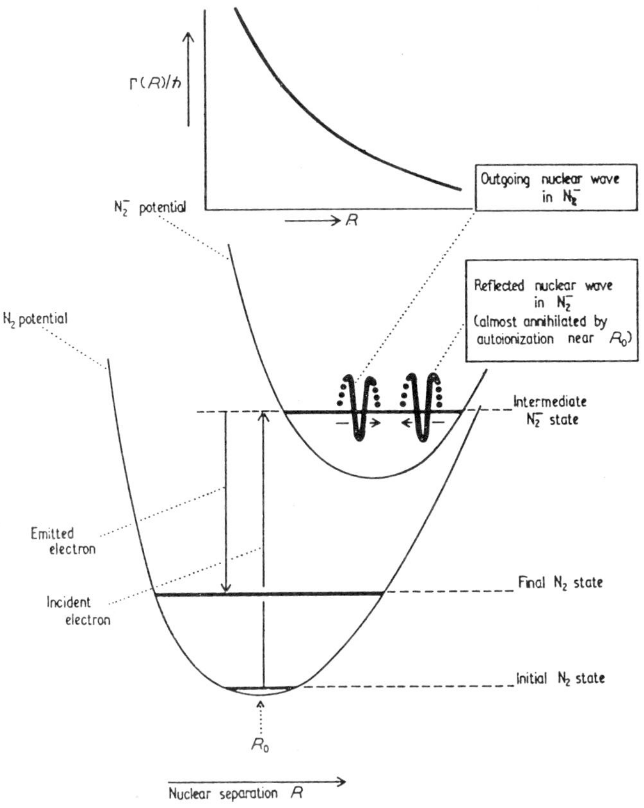

Fig. 11. A schematic picture of the potential energy curves and nuclear wave functions in the boomerang model of vibrational excitation (figure 2 from Birtwistle and Herzenberg [135]).

the corresponding eigenvalues are complex. This corresponds to a resonance state of the electron-molecule system. We write in this case

$$W_n(\underline{R}) = E_n(\underline{R}) - \tfrac{1}{2}\, i\, \Gamma_n(\underline{R}) \tag{120}$$

where the resonance position $E_n(\underline{R})$ and resonance width $\Gamma_n(\underline{R})$ are both real and $\Gamma_n(\underline{R})$ is positive.

If only one resonance is important at the energy of interest then the complete collision wave function including the nuclear motion can be written in accordance with the Born Oppenheimer approximation as

$$\Psi_i = \psi_n^r\, \eta_{ni}(R) + \psi_i^P \tag{121}$$

where the suffix i stands for the rotational, vibrational and electronic quantum numbers defining the initial state. The wave function Ψ_i^P is orthogonal to ψ_n^r with respect to the electronic coordinates and contains the incident wave. The wave function $\eta_{ni}(\underline{R})$ varies rapidly with $\underline{R}$ and describes the nuclear motion. In order to derive an equation for η_{ni} we substitute eq. (121) into eq. (15) and project onto the function ψ_n^r. If we assume that the resonant state is slowly varying with $\underline{R}$ in accordance with the Born Oppenheimer approximation and we remember that H_m+T+V = H + $T_{\underline{R}}$, we find that η_{ni} satisfies the equation

$$(T_{\underline{R}} + W_n(\underline{R}) - E)\,\eta_{ni}(\underline{R}) = - \langle \psi_n^r | H | \Psi_i^P \rangle_e \qquad (122)$$

The integral on the right hand side of this equation involves only the electronic coordinates and goes over an internal region of space where the colliding electron is inside the charge distribution of the molecule.

In practise it is often a good approximation to write

$$\langle \psi_n^r | H | \Psi_i^P \rangle_e = \zeta_{ni}(\underline{R})\chi_i(\underline{R}) \qquad (123)$$

where ζ_{ni} is the so-called "entry amplitude" from the initial state into the resonant state and χ_i is the initial vibrational wave function of the target. We can write a solution of eq. (122) in the form

$$\eta_{ni}(\underline{R}) = - \int G_n(\underline{R},\underline{R}')\,\zeta_{ni}(\underline{R}')\,\chi_i(\underline{R}')\,d\underline{R}' \qquad (124)$$

where $G_n(\underline{R},\underline{R}')$ is the Green's function of the operator on the left hand side of eq. (122). The cross section for vibrational excitation is then obtained by calculating the decay of the resonant state into the final vibrational state. The corresponding amplitude for a transition from an initial state i to a final state j is given by

$$f_{ji} = - \int\int \chi_j^*(\underline{R})\zeta^*_{nj}(\underline{R})\,G_n(\underline{R},\underline{R}')\,\zeta_{ni}(\underline{R}')\,\chi_i(\underline{R}')\,d\underline{R}\,d\underline{R}' \,. \qquad (125)$$

Dubé and Herzenberg [15] have applied this theory to calculate $e^- - N_2$ vibrational excitation cross sections. We have already shown some of their results, compared with experiment in figure 1. The theory assumes that there is only one resonance state of N_2^- with $^2\Pi_g$ symmetry which contributes to the cross section and the real and imaginary parts of the corresponding potential energy curve is written in terms of six adjustable parameters. The potential energy curve obtained in this way is in good agreement with ab-initio calculations. Alternatively ab-initio calculations of complex potential energy curves can be input into this boomerang model to yield vibrational excitation cross sections.

Dissociative attachment cross sections can be obtained by a straightforward extension of the above theory. The electron is captured as before into an intermediate resonance state. The nuclei then move apart along the resonance potential curve. If they separate to the point where the resonance curve crosses the neutral molecule curve and becomes real, before the resonance decays then dissociative attachment occurs. Otherwise we have vibrational excitation as discussed above. The cross section for dissociative attachment is obtained by calculating the flux of negative ions at large internuclear distances giving

$$\sigma_{DA} = \frac{2\pi^2}{k_i}\,\frac{K}{\mu}\,\lim_{R\to\infty}|a(R)|^2 \qquad (126)$$

where k_i and K are the wave numbers describing the motion of the incident electron and outgoing ion in the centre-of-mass system, μ is the reduced mass of the nuclei and a(R) is the amplitude of the outgoing-wave solution of eq. (122).

We conclude this discussion by noting that the resonance theory of vibrational excitation and dissociative attachment based on eq. (122) is clearly only approximate. A detailed discussion of the validity of the approximations inherent in the derivation of this equation has been given by Herzenberg [134]. One important modification in a more complete theory is that the potential $W_n(R)$ becomes non-local and energy dependent as well as complex. The locality of the potential implies that the electronic wave function follows the nuclei adiabatically when all the electrons are close to the molecule. Bardsley [137] has formulated a more complete theory which gives rise to a non-local potential, and he shows that the potential becomes local if the energy of the incident electron outside the molecule is so high that several vibrational states may be excited. However the question of the non-locality of the potential is still one for discussion.

As an example of an application of the theory of dissociative attachment we show in figure 12 the dissociative attachment cross section for F_2 defined by

$$e^- + F_2 \rightarrow F + F^- \tag{127}$$

and calculated by Hazi et al [138] compared with semi-empirical calculations and with experiment. In this calculation the resonance parameters were extracted from a large-scale CI calculation using the Stieltjes-moment theory [140], and the wave equation (122) for the dissociating nuclei solved using a non-local potential which avoids the local potential approximation used in the boomerang model. The ab initio calculations of Hazi et al [138] are about a factor of 1.5 above experiment which can be regarded as satisfactory in view of the sensitivity of the results to the detailed behaviour of the potential energy curves.

We conclude this section on resonance theory methods by mentioning an ab-initio approach using Feshbach projection operators [141]. The basic idea is to extract the rapid variations of the T-matrix caused by resonances, virtual states or bound states such that the remaining "background" T-matrix is weakly dependent on energy and on internuclear distance and can be treated in the adiabatic-nuclei approximation. The formalism yields a simple prescription for treating the "resonant" part of the T-matrix essentially exactly thus taking into account non-Born Oppenheimer effects.

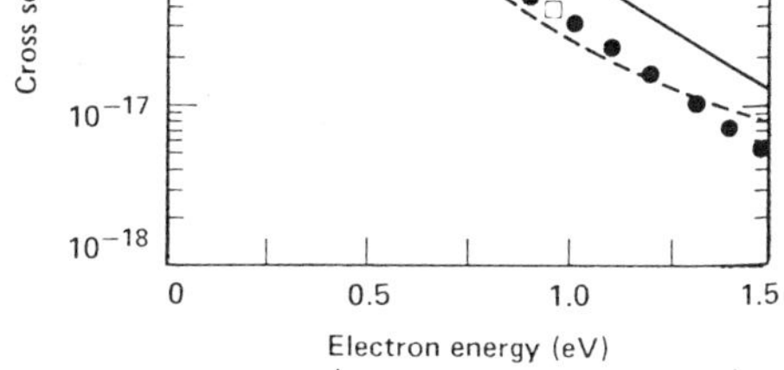

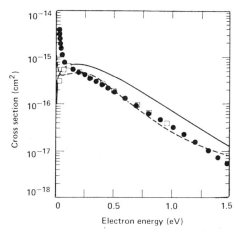

Fig. 12. Dissociative attachment cross section for the ground vibrational state of F_2. Solid line, theory of Hazi [138]; dashed line, semi-empirical calculation of Hall [139]; squares, semi-empirical calculations of Bardsley et al; solid circles, experiment of Chantry (fig.2 of Hazi et al [138]).

We introduce the projection operators

$$P = \int |\phi_k><\phi_k| k dk d\Omega_k \qquad Q = |\phi_d><\phi_d| \tag{128}$$

where ϕ_d is a square integrable electronic state which approximates the bound state or resonance and ϕ_k are continuum states which are orthogonal to ϕ_d. We assume that ϕ_d changes only slowly when the internuclear distance is varied hence it is quasi-diabatic. We note that these projection operators are closely related to those introduced earlier by eqs. (32) and (37) except that Q now projects onto a single square integrable function. Domcke et al [142,143,144] then show that the multi-channel resonant T-matrix is exactly given by the formal operator expression

$$T_{res}(k_f, v'; k_i, v) = <v'|V_{kf}(R)\,(E - \mathcal{H})^{-1}\,V^*_{ki}(R)|v> \tag{129}$$

where

$$\mathcal{H} = H_o + \varepsilon_d(R) + \Delta(R, E - H_o) - \tfrac{1}{2}\, i\Gamma(R, E - H_o) \tag{130}$$

Here H_o is the vibrational Hamiltonian of the target and

$$V_k(R) = <\phi_k|H|\phi_d> \tag{131}$$

$$\varepsilon_d(R) = <\phi_d|H|\phi_d> \tag{132}$$

$$\Gamma(R,E) = 2\pi \int |V_k(R)|^2\, d\Omega_k \tag{133}$$

$$\Delta(R,E) = \frac{1}{2\pi}\, P \int \frac{\Gamma(R,E')}{E-E'}\, dE' \quad , \tag{134}$$

H being the electronic Hamiltonian. Methods for the ab initio calculation of the basic qualities ε_d, Γ and Δ have been given by Hazi [140] and Berman and Domcke [145]. This theory is equivalent to the non-local-complex-potential theory formulated by Bardsley [137] and others.

This projection operator formalism has been applied to a variety of resonance phenomena including shape resonances close to threshold, threshold peaks in polar molecules and threshold effects [141] where we note that threshold effects are included provided that the energy dependence of V_k, Γ and Δ are properly taken into account.

R-Matrix Method

The fixed-nuclei R-matrix method discussed earlier has been extended to treat vibrational excitation and dissociative attachment by Schneider et al [146]. The interaction region is again partitioned into two regions but now the internal region is taken in the case of diatomic molecules, to be a hypersphere defined by $0 \leq r \leq a$ and $A_{in} \leq R \leq A_{out}$ as illustrated in figure 13. Here r and a are defined as in the fixed nuclei approximation (see figure 9), R is the internuclear separation, A_{in} is chosen to encircle the nuclear repulsion singularity in the potential and A_{out} is chosen such that those target vibrational states which are to be included in the calculation have negligible amplitude for $R \geq A_{out}$. For $r > a$ the system separates into an electron plus residual molecule as discussed earlier in the fixed-nuclei approximation, while for $R > A_{out}$ the system separates into an atom plus a negative ion corresponding to dissociative attachment.

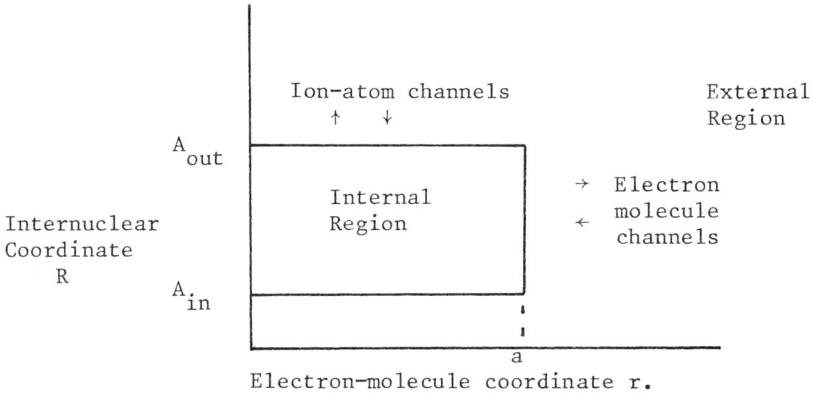

Fig. 13. Partitioning of space in the non-adiabatic R-matrix method.

We introduce the following expansion basis in the internal region

$$\Psi = \sum_i \psi_i \, \eta_i \, (\underline{R}) \tag{135}$$

where the ψ_i are the R-matrix basis states defined by eqs. (85) and (86) for each value of the internuclear distance and η_i are wave functions describing the nuclear motion. This equation replaces eq. (121) in the resonance theory approach. In order to obtain equations for the functions η_i we substitute eq. (135) into eq. (15) and project onto the functions ψ_i. If we assume, in accordance with the Born-Oppenheimer approximation, that the nuclear kinetic energy operator acts only on the functions $\eta_i(R)$ we obtain

$$(T_{\underline{R}} + E_i(R) - E) \, \eta_i(\underline{R}) = \langle \psi_i | L_b | \Psi \rangle \tag{136}$$

where the $E_i(R)$ are defined by eq. (86) and the Bloch operator is defined by eq. (87). We note that in some applications some eigenvalues $E_i(R)$ are found to cross as R is varied. In this case, additional terms must be included in eq. (136) corresponding to the action of $T_{\underline{R}}$ on the R-matrix basis states $\psi_i(R)$ (eg see Gillan et al [65]). In addition, if dissociative attachment is being considered, then the Bloch operator L_b in eq. (136) must be augmented by an additional term allowing for dissociation into $(A+B^-)$-type channels. Schneider et al [146] show that this leads to additional terms coupling the electronic and vibrational channels with the dissociating channels. We will not consider such extensions here.

It is interesting to compare eq. (136) with the corresponding eq. (122) obtained using the resonance theory approach. We see that the potential on the left-hand-side of the equation is now real rather than complex and the surface term replaces the "entry amplitude" on the right-hand-side of the equations.

Electronic and vibrational excitation cross sections can be obtained by projecting the total wave function defined by eq. (135) onto the appropriate channel functions and evaluating it on the surface of the internal region. We obtain

$$\langle \Phi_i \, \chi_{iv} | \Psi \rangle_{r=a} = \sum_k \langle \Phi_i \, \chi_{iv} | \psi_k \, \eta_k \rangle_{r=a} \, . \tag{137}$$

We substitute for η_k from eq. (136) by introducing the Green's function

$G_k(\underline{R},\underline{R}')$ for the corresponding operator on the left-hand-side of this equation. This yields

$$\langle\Phi_i\ \chi_{iv}|\Psi\rangle_{r=a} = \sum_{i'v'} \mathcal{R}_{iv\ i'v'} \left[\left(\frac{d}{dr} - \frac{b}{r}\right) \langle\Phi_{i'}\chi_{i'v'}|\Psi\rangle\right]_{r=a} \tag{138}$$

where we have introduced a generalised R-matrix by the equation

$$\mathcal{R}_{iv\ i'v'} = \frac{1}{2a} \sum_k \langle w_{ik}(\underline{R})\chi_{iv}(\underline{R})\ G_k(\underline{R},\underline{R}')\chi_{i'v'}(\underline{R}')\ w_{i'k}(\underline{R}')\rangle \tag{139}$$

and where the surface amplitudes w_{ik} are defined by eq. (94). If we neglect the kinetic energy operator $T_{\underline{R}}$ in eq. (136) then

$$G_k(\underline{R},\underline{R}') = \frac{\delta(\underline{R}-\underline{R}')}{E_k(\underline{R})-E} \tag{140}$$

and eq. (138) reduces to

$$\mathcal{R}_{iv\ i'v'} = \langle\chi_{iv}(\underline{R})\ R_{ii'}\ \chi_{i'v'}(\underline{R})\rangle \tag{141}$$

where $R_{ii'}$ is the fixed-nuclei R-matrix introduced earlier by eq. (93). Comparing this then with eq. (42) we see that after projecting onto the rotational eigenfunctions, eq. (141) corresponds to the adiabatic-nuclei approximation for the R-matrix.

Once the generalised R-matrix $\mathcal{R}_{iv\ i'v'}$ has been determined the final step in the calculation is to solve the collision problem in the external region. The relevant equations which must be solved in this region are the hybrid equations coupling the electronic and vibrational levels of the molecule. These equations have already been discussed in connection with eq. (116) where we note that the coupling potentials are local and behave as r^{-n} in this region.

This theory has been used to calculate vibrational excitation cross section in e-N_2 scattering by Schneider et al [147] and by Morgan [148]. We present in figure 14 Morgan's vibrational excitation cross sections compared with experimental measurements of Allan [149]. The qualitative features of the $0 \rightarrow v$ excitation cross sections for $v = 1$ to 17 are represented well by the theory but the theory increasingly underestimates the magnitude of the cross section as v increases. Thus if we define

$$R_v = \frac{[\sigma(0\rightarrow v)]_{max}}{[\sigma(0\rightarrow 1)]_{max}} \tag{142}$$

then R_{17} (experiment) $= 0.43.10^{-6}$ and R_{17} (theory) $= 0.39.10^{-8}$. Nevertheless, the theoretical result which is based on a completely ab-initio theory with no adjustable parameters is encouraging.

Energy Modified Adiabatic Approximation

We conclude this section by mentioning an interesting way of extending the adiabatic-nuclei approximation introduced by Nesbet [150]. In this approach, called the energy modified adiabatic (EMA) approximation the S-matrix elements connecting the vibrational states are defined by

$$S_{iv,i'v'} = \langle\chi_{iv}|S_{ii'}\ (E - H_i;\ R)|\chi_{i'v'}\rangle \tag{143}$$

where $S_{ii'}(E - H_i;\ R)$ is the S-matrix calculated in the fixed-nuclei approximation at the internuclear separation R at an energy defined by the operator $H_i = E_i(R) + T_{vib}$. This has the effect of including the

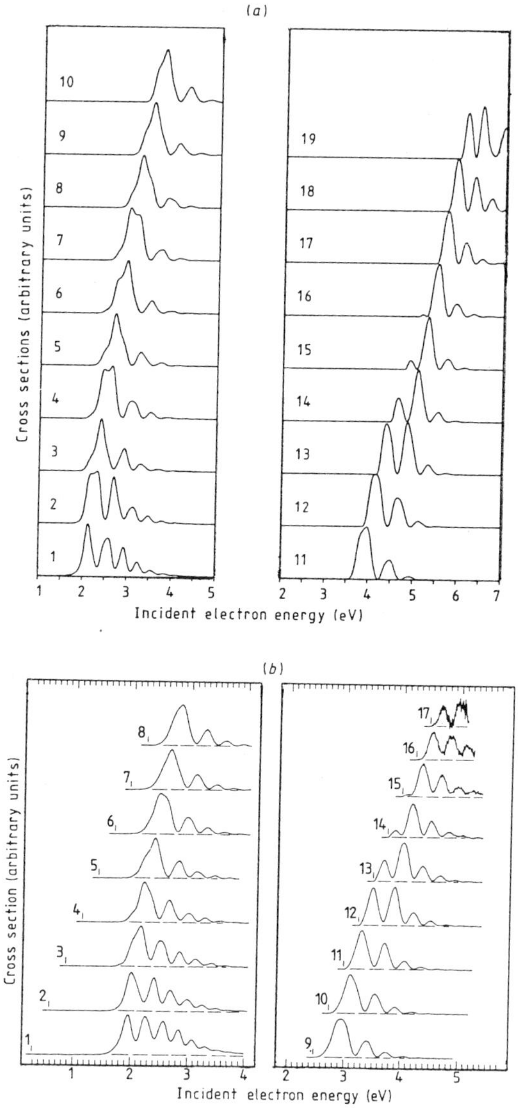

Fig. 14. Vibrational excitation cross sections for 0→v transitions in $e^- -N_2$ scattering. (a) R-matrix calculation of Morgan [148], (b) experiment of Allan [149] (fig.2 of Morgan [148])

internal energy of the target and has been successfully applied to describe the observed structure in the vibrational excitation cross section in $e^- -N_2$ scattering.

FURTHER ILLUSTRATIVE RESULTS

In this section we present a few further results which illustrate the theory and calculational methods presented in the previous sections.

$e^- - H_2^+$ collisions and H_2 photoionisation

The hydrogen molecular ion is the simplest molecular target from which we can scatter electrons. Furthermore, the $e - H_2^+$ states are also reached by photoionising the neutral hydrogen molecule. The following closely related processes are of interest:

Elastic scattering,

$$e^- + H_2^+ \rightarrow e^- + H_2^+ , \tag{144}$$

vibrational excitation,

$$e^- + H_2^+(v) \rightarrow e^- + H_2^+(v') , \tag{145}$$

electronic excitation,

$$e^- + H_2^+ (X^2\Sigma_g^+) \rightarrow e^- + H_2^+(A^2\Sigma_u^+) \tag{146}$$

dissociative recombination,

$$e^- + H_2^+ \rightarrow H + H \tag{147}$$

three body dissociation,

$$e^- + H_2^+ \rightarrow H + H^+ + e^- \tag{148}$$

photoionisation,

$$h\nu + H_2 \rightarrow H + H \tag{149}$$

vibrationally resolved photoionisation,

$$h\nu + H_2(v) \rightarrow H_2^+(v') + e^- \tag{150}$$

and dissociative photoionisation

$$h\nu + H_2 \rightarrow H + H^+ + e^- . \tag{151}$$

All of these processes go through common intermediate continuum states of H_2. These states may be resonant or non-resonant. We present in figure 15 the relevant potential curves for H_2 and H_2^+. The Rydberg resonance converging to the first excited $A^2\Sigma_u^+$ state of H_2^+ have a profound effect on all of the above processes.

Calculations on $e - H_2^+$ collisions have been carried out using a number of approximations discussed earlier in these lectures. We present in figure 16 the positions of the three lowest resonances with $^1\Sigma_g^+$ symmetry and the width of the lowest resonance. All calculations for the position of the lowest resonance are in good agreement, however there are considerable discrepancies in the calculated width. In particular the width obtained by the older calculations of Bottcher and Docken [157] are in clear disagreement with the rest at large values of R. The importance of these resonance curves is that they control the dissociative recombination process denoted by eq. (147). The corresponding $^1\Sigma_u^+$ and $^1\Pi_u$ resonances also play an important role in H_2 photoionisation. As an example we show in figure 17 the asymmetry parameter β defined by eq. (53) for two H_2 geometries calculated using the R-matrix theory with the inclusion of two electronic states in the final state expansion [151]. This work is in good

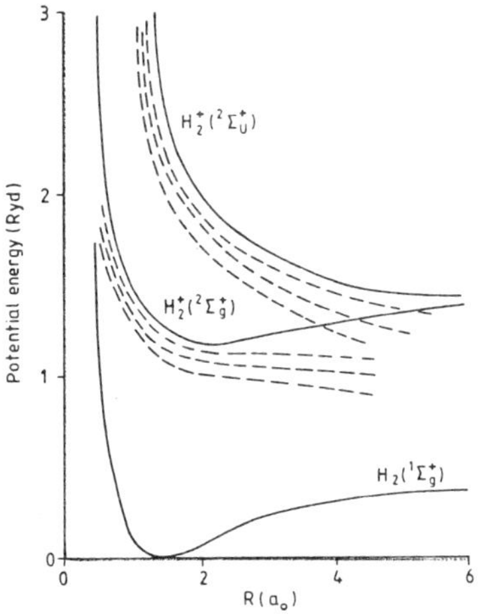

Fig. 15. Potential curves for H_2 and H_2^+. The dashed lines indicate schematically the H_2^* Rydberg states converging to $H_2^+(X^2\Sigma_g^+)$ and $H_2^+(A^2\Sigma_u^+)$ (Figure 1 from Tennyson et al [151]).

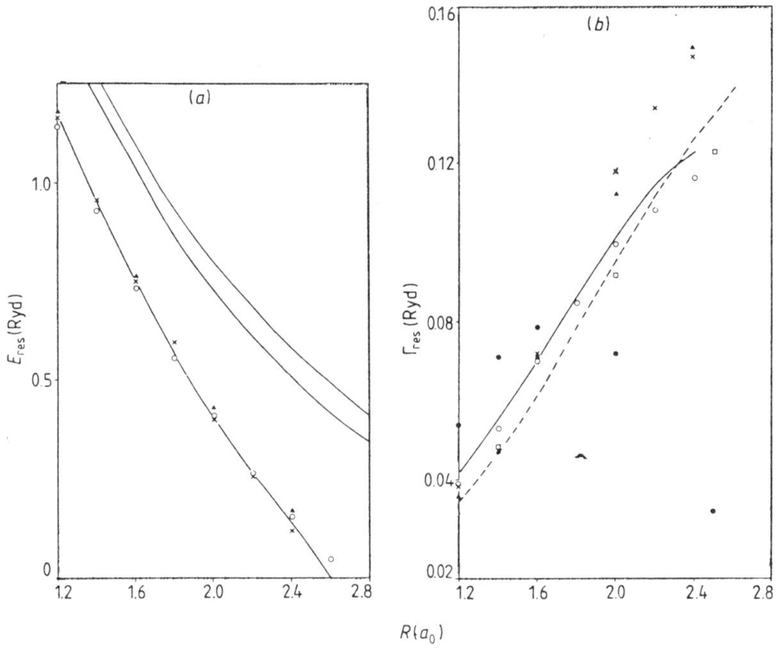

Fig. 16. (a) Position of the lowest three resonances of $^1\Sigma_g^+$ symmetry as a function of H_2^+ separation Full curve, R-matrix calculation including polarisation [152]; Broken curve, R-matrix calculation excluding polarisation [152]; *, Tagaki and Nakamura [153]; O, Sato and Hara [154];▲ , LA calculation of Collins and Schneider [155]; □, Hazi et al [156];● , Bottcher and Docken [157] (Figure 1 from Tennyson and Noble [152]).

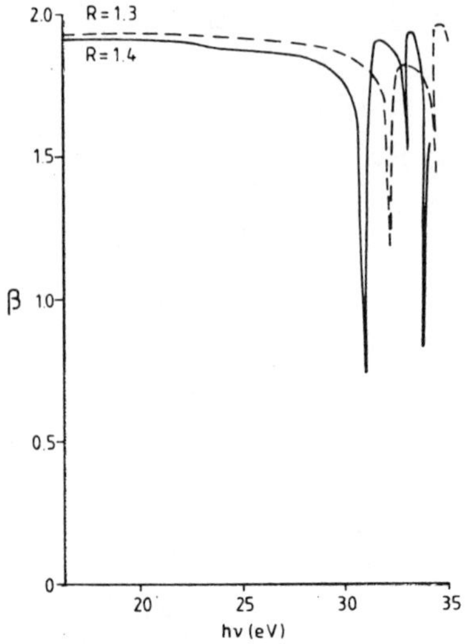

Fig. 17. Asymmetry parameter β as a function of the photon energy for two H_2 geometries (Figure 5 from Tennyson et al [151]).

agreement with other recent calculations but in disagreement with experiment which finds a lower value of β close to threshold [158]. Further work needs to be carried out to understand this discrepancy.

$e^- - H_2$ Collision

The next simplest system to consider is molecular hydrogen. Recently state-of-the-art ab-initio calculations have been carried out using the R-matrix method [159], the LA method [160] and the Schwinger variational method [161] for the electronic excitation process

$$e^- + H_2(^1\Sigma_g^+) \rightarrow e^- + H_2(^3\Sigma_u^+) \quad . \tag{152}$$

We present these results in figure 18 where they are compared with LA calculations which omitted certain relaxation effects (this corresponds to older calculations by Chung and Lin [162]) and with experiment. The three calculations which retain two electronic states in the expansion (together with additional correlation functions in the R-matrix case) are in excellent agreement with each other and with experiment. This calculation shows that the three methods will give equivalent results when equivalent approximations are made.

Photoionisation of N_2

Photoionisation leading to the $X^2\Sigma_g^+$ state of N_2^+ is of primary interest due to the appearance of a shape resonance in the cross section. In the one electron picture we have

$$h\nu + N_2 \rightarrow N_2^*(3\sigma_g^{-1}\, k\sigma_u\ {}^1\Sigma_u^+) \rightarrow N_2^+(X^2\Sigma_g^+) + e^- \tag{153}$$

where the intermediate resonance state is mainly f-wave ($\ell = 3$). Such resonances are a common feature of molecular photoionisation cross sections as discussed by Dehmer et al [18]. Also we note that this $^1\Sigma_u^+$ resonance in N_2 is the analogue of the $^2\Sigma_u^+$ resonance at 25 eV in $e^- - N_2$ scattering shown in figure 5.

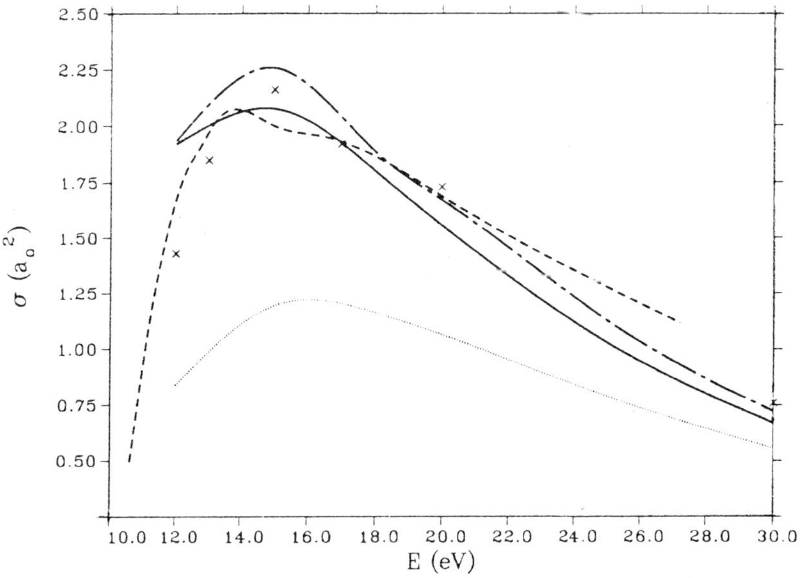

Fig. 18. Comparison of theory and experiment for $^3\Sigma_u^+$ excitation in $e^- - H_2$ collisions. dashed line, R-matrix method [159]; solid line LA method [160]; chain line; Schwinger method [161]; dots, LA method omitting relaxation [162]; crosses; experiment [163] (Figure 3 from Collins and Schneider [37]).

Several calculations of this photoionisation cross section have been made. We show in figure 19 calculations by Lucchese et al [164] in the dipole length and velocity approximations using both a Hartree-Fock and a CI initial state compared with experiment. The final state wave function was obtained using a single-centre iterative Schwinger variation method. The peak in the cross section at a photon energy of about 25 eV is clearly visible.

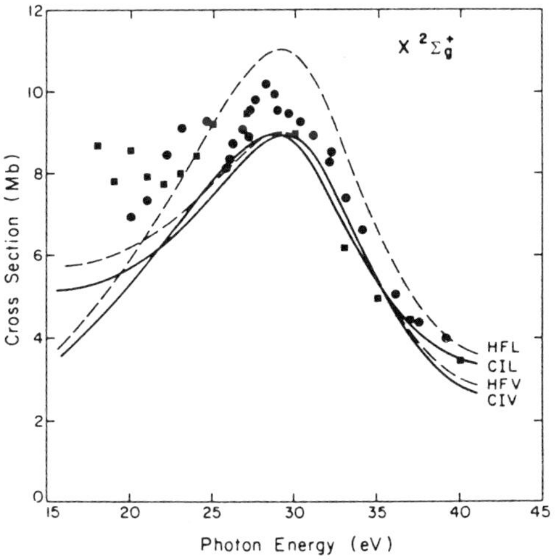

Fig. 19. Photoionisation cross section of N_2 leading to $N_2^+(X^2\Sigma_g^+)$. Calculations by Lucchese et al [164]; ■, expt. of Plummer et al [165]; ●, expt. of Hamnett et al [166] (Figure 4 from Lucchese et al [164]).

This resonance can also affect other photoionisation channels as pointed out by Stephens and Dill [167]. As an example, the asymmetry parameter β was calculated for photoionisation of N_2 leading to the $2\sigma_u^{-1}$ $B^2\Sigma_u^+$ state including coupling to the $3\sigma_g^{-1}$ $X\ ^2\Sigma_g^+$ state by Basden and Lucchese [168]. Their results using the $\tilde{C}$-functional method [102,107,164] compared with experiment are shown in figure 20. The observed oscillation in β is due to the coupling with the $^1\Sigma_u^+$ resonance and is well described by the theory.

Channel coupling effects can also be described by multichannel quantum-defect theory (MQDT). As an example, Raoult et al [172] have applied the MQDT approach to the study of autoionisation in the Hopfield series of molecular nitrogen. This is a very active area of research providing an analysis of extensive experimental data in terms of a few parameters which may be calculated from first principles or fitted to the data.

e^- - CH_4 Collisions

As one example of electron scattering by a polyatomic molecule we present recent integrated elastic cross section calculations carried out by Gianturco and Scialla [48]. We present in figure 21 a summary of the theoretical models compared with experiment. The full line in this figure refer to the SCE model with an orthogonality constraint discussed earlier in these lectures, while the dotted and dashed curves refer respectively to the HFEGE plus orthogonality with and without an asymptotic adjustment to the energy as discussed in connection with eq. (56). Finally the chained curve refers to a modified orthogonalised SCE model. It is interesting to see that both semi-classical models are close to the elastic experimental data indicated by squares and crosses of Lohmann and Buckman [173] and Ferch et al [174]. The experimental points represented by triangles refer to total cross section measurements of James [175] . It is also interesting to note that the above model also fares reasonably well even for more complicated systems like SiH_4.

We see from figure 21 that the cross section exhibits a broad peak in the 7 - 8 eV region. This is due to a shape resonance in the t_2 symmetry state.

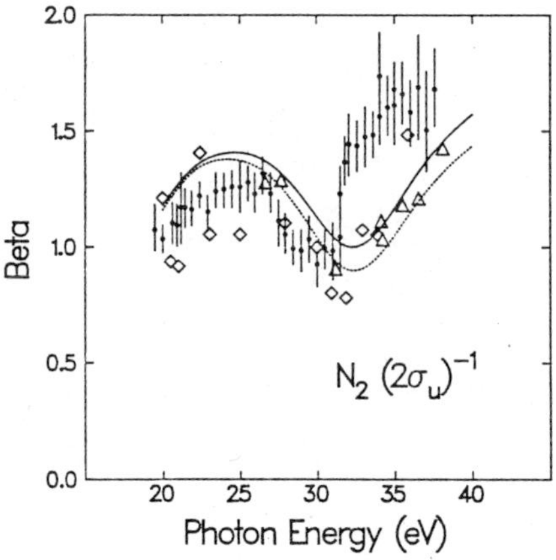

Fig. 20. Photoelectron asymmetry parameter for N_2 photoionisation leading to the $2\sigma_u^{-1}$ $B\ ^2\Sigma_u^+$ state of N_2^+. ——— , dipole velocity approx. [168]; - - - -, dipole length approx. [168];● , expt. of Southworth et al [169];△ , expt of Adam et al [170];◇, expt of Marr et al [171]. (Figure 1 from Basden and Lucchese [168]).

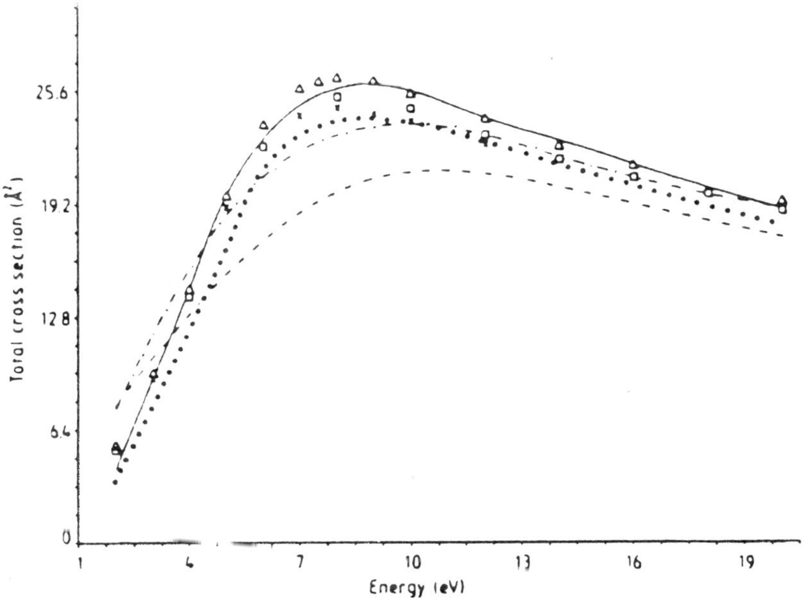

Fig. 21. Integrated elastic cross section for e^--CH_4 collisions with different exchange and correlation potentials discussed in the text (Figure 3 from Gianturco and Scilla [48]).

ACKNOWLEDGMENTS

It is a pleasure to acknowledge many valuable discussion on the subject matter of these lectures with colleagues at the Queen's University of Belfast, at the SERC Daresbury Laboratory and at the University of London.

REFERENCES

[1] Y. Itikawa, Phys. Reports 46, 117 (1978).
[2] P. G. Burke, Adv. At. Mol. Phys. 15, 471 (1979).
[3] N. F. Lane, Rev. Mod. Phys. 52, 29 (1980).
[4] D. W. Norcross and L. A. Collins, Adv. At. Mol. Phys. 18, 341 (1982).
[5] D. G. Thompson, Adv. At. Mol. Phys. 19, 309 (1983).
[6] F. A. Gianturco and A. Jain, Phys. Reports 143, 347 (1986).
[7] M. A. Morrison, Adv. At. Mol. Phys. 23 to be published (1987).
[8] S. Trajmar, D. F. Register and A. Chutjian, Phys. Reports, 97, 219,(1983).
[9] T. N. Rescigno, B. V. McKoy and B. I. Schneider (eds) Electron-Molecule and Photon-Molecule Collisions (Plenum Press, New York, 1979).
[10] F. A. Gianturco and G. Stefani (eds) Wavefunctions and Mechanisms for Electron Scattering Processes (Springer-Verlag, New York, Berlin, 1984).
[11] S. P. McGlynn, G. L. Findlay and R. H. Huebner (eds) Photophysics and Photochemistry in the Vacuum Ultraviolet (Dortrecht: Reidel Publishing Co., 1985).
[12] I. Shimamura and K. Takayanagi (eds) Electron-Molecule Colisions (Plenum Press, New York 1984).
[13] L. C. Pitchford, B. V. McKoy, A. Chutjian and S. Trajmar (eds). Swarm Studies and Inelastic Electron-Molecule Collisions (Springer-Verlag New York, Berlin, 1987).
[14] P. G. Burke in Atomic and Molecular Collision Theory, ed. F.A.Gianturco (Plenum Press, New York 1982) p. 69.
[15] L. Dubé and A. Herzenberg, Phys. Rev. A20, 194 (1979).
[16] H. Ehrhardt and K. Willman, Z. Phys. 204, 462, 1967.
[17] G. J. Schulz, Rev. Mod. Phys. 45, 423 (1973).

[18] J. L. Dehmer, A. C. Parr and S. H. Southworth, in Handbook on Synchrotron Radiation, Volume II, ed. G. V. Marr (North Holland Amsterdam 1986).
[19] K. Rohr and F. Linder, J. Phys. B. (Atom. Molec. Phys.) 9, 2521, (1976).
[20] P. G. Burke, N. Chandra and F. A. Gianturco, J. Phys. B. (Atom. Molec. Phys.) 5, 2212, (1972).
[21] H.S.W. Massey, Proc. Camb. Phil. Soc., 28, 99, (1932).
[22] D. W. Norcross and N. T. Padial, Phys. Rev. A25, 226, (1982).
[23] A. M. Arthurs and A. Dalgarno, Proc. Roy. Soc. A304, 465, (1960).
[24] H. C. Stier, Z. Phys. 76, 439, (1932).
[25] J. B. Fisk, Phys. Rev., 49, 167, (1936).
[26] M. E. Rose, in Elementary Theory of Angular Momentum (John Wiley, New York, 1957).
[27] D. M. Chase, Phys. Rev., 104, 838, (1956).
[28] A. Altshuler, Phys. Rev., 107, 114, (1957).
[29] E. S. Chang and U. Fano, Phys. Rev. A6, 173, (1972).
[30] L. A. Collins and D. W. Norcross, Phys. Rev. A18, 478, (1978).
[31] E. S. Chang and A. Temkin, J. Phys. Soc. Japan, 29, 172, (1970).
[32] M. A. Morrison, A. N. Feldt and B. C. Saha, Phys. Rev. A30, 2811, (1984).
[33] P. G. Burke, C. J. Noble and S. Salvini, J. Phys. B. (Atom. Molec. Phys.) 16, L113, (1983).
[34] R. E. Kennerly, Phys. Rev. A21, 1876, (1980).
[35] C. N. Yang, Phys. Rev., 74, 764, (1948).
[36] N. A. Cherepkov, Adv. At. Mol. Phys. 19, 395, (1983).
[37] L. A. Collins and B. I. Schneider, in Electron Molecule Scattering and Photoionisation, eds. P. G. Burke and J. B. West (Plenum Press, New York, 1988).
[38] K. R. Hoffman, M. S. Dababneh, Y. S. Hsieh, W. E. Kappila, V. Poi, J. H. Smart and T. S. Stein, Phys. Rev. A25, 1393, (1982).
[39] G. Dalba, P. Fornasini, I. Lazzizzea, G. Ranier and A. Zecca, J. Phys. B. (Atom. Molec. Phys.) 13, 2839, (1980).
[40] S. Hara, J. Phys. Soc. Japan, 22, 710, (1967).
[41] P. G. Burke and N. Chandra, J. Phys. B. (Atom. Molec. Phys.) 5, 1696 (1972).
[42] S. Salvini and D. G. Thompson, J. Phys. B. (Atom. Molec. Phys.) 14, 3797, (1981).
[43] A. Jain and D. G. Thompson, J. Phys. B. (Atom. Molec. Phys.) 15, L631, (1982).
[44] A. Jain and D. G. Thompson, J. Phys. B. (Atom. Molec. Phys.) 16, 3077, (1983).
[45] A. Jain and D. G. Thompson, J. Phys. B. (Atom. Molec. Phys.) 16, L347, (1983).
[46] A. Jain and D. G. Thompson, J. Phys. B. (Atom. Molec. Phys.) 17, 443, (1984).
[47] J. B. Furness and I. E. McCarthy, J. Phys. B. (Atom. Molec. Phys.) 6, 2280, (1973).
[48] F. A. Gianturco and S. Scialla, in Electron Molecule Scattering and Photoionisation, eds. P. G. Burke and J. B. West (Plenum Press, New York, 1988).
[49] N. Abusalbi, R. A. Eades, T. Namin , D. Thirnmalai, D. A. Dixon and D. G. Truhlar, J. Chem. Phys., 78, 1213, (1983).
[50] T. N. Rescigno and A. E. Orel, Phys. Rev., A23, 1134 (1981).
[51] B. I. Schneider and L. A. Collins, Phys. Rev. A24, 1264, (1981).
[52] F. A. Gianturco and D. G. Thompson, J. Phys. B. (Atom. Molec. Phys.) 9, L383, (1976).
[53] T. L. Gibson and M. A. Morrison, Phys. Rev. A29, 2497, (1984).
[54] A. Temkin, Phys. Rev. 107, 1004, (1957).
[55] A. Temkin and K. V. Vasavada, Phys. Rev., 160, 109, (1967).
[56] N. F. Lane and R.J.W. Henry, Phys. Rev., 173, 183, (1968).
[57] J. K. O'Connell and N. F. Lane, Phys. Rev., A27, 1893, (1983).

[58] N. T. Padial and D. W. Norcross, Phys. Rev., A29, 1742, (1984).
[59] F. A. Gianturco, A. Jain and L. C. Pantano, J. Phys. B. (Atom. Molec. Phys.) 20, 571, (1987).
[60] R. J. Damburg and E. Karule, Proc. Phys. Soc. 90, 637, (1967).
[61] P. G. Burke and J.F.B. Mitchell, J. Phys. B, (Atom. Molec. Phys.) 7, 665, (1974).
[62] B. I. Schneider in Electronic and Atomic Collisions ed. G. Watel (North Holland, Amsterdam, 1978) p. 257.
[63] R. K. Nesbet, C. J. Noble, L. A. Morgan and C. A. Weatherford, J. Phys. B. (Atom. Molec. Phys.) 17, L891 (1984).
[64] C. J. Gillan, C. J. Noble and P. G. Burke, J. Phys. B. (Atom. Molec. Phys.) to be published (1987).
[65] C. J. Gillan, O. Nagy, P. G. Burke, L. A. Morgan and C. J. Noble, J. Phys. B. (Atom. Molec. Phys.) 20, 4585 (1987).
[66] R. K. Nesbet, J. Chem. Phys. 40, 3619 (1964).
[67] J. S. Bell and E. J. Squires, Phys. Rev. Letters, 3, 96, (1959).
[68] A. Klonover and U. Kaldor, Chem. Phys. Lett., 51, 321, (1977).
[69] A. Klonover and U. Kaldor, J. Phys. B. (Atom. Molec. Phys.) 11, 1623, (1978).
[70] A. Klonover and U. Kaldor, J. Phys. B. (Atom. Molec. Phys.) 12, 323, (1979).
[71] A. Klonover and U. Kaldor, J. Phys. B. (Atom. Molec. Phys.) 12, L61, (1979).
[72] E. Ficocelli Varracchio and U. T. Lamanna, to be published (1987).
[73] B. I. Schneider and L. A. Collins, Phys. Rev. A30, 95, (1984).
[74] P. G. Burke and A. L. Sinfailam, J. Phys. B. (Atom. Molec. Phys.), 3, 641, (1970).
[75] N. Chandra, Comput, Phys. Commun., 5, 417, (1973).
[76] G. Raseev, Comput, Phys. Commun., 20, 275, (1980).
[77] W. N. Sams and D. J. Kouri, J. Chem. Phys., 51, 4809, (1969).
[78] E. R. Smith and R.J.W. Henry, Phys. Rev., A7, 1585, (1973).
[79] M. A. Morrison, in Electron-Molecule and Photon-Molecule Collisions, eds., T. N. Rescigno, B.V. McKoy and B. I. Schneider (Plenum Press, New York, 1979) p.15.
[80] L. A. Collins, W. D. Robb and M. A. Morrison, J. Phys. B. (Atom. Molec. Phys.) 11, L777, (1978).
[81] W. D. Robb and L. A. Collins, Phys. Rev., A22, 2474, (1980).
[82] D. W. Norcross in Atoms in Astrophysics eds. P. G. Burke, W. B. Eissner, D. G. Hummer and I. C. Percival (Plenum Press, New York, 1983) p. 55.
[83] C. J. Noble and R. K. Nesbet, Comput. Phys. Commun. 33, 399 (1984).
[84] M. J. Seaton, J. Phys. B. (Atom. Molec. Phys.) 7, 1817, (1974).
[85] L. A. Collins and B. I. Schneider, Phys. Rev. 24, 2387, (1981).
[86] B. I. Schneider and L. A. Collins, Phys. Rev. A33, 2970, (1986).
[87] H. Feshbach, Ann. Phys. (N.Y.) 5, 357, (1958).
[88] H. Feshbach, Ann. Phys. (N.Y.) 19, 287, (1962).
[89] E. P. Wigner, Phys. Rev. 70, 15, (1946).
[90] E. P. Wigner, Phys. Rev., 70, 606, (1946).
[91] E. P. Wigner and L. Eisenbud, Phys. Rev., 72, 29, (1947).
[92] P. G. Burke, A. Hibbert and W. D. Robb, J. Phys. B. (Atom. Molec. Phys.) 4, 153, (1971).
[93] P. G. Burke and W. D. Robb, Adv. Atom. Molec. Phys., 11, 143, (1975).
[94] B. I. Schneider, Chem. Phys. Lett., 31, 237, (1975).
[95] B. I. Schneider, Phys. Rev., A11, 1957, (1975).
[96] P. G. Burke, I. Mackay and I. Shimamura, J. Phys. B., (Atom. Molec. Phys.), 10, 2497, (1977).
[97] C. J. Noble, Daresbury Laboratory Report DL/SCI/TH33T (1982).
[98] C. Bloch, Nucl. Phys., 4, 503, (1957).
[99] P.J.A. Buttle, Phys. Rev., 160, 691, (1967).
[100] J. Tennyson, P. G. Burke and K. A. Berrington, Comput. Phys. Commun to be published (1987).
[101] K. Takatsuka and V. McKoy, Phys. Rev. A24, 2473, (1981).

[102] K. Takatsuka and V. McKoy, Phys. Rev., A30, 1734, (1984).
[103] T. L. Gibson, M.A.P. Lima, V. McKoy and W. M. Huo, Phys. Rev., A35, 2473, (1987).
[104] W. M. Huo, M.A.P. Lima, T. L. Gibson and V. McKoy, Phys. Rev. A. to be published (1987).
[105] C. Lovelace, Phys. Rev., 135, B1225, (1964).
[106] S. K. Adikari and I. H. Sloan, Phys. Rev., C11, 1133, (1975).
[107] M. T. Lee, K. Takatsuka and B.V.McKoy, J. Phys. B. (Atom. Molec. Phys.) 14, 4115, (1981).
[108] D. Dill and J. L. Dehmer, J. Chem. Phys., 61, 692, (1974).
[109] J. L. Dehmer and D. Dill in Electron-Molecule and Photon-Molecule Collisions eds. T. N. Rescigno, B.V.McKoy and B. I. Schneider (Plenum Press, New York, 1979) p. 225.
[110] P. Ziesche and W. John, J. Phys. B. (Atom. Molec. Phys.) 9, 333, (1976).
[111] K. H. Johnson in Advances in Quantum Chemistry ed P. O. Löwdin (Academic Press, New York, 1973) p. 143.
[112] P. W. Langhoff, Chem. Phys. Letters, 22, 60, (1973).
[113] P. W. Langhoff in Electron-Molecule and Photon-Molecule Collisions eds. T. N. Rescigno, V. McKoy and B. I. Schneider (Plenum Press, New York 1979) p. 183.
[114] A. U. Hazi and H. S. Taylor, Phys. Rev., A1, 1109, (1970).
[115] Y. K. Ho, Phys. Rep., 99, 1, (1983).
[116] S. Hara, J. Phys. Soc. Japan, 27, 1009, (1969).
[117] K. L. Bell, J. Phys. B. (Atom. Molec. Phys.) 14, 2895, (1981).
[118] M. A. Crees and D. L. Moores, J. Phys. B. (Atom. Molec. Phys.) 10, L225, (1977).
[119] S. Hara and S. Ogata, J. Phys. B. (Atom. Molec. Phys.) 18, L59, (1985).
[120] S. Hara, J. Phys. B. (Atom. Molec. Phys.) 18, 3759, (1985).
[121] S. Hara, H. Sato, S. Ogata and N. Tamba, J. Phys. B. (Atom. Molec. Phys.) 19, 1177, (1986).
[122] E. C. Sullivan and A. Temkin, Comput. Phys. Commun., 25, 97, (1982).
[123] K. Onda and A. Temkin, Phys. Rev. A, 28, 621, (1983).
[124] C. A. Weatherford, K. Onda and A. Temkin, Phys. Rev. A, 31, 3620, (1985).
[125] C. A. Weatherford, F. B. Brown and A. Temkin, Phys. Rev. A, 35, 4561, (1987).
[126] N. Chandra and A. Temkin, Phys. Rev. A, 13, 188, (1976).
[127] N. Chandra and A. Temkin, Phys. Rev. A, 14, 507, (1976).
[128] A. Temkin in Electronic and Atomic Collisions eds. N. Oda and K. Takayanagi (North Holland, Amsterdam 1980) p. 95.
[129] B. H. Choi and R. T. Poe, Phys. Rev. A, 16, 1821,(1977).
[130] B. H. Choi and R. T. Poe, Phys. Rev. A, 16, 1831, (1977).
[131] A. Herzenberg and F. Mandl, Proc. Roy. Soc., A270, 48, (1962).
[132] J. N. Bardsley and F. Mandl, Rep. Prog. Phys., 31, 471, (1968).
[133] A. Herzenberg, J. Phys. B. (Atom. Molec. Phys.) 1, 548, (1968).
[134] A. Herzenberg, in Electron-Molécule Collisions eds. I. Shimamura and K. Takayanagi (Plenum Press, New York 1984) p. 191.
[135] D. T. Birtwistle and A. Herzenberg, J. Phys. B. (Atom. Molec. Phys.) 4, 53, (1971).
[136] A.J.F. Siegert, Phys. Rev., 56, 750, (1939).
[137] J. N. Bardsley, J. Phys. B. (Atom. Molec. Phys.) 1, 349, 365, (1968).
[138] A. U. Hazi, A. E. Orel and T. N. Rescigno, Phys. Rev. Letters, 46, 918, (1981).
[139] R. J. Hall, J. Chem. Phys., 68, 1803, (1978).
[140] A. U. Hazi in Electron-Atom and Electron-Molecule Collisions ed. J. Hinze (Plenum Press, New York 1983), 103.
[141] W. Domcke in Swarm Studies and Inelastic Electron Molecule Collisions eds. L. C. Pitchford, B. V. McKoy, A. Chutjian and S. Trajmar (Springer Verlag, New York, Berlin, 1987) p. 205.
[142] W. Domcke and L. S. Cederbaum, J. Phys. B. (Atom. Molec. Phys.) 13, 2829, (1980).

[143] M. Berman, H. Estrada, L. S. Cederbaum and W. Domcke, Phys. Rev. A, 28, 1363, (1983).
[144] C. Münsdel and W. Domcke, J. Phys. B. (Atom. Molec. Phys.) 17, 3593, (1984).
[145] M. Berman and W. Domcke, Phys. Rev. A, 29, 2485, (1984).
[146] B. I. Schneider, M. Le Dourneuf and P. G. Burke, J. Phys. B. (Atom. Molec. Phys.) 12, L365, (1979).
[147] B. I. Schneider, M. Le Dourneuf and Vo Ky Lan, Phys. Rev. Letters, 43, 1926, (1979).
[148] L. A. Morgan, J. Phys. B. (Atom. Molec. Phys.) 19, L439, (1986).
[149] M. Allan, J. Phys. B. (Atom. Molec. Phys.) 18, 4511, (1985).
[150] R. K. Nesbet, Phys. Rev. A, 19, 551, (1979).
[151] J. Tennyson, C. J. Noble and P. G. Burke, Int. J. Quantum. Chem., 29, 1033, (1986).
[152] J. Tennyson and C. J. Noble, J. Phys. B. (Atom. Molec. Phys.) 18, 155, (1986).
[153] H. Tagaki and H. Nakamura, Phys. Rev. A, 27, 691, (1983).
[154] H. Sato and S. Hara (1983) quoted by Tagaki and Nakamura [153].
[155] L. A. Collins and B. I. Schneider, Phys. Rev. A27, 101, (1983).
[156] A. U. Hazi, C. Derkits and J. N. Bardsley, Phys. Rev. A, 27, 1319, (1983).
[157] C. Bottcher and K. Docken, J. Phys. B. (Atom. Molec. Phys.), 7, L5, (1974).
[158] G. V. Marr, private communication (1987).
[159] K. L. Baluja, C. J. Noble and J. Tennyson, J. Phys. B. (Atom. Molec. Phys.), 18, L851, (1985).
[160] B. I. Schneider and L. A. Collins, J. Phys. B. (Atom. Molec. Phys.), 18, L857, (1985).
[161] M.A.P. Lima, T. L. Gibson, W. M. Huo and V. McKoy, J. Phys. B. (Atom. Molec. Phys.) 18, L865, (1985).
[162] S. Chung and C. C. Lin, Phys. Rev. A17, 1874, (1978).
[163] H. Nishimura and A. Danjo, J. Phys. Soc. Japan, 55, 3031, (1986).
[164] R. R. Lucchese, G. Raseev and V. McKoy, Phys. Rev. A25, 2572, (1982).
[165] E. W. Plummer, T. Gustafsson, W. Gudat and D. E. Eastman, Phys. Rev., A15, 2339, (1977).
[166] A. Hammett, W. Stoll and C. E. Brian, J. Electron. Spec. Relat. Phenom., 8, 367, (1976).
[167] J. A. Stephens and D. Dill, Phys. Rev., A31, 1968, (1985).
[168] B.Basden and R. L. Lucchese, Phys. Rev. A34, 5158, (1986).
[169] S. H. Southworth, A. C. Parr, J. E. Hardis and J. L. Dehmer, Phys. Rev., A33, 1020, (1986).
[170] M. Y. Adam, P. Morin, P. Lablanquie and I. Nenmer, unpublished.
[171] G. V. Marr, J. M. Morton, R. M. Holmes and D. G. McCoy, J. Phys. B. (Atom. Molec. Phys.), 12, 43, (1979).
[172] M. Raoult, H.Le Rouzo, G. Raseev and H. Lefebvre-Brian, J. Phys. B. (Atom. Molec. Phys.), 16, 4601, (1983).
[173] B. Lohmann and S. J. Buckman, J. Phys. B. (Atom. Molec. Phys.) 19, 2565, (1986).
[174] J. Ferch, B. Granitz and W. Raith, J. Phys. B. (Atom. Molec. Phys.) 18, L445, (1985).
[175] R. K. Jones, J. Chem. Phys., 82, 5424, (1985).

ELECTRON AND PHOTON COLLISIONS WITH ATOMS

C.J. Joachain

Physique Théorique, Université Libre de Bruxelles, Belgium
and Institut de Physique Corpusculaire, Université de
Louvain, Louvain-La-Neuve, Belgium

ABSTRACT

The theoretical methods used to analyze electron and photon collisions with atoms are reviewed. First the theory of electron-atom collisions is discussed, beginning with elastic scattering and excitation, and followed by electron impact ionization. A survey is then given of the theory of atomic photoionization.

1. INTRODUCTION

Electron and photon collisions with atoms have attracted considerable interest since the early days of atomic physics, because such processes provide the means of investigating the dynamics of several-particle systems at a fundamental level. In addition, a detailed understanding of these collision phenomena is required in several other fields of physics such as astrophysics, plasma physics and laser physics.

In recent years a number of important advances have occured on the experimental side. These include absolute measurements of cross sections, experiments using coincidence techniques, polarized beams or targets, incident positrons, synchrotron radiation or lasers. Many of these experiments provide stringent tests of the theory and have stimulated the development of new theoretical methods. The aim of these lectures is to give a survey of the theory, with particular emphasis on simple atomic systems for which the dynamics of the collision can be analyzed in detail. Section 2 is devoted to elastic scattering and excitation in electron-atom collisions, while Section 3 deals with electron impact ionization of atoms. Finally, in Section 4 we discuss photoionization. Multiphoton processes and laser-assisted electron-atom collisions will not be considered, as these topics will be discussed in the lectures of M. Gavrila. Atomic units (a u) will be used, unless otherwise stated.

2. ELECTRON-ATOM COLLISIONS. ELASTIC SCATTERING AND EXCITATION

2.1. The low energy region

We begin our discussion of electron-atom collisions by considering elastic scattering and excitation at low energies, such that the velocities of the incident and scattered electrons are less than the average velocities of the target electrons playing an active role in the collision.

2.1.1. The wave equation, scattering amplitudes and cross sections

The Schrödinger wave equation describing the collision of an electron with a target atom or ion containing N electrons and having nuclear charge Z is

$$H_{N+1}\,\Psi = E\,\Psi \tag{2.1}$$

where E is the total energy of the electron-atom (ion) system and the (N+1)- electron Hamiltonian of this system is given by

$$H_{N+1} = \sum_{i=o}^{N}\left(-\frac{1}{2}\nabla_i^2 - \frac{Z}{r_i}\right) + \sum_{i>j=o}^{N}\frac{1}{r_{ij}} \tag{2.2}$$

with $r_{ij} = |\mathbf{r}_i-\mathbf{r}_j|$. Here $\mathbf{r}_i$ and $\mathbf{r}_j$ are the vector coordinates of the electrons i and j, respectively. The origin of coordinates is the target nucleus, assumed to have infinite mass. In writing the Hamiltonian (2.2) we have neglected all relativistic effects, a restriction which will be relaxed below. We also introduce the target eigenstates ψ_n such that

$$H_N \psi_n = w_n \psi_n \tag{2.3}$$

where H_N is the Hamiltonian (2.2), but with the summation starting at the index (1) instead of (o), and w_n are the target eigenenergies. The eigenkets of H_N will be denoted by $|n\rangle$.

Let us now consider a scattering process of the type

$$e^- + A(i) \rightarrow e^- + A(j) \tag{2.4}$$

in which an electron is incident upon the target atom A in the initial eigenstate ψ_i, and in the final state a scattered electron emerges and the target A is left in the eigenstate ψ_j. From energy conservation we have

$$E = \frac{1}{2}k_i^2 + w_i = \frac{1}{2}k_j^2 + w_j \tag{2.5}$$

We look for a solution of the Schrödinger equation (2.1) corresponding to an electron in spin state $\chi_{1/2m_i}$ incident upon a target atom or ion in state ψ_i and scattered into any combination of spin states $\chi_{1/2m_j}$ and target states ψ_j. The asymptotic form of the wave function in the case of a neutral target is

$$\Psi_i^{(+)} \underset{r\to\infty}{\rightarrow} \psi_i\chi_{\frac{1}{2}m_i} e^{ik_i z} + \sum_j \psi_j\chi_{\frac{1}{2}m_j} f_{ji}(\theta,\phi) \frac{e^{ik_j r}}{r} \tag{2.6}$$

where the subscript i on $\Psi_i^{(+)}$ refers to the initial state and the superscript (+) indicates that $\Psi_i^{(+)}$ satisfies outgoing spherical waves boundary conditions[1]. In the above equation $(r\ \theta\ \phi)$ are the spherical polar coordinates of the unbound electron and $f_{ji}(\theta,\phi)$ is the scattering amplitude for the transition from state $\psi_i\chi_{1/2m_i}$ to state $\psi_j\chi_{1/2m_j}$. The second term on the right of (2.6) contains contributions at infinity from all atomic states corresponding to open channels, for which $k_j^2 \geqslant o$. The remaining states corresponding to closed channels such that $k_j^2 < o$ give decaying exponential terms which do not contribute to the asymptotic form. These terms, however, play an important role when the projectile electron is within or near the target, giving rise for example to intermediate resonance states. Finally, when the target is an ion one must modify the exponents in (2.6) by including the logarithmic phase factors due to the distortion by the Coulomb field. Since this modification introduces no essential complication we shall not write these logarithmic phase factors explicitly in what follows.

The differential cross section for a transition $|\mathbf{k}_i,\psi_i\chi_{1/2m_i}\rangle \rightarrow |\mathbf{k}_j,\psi_j\chi_{1/2m_j}\rangle$ can be obtained in the usual way by calculating the incident and outgoing fluxes. It is given by

$$\frac{d\sigma_{ji}}{d\Omega} = \frac{k_j}{k_i}|f_{ji}|^2 \tag{2.7}$$

and the total (integrated) cross section is obtained by integrating over all scattering angles.

The fundamental problem to be solved is to calculate the scattering amplitudes for the processes (2.4). At sufficiently low incident electron energies, only a finite number of target states can be excited (i.e. only a finite number of channels are open) and it is possible to represent explicitly all these open channels in the total electron-atom wave function $\Psi_i^{(+)}$. We shall now discuss several methods which exhibit this feature, following the treatment of Burke and Joachain[2].

2.1.2. The two-electron problem

In order to introduce the basic concepts in a simple way, we shall first consider the two-electron problem corresponding to the collision of an electron with atomic hydrogen or with a hydrogen-like ion of nuclear charge Z. Clearly our discussion will also be appropriate for describing the scattering of an electron by an atom or ion having one active electron outside inert closed shells. Since the non-relativistic Hamiltonian H_2 defined by (2.2) with N=1 does not contain the electron spin coordinates, the space and spin parts of the total wave function $\Psi_i^{(+)}$ can be separated. Denoting by $\underset{\sim}{r}_o$ and $\underset{\sim}{r}_1$ the position vectors of the two electrons, the space part of $\Psi_i^{(+)}$ satisfies the Schrödinger equation

$$H_2\,\Psi_i^{(+)}(\underset{\sim}{r}_o,\underset{\sim}{r}_1) = E\,\Psi_i^{(+)}(\underset{\sim}{r}_o,\underset{\sim}{r}_1) \tag{2.8a}$$

where

$$H_2 = -\frac{1}{2}\nabla_o^2 - \frac{1}{2}\nabla_1^2 - \frac{Z}{r_o} - \frac{Z}{r_1} + \frac{1}{r_{o1}} \tag{2.8b}$$

while the space part of the target eigenfunction corresponding to the eigenenergy w_j, $\psi_j(\underset{\sim}{r}) \equiv \langle \underset{\sim}{r}|j\rangle$, satisfies the equation

$$\left(-\frac{1}{2}\nabla^2 - \frac{Z}{r} - w_j\right)\psi_j(\underset{\sim}{r}) = 0 \tag{2.9}$$

Let us examine the asymptotic form of the solution of eq.(2.8). We consider initially an asymmetric solution where the incident electron is labelled o and the scattered electron is labelled either o corresponding to direct scattering or 1 corresponding to exchange scattering. Hence we can write

$$\Psi_i^{(+)}(\underset{\sim}{r}_o,\underset{\sim}{r}_1) \underset{r_o\to\infty}{\to} \psi_i(\underset{\sim}{r}_1)e^{ik_iz_o} + \sum_j \psi_j(\underset{\sim}{r}_1) f_{ji}(\theta_o,\phi_o)\frac{e^{ik_jr_o}}{r_o} \tag{2.10a}$$

and

$$\Psi_i^{(+)}(\underset{\sim}{r}_o,\underset{\sim}{r}_1) \underset{r_1\to\infty}{\to} \sum_j \psi_j(\underset{\sim}{r}_o) g_{ji}(\theta_1,\phi_1)\frac{e^{ik_jr_1}}{r_1} \tag{2.10b}$$

where f_{ji} and g_{ji} are called the direct and the exchange amplitudes, respectively. Unlike the amplitude $\mathcal{f}_{ji}$ of eq.(2.6) they do not explicitly involve the spin of the unbound electron.

Wave function expansions

Suppose that we expand the wave function $\Psi_i^{(+)}(\underset{\sim}{r}_o,\underset{\sim}{r}_1)$ in terms of the target eigenfunctions. Omitting the subscript i and superscript (+) for notational simplicity, we write

$$\Psi(\underset{\sim}{r}_o,\underset{\sim}{r}_1) = \sum_i F_i(\underset{\sim}{r}_o)\psi_i(\underset{\sim}{r}_1) \tag{2.11a}$$

or

$$\Psi(\underset{\sim}{r}_o,\underset{\sim}{r}_1) = \sum_i G_i(\underset{\sim}{r}_1)\psi_i(\underset{\sim}{r}_o) \tag{2.11b}$$

where the summation includes an integration over the continuum. Castillejo et al.[3] have shown that these expansions can only satisfy both boundary conditions (2.10) if suitable paths are chosen around singularities that occur in $F_i(\underset{\sim}{r}_o)$ and $G_i(\underset{\sim}{r}_1)$ in the continuum part of the expansion. In order to avoid the difficulties associated with these singularities, one can consider the alternative expansion

$$\Psi(\underset{\sim}{r}_o,\underset{\sim}{r}_1) = \sum_i \left[F_i(\underset{\sim}{r}_o)\psi_i(\underset{\sim}{r}_1) + \psi_i(\underset{\sim}{r}_o)G_i(\underset{\sim}{r}_1)\right] \tag{2.12}$$

Clearly the functions F_i and G_i are not uniquely defined in the above equation, since $\Psi(\underset{\sim}{r}_o,\underset{\sim}{r}_1)$ is unchanged by the substitutions

$$F_i \rightarrow \bar{F}_i = F_i + \sum_j a_{ij}\psi_j \ , \ G_i \rightarrow \bar{G}_i = G_i - \sum_j a_{ji}\psi_j \tag{2.13}$$

where the coefficients a_{ij} are arbitrary constants. This arbitrariness can be used to eliminate the singularities. Indeed, by choosing the constants a_{ij} so that the functions $\bar{F}_j$ and $\bar{G}_j$ are orthogonal to atomic states of lower energies, namely

$$\langle\psi_i|\bar{F}_j\rangle = \langle\psi_i|\bar{G}_j\rangle = 0 \ , \ w_i < w_j \tag{2.14}$$

one can prove that for open (energetically available) channels not involving ionization, the asymptotic forms of the functions $\bar{F}_j$ and $\bar{G}_j$ are given by

$$\bar{F}_j(\underset{\sim}{r}) \underset{r\to\infty}{\rightarrow} \delta_{ji} e^{ik_j z} + f_{ji}(\theta,\phi)\frac{e^{ik_j r}}{r} \tag{2.15a}$$

and

$$\bar{G}_j(\underset{\sim}{r}) \underset{r\to\infty}{\rightarrow} g_{ji}(\theta,\phi)\frac{e^{ik_j r}}{r} \tag{2.15b}$$

In the closed channels ($k_j^2 < 0$) the functions $\bar{F}_j$ and $\bar{G}_j$ vanish asymptotically.

In order to achieve maximum simplicity in the form of the wave function expansion, we can take advantage of the symmetry of the Hamiltonian (2.8b) with respect to exchange of the spatial coordinates of the two electrons. Thus we can look for symmetric (+) and antisymmetric (-) solutions of the Schrödinger equation (2.8a) and write for the space part of the total wave function the expansion

$$\Psi^{\pm}(\underset{\sim}{r}_o,\underset{\sim}{r}_1) = \sum_i \left[F_i^{\pm}(\underset{\sim}{r}_o)\psi_i(\underset{\sim}{r}_1) \pm F_i^{\pm}(\underset{\sim}{r}_1)\psi_i(\underset{\sim}{r}_o)\right] \tag{2.16}$$

where the functions $F_i^{\pm} = \overline{F}_i \pm \overline{G}_i$ do not contain singularities.

Now, according to the Pauli exclusion principle, the full wave function $\Psi(\underline{x}_o,\underline{x}_1)$ of the system, where $\underline{x}_i \equiv (\underset{\sim}{r}_i,\sigma_i)$ represents the space and spin coordinates of electron i, must be antisymmetric with respect to interchange of the space and spin coordinates of the two electrons. Let $\underset{\sim}{S}$ be the total spin operator and let us introduce a singlet spin function χ_{oo} which is an eigenfunction of the operators $\underset{\sim}{S}^2$ and $\underset{\sim}{S}_z$, corresponding respectively to the quantum numbers S=0 and M_S=0. That is

$$\chi_{oo}(\sigma_o,\sigma_1) = \frac{1}{\sqrt{2}}\left[\alpha(\sigma_o)\beta(\sigma_1) - \alpha(\sigma_1)\beta(\sigma_o)\right] \tag{2.17}$$

where α and β are the basic spinors

$$\alpha = \begin{pmatrix}1\\0\end{pmatrix} \quad , \quad \beta = \begin{pmatrix}0\\1\end{pmatrix} \tag{2.18}$$

The spin function $\chi_{oo}(\sigma_o,\sigma_1)$ is obviously antisymmetric with respect to interchange of the two spin coordinates σ_o and σ_1. We also introduce three triplet spin eigenfunctions of the operators $\underset{\sim}{S}^2$ and S_z corresponding to S=1 and M_S+1,0,-1, namely

$$\begin{aligned}
\chi_{11}(\sigma_o,\sigma_1) &= \alpha(\sigma_o)\alpha(\sigma_1)\\
\chi_{10}(\sigma_o,\sigma_1) &= \frac{1}{\sqrt{2}}\left[\alpha(\sigma_o)\beta(\sigma_1) + \alpha(\sigma_1)\beta(\sigma_o)\right]\\
\chi_{1-1}(\sigma_o,\sigma_1) &= \beta(\sigma_o)\beta(\sigma_1)
\end{aligned} \tag{2.19}$$

These functions are symmetric with respect to interchange of the two spin coordinates. For singlet (S=0) scattering we can therefore write the total wave function as the product

$$\Psi^+(\underline{x}_o,\underline{x}_1) = \Psi^+(\underset{\sim}{r}_o,\underset{\sim}{r}_1)\chi_{oo}(\sigma_o,\sigma_1) \tag{2.20}$$

For triplet (S=1) scattering we have instead

$$\Psi^-(\underline{x}_o,\underline{x}_1) = \Psi^-(\underset{\sim}{r}_o,\underset{\sim}{r}_1)\chi_{1M_S}(\sigma_o,\sigma_1), \ M_S = -1,0,1 \tag{2.21}$$

Let us now return to the expansion (2.16). Using eqs. (2.15) and the fact that $F_j^{\pm} = \overline{F}_j \pm \overline{G}_j$, we find that for open channels not involving ionization one has asymptotically

$$F_j^{\pm}(\underset{\sim}{r}) \underset{r\to\infty}{\to} \delta_{ji} e^{ik_j z} + f_{ji}^{\pm}(\theta,\phi) \frac{e^{ik_j r}}{r} \tag{2.22}$$

where the scattering amplitudes

$$f_{ji}^{\pm} = f_{ji} \pm g_{ji} \tag{2.23}$$

are called respectively the singlet (+) and triplet (-) scattering amplitudes. For closed channels the functions $F_j^{\pm}(\underset{\sim}{r})$ vanish asymptotically From eq. (2.7) and the above equations it follows that for an unpolarized system (i.e. for random spin orientations) the differential cross section for a transition $|\underset{\sim}{k}_i,\psi_i\rangle \to |\underset{\sim}{k}_j,\psi_j\rangle$ is given by

$$\frac{d\sigma_{ji}}{d\Omega} = \frac{k_j}{k_i}\left(\frac{1}{4}|f_{ji} + g_{ji}|^2 + \frac{3}{4}|f_{ji} - g_{ji}|^2\right) \tag{2.24}$$

Since the functions $F_i^{\pm}(\underset{\sim}{r})$ contribute to the asymptotic form only for open channels, it is possible to rewrite the expansion (2.16) in the alternative form

$$\Psi^{\pm}(\underset{\sim}{r}_o,\underset{\sim}{r}_1) = \sum_i \left[F_i^{\pm}(\underset{\sim}{r}_o)\psi_i(\underset{\sim}{r}_1) \pm F_i^{\pm}(\underset{\sim}{r}_1)\psi_i(\underset{\sim}{r}_o)\right] + \sum_i c_i^{\pm}\chi_i^{\pm}(\underset{\sim}{r}_o,\underset{\sim}{r}_1) \tag{2.25}$$

where the first summation only runs over target eigenstates corresponding to open channels and the $\chi_i^{\pm}(\underset{\sim}{r}_o,\underset{\sim}{r}_1)$ are square integrable correlation functions, which are either symmetric (+) or antisymmetric (-) with respect to the interchange of the spatial coordinates of the two electrons. We remark that if the total energy $E<o$, so that only a finite number of channels are open, then all the target eigenstates ψ_i which are energetically accessible can be included in the first expansion. If the correlation functions $\chi_i^{\pm}$ are chosen to span the space orthogonal to that corresponding to the first expansion, then equation (2.25) provides an exact representation of the total wave function.

Another expansion of the wave functions $\Psi^{\pm}(\underset{\sim}{r}_o,\underset{\sim}{r}_1)$ which has proved useful is the pseudo-state expansion proposed by Burke and Schey[4] and Damburg and Karule[5]. In this case the sum over energetically accessible target eigenstates is supplemented by a sum over pseudo-states, namely

$$\Psi^{\pm}(\underset{\sim}{r}_o,\underset{\sim}{r}_1) = \sum_i \left[F_i^{\pm}(\underset{\sim}{r}_o)\psi_i(\underset{\sim}{r}_1) \pm F_i^{\pm}(\underset{\sim}{r}_1)\psi_i(\underset{\sim}{r}_o)\right]$$
$$+ \sum_i \left[H_i^{\pm}(\underset{\sim}{r}_o)\overline{\psi}_i(\underset{\sim}{r}_1) \pm H_i^{\pm}(\underset{\sim}{r}_1)\overline{\psi}_i(\underset{\sim}{r}_o)\right] \tag{2.26}$$

The pseudo-states $\overline{\psi}_i$ are not target eigenstates and there is no unique prescription for determining them. For example, they can be chosen to optimize the representation of the short-range correlation effects or long-range polarization effects occuring in the transitions under consideration. They can also be chosen to diagonalize the target Hamiltonian

$$\langle\bar{\psi}_i|-\frac{1}{2}\nabla^2 - \frac{Z}{r}|\bar{\psi}_j\rangle = \bar{w}_i\ \delta_{ij} \tag{2.27}$$

and to be orthogonal to the target eigenstates ψ_i retained in the first expansion on the right of (2.26). If the pseudo-states $\bar{\psi}_i$ are expanded in the basis set of the target eigenstates, namely

$$\bar{\psi}_i = \sum_j b_{ij}\ \psi_j \tag{2.28}$$

this summation usually includes a contribution from the continuum. As a result the continuum will then be partially represented by the expansion (2.26).

Finally, it is also possible to write down an expansion of the wave functions $\Psi^{\pm}(\underset{\sim}{r}_o,\underset{\sim}{r}_1)$ in which target eigenstates ψ_i, pseudo-states $\bar{\psi}_i$ and square integrable correlation functions $\chi_i^{\pm}$ are all retained. We shall still represent such expansions formally by eq.(2.25), provided that the first summation is assumed to include both target eigenstates and pseudo-states.

Coupled equations

We shall now derive the equations satisfied by the unknown functions $F_i^{\pm}$. For this purpose it is convenient to use the projection operator formalism of Feshbach[6]. We suppose that the space part of the wave function is expanded according to (2.25). Let P be an operator which projects onto the target eigenstates and pseudo-states ψ_i in (2.25) and Q as the operator which projects onto the quadratically integrable functions $\chi_i^{\pm}$. Burke and Taylor[7] have shown that there is no loss of generality in taking the functions $\chi_i^{\pm}$ to be orthogonal to the functions ψ_i, so that

$$PQ = QP = 0 \tag{2.29}$$

In addition, since P and Q are projection operators, we have

$$P^2 = P \quad , \quad Q^2 = Q \tag{2.30}$$

Assuming that the two expansions in (2.25) are complete so that

$$P + Q = I \tag{2.31}$$

we can rewrite the Schrödinger equation (2.8a) as

$$(H - E)\ (P + Q)\Psi = 0 \tag{2.32}$$

where we have omitted the subscript and superscript on H and Ψ for notational simplicity. Acting on this equation with P and Q and using eqs. (2.29), (2.30) and (2.31) we find that

$$(H_{PP} - E)P\Psi + H_{PQ}\, Q\Psi = 0 \tag{2.33a}$$

and

$$(H_{QQ} - E)Q\Psi + H_{QP}P\Psi = 0 \tag{2.33b}$$

where $H_{PP} = PHP$, $H_{QQ} = QHQ$, $H_{PQ} = PHQ$ and $H_{QP} = QHP$. Solving (2.33b) for $Q\Psi$ and substituting into (2.33a), we obtain

$$(H_{PP} + H_{PQ} \frac{1}{E - H_{QQ}} H_{QP} - E\,)P\Psi = 0 \tag{2.34}$$

which are a set of coupled equations for the functions $F_i^{\pm}(\underset{\sim}{r})$. The equations obtained in the absence of the functions $\chi_i^{\pm}$, namely

$$(H_{PP} - E)P\Psi = 0 \tag{2.35}$$

are often called the close coupling equations. The additional object

$$\mathcal{V} = H_{PQ} \frac{1}{E - H_{QQ}} H_{QP} \tag{2.36}$$

which is present in (2.34) allows for propagation in the closed channel space; we shall see in Section 2.2 that it is closely related to the optical potential.

To obtain the explicit form of the coupled equations (2.34) we write eqs. (2.33) in terms of the functions ψ_i and $\chi_i^{\pm}$. That is

$$\int \psi_i^{\star}(\underset{\sim}{r}) \; (H - E)\Psi^{\pm}(\underset{\sim}{r}_o,\underset{\sim}{r}_1)d\underset{\sim}{r} = 0 \tag{2.37a}$$

and

$$\int \chi_i^{\pm\star} (\underset{\sim}{r}_o,\underset{\sim}{r}_1) \; (H - E)\Psi^{\pm}(\underset{\sim}{r}_o,\underset{\sim}{r}_1)d\underset{\sim}{r}_o d\underset{\sim}{r}_1 = 0 \tag{2.37b}$$

The first of these equations can be simplified using eq. (2.9) and (2.27) which are satisfied respectively by the target eigenstates and pseudostates. In what follows we shall suppose for simplicity that only target eigenstates are retained in the first expansion in eq. (2.25). The second equation (2.37) can be used to eliminate the coefficients $c_i^{\pm}$ in the expansion (2.25). Assuming that the square integrable functions $\chi_i^{\pm}$ are chosen to diagonalize H so that

$$<\chi_i^{\pm}|\,H\,|\chi_j^{\pm}> = \varepsilon_i^{\pm}\;\delta_{ij} \tag{2.38}$$

one finds that the functions $F_i^{\pm}(\mathbf{r})$ satisfy the coupled integro-differential equations

$$(\nabla^2 + k_i^2)\, F_i^{\pm}(\mathbf{r}) = 2\sum_j (V_{ij} + W_{ij}^{\pm} + X_{ij}^{\pm}) F_j^{\pm}(\mathbf{r}) \tag{2.39}$$

where V_{ij} is a direct potential matrix given by

$$\begin{aligned} V_{ij}(\mathbf{r}) &= \int \psi_i^{\star}(\mathbf{r}')\left(-\frac{Z}{r} + \frac{1}{|\mathbf{r}-\mathbf{r}'|}\right)\psi_j(\mathbf{r}')d\mathbf{r}' \\ &= -\frac{Z}{r}\delta_{ij} + \int \psi_i^{\star}(\mathbf{r}')\frac{1}{|\mathbf{r}-\mathbf{r}'|}\psi_j(\mathbf{r}')d\mathbf{r}' \end{aligned} \tag{2.40}$$

and $W_{ij}^{\pm}$ is a non-local exchange potential such that

$$W_{ij}^{\pm} F_j^{\pm}(\mathbf{r}) = \pm \int \psi_i^{\star}(\mathbf{r}') \left(\frac{1}{|\mathbf{r}-\mathbf{r}'|} + w_i + w_j - E\right)\psi_j(\mathbf{r}) F_j^{\pm}(\mathbf{r}')d\mathbf{r}' \tag{2.41}$$

The non-local correlation potential $X_{ij}^{\pm}$ appearing in (2.39) is found from (2.36) and (2.38) to be such that

$$X_{ij}^{\pm} F_j^{\pm}(\mathbf{r}) = 2\sum_k \int Z_{ik}^{\pm\star}(\mathbf{r}) \frac{1}{E-\varepsilon_k^{\pm}} Z_{jk}(\mathbf{r}')F_j^{\pm}(\mathbf{r}')d\mathbf{r}' \tag{2.42}$$

where

$$Z_{ij}^{\pm\star}(\mathbf{r}) = \int \psi_i^{\star}(\mathbf{r}')\,(H - E)\,\chi_j^{\pm}(\mathbf{r},\mathbf{r}')d\mathbf{r}' \tag{2.43}$$

The coupled equations (2.39) are the basic equations to be solved for the analysis of low-energy electron scattering by one-electron atoms (ions). If only the initial state of the target is retained on the right of (2.25), eqs. (2.39) reduce to the single equation

$$(\nabla^2 + k_i^2)F_i^{\pm}(\mathbf{r}) = 2(V_{ii} + W_{ii}^{\pm})F_i^{\pm}(\mathbf{r}) \tag{2.44}$$

which is known as the static-exchange approximation. The direct potential V_{ii} is called the static potential. For example, in the case of the ground state (1s) of atomic hydrogen, one readily finds that the static potential is given by

$$V_{1s,1s}(r) = -\left(1+\frac{1}{r}\right)e^{-2r} \tag{2.45}$$

The static-exchange approximation is clearly of limited validity. Firstly, it only describes elastic scattering and even for this process it neglects the polarization of the target by the free electron. Secondly, at higher energies it does not take into account the loss of flux into other channels. Nevertheless, it has been widely used as a starting point for more accurate calculations.

When more than one target state is retained in the first expansion on the right of (2.25) one obtains the "close-coupling approximation". Thus, assuming that M target states are kept in this expansion, one obtains the close-coupling equations

$$(\nabla^2 + k_i^2)F_i^{\pm}(\underset{\sim}{r}) = 2\sum_{j=1}^{M}(V_{ij} + W_{ij}^{\pm})F_j^{\pm}(\underset{\sim}{r}) \qquad (2.46)$$

The close coupling method is a good approximation for strong transitions between low-lying states whose energies are well separated from all other states. In other cases the convergence is often slow and the more general expansion (2.25) must be used, leading to the full coupled equations (2.39).

In order to solve the basic equations (2.39), a partial wave decomposition must be carried out. We shall return below to this point, after generalizing our results to the case of complex target atoms and ions.

2.1.3. Electron collisions with complex atoms and ions

For systems with three or more electrons it is not possible to write down the total wave function of the problem in the simple forms (2.20) or (2.21). Nevertheless, in order to avoid singularities in the continuum part of the expansion of the wave function analogous to those discussed above for electron scattering by one-electron atoms (ions), it is necessary to look for an expansion which is antisymmetric with respect to the interchange of the space and spin coordinates of all the electrons. By analogy with the expansion (2.25) used for the two-electron problem, we shall write this expansion in the form

$$\Psi(\underline{x}_o,\underline{x}_1,\ldots\underline{x}_N) = \mathcal{A}\sum_i F_i(\underline{x}_o)\psi_i(\underline{x}_1,\ldots\underline{x}_N) + \sum_i c_i\chi_i(\underline{x}_o,\underline{x}_1,\ldots\underline{x}_N) \qquad (2.47)$$

where $\mathcal{A}$ is the antisymmetrization operator. The first summation goes over a limited number of target eigenstates and may also include some pseudo-states, while the second summation is over a set of square integrable functions χ_i, each of them being antisymmetric with respect to the interchange of the space and spin coordinates of the N+1 electrons. In practice, the target eigenstates and pseudo-states cannot be obtained exactly, but instead are generally expanded in a configuration - interaction basis.

By substituting the expansion (2.47) into the Schrödinger equation (2.1) and projecting onto the target functions ψ_i and the square-integrable functions χ_i, one obtains for the functions F_i a set of coupled integro-differential equations analogous to the equations (2.39) derived above for the two-electron problem.

2.1.4. Partial wave analysis

The partial wave decomposition of the basic coupled equations of the theory can be performed by remembering that the non-relativistic (N+1)-electron Hamiltonian (2.2) commutes with the square of the total orbital angular momentum operator $\underset{\sim}{L}^2$, with the square of the total spin angular momentum operator $\underset{\sim}{S}^2$, and also with the operators L_z, S_z and the parity operator π. Thus we can form solutions $\Psi^\Gamma(\underline{x}_o,\underline{x}_1,\ldots\underline{x}_N)$ of the Schrödinger equation which are eigenfunctions of these operators belonging to the eigenvalues $\Gamma \equiv LSM_LM_S\pi$, where π is the parity quantum number. These eigenfunctions Ψ^Γ can be expanded in analogy with (2.47) as

$$\Psi^\Gamma(\underline{x}_o,\underline{x}_1,\ldots\underline{x}_N) = \mathcal{A}\sum_i F_i^\Gamma(r_o)\psi_i^\Gamma(\underline{x}_1,\ldots\underline{x}_N;\hat{\underset{\sim}{r}}_o,\sigma_o)+\sum_i c_i^\Gamma\chi_i^\Gamma(\underline{x}_o,\underline{x}_1,\ldots\underline{x}_N) \tag{2.48}$$

where the channel functions ψ_i^Γ are constructed by combining the target eigenfunctions with the angular and spin functions of the scattered electron to form eigenstates of the operators $\underset{\sim}{L}^2$, $\underset{\sim}{S}^2$, L_z, S_z and π corresponding to the eigenvalues $\Gamma \equiv LSM_LM_S\pi$.

By substituting (2.48) into the Schrödinger equation (2.1), projecting onto the channel functions ψ_i^Γ and onto the square-integrable functions χ_i^Γ and after eliminating the coefficients c_i^Γ one obtains for the reduced radial functions F_i^Γ coupled integro-differential equations of the form

$$\left(\frac{d^2}{dr^2} - \frac{\ell(\ell+1)}{r^2} + k_i^2\right)F_i^\Gamma(r) = 2\sum_j(V_{ij}^\Gamma+W_{ij}^\Gamma+X_{ij}^\Gamma)F_j^\Gamma(r) \tag{2.49}$$

where ℓ is the orbital angular momentum of the free electron, and the potentials V_{ij}^Γ, W_{ij}^Γ and X_{ij}^Γ are partial-wave decompositions of the direct, exchange and correlation potentials. The equations (2.49) were first derived for e^--H scattering - in the absence of the square - integrable functions - by Percival and Seaton[8]. It should be noted that the potentials V_{ij}^Γ, W_{ij}^Γ and X_{ij}^Γ are independent of M_L and M_S, so that one must only solve eqs.(2.49) for the required $LS\pi$ symmetries.

In general, the expressions giving V_{ij}^Γ, W_{ij}^Γ and X_{ij}^Γ are very cumbersome, so that in practice these potentials are determined by computer programs. It is worth noting, however, that the direct potential V_{ij}^Γ is local and has the asymptotic form

$$V_{ij}^\Gamma(r) = -\frac{(Z-N)}{r}\delta_{ij} + \sum_{\lambda=1}^{\lambda_{max}} b^\lambda r^{-\lambda-1} \quad , \quad r > a \tag{2.50}$$

where b^λ are coefficients and a is the range beyond which the bound orbitals in ψ_i^Γ and ψ_j^Γ are negligible. The dominant non-Coulombic long-range potential is the dipole (r^{-2}) potential which couples target states between which dipole transitions are allowed. It gives rise, in second order, to an equivalent diagonal (i=j) polarization potential V_{pol} having the asymptotic form[3]

$$V_{pol}(r) \underset{r\to\infty}{\to} -\frac{\alpha}{2r^4} \tag{2.51}$$

where α is the dipole polarizability of the target in the state i considered.

2.1.5. The S-matrix, K-matrix and T-matrix

Let us assume that for a given set of quantum numbers Γ there are M coupled channels, corresponding to retaining a finite number of target states and pseudo-states in the expansion (2.48). Then both i and j in (2.49) will take the values 1,2,...M and the M coupled equations (2.49) will in general have 2M linearly independent solutions. The requirement that the physically acceptable reduced radial functions vanish at the origin reduces this number to M linearly independent solutions. We shall therefore introduce a second subscript on the functions F_i^Γ, and denote these solutions by F_{ji}^Γ, with i,j = 1,2,...M. Thus

$$F_{ji}^\Gamma(o) = 0 \quad , \quad i,j = 1,2,\ldots M \tag{2.52}$$

Moreover, for large r one has in the open channels ($k_j^2 \geqslant o$)

$$F_{ji}^\Gamma(r) \underset{r\to\infty}{\to} k_j^{-1/2}(\delta_{ji}e^{-i\Theta_j} - S_{ji}^\Gamma e^{i\Theta_j}) \tag{2.53a}$$

while in the closed channels ($k_j^2 < o$)

$$F_{ji}^\Gamma(r) \underset{r\to\infty}{\to} 0 \tag{2.53b}$$

The quantity Θ_j which appears in (2.53a) is given by

$$\Theta_j = k_j r - \frac{1}{2}\ell_j\pi + \frac{\mathcal{Z}}{k_j}\,\ell n(2k_j r) + \sigma_j \tag{2.54}$$

where $\mathcal{Z}$ = Z-N is the "residual charge" and $\sigma_j = \arg\Gamma(\ell_j+1 - i\mathcal{Z}/k_j)$ is the Coulomb phase shift. The objects S_{ji}^Γ in (2.53a) are the elements of the S-matrix. Assuming that there are M_a open channels, we see that the S-matrix defined by (2.53a) has dimension $M_a x M_a$. It is not difficult to prove that the S-matrix is symmetric (from time reversal invariance) and unitary (from probability conservation).

Instead of the complex solutions defined by the boundary conditions (2.53), it is often convenient to introduce real solutions defined by the boundary conditions

$$F_{ji}^\Gamma(r) \underset{r\to\infty}{\to} k_j^{-1/2}(\delta_{ji}\sin\Theta_j + K_{ji}^\Gamma\cos\Theta_j), j=1,2,\ldots M_a, \tag{2.55a}$$

$$F^{\Gamma}_{ji}(r) \underset{r\to\infty}{\to} 0 \ , \ j=M_a+1,\ldots M \tag{2.55b}$$

with $i=1,2,\ldots M$. These solutions are linear combinations of those defined by (2.53). The objects K^{Γ}_{ji} are the elements of the K-matrix, which is related to the S-matrix by

$$\underline{S} = \frac{\underline{I} + i\underline{K}}{\underline{I} - i\underline{K}} \tag{2.56}$$

Since the S-matrix is symmetric it follows that the K-matrix is also symmetric. Moreover, equation (2.56) together with the unitarity of the S-matrix and the symmetry of the K-matrix implies that the K-matrix is real.

Following Kato[9], one can also introduce a generalized K-matrix by requiring the functions F^{Γ}_{ji} to satisfy the asymptotic boundary conditions

$$F^{\Gamma\tau}_{ji}(r) \underset{r\to\infty}{\to} k_j^{-1/2}\left[\delta_{ji}\sin(\Theta_j+\tau_j)+K^{\Gamma\tau}_{ji}\cos(\Theta_j+\tau_j)\right] \tag{2.57a}$$

$$F^{\Gamma\tau}_{ji}(r) \underset{r\to\infty}{\to} 0 \ , \ j=M_a+1,\ldots M \tag{2.57b}$$

where $i=1,2,\ldots M$ and the real parameters τ_j can be chosen arbitrarily in each open channel. The M_a solutions $\underline{F}^{\Gamma\tau}$ defined by (2.57a) are linear combinations of the M_a solutions $\underline{F}^{\Gamma}$ defined by (2.55a). The flexibility in the choice of τ_j is useful when approximations are made to calculate the K-matrix, as we shall see shortly.

Finally, one can define solutions satisfying the outgoing wave boundary conditions

$$F^{\Gamma}_{ji}(r) \underset{r\to\infty}{\to} k_j^{-1/2}(\delta_{ji}\sin\Theta_j + \frac{1}{2i}e^{i\Theta_j}T^{\Gamma}_{ji}),j=1,2,\ldots M_a \tag{2.58a}$$

$$F^{\Gamma}_{ji}(r) \underset{r\to\infty}{\to} 0 \ , \ j = M_a+1,\ldots M \tag{2.58b}$$

with $i = 1,2,\ldots M$. The objects T^{Γ}_{ji} are the elements of the T-matrix, which is related to the S-matrix by

$$\underline{S} = \underline{I} + \underline{T} \tag{2.59}$$

2.1.6. Scattering amplitudes, cross sections and collision strengths

The scattering amplitudes f_{ji} defined by (2.7) - and hence the differential and total cross sections - can readily be related[2] to the S-matrix (or the K or T-matrices). As an example, the total cross section for a transition (i→j) - obtained by averaging the differential cross section (2.7) over the initial magnetic quantum numbers, summing over the final magnetic quantum numbers and integrating over all scattering angles of the scattered electron - is given by

$$\sigma_{tot}(i\rightarrow j) = \frac{\pi}{k_i^2} \sum_{LS\pi} \frac{(2L+1)(2S+1)}{2(2L_i+1)(2S_i+1)} \sum_{\ell_i \ell_j} |T^{\Gamma}_{ji}|^2 \qquad (2.60)$$

where ℓ_i and ℓ_j are the orbital angular momentum quantum numbers of the incident and scattered electron, respectively.

A further quantity which is often used in applications is the collision strength. It is defined in terms of the total cross section for a transition (i→j) by the relation

$$\Omega(i,j) = \frac{\omega_i k_i^2}{\pi} \sigma_{tot}(i\rightarrow j) \qquad (2.61)$$

where $\omega_i = (2L_i+1)(2S_i+1)$ is the statistical weight of the initial state We remark that the collision strength is a dimensionless quantity and is symmetric : $\Omega(i,j) = \Omega(j,i)$.

Most of the recent results emerging from the coupled equations (2.49) have been obtained by using one of four methods : the variational method [10-12], the R-matrix method [13,14], the reduction of these equations to a system of linear algebraic equations[15] and the non-iterative integral equations method [16]. These methods and the associated computer programs have been reviewed by Burke and Eissner[17]. We shall only discuss here the variational and R-matrix methods.

2.1.7. The variational method

Among the various variational principles which have been proposed for scattering problems, we shall select the Hulthén-Kohn method[18,19] which has been used extensively in low-energy electron-atom collisions. We begin by rewriting the coupled integro-differential equations (2.49) in the form

$$\sum_k L^{\Gamma}_{jk} F^{\Gamma}_k = 0 \qquad (2.62)$$

where we have transferred all the terms of (2.49) into the left-hand side in order to define the operator L^{Γ}_{jk}.

Let us consider the functional

$$I^{\Gamma\tau}_{ji} = \sum_{k\ell} \int_0^{\infty} F^{\Gamma\tau}_{kj}(r) L^{\Gamma}_{k\ell} F^{\Gamma}_{\ell i}(r)dr \qquad (2.63)$$

where the radial functions satisfy the generalized K-matrix boundary conditions (2.57). It is clear that $I^{\Gamma\tau}_{ji} = 0$ if the functions are exact solutions of the equations (2.62). Suppose that we allow the functions $F^{\Gamma\tau}_{ji}$ to undergo small variations $\delta F^{\Gamma\tau}_{ji}$ about the exact solutions of (2.62), satisfying the boundary conditions

$$\delta F^{\Gamma\tau}_{ji}(o) = 0 \quad , \tag{2.64a}$$

$$\delta F^{\Gamma\tau}_{ji}(r) \underset{r\to\infty}{\to} k_j^{-1/2} \cos(\Theta_j+\tau_j)\,\delta K^{\Gamma\tau}_{ji} \tag{2.64b}$$

Using Green's theorem, one then obtains the Hulthén-Kohn variational principle

$$\delta(\underline{I}^{\Gamma\tau} + \underline{K}^{\Gamma\tau}) = 0 \tag{2.65}$$

We remark that this variational principle is not unique since the parameters τ_j in (2.57) can be chosen arbitrarily.

A variational estimate of the K-matrix can now be obtained as follows. Let us assume that we have constructed trial functions $\underline{F}_t^{\Gamma\tau}$ satisfying the conditions (2.57) and (2.64), and corresponding to a trial K-matrix $\underline{K}_t^{\Gamma\tau}$. Substituting $\underline{F}_t^{\Gamma\tau}$ into the right-hand side of (2.63) to obtain $\underline{I}_t^{\Gamma\tau}$, and then using the variational principle (2.63) yields the variational estimate

$$[\underline{K}^{\Gamma\tau}] = \underline{K}_t^{\Gamma\tau} + \underline{I}_t^{\Gamma\tau} \tag{2.66}$$

which is correct up to terms of second order. If the trial functions depend on m parameters $c_1^{\Gamma\tau},\ldots c_m^{\Gamma\tau}$, the variation with respect to these parameters and with respect to the K-matrix elements $K^{\Gamma\tau}_{ji}$ yields the system of (m+1) equations

$$\frac{\partial I^{\Gamma\tau}_{ji}}{\partial c_k^{\Gamma\tau}} = 0 \quad , \quad k = 1,2,\ldots,m$$

$$\frac{\partial I^{\Gamma\tau}_{ji}}{\partial K^{\Gamma\tau}_{ji}} = -1 \tag{2.67}$$

If the trial functions $\underline{F}_t^{\Gamma\tau}$ depend linearly on the parameters $c_1^{\Gamma\tau},\ldots c_m^{\Gamma\tau}$, then eqs (2.67) form a system of (m+1) linear equations which are readily solved to yield the wave function and the first order K-matrix elements. The variational (second order) estimate of the K matrix is then obtained by using equation (2.66).

Thus far we have not discussed the choice of the parameters τ_j. Clearly each choice of τ_j gives a different variational estimate of the

K-matrix. This freedom in the choice of the parameters τ_j can be used to avoid the anomalous singularities[20] which can arise when one solves the system of equations (2.67). For example, in the "anomaly-free" (AF) method the Kohn method with $\tau_j=0$ or the "inverse Kohn" (also called the Rubinow[21]) method with $\tau_j=\pi/2$ are used in order to avoid these singularities.

As an illustration, we show in Fig.1 the differential cross section for e^--H(1s) elastic scattering at an incident electron energy of 4.89 eV.

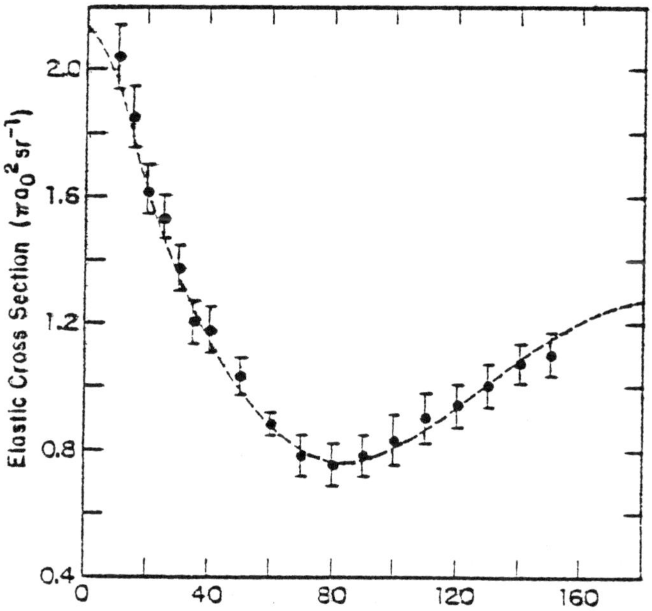

Fig. 1. The differential cross section for e^--H(1s) elastic scattering at an incident electron energy of 4.89 eV. The theoretical results (dashed line) are obtained from variational calculations of the $\ell=0,1$ and 2 phase shifts. The dots represent the absolute measurements of Williams[25].

The theoretical results, obtained from accurate variational calculations of the $\ell=0,1$ and 2 phase shifts (performed respectively by Schwartz[22], Armstead[23] and Register and Poe[24], are seen to be in excellent agreement with the absolute experimental data of Williams[25].

As a second example, we show in Fig. 2 the total and momentum transfer cross sections for elastic electron-helium scattering. The matrix variational results of Nesbet[26] are seen to agree very well with the experimental data of Andrick and Bitsch[27], Kauppila et al[28] and Crompton et al[29]. Also shown in Fig. 2 are the results of O'Malley et al[30], obtained by using the R-matrix method, to which we now turn our attention.

2.1.8. The R-matrix method

Originally introduced by Wigner and Eisenbud[31] in the theory of nuclear reactions, the R-matrix method was first applied to atomic

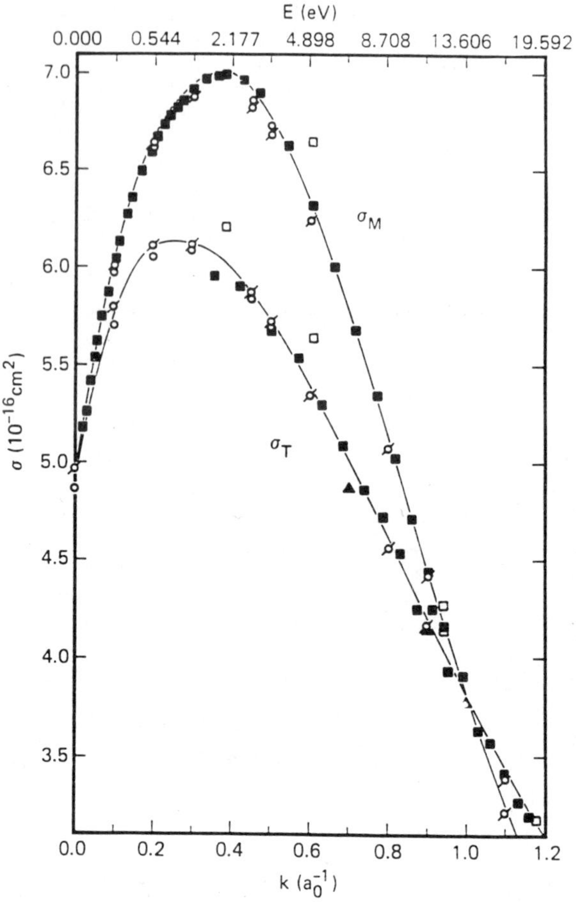

Fig. 2. The total cross section (σ_T) and momentum transfer cross section (σ_M) for elastic electron-helium scattering. Theoretical results: —— matrix variational calculations of Nesbet[26]; ϕ calculated and o estimated R-matrix values of O'Malley et al[30]. Experimental data : □ Andrick and Bitsch[27], ■ Kauppila et al[28] (σ_T); Crompton et al[29] (σ_M). From ref. 12.

collisions by Burke et al[32,33]. The basic idea is that the dynamics of the projectile-target system depends on the relative distance r of the colliding electron and the atomic nucleus, so that it is convenient to partition configuration space into two regions by a sphere of radius r=a, as illustrated in Fig. 3. The radius a is chosen so that the charge distribution of the target states of interest is contained within the sphere r=a. In the internal region (r<a) the direct, exchange and correlation potentials in equations (2.49) are all non-zero and the (N+1) electron complex consisting of the projectile electron and the N-electron target has many of the properties of a bound state. Hence a configuration-interaction expansion of the total wave function similar to that used in bound state calculations is appropriate in the internal region. By contrast, in the external region (r>a) the exchange and correlation effects between the unbound electron and the target can be neglected, and the unbound electron moves in the direct potential which has achieved its asymptotic form (2.50). The collision problem can then be readily solved in this region by using asymptotic methods. As we shall see shortly, the link between the solutions in the two regions is

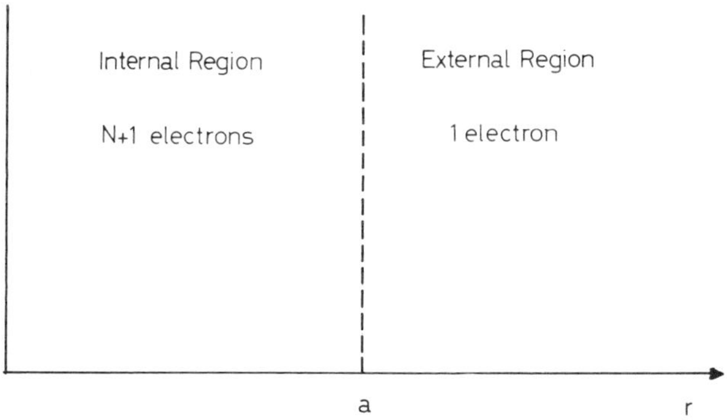

Fig. 3. Partitioning of configuration space in the R-matrix method

provided by the R-matrix.

Following the treatment of Burke[14] we assume, in analogy with (2.48), that the total wave function of the (N+1) electron-atom system can be expanded in the internal region in terms of the basis

$$\Psi_k^\Gamma(\underline{x}_o,\underline{x}_1,\ldots\underline{x}_N) = \mathcal{A}\sum_{ij} a_{ijk}^\Gamma u_j^\Gamma(r_o)\psi_i^\Gamma(\underline{x}_1,\ldots\underline{x}_N,\hat{r}_o,\sigma_o)$$

$$+\sum_i b_{ik}^\Gamma \chi_i^\Gamma(\underline{x}_o,\underline{x}_1,\ldots\underline{x}_N) \quad (2.68)$$

In this equation the channel functions ψ_i^Γ and the square-integrable functions χ_i^Γ are the same as in (2.48). The u_j^Γ are radial basis functions describing the motion of the free electron which do not vanish on the boundary of the inner region. The expansion coefficients a_{ijk}^Γ and b_{ij}^Γ are determined by diagonalising the operator $H_{N+1} + L_{N+1}$ in the internal region, namely

$$\langle\Psi_k^\Gamma|H_{N+1} + L_{N+1}|\Psi_{k'}^\Gamma\rangle = E_k^\Gamma\,\delta_{kk'} \quad (2.69)$$

where H_{N+1} is the (N+1)-electron Hamiltonian (2.2) and

$$L_{N+1} = \frac{1}{2}\sum_i\sum_j |\psi_i^\Gamma\rangle\,\delta(r_j-a)\left(\frac{d}{dr_j} - \frac{b}{r_j}\right)\langle\psi_i^\Gamma| \quad (2.70)$$

is the surface operator introduced by Bloch[34], which guarantees that $H_{N+1} + L_{N+1}$ is Hermitian in the internal region. In the above equation b is an arbitrary parameter, usually chosen to be zero.

The Schrödinger equation in the inner region can be written as

$$(H_{N+1} + L_{N+1} - E)\Psi^\Gamma = L_{N+1}\,\Psi^\Gamma \quad 52.71)$$

and has the formal solution

$$\Psi^{\Gamma} = \frac{1}{H_{N+1} + L_{N+1} - E} L_{N+1} \Psi^{\Gamma} \qquad (2.72)$$

Expanding the operator $(H_{N+1} + L_{N+1} - E)^{-1}$ in terms of the eigenfunctions defined by (2.69), we have

$$|\Psi^{\Gamma}\rangle = \sum_k \frac{|\Psi_k^{\Gamma}\rangle\langle\Psi_k^{\Gamma}|L_{N+1}|\Psi^{\Gamma}\rangle}{E_k^{\Gamma} - E} \qquad (2.73)$$

We now project this equation onto the channel functions ψ_i^{Γ} and evaluate it on the boundary $r_o=a$. We introduce the surface amplitudes w_{ik}^{Γ} by

$$w_{ik}^{\Gamma} = \langle \psi_i^{\Gamma}|\Psi_k^{\Gamma}\rangle_{r_o=a} \qquad (2.74)$$

and the reduced radial wave functions F_i^{Γ} by

$$F_i^{\Gamma}(r_o) = \langle\psi_i^{\Gamma}|\Psi^{\Gamma}\rangle \qquad (2.75)$$

where in both equations (2.74) and (2.75) the integrals are taken over all space and spin coordinates except the radial coordinate r_o. We then obtain

$$F_i^{\Gamma}(a) = \sum_j R_{ij}^{\Gamma}(E) \left[a \frac{dF_j^{\Gamma}}{dr_o} - bF_j^{\Gamma} \right]_{r_o=a} \qquad (2.76)$$

where

$$R_{ij}^{\Gamma}(E) = \frac{1}{2a} \sum_k \frac{w_{ik}^{\Gamma} w_{jk}^{\Gamma}}{E_k^{\Gamma} - E} \qquad (2.77)$$

is the R-matrix.

Equations (2.76) and (2.77) are the basic equations of the R-matrix method. The R-matrix is obtained at all energies by diagonalizing the operator $H_{N+1} + L_{N+1}$ once to determine the eigenvalues E_k^{Γ} and the corresponding surface amplitudes w_{ik}^{Γ}. The logarithmic derivative of the reduced radial wave function on the boundary is then given by (2.76). This is then matched to the solution in the outer region, giving the K-matrix, S-matrix or T-matrix. In practice, the expansion (2.77) is slowly convergent, and a correction for the omitted far-away poles suggested by Buttle[35] must be included.

We show in Fig. 4 the 1^1S-2^3S excitation cross section in helium calculated by Freitas et al[36] by using the R-matrix method and including the first five target eigenstates (dashed line) and the first eleven target eigenstates (solid curve) in the expansion (2.68). The two calculations agree near threshold, but diverge near the n=3 thresholds where the eleven state calculations exhibit substantial resonance structure. The eleven state total metastable excitation cross sections are in excellent agreement with the measurements of Buckman et al[38] near these resonances.

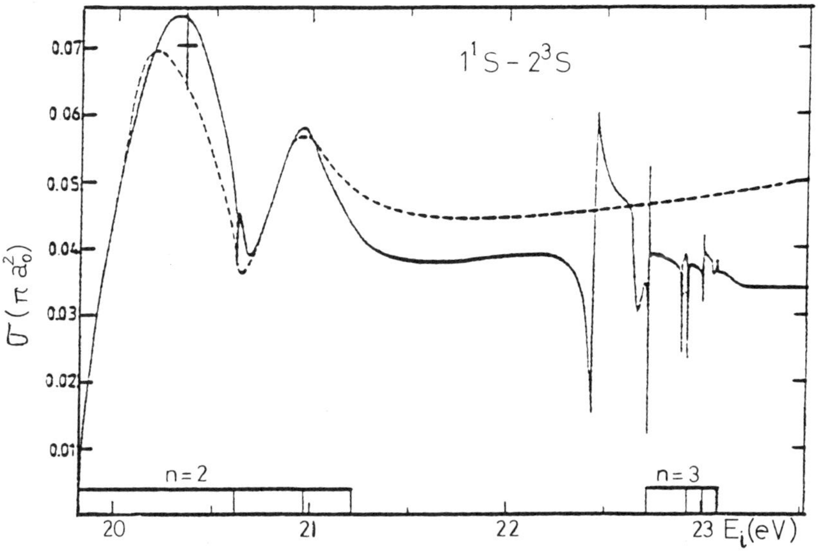

Fig. 4. The electron-helium 1^1S-2^3S excitation cross section calculated by using the R-matrix method in a five state (---) and an eleven state (——) approximation[36]. The cross represents the measurement of Johnston and Burrow[37]. From ref. 36.

2.1.9. Resonances

It is important at this point to stress the fundamental role of resonances in low-energy electron-atom (ion) scattering. The general theory of such resonances has been discussed by Burke[39] and resonances in electron-atom and electron-ion collisions have been reviewed by Golden[40] and Henry[41].

It is useful to distinguish two categories of resonances, namely open-channel (or shape) resonances and closed-channel (or Feshbach) resonances. Open-channel resonances occur when the interaction between the incident electron and the target is such that a quasi-bound intermediate state of the electron-target system can be formed without changing the target configuration. The effective potential between the projectile electron and the target then has a characteristic shape, with an inner well and an outer barrier, so that open-channel resonances are also referred to as shape resonances. We remark that shape resonances occur above the threshold of the channels to which they are most strongly coupled, and that they may occur even if only a single channel is involved.

On the other hand, closed-channel or Feshbach resonances occur for incident energies just below the threshold for the excitation of some state. This may happen when a compound, autoionizing state of the (N+1)-electron system (projectile electron plus target) coincides with an electron in the continuum of the N-electron system. This autoionizing or doubly excited state will either decay radiatively to a bound state of the (N+1)-electron system, or it will autoionize into the continua associated with the target N-electron states. In the latter instance we have a closed-channel resonant behaviour. Since open-channel resonances are above the threshold energies of the target states to which they are most strongly coupled (so that there is an additional decay mode nearby which is open), they are usually much broader than closed-channel resonances, which lie below the threshold corresponding to the target states with which they couple strongly. The projection-operator formalism of Feshbach[6] outlined above is particularly convenient to study the closed-channel resonances.

2.1.10. Quantum defect theory

The fact that the R-matrix is a real meromorphic function of the energy, with simple poles on the real axis, provides the basis of the quantum defect theory[42-44], in which the description of an electron in the field of a positive ion is made in terms of analytically known functions of the energy. In this way electron-ion scattering can be analyzed in terms of a few parameters. In particular, if the K-matrix elements are calculated from the coupled integro-differential equations at a few energies above threshold, they can be extrapolated yielding an infinite series of resonances below threshold.

2.1.11. Relativistic effects

Before leaving the low-energy domain, let us consider briefly relativistic effects. These can become significant when Z increases. For atoms (ions) with small and intermediate Z values, the scattering calculations can be first performed in LS coupling; the K matrices are subsequently recoupled to yield transitions between fine-structure levels. On the other hand, for atoms or ions with high Z-values relativistic terms must be kept in the Hamiltonian. This can be done to lowest order by using the Breit-Pauli Hamiltonian

$$H_{N+1}^{BP} = H_{N+1} + H_{N+1}^{REL} \tag{2.78}$$

where H_{N+1} is the non-relativistic Hamiltonian (2.2) and H_{N+1}^{REL} contains the one-body spin-orbit, Darwin and mass-correction terms. Extensions of the non-relativistic calculations based on the Breit-Pauli Hamiltonian (2.78) have been made[45-47]. Alternatively, programs for calculating electron-atom (ion) cross sections have been written[48,49] based on the Dirac Hamiltonian

$$H_{N+1}^{D} = \sum_{i=o}^{N} (c\boldsymbol{\alpha}\cdot\mathbf{p}_i + \beta c^2 - \frac{Z}{r_i}) + \sum_{i>j=o}^{N} \frac{1}{r_{ij}} \tag{2.79}$$

where $\boldsymbol{\alpha}$ and β are the usual Dirac matrices and c is the velocity of light.

As an example, we consider e^--Hg excitation, for which R-matrix calculations have been performed using the Breit-Pauli Hamiltonian (2.78) and including the $6s^2\ ^1S^o$, $6s6p\ ^3P^o_{0,1,2}$ and $6s6p\ ^1P^o_1$ target states in the expansion[50,51]. We show in Fig. 5 the integrated Stokes parameter[52] η_1 for the transition $6s6p\ ^3P^o_1 \to 6s^2\ ^1S^e_0$ in mercury, excited by polarized electrons.

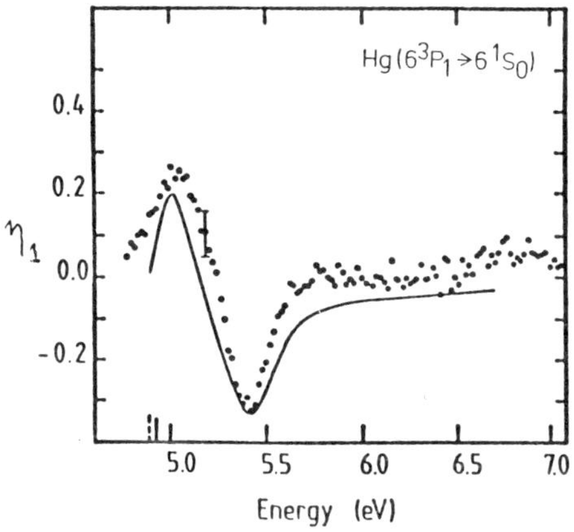

Fig. 5. The integrated Stokes parameter η_1 for the transition $6s6p\ ^3P^o_1 \to 6s^2\ ^1S^e_0$ in Hg, produced by polarized electron impact. The linear polarization η_1 is obtained with transverse electron polarization and the photon detector in the y direction. The light polarization is normalized to the electron polarization P_y. Theoretical curve (——) : Bartschat et al[51]. Experimental data : Wolcke et al[52]. From ref. 51.

It should be noted that the Stokes parameters provide stringent tests of spin-dependent effects. In particular, if all spin-dependent effects were negligible then η_1 would vanish identically. The results close to threshold also depend on the location of the $6s5p^2\ ^2D_{3/2}$ e^- - Hg resonance. The good agreement between theory[51] and experiment[52] seen in Fig. 5 shows that the three main effects of strong channel coupling, electron exchange and the spin-orbit interaction are correctly taken into account by the theory. A further discussion of these effects, together with results for other heavy atoms have been given by Bartschat and Burke[53].

2.2. The region of intermediate and high energies

Let us now consider electron-atom (ion) elastic scattering and excitation at impact energies larger than the ionization energy of the target. In this case an infinite number of channels is open and it is not possible to represent all these open channels explicitly in the total electron-atom wave function. In this Section we shall examine various theoretical methods which have been proposed to deal with fast electron-atom (ion) collisions. Detailed discussions of several of these methods may be found in the review articles of Bransden and McDowell[54], Byron and Joachain[55] and Walters[56]. We assume that the electron-atom system is described by the non-relativistic Hamiltonian (2.2). We shall begin by analyzing the "high-energy" domain which extends from several times the ionization threshold upwards. The case of "intermediate energies" will be discussed subsequently.

2.2.1. The Born series

In the high-energy region it is reasonable to try an approach based on perturbation theory. We shall first discuss direct collisions. The free motion of the projectile electron and the target is described by the direct arrangement channel Hamiltonian

$$H_d = K + H_N \tag{2.80}$$

where K is the kinetic energy operator of the projectile and H_N is the target Hamiltonian having eigenenergies w_n and eigenkets $|n\rangle$. The full Hamiltonian of the system is

$$H_{N+1} = H_d + V_d \tag{2.81}$$

where V_d is the interaction between the electron and the target in the initial (direct) arrangement channel, namely

$$V_d = -\frac{Z}{r_o} + \sum_{j=1}^{N} \frac{1}{r_{oj}} \tag{2.82}$$

with $r_{oj} = |\underset{\sim}{r}_o - \underset{\sim}{r}_j|$ and we recall that $\underset{\sim}{r}_o$ denotes the coordinates of the projectile electron. For simplicity we shall concentrate here on the case of neutral target atoms, for which N = Z.

We now write the Born series for the direct scattering amplitude as

$$f = \sum_{n=1}^{\infty} \bar{f}_{Bn} \tag{2.83}$$

where the nth Born term $\bar{f}_{Bn}$ contains n times the interaction V_d and (n-1) times the direct Green's operator[1]

$$G_d^{(+)} = \frac{1}{E-H_d + i\varepsilon} \quad , \quad \varepsilon \to o^+ \tag{2.84}$$

The first term $\bar{f}_{B1}$ is the familiar first Born amplitude, which has been calculated for a large number of transitions. We simply recall here that if $E_i = k_i^2/2$ denotes the incident electron energy, the total (integrated) first Born cross sections fall off like E_i^{-1} at high energies for "spectroscopically forbidden" transitions (such as S-S or S-D transitions), while for "spectroscopically allowed" transitions (such as S-P transitions) they decrease like $E_i^{-1} \log E_i$ for large E_i. This behaviour may be traced to the infinite range of the Coulomb interaction.

One of the basic goals of the theory is to obtain systematic improvements over the first Born approximation. To this end, we first examine the second Born term $\bar{f}_{B2}$. For the direct transition $|\underset{\sim}{k}_i, i\rangle \to |\underset{\sim}{k}_f, f\rangle$, we have [1]

$$\bar{f}_{B2} = 8\pi^2 \sum_n \int d\underset{\sim}{q} \, \frac{\langle \underset{\sim}{k}_f, f|V_d|\underset{\sim}{q}, n\rangle\langle \underset{\sim}{q}, n|V_d|\underset{\sim}{k}_i, i\rangle}{q^2-k_i^2+2(w_n-w_i)-i\varepsilon} \quad , \quad \varepsilon \to o^+ \tag{2.85}$$

A large amount of work has been devoted in recent years to the evaluation of $\bar{f}_{B2}$. A few exact calculations have been performed for the simplest transitions in atomic hydrogen [57-60], but in general $\bar{f}_{B2}$ must be evaluated approximatively. At sufficiently high energies, a good approximation to $\bar{f}_{B2}$ is obtained by replacing the energy differences $(w_n - w_i)$ by an average excitation energy $\bar{w}$, so that the sum on the intermediate target states can be done by closure. An improvement over this procedure consists in evaluating exactly the first few terms in the sum, while treating the remaining states by closure or by using pseudo-states. This method has been widely used in recent years, along with further improvements[55,56,59].

The behaviour of the terms $\bar{f}_{Bn}$ as a function of (large) k_i and of the magnitude Δ of the momentum transfer $\underset{\sim}{\Delta} = \underset{\sim}{k}_i - \underset{\sim}{k}_f$ has been analyzed by Byron and Joachain[55,61] for a target atom whose initial state is an S-state. Their results for direct elastic scattering (where $|\underset{\sim}{k}_i|=|\underset{\sim}{k}_f|=k$) are summarized in Table 1. This Table also shows the dependence of the corresponding terms of the Glauber series, which will be considered shortly.

Looking at Table 1, we see that for direct elastic scattering the dominant contribution to the scattering amplitude at large k is given by the first Born term $\bar{f}_{B1}$ at all momentum transfers. At small Δ the real part of the second Born term, Re $\bar{f}_{B2}$, is governed by polarization effects and gives the dominant correction (of order k^{-1}) to the first Born differential cross section $d\sigma_{B1}/d\Omega$. The imaginary part of the second Born term, Im $\bar{f}_{B2}$, gives at small Δ a correction to $d\sigma_{B1}/d\Omega$ of order $(\log k)^2/k^2$ taking into account absorption effects due to the loss of flux into the open channels. At large Δ the terms $\bar{f}_{Bn}$ $(n \geq 2)$ of the direct elastic amplitude are dominated by processes in which the atom remains in its initial state $|i\rangle$ in all intermediate states; this reflects the fact that at large angles elastic scattering is governed by the static potential $V_{ii} = \langle i|V_d|i\rangle$. For example, we have represented in Fig. 6 the diagram corresponding to the dominant contribution to $\bar{f}_{B2}$ at large momentum transfers. This dominant contribution is obtained by keeping in the sum on intermediate target states on the right of (2.85) the only term n=i for which the target atom remains in the initial state $|i\rangle$; we shall call this term $\bar{f}_{B2}$ (i).

The situation is different for direct inelastic collisions where for large Δ the first Born term falls off rapidly and the Born series is dominated by the second Born term $\bar{f}_{B2}$. This is illustrated in Table 2 for inelastic S-S transitions, and may be understood as follows. Large momentum transfer collisions can only take place if the projectile electron collides with the much heavier atomic nucleus. Now, for inelastic scattering the orthogonality of the target initial and final states, $\langle f|i\rangle = 0$, removes the electron-nucleus interaction term $- Z/r_0$ from the first Born amplitude $\bar{f}_{B1}$. As a result, the first Born approximation differential cross section for high-energy large angle inelastic scattering is orders of magnitude too small. The fact that $\bar{f}_{B2}$ falls off more slowly than $\bar{f}_{B1}$ for inelastic collisions at large Δ is due to the possibility of off-shell elastic scattering in the intermediate states $|i\rangle$ and $|f\rangle$ (the initial and final target states of the inelastic transition), where the projectile electron can experience the Coulomb potential of the nucleus. The corresponding two contributions to $\bar{f}_{B2}$, which we shall denote respectively by $\bar{f}_{B2}(i)$ and $\bar{f}_{B2}(f)$, are obtained by keeping in the sum on intermediate states in (2.85) the terms n=i or n=f, respectively. The two diagrams representing these contributions are shown in Fig. 7. We remark that since the small momentum transfer region $(\Delta \lesssim 1)$ - where the first Born approximation is accurate for large $k(= \underset{\sim}{k}_i)$ - corresponds to angles $\theta \lesssim k^{-1}$, the angular region in which the first Born approximation is valid shrinks with increasing energy. However, because the dominant contribution to total (integrated) cross sections comes from the region $\Delta \lesssim 1$, the first Born values for total cross sections are reliable at sufficiently high energies.

TABLE 1

Dependence of various terms of the Born and Glauber series for the direct scattering amplitude corresponding to elastic scattering by an atom in an S-state, as a function of (large) $k = |\underset{\sim}{k}_i| = |\underset{\sim}{k}_f|$ and of $\Delta = |\underset{\sim}{k}_i - \underset{\sim}{k}_f|$. The dominant contributions are framed. The terms located above the dashed line contribute through order k^{-2} to the differential cross section

Order of Pert. Theory	Term	Small Δ ($\Delta < k^{-1}$)	Interm. Δ ($k^{-1} < \Delta < 1$)	Large Δ ($\Delta > k$)
First	$\bar{f}_{B1} = \bar{f}_{G1}$	$\boxed{1}$	$\boxed{1}$	$\boxed{\Delta^{-2}}$
Second	Re $\bar{f}_{B2}$	k^{-1}	k^{-2}	$k^{-2}\ \Delta^{-2}$
	Re $\bar{f}_{G2}$	0	0	0
	Im $\bar{f}_{B2}$	k^{-1} log k	k^{-1}	$k^{-1}\ \Delta^{-2}$ log Δ
	Im $\bar{f}_{G2}$	k^{-1} log Δ	k^{-1}	$k^{-1}\ \Delta^{-2}$ log Δ
Third	Re $\bar{f}_{B3}$	k^{-2}	k^{-2}	$k^{-2}\ \Delta^{-2}\ \log^2 \Delta$
	Re $\bar{f}_{G3}$	k^{-2}	k^{-2}	$k^{-2}\ \Delta^{-2}\ \log^2 \Delta$
	- - - - - - - -	- - - - - - - -	- - - - - - - -	- - - - - - - -
	Im $\bar{f}_{B3}$	k^{-3}	k^{-3}	$k^{-3}\ \Delta^{-2}$ log Δ
	Im $\bar{f}_{G3}$	0	0	0
n^{th} $(n > 3)$	$\bar{f}_{Bn}$	$(ik)^{1-n}$	$(ik)^{1-n}$	$(ik)^{1-n}\Delta^{-2}\log^{n-1}\Delta$
	$\bar{f}_{Gn}$	$(ik)^{1-n}$	$(ik)^{1-n}$	$(ik)^{1-n}\Delta^{-2}\log^{n-1}\Delta$

The terms $\bar{g}_{Bn}$ of the Born series for exchange scattering are much more difficult to analyze than the direct Born terms $\bar{f}_{Bn}$. We simply mention that for large k the term $\bar{g}_{B2}$ falls off more slowly than g_{B1}, except for elastic exchange scattering at small Δ, where the Bonham - Ochkur amplitude[62,63] g_{Och}, which is the leading piece of $\bar{g}_{B1}$, is of order k^{-2}. It is also interesting to note that if we know all the direct amplitudes half-off-shell, then the exchange amplitude for a transition $|\underset{\sim}{k}_i, i\rangle \to |\underset{\sim}{k}_f, f\rangle$, which we write in an explicit notation as $g(\underset{\sim}{k}_f, f; \underset{\sim}{k}_i, i)$, may be obtained by quadratures via the relation[64]

$$g(\underset{\sim}{k}_f, f; \underset{\sim}{k}_i, i) = \bar{g}_{B1}(\underset{\sim}{k}_f, f; \underset{\sim}{k}_i, i) + (2\pi^2)^{-1} \sum_n \int d\underset{\sim}{q}\ \frac{\bar{g}_{B1}(\underset{\sim}{k}_f, f; \underset{\sim}{q}, n) f(\underset{\sim}{q}, n; \underset{\sim}{k}_i, i)}{q^2 - k_i^2 + 2(w_n - w_i) - i\varepsilon}, \quad \varepsilon \to 0^+ \tag{2.86}$$

where $\bar{g}_{B1}(\underset{\sim}{k}_f, f; \underset{\sim}{k}_i, i)$ is the on-shell first Born exchange amplitude, $\bar{g}_{B1}(\underset{\sim}{k}_f, f; \underset{\sim}{q}, n)$ is a half-off-shell first Born exchange amplitude and $f(\underset{\sim}{q}, n; \underset{\sim}{k}_i, i)$ is a half-off-shell direct amplitude.

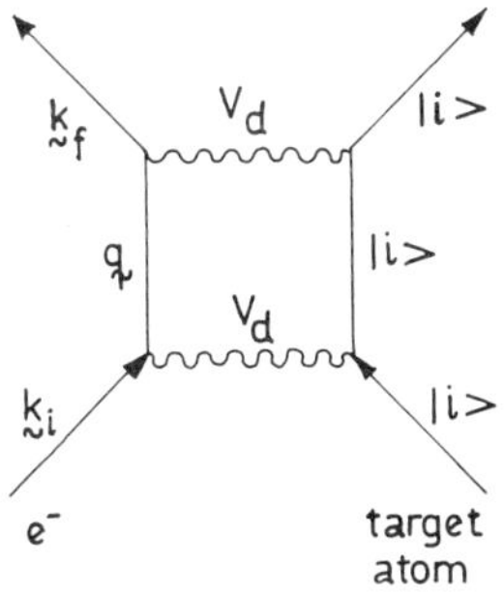

Fig. 6. Diagram representing the dominant contribution $\bar{f}_{B2}(i)$ to the second Born term $\bar{f}_{B2}$ at large momentum transfers for an elastic scattering process. The target atom remains in the initial state $|i\rangle$ in the intermediate state. The vector $\underset{\sim}{q}$ is the intermediate momentum of the projectile and V_d is the direct target-projectile interaction (2.82).

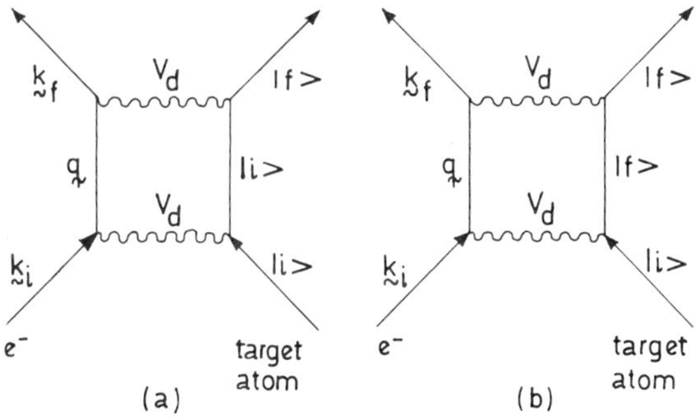

Fig. 7. The two diagrams representing the dominant contributions to $\bar{f}_{B2}$ at large momentum transfer for an inelastic transition $|\underset{\sim}{k}_i, i\rangle \rightarrow |\underset{\sim}{k}_f, f\rangle$. Fig. 7(a) corresponds to the contribution $\bar{f}_{B2}(i)$ in which the target atom remains in the initial state $|i\rangle$ in the intermediate state. Fig. 7(b) represents the contribution $\bar{f}_{B2}(f)$ in which the intermediate state of the target atom is the final state $|f\rangle$.

TABLE 2

Dependence of various terms of the Born and Glauber series for the direct scattering amplitude corresponding to inelastic (S-S) transitions as a function of (large) $k = |\underset{\sim}{k}_i|$ and of $\Delta = |\underset{\sim}{k}_i - \underset{\sim}{k}_f|$. The dominant contributions are framed. The terms located above the dashed line contribute through order k^{-2} to the differential cross section.

Order of Pert. Theory	Term	Small Δ ($\Delta < k^{-1}$)	Interm. Δ ($k^{-1} < \Delta < 1$)	Large Δ ($\Delta > k$)
First	$\bar{f}_{B1} = \bar{f}_{G1}$	$\boxed{1}$	$\boxed{1}$	Δ^{-6}
Second	Re $\bar{f}_{B2}$	k^{-1}	k^{-2}	$k^{-2}\,\Delta^{-2}$
	Re $\bar{f}_{G2}$	0	0	0
	Im $\bar{f}_{B2}$	k^{-1} log k	k^{-1}	$\boxed{k^{-1}\,\Delta^{-2}}$
	Im $\bar{f}_{G2}$	k^{-1} log Δ	k^{-1}	$\boxed{k^{-1}\,\Delta^{-2}}$
Third	Re $\bar{f}_{B3}$	k^{-2}	k^{-2}	$k^{-2}\,\Delta^{-2}$ log Δ
	Re $\bar{f}_{G3}$	k^{-2}	k^{-2}	$k^{-2}\,\Delta^{-2}$ log Δ
	- - - - - -	- - - - - -	- - - - - -	- - - - - -
	Im $\bar{f}_{B3}$	k^{-3}	k^{-3}	$k^{-3}\,\Delta^{-2}$ log Δ
	Im $\bar{f}_{G3}$	0	0	0
n^{th} ($n > 3$)	$\bar{f}_{Bn}$	$(ik)^{1-n}$	$(ik)^{1-n}$	$(ik)^{1-n}\,\Delta^{-2}\,\log^{n-2}\Delta$
	$\bar{f}_{Gn}$	$(ik)^{1-n}$	$(ik)^{1-n}$	$(ik)^{1-n}\,\Delta^{-2}\,\log^{n-2}\Delta$

2.2.2. The Glauber approximation

The accurate evaluation of the higher Born terms $\bar{f}_{Bn}$ ($n \geq 3$) is beyond the limit of present expertise, but fortunately, useful information about these terms can be obtained by using eikonal methods [65]. We shall begin by discussing the Glauber approximation[66], which is the simplest multi-particle generalization of the potential scattering eikonal approximation of Moliere[67]. For a direct collision $|\underset{\sim}{k}_i, i\rangle \rightarrow |\underset{\sim}{k}_f, f\rangle$, the Glauber scattering amplitude is given by

$$f_G = \frac{k_i}{2\pi i} \int d^2 \underset{\sim}{b}_o \exp(i\underset{\sim}{\Delta}\cdot\underset{\sim}{b}_o) \langle f | \left\{ \exp\left[\frac{i}{k_i}\,\chi_o(\underset{\sim}{b}_o, X)\right] - 1 \right\} | i \rangle \tag{2.87}$$

where the symbol X denotes the ensemble of the target coordinates and we use a cylindrical coordinate system, with $\underset{\sim}{r}_o = \underset{\sim}{b}_o + z_o\hat{z}$. The Glauber phase χ_o is given in terms of the direct interaction (2.82) by

$$\chi_o = -\int_{-\infty}^{+\infty} V_d(\underset{\sim}{b}_o, z_o, X)\, dz_o, \tag{2.88}$$

the integration being performed along a z-axis perpendicular to Δ and lying in the scattering plane.

Detailed discussions of the Glauber approximation may be found in the review articles of Joachain and Quigg[68] and Byron and Joachain[55]. We first note that the Glauber approach may be viewed as an eikonal approximation to a "frozen target" model proposed by Chase[69], in which closure is used with an average excitation energy $\bar{w} = 0$. Interesting insight into the properties of the Glauber method may also be gained by expanding the Glauber amplitude (2.87) in powers of V_d, namely

$$f_G = \sum_{n=1}^{\infty} \bar{f}_{Gn} \tag{2.89}$$

where

$$\bar{f}_{Gn} = \frac{k_i}{2\pi i}\frac{1}{n!}\left(\frac{i}{k_i}\right)^n \int d^2b_o \exp(i\Delta \cdot b_o) \langle f | \chi_o^n(b_o, X) | i \rangle \tag{2.90}$$

and comparing the nth Glauber term $\bar{f}_{Gn}$ with the corresponding nth Born term $\bar{f}_{Bn}$. Because of our choice of z-axis we have $\bar{f}_{B1}=\bar{f}_{G1}$. For $n \geq 2$, the terms $\bar{f}_{Gn}$ are alternately real or purely imaginary, while the corresponding Born terms are complex. This special feature of the Glauber amplitude leads to several defects such as i) the absence in the elastic scattering case of the important term Re $\bar{f}_{B2}$ which accounts for polarization effects at small Δ and ii) identical cross sections for electron- and positron- atom scattering. Other deficiencies of the Glauber amplitude (2.87) include i) a logarithmic divergence for elastic scattering in the forward direction, which is due to the choice $\bar{w}=0$ made in obtaining the Glauber amplitude and may be traced to the behaviour of $\bar{f}_{G2}$ at $\Delta=0$, as shown in Table 1 and ii) a poor description of inelastic collisions involving non-spherically symmetric states. Nevertheless, the Glauber amplitude (2.87) has one attractive property : it includes terms from all orders of perturbation theory in such a way as to ensure unitarity. We shall come back to this point below.

2.2.3. The eikonal-Born series method

The eikonal-Born series (EBS) method, introduced by Byron and Joachain[65], is based on an analysis of the terms of the Born series (2.83) and of the Glauber series (2.89), the aim being to obtain a consistent expansion of the scattering amplitude in powers of k^{-1}. The main results are summarized in Tables 1 and 2 for direct elastic and inelastic (S-S) transitions, respectively. We see that for these processes the Glauber term $\bar{f}_{Gn}$ gives in each order of perturbation theory the leading piece of the corresponding Born term (for large k) for all momentum transfers, except in second order where the long range of the Coulomb potential is responsible for the anomalous behaviour of $\bar{f}_{G2}$ at small Δ. We also remark from Tables 1 and 2 that neither the second Born amplitude $f_{B2} = \bar{f}_{B1} + \bar{f}_{B2}$ nor the Glauber amplitude f_G are correct through order k^{-2}. In fact, a consistent calculation of the direct scattering amplitude through that order requires the terms f_{B1}, $\bar{f}_{B2}$ and Re $\bar{f}_{B3}$ (or $\bar{f}_{G3}$). Since Re $\bar{f}_{B3}$ is very difficult to evaluate, and because $\bar{f}_{G3}$ is a good approximation to Re $\bar{f}_{B3}$ for large enough k, one can use $\bar{f}_{G3}$ in place of Re $\bar{f}_{B3}$. Thus one obtains in this way the eikonal-Born series direct scattering amplitude

$$f_{EBS} = \bar{f}_{B1} + \bar{f}_{B2} + \bar{f}_{G3} \tag{2.91}$$

We note that the EBS' amplitude [70] (also called the modified Glauber amplitude)

$$f_{EBS'} = f_G - \bar{f}_{G2} + \bar{f}_{B2}$$
$$= \bar{f}_{B1} + \bar{f}_{B2} + \sum_{n=3}^{\infty} \bar{f}_{Gn} \tag{2.92}$$

also gives a consistent picture of the direct scattering amplitude through order k^{-2}. In addition, exchange effects are taken into account in the EBS theory by keeping the relevant terms in the Born series for the exchange amplitude. For example, in the case of elastic scattering the Bonham-Ochkur amplitude g_{Och}, which is of order k^{-2} for large k and fixed Δ, must be taken into account in order to perform a consistent calculation of the differential cross section through order k^{-2}.

The EBS method is very successful when perturbation theory converges rapidly, namely at high energies, small and intermediate angles and for light atoms. This is illustrated in Fig.8, where the differential cross

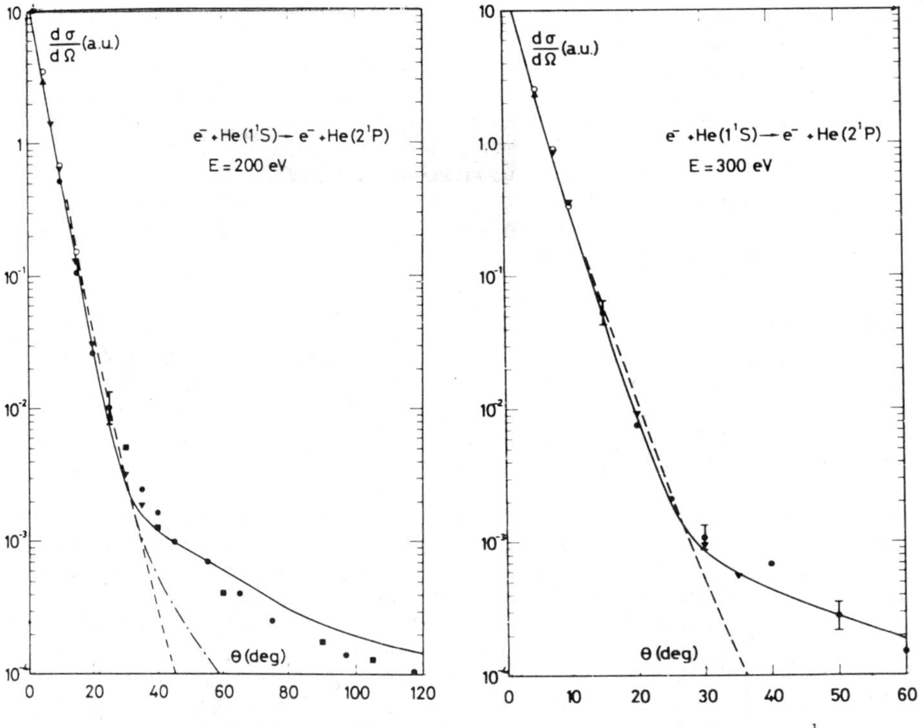

Fig. 8. Differential cross section for the transition $e^- + He(1^1S) \rightarrow e^- + He(2^1P)$ at incident electron energies (a) of 200 eV and (b) of 300 eV. Theoretical curves : --- first Born approximation, -.- : Glauber approximation, —— EBS calculation[71]. Experimental data : ▲ Chamberlain et al[72]; ■ Opal and Beaty[73], ● Suzuki and Takayanagi[74], ▾ Dillon and Lassettre[75]. From ref. 71.

section for the excitation of the 2^1P state of helium is shown at incident electron energies of 200 eV and 300 eV. This figure also shows the failure of the first Born approximation for this inelastic process, in agreement with the discussion of Section 2.2.1. Since we are dealing with an S-P transition, the Glauber approximation is also expected to be inaccurate outside the small-angle region (which is essentially controlled by $\bar{f}_{B1}$ in the present case), and this is confirmed by the examination of Fig. 8(a).

When perturbation theory is more slowly convergent, improvements over the EBS method are necessary. These can be obtained by constructing methods which include terms from all orders of perturbation theory in a unitary way. It is to this question that we now turn our attention.

2.2.4. The unitarized EBS method

In the case of potential scattering, Wallace[76] has obtained improvements over the eikonal approximation by writing down in a systematic way the corrections to the eikonal phase. Detailed studies of the relationships between the terms of the Born, eikonal and Wallace series for potential scattering have been made by Byron, Joachain and Mund[77]. In the light of these developments, Byron, Joachain and Potvliege[78,79] have proposed a generalization of the potential scattering Wallace amplitude to the multi-particle case. For the direct transition $|\mathbf{k}_i, i\rangle \rightarrow |\mathbf{k}_f, f\rangle$ this multi-particle Wallace amplitude is given by[79]

$$f_W = \frac{k_i}{2\pi i} \int d^2 b_o \exp(i\boldsymbol{\Delta}.\mathbf{b}_o) \langle f | \left\{ \exp[\frac{i}{k_i} \chi_o(\mathbf{b}_o, X)] [1 + \frac{i}{k_i^3} \chi_1(\mathbf{b}_o, X)] - 1 \right\} | i \rangle \qquad (2.93)$$

where χ_o is the Glauber phase (2.88), while the Wallace correction χ_1, which is of second order in the interaction potential V_d, reads

$$\chi_1(\mathbf{b}_o, X) = \frac{1}{2} \int_{-\infty}^{+\infty} (\boldsymbol{\nabla}\chi_+) \cdot (\boldsymbol{\nabla}\chi_-) dz_o \qquad (2.94a)$$

with

$$\chi_+(\mathbf{b}_o, z_o, X) = -\int_{-\infty}^{z_o} V_d(\mathbf{b}_o, z', X) dz', \quad \chi_-(\mathbf{b}_o, z_o, X) = -\int_{z_o}^{\infty} V_d(\mathbf{b}_o, z', X) dz' \qquad (2.94b)$$

The terms of the Wallace series, obtained by expanding f_W in powers of V_d, will be called $\bar{f}_{Wn}$.

At this point, we remark that since the excitation energies in both the initial and final channels have been set equal to zero, the long-range polarization effects will be missing from f_W and the quantity Im $\bar{f}_{W2}$ (=Im $\bar{f}_{G2}$) will diverge at Δ=0 for elastic scattering. Following Byron, Joachain and Potvliege[78], we therefore construct a new amplitude with $\bar{f}_{B2}$ inserted in the place of $\bar{f}_{W2}$. That is,

$$f_{UEBS} = f_W - \bar{f}_{W2} + \bar{f}_{B2} \qquad (2.95)$$

At small momentum transfers, where higher-order terms of perturbation theory are unimportant, we retrieve the EBS amplitude (2.91) by keeping the terms through order k^{-2} in (2.95). On the other hand the terms $\bar{f}_{B2}$ and $\bar{f}_{W2}$ differ negligibly at large Δ, so that at large angles f_{UEBS} will differ little from f_W and will provide a more accurate value of the direct scattering amplitude than the EBS' amplitude (2.92). Moreover, the amplitude f_{UEBS} will be nearly unitary at all angles. It is this " all-order" amplitude which is called the "unitarized EBS" amplitude. It may be used in equation (2.86) to obtain a corresponding " all-order" exchange amplitude.

Applications of the UEBS method have been made thus far to elastic scattering and to the 1s-2s and 1s-2p transitions in atomic hydrogen, for both electron and positron impact[64,78,79]. As an example, we compare in Fig. 9 the UEBS magnetic sublevel differential cross sections for excitation of the 2pm states, at an incident electron energy of 54.4 eV

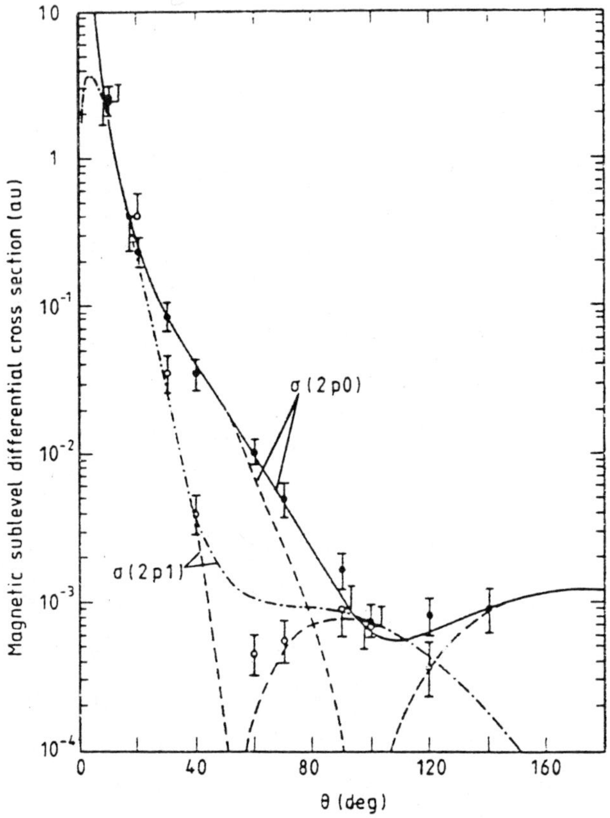

Fig. 9. Differential cross section for the excitation of the 2pm sublevels of atomic hydrogen by electron impact at an incident electron energy of 54.4 eV. Theoretical curves : ___ UEBS results[79] for the excitation of the 2po state; -.- UEBS results[79] for the excitation of the 2pl (or 2p-1) state; --- UEBS results without exchange. The experimental data are those of Williams[80]; the black dots (•) correspond to the excitation of the 2po state and the open circles (o) to the excitation of the 2pl (or 2p-1) state. From ref. 79.

with the experimental data of Williams[80]. The agreement between theory and experiment is seen to be good. It is worth noting from Fig.9 that at this rather low impact energy the magnetic sublevel differential cross sections are very sensitive to exchange effects in certain angular regions.

2.2.5. Optical potentials

The original idea of the optical potential method is to analyze the elastic scattering of a particle from a complex target by replacing the complicated interactions between the projectile and the target particles by an optical potential (or pseudo- potential) in which the incident particle moves[1,81]. Once the optical potential V_{opt} is determined, the original many-body elastic scattering problem reduces to a one-body situation. However, this reduction is in general a difficult task, and approximation methods are necessary.

Let us begin by considering direct elastic scattering, the corresponding direct part of the optical potential being denoted by V^d_{opt}. Formal expressions for V^d_{opt} are readily obtained by using the Feshbach projection operator formalism[6]. Let

$$P = |i\rangle\langle i| \tag{2.96}$$

be an operator projecting on the initial target state $|i\rangle$, and let $Q = I-P$. Remembering that the total Hamiltonian of the system is given by $H_{N+1} = H_d + V_d = K + H_N + V_d$, and that the operators P and Q commute with H_d, we may write equation (2.34) in the form

$$(K + V^d_{opt} - E_i)\, P\,\Psi = 0 \tag{2.97}$$

where $E_i = E-w_i = k_i^2/2$ is the projectile energy and we have used the fact that $H_N|i\rangle = w_i|i\rangle$. The direct optical potential V^d_{opt} appearing in the above equation is given by

$$V^d_{opt} = PV_dP + \mathcal{V} \tag{2.98}$$

where PV_dP is the potential V_d acting in P space and $\mathcal{V}$ is the operator given by (2.36).

At intermediate and high energies it is appropriate to make a perturbative expansion of (2.98) in powers of V_d. That is,

$$V^d_{opt} = V^{(1)} + V^{(2)} + V^{(3)} + \ldots \tag{2.99}$$

where the first order term is $V^{(1)} = PV_dP$, the second order term reads

$$V^{(2)} = PV_dQ\,\frac{1}{E-H_d+i\varepsilon}\,QV_dP\ , \quad \varepsilon\to o^+ \tag{2.100}$$

ans so on. With P given by (2.96), we see that $V^{(1)} = \langle i|V_d|i\rangle$ is just the static potential, while a more explicit expression of $V^{(2)}$ is given by

$$V^{(2)} = \sum_{n\neq i} \frac{\langle i|V_d|n\rangle\ \langle n|V_d|i\rangle}{k_i^2/2 - K - (w_n-w_i)+i\varepsilon}\ ,\varepsilon\to o^+ \tag{2.101}$$

The static potential $V^{(1)} = \langle i|V_d|i\rangle$ is readily evaluated for simple target atoms, or when an independent particle model is used to describe the target state $|i\rangle$, which we assume here to be spherically symmetric. We remark that $V^{(1)}$ is real and of short range, and hence does not account for polarization and absorption effects which play an important role in the energy range considered here. However, for small values of the projectile coordinate r_o, the static potential correctly reduces to the Coulomb interaction $-Z/r_o$ acting between the projectile electron and the target nucleus, and hence gives a good account of large angle direct scattering, as we already noted in Section 2.2.1.

Although the second and higher order terms of the direct optical potential are in general complicated, non-local, complex operators, at sufficiently high energies local approximations to them can be obtained[59,82-85]. For example, Byron and Joachain[59] converted the lowest order terms of perturbation theory, calculated by using the EBS method for elastic scattering, into an ab-initio optical potential. The second order part $V^{(2)}$ of the direct optical potential may then be written approximately as

$$V^{(2)} = V_{pol} + i\, V_{abs} \tag{2.102}$$

where V_{pol} and V_{abs} are real and central but energy-dependent. The term V_{pol} (which falls off like r_o^{-4} at large r_o) accounts for dynamic polarization effects and iV_{abs} for absorption effects due to loss of flux from the incident channel. The leading contribution of the third order potential $V^{(3)}$ has also been obtained for $e^{\pm}$-H(1s) scattering[59] and for $e^{\pm}$-H(2s), $e^{\pm}$-He(2^1S) and $e^{\pm}$-He(2^3S) scattering[86].

Having obtained a local approximation for V^d_{opt}, exchange effects may be taken into account by using a local exchange pseudo-potential V^{ex}_{opt}, which is usually taken to be an approximation to static - exchange[87-90]. The full optical potential V_{opt} containing the direct and exchange parts, is then treated in a unitary, essentially exact partial wave manner. It is worth noting that in performing such an exact treatment of the (approximate) optical potential one generates approximations to all terms of perturbation theory. In particular, the static potential is taken care of exactly, a feature which is an important advantage for large angle scattering.

As an illustration of this optical model theory, we compare in Fig. 10 the theoretical predictions[84,85] with the differential cross sections recently measured by Hyder et al[91] for $e^{\pm}$-argon scattering at an incident energy of 300 eV. For incident positrons the direct potential V_d has of course the opposite sign of that written in equation (2.82) for incident electrons, and in addition there are no exchange effects. It should be noted that the positron measurements constitute an additional test of the theory. A detailed discussion of the angular distributions for $e^{\pm}$-argon scattering has been given recently by Joachain and Potvliege[92] and a review of fast positron-atom collisions may be found in ref.93.

The optical potential formalism outlined above is readily generalized to the case in which we are interested in transitions between a certain number M of target states. Equations (2.97) - (2.100) are still valid, provided the expression (2.96) of P is now replaced by

$$P = \sum_{n=1}^{M} |n\rangle\langle n| \tag{2.103}$$

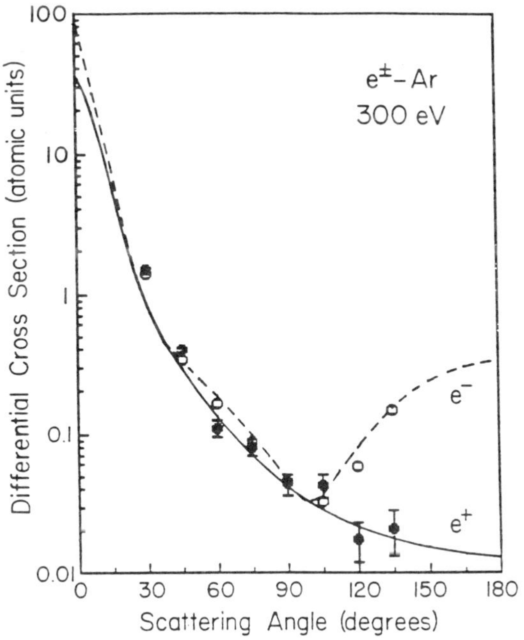

Fig. 10. Differential cross section for electron (---) and positron(——) elastic scattering by argon at 300 eV, calculated by Joachain et al[84,85] from an ab-initio optical model theory. The experimental data of Hyder et al[91] are : o electron scattering, • positron scattering. From ref.91.

Of course, equation (2.97) now represents a set of coupled equations, and the optical potential is a potential matrix. In particular, the second-order potential (SOP) method of Bransden and Coleman[94] is an approximation to these coupled equations, in which the optical potential is treated to second order in V_d. For example, in the case of the two-electron system, the coupled equations of the SOP method read

$$(\nabla^2+k_i^2)F_i^{\pm}(\underset{\sim}{r}) = 2\sum_{j=1}^{M}(V_{ij}+W_{ij}^{\pm}+V_{ij}^{(2)})F_j^{\pm}(\underset{\sim}{r}) \tag{2.104}$$

where we have used the notation of (2.46). In the above equation $V_{ij}^{(2)}$ is a second order potential matrix which accounts approximately for the coupling with the states $n\geqslant M+1$ omitted from the close-coupling expansion. Neglecting exchange, it is given by

$$V_{ij}^{(2)} = \sum_{n\geqslant M+1} \frac{\langle i|V_d|n\rangle\langle n|V_d|j\rangle}{k_j^2/2-K-(w_n-w_j)+i\varepsilon}\ , \quad \varepsilon\to 0^+ \tag{2.105}$$

which is the generalization of the one-channel direct second order potential (2.101). A detailed discussion of the second order potential method may be found in the review article of Bransden and McDowell[54]. Among the various applications of the method, we mention in particular the study of elastic scattering, 1s-2s and 1s-2p excitation in e^--H collisions performed by Bransden et al.[95].

McCarthy et al[96,97] have studied approximations to the optical potential going beyond the second order approximation, and making allowance for exchange. Their method is often called the "coupled-channel optical model" (CCOM).

2.2.6. Distorted waves

Distorted wave treatments are characterized by the fact that the interaction is broken in two parts, one which is treated exactly and the other which is handled by perturbation theory[1]. This separation is dictated by the physics of the problem, and consequently many kinds of distorted wave methods have been applied to electron-atom (ion) collisions. Detailed discussions of distorted-wave treatments may be found in the reviews of Bransden and McDowell[54] and Henry[41]. We mention in particular the well-known distorted wave Born approximation[1] (DWBA), the distorted wave second Born approximation[60,98,99] (DWSBA) and the pseudo-state close-coupling plus distorted wave second Born approximation[100].

2.2.7. Modified low-energy methods

The most elaborate methods discussed above can usually be applied down to relatively low impact energies, but as the projectile energy is decreased to be only somewhat larger than the ionization energy of the target, other approximation schemes are necessary. Instead of pushing down the validity of high-energy methods, one can try to extend the low-energy approaches discussed in Section 2.1 to that "sub-intermediate" energy region. For example, Burke and Webb[101] and Callaway and Wooten[102] have generalized the pseudo-state approach to the case where the pseudo-states can carry away flux into the open channels (including the ionizing ones). Extensive calculations have been made by Callaway et al[103] for e^--H scattering and by Willis and McDowell[104] for e^--He collisions. Unfortunately, this method presents undesirable features due to the unphysical pseudo-resonances associated with the pseudo-states, although Burke et al[105] have shown that useful information can be obtained by performing a suitable averaging of the scattering amplitudes over the pseudo-resonances. Another difficulty of the pseudo-state method is to define a satisfactory pseudo-state basis, particularly for complex targets.

There has also been much interest in methods which attempt to approximate the wave function in a limited region of configuration space by a combination of square-integrable (L^2) functions, avoiding the explicit representation of the scattering boundary conditions, and devising indirect means of extracting the scattering parameters. These L^2 methods which have been reviewed by Reinhardt[106] and Broad[107], have been successful in treating the elastic scattering of electrons and positrons by hydrogen atoms, but numerical instabilities have been encountered in the evaluation of inelastic scattering parameters. Bransden and Stelbovics[108] have proposed to expand the closed-channel part of the optical potential on an L^2 basis, using a method suggested by the work of Heller et al[109,110], Broad[111] and Reading et al[112]. This approach was applied successfully to a model problem of two coupled S channels in which the first is treated explicitly (P space) and the second (Q space) is described by the closed-channel part of an optical potential represented on an L^2 basis. However, extensions of this approach to more realistic electron-atom problems (involving more channels) has proved difficult due to numerical instabilities[113].

Recently, Burke et al[114] have proposed a new R-matrix method with the aim of obtaining accurate scattering amplitudes in the "sub-intermediate" energy region. Indeed, while the expansion (2.68) gives precise results for low electron impact energies where only a small number of channels are open, it is less satisfactory at impact energies such that an infinite number of channels are open. In the new R-matrix method the calculation of the scattering of an electron by an (N+1)-electron target is carried out in two steps. In the first one a complete set of (N+1)-electron bound and continuum basis states are determined by considering the collision of an electron with an N-electron target. This basis, denoted by Ψ_k^Γ is defined by equations (2.68) and (2.69). In the second step these states are then used as target states for the (N+2)-electron system. The appropriate basis in this case is defined by

$$\theta_k^\Gamma = \sum_{ij} \alpha_{ijk}^\Gamma u_j^\Gamma(r_o)\bar{\Psi}_k^\Gamma + \sum_i \beta_{ik}^\Gamma \chi_i^\Gamma \qquad (2.105)$$

where the $\bar{\Psi}_k^\Gamma$ are channel functions formed from the Ψ_k^Γ, the u_j are continuum basis functions and the χ_i are (N+2)-electron square integrable functions. The expansion coefficients α_{ijk} and β_{ik} are obtained by diagonalizing $H_{N+2} + L_{N+2}$ in the internal region,

$$\langle\Theta_k^\Gamma|H_{N+2} + L_{N+2}|\Theta_{k'}^\Gamma\rangle = E_{k,N+2}^\Gamma \delta_{kk'} \qquad (2.106)$$

where H_{N+2} is the (N+2)-electron Hamiltonian and L_{N+2} the appropriate Bloch operator.

In this new version of the R-matrix method the partitioning of configuration space is such that the internal region is divided into two sub-regions (see Fig. 11). The radius a_1 of the inner sub-region is chosen to just envelope the bound states ψ_i^Γ of the N-electron target

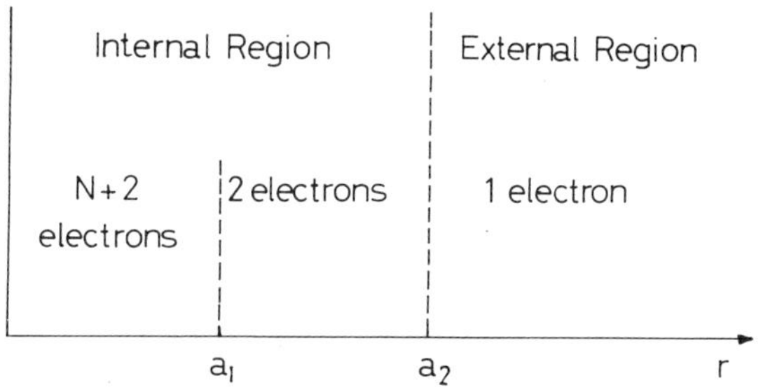

Fig. 11. Partitioning of configuration space in the new R-matrix method. From ref. 114.

retained in expansion (2.68). The radius a_2 of the outer sub-region is chosen to just envelope the bound states $\bar{\Psi}_k^\Gamma$ of the (N+1)-electron target retained in expansion (2.105). In the region $a_1 \leq r \leq a_2$ only the valence electron of the (N+1)-electron target and the scattered electron are present; both of these are represented by continuum wave functions.

This new version of the R-matrix theory has the additional possibility of representing ionization processes as well as excitation processes. It has been combined[114] with the T-matrix averaging technique of Burke et al[105] to obtain accurate ^1S elastic e^--H(1s) cross sections from threshold to 60 eV.

2.2.8. Tests of the theory

We shall conclude this survey of elastic scattering and excitation at intermediate and high energies by discussing some stringent tests of the theory.

Positron scattering

A first test, which has already been mentionned above, is to investigate the differences between electron and positron scattering. We recall in this connection that the first Born or Glauber approximations give identical cross sections for incident electrons or positrons, so that more sophisticated methods are required to explore these differences. For example, we have shown in Section 2.2.5 that theoretical predictions based on an ab-initio optical model[84,85] have been confirmed by the elastic $e^{\pm}$-Ar angular distributions recently measured by Hyder et al[91]. Clearly inelastic angular distributions for incident positrons would also be of great interest. We also remark that the difference $\Delta\sigma$ between the total (complete) cross section for electron scattering, $\sigma_{tot}(e^-)$ and that for positron scattering, $\sigma_{tot}(e^+)$ namely

$$\Delta\sigma = \sigma_{tot}(e^-) - \sigma_{tot}(e^+) \tag{2.106}$$

is also a delicate test of theoretical methods, since this quantity depends on third and higher odd order contributions to the imaginary part of the direct elastic forward (θ=0) amplitude, and on contributions to the imaginary part of the exchange elastic forward amplitude. A detailed comparison between measurements of total cross sections for intermediate-energy $e^{\pm}$-He, $e^{\pm}$-Ne and $e^{\pm}$-Ar collisions with theoretical calculations has been made by Kauppila et al[115] and Stein and Kauppila[116].

Polarized electrons

A second stringent test of the theory is provided by experiments using polarized electrons and (or) targets[52]. Such experiments will be discussed during this meeting by J. Kessler, so that I shall only consider here a simple example : the elastic scattering of electrons by spin - 1/2 atoms, in the absence of spin-orbit interaction (e.g. in atomic hydrogen or in the light alkalis). Let us first assume that the incident electrons and the target atoms are completely polarized. Denoting the spin directions by the symbols ↑ and ↓ , the scattering of electrons with spin parallel or antiparallel to the atomic spin leads to the following possibilities[1] :

Process	Differential cross section	
$e\uparrow + A\uparrow \rightarrow e\uparrow + A\uparrow$	$\lvert f-g\rvert^2$	(2.107a)
$e\uparrow + A\downarrow \rightarrow e\uparrow + A\downarrow$	$\lvert f\rvert^2$	(2.107b)
$e\uparrow + A\downarrow \quad e\downarrow + A\uparrow$	$\lvert g\rvert^2$	(2.107c)

where f is the direct amplitude, g is the exchange amplitude and we have made use of the fact that $k_j/k_i=1$ for elastic scattering. In contrast with experiments with unpolarized particles, which generally result in the measurement of spin-averaged cross sections [see (2.24)], experiments using polarized beams and targets yield separate information about direct and exchange scattering. In fact, it is sufficient for the observation of the individual cross sections listed in (2.107) to perform simpler experiments, in which either the electrons or the target atoms are initially polarized[52]. Furthermore, in practice one cannot use fully polarized electron beams or targets but this does not introduce any basic difficulty. Indeed, by using the density matrix formalism[1,52,117], it is easy to show that a partially polarized electron beam with a degree of polarization P_e can be considered to be made up of a totally polarized fraction and an unpolarized fraction in the ratio $P_e/(1-P_e)$.

Among the various possible experiments using polarised electron beams and (or) targets[118], let us consider one in which polarized electrons are scattered by polarized hydrogen atoms, and the scattered electron intensity is measured for the electron and target spins anti-parallel (↑↓) and parallel (↑↑). Such an experiment has become feasible because of the development of improved polarized electron sources and of sources of spin-polarized hydrogen atoms[119-123]. The information provided by the experiment can be expressed in the form of the measured asymmetry.

$$\Delta = P_e\, P_a\, A \tag{2.108}$$

where P_e and P_a are the degrees of polarization of the incident electron and of the atomic electron, respectively, and A is the asymmetry parameter, defined for a transition from target state $|i\rangle$ to $|j\rangle$ by

$$A = \frac{\sigma_{ji}(\uparrow\downarrow) - \sigma_{ji}(\uparrow\uparrow)}{\sigma_{ji}(\uparrow\downarrow) + \sigma_{ji}(\uparrow\uparrow)} \tag{2.109}$$

Here $\sigma_{ji}(\uparrow\downarrow)$ and $\sigma_{ji}(\uparrow\uparrow)$ are the cross sections for the process $i \to j$ when the incident electron and target atom spins are antiparallel and parallel, respectively. These cross sections can be either differential or total (integrated). In the latter case we have an integrated asymmetry parameter which contains of course less information. We note that if $\sigma_{ji}(S=0)$ denotes the singlet cross section for the process $i \to j$, obtained from the singlet scattering amplitude f^{+}_{ji} of equation (2.23), and if $\sigma_{ji}(S=1)$ is the corresponding triplet cross section - obtained from the triplet scattering amplitude f^{-}_{ji}, we have

$$A = \frac{\sigma_{ji}(S=0) - \sigma_{ji}(S=1)}{\sigma_{ji}(S=0) + 3\sigma_{ji}(S=1)} = \frac{1 - r_{ji}}{1 + 3\, r_{ji}} \tag{2.110a}$$

where

$$r_{ji} = \frac{\sigma_{ji}(S=1)}{\sigma_{ji}(S=0)} \tag{2.110b}$$

is the ratio of the triplet to the singlet cross section for the transition $i \to j$. From (2.110) we see that the lower bound of A is equal to $-1/3$ and is reached for pure triplet scattering.

As an example, we compare in Fig. 12 the experimental data of Fletcher et al[121] with theoretical calculations of Mc Dowell et al[124] for the (differential) asymmetry parameter A corresponding to e^--H(1s)

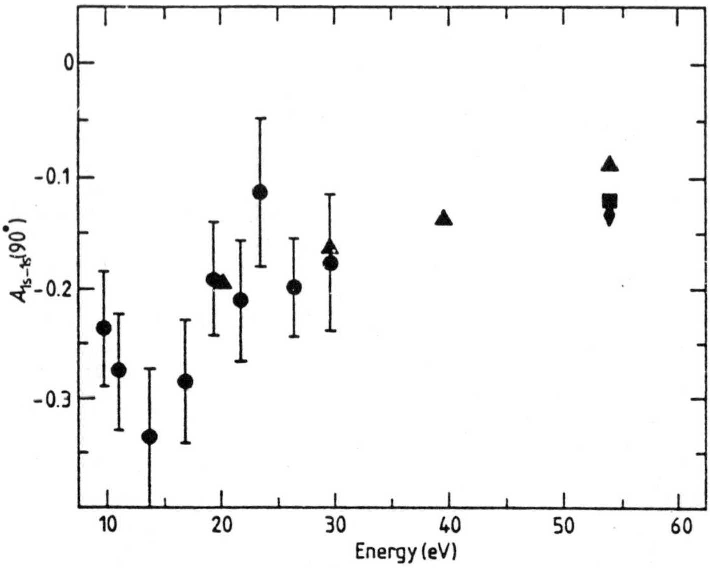

Fig. 12. The asymmetry parameter A for e^--H(1s) elastic scattering at the scattering angle $\theta=90°$, as a function of the incident electron energy. Theoretical results of McDowell et al[124] : ▲ static-exchange calculation; ■ six-state close-coupling calculation, ♦ UEBS calculation. Experimental data : ● , Fletcher et al[121]. From ref. 124.

elastic scattering at a scattering angle $\theta=90°$. The agreement between theory and experiment is seen to be encouraging. It would be very interesting to have measurements at higher energies for this process, and also for inelastic transitions.

Electron-photon coincidence experiments

A third exacting test of the theory is provided by electron-photon coincidence experiments, in which the target atoms are excited by electron impact, and the scattered electrons are observed in coincidence with the photons emitted by the excited atoms. Such experiments have been reviewed recently by Slevin[125]. As a first example, we shall consider the simple case of the excitation of the 2^1P states of helium from the ground state 1^1S, for which the first electron-photon coincidence experiment was performed[126]. At intermediate energies, the collision takes place in a time of the order of 10^{-14}s, and the atoms decay with a mean life of about 10^{-9}s, so that excitation and decay can be treated as independent processes. Neglecting spin-dependent forces, the excitation into the 2^1P states can therefore be described as a coherent superposition of excitations into degenerate magnetic sublevels [128], and immediately after the collision the eigenstate $|\psi(2^1P)\rangle$ of the excited atom may be written as

$$|\psi(2^1P)\rangle = a(1)|11\rangle + a(0)|10\rangle + a(-1)|1-1\rangle \tag{2.111}$$

where the (complex) amplitudes a_M are functions of the incident electron energy and the electron scattering angles (θ,ϕ); they describe the excitation of the magnetic sublevels $|LM\rangle$ (with L=1 in the present case). We note that the ϕ-dependence of a(M) can be factored out :

$$a(M) = a_M\, e^{-iM\phi} \tag{2.112}$$

where we are using the "collision frame" in which the direction of the incident electron beam is taken as the z-axis.

Since observables are related to the quantities $|\psi|^2$, the results of experimental observation on the atomic ensemble under consideration can be expressed in terms of all possible bilinear combinations of the amplitudes, a_M, $a_M^\star$. These quantities can be conveniently written in terms of the 3x3 density matrix[117]

$$\rho = \begin{pmatrix} |a_1|^2 & a_1\, a_o^\star & a_1\, a_{-1}^\star \\ a_o\, a_1^\star & |a_o|^2 & a_o\, a_{-1}^\star \\ a_{-1}\, a_1^\star & a_{-1}\, a_o^\star & |a_{-1}|^2 \end{pmatrix} \tag{2.113}$$

which completely describes the state of the detected atomic subensemble immediately after the excitation. We see that the diagonal elements of ρ are given by

$$\rho_{MM} = |a_M|^2 = \sigma_M \tag{2.114}$$

where $\sigma_M \equiv d\sigma_M/d\Omega$ is the differential cross section for excitation of the state $|1M\rangle$. We remark that if f_M denotes the scattering amplitude for the excitation of the $|1M\rangle$ state from the ground state, defined in accordance with (2.7), then $|a_M|^2$ and $|f_M|^2$ only differ by a factor of k_j/k_i. We also note that the trace of ρ gives the differential cross section $\sigma \equiv d\sigma/d\Omega$ summed over all magnetic substates :

$$\mathrm{Tr}\rho = \sum_M \sigma_M = \sigma \tag{2.115}$$

Because the density matrix ρ is Hermitian, not all its elements are independent. Furthermore, reflection invariance leads to additional contraints. As a result, the density matrix (2.113) can be specified by three independent parameters, a fact which can also be deduced by recollecting that the overall phase of $\psi(2^1P)$ is unobservable, and that reflection invariance implies that $a_{-1}=-a_1$. Setting $a_o=|a_o|\exp(i\alpha_0)$ and $a_1=|a_1|\exp(i\alpha_1)$, the three independent parameters can be chosen to be σ, λ and χ, where [127]

$$\sigma = \sigma_o + 2\sigma_1 \quad , \quad \lambda = \sigma_o/\sigma \quad , \quad \chi = \alpha_1 - \alpha_o \tag{2.116}$$

The two parameters λ and χ represent the off-diagonal elements of the density matrix ρ; these elements are combinations of the scattering amplitudes with different magnetic quantum numbers and take into account the interference between the different magnetic sublevels. As an example, we show in Fig. 13 the λ and |χ| parameters as a function of the electron

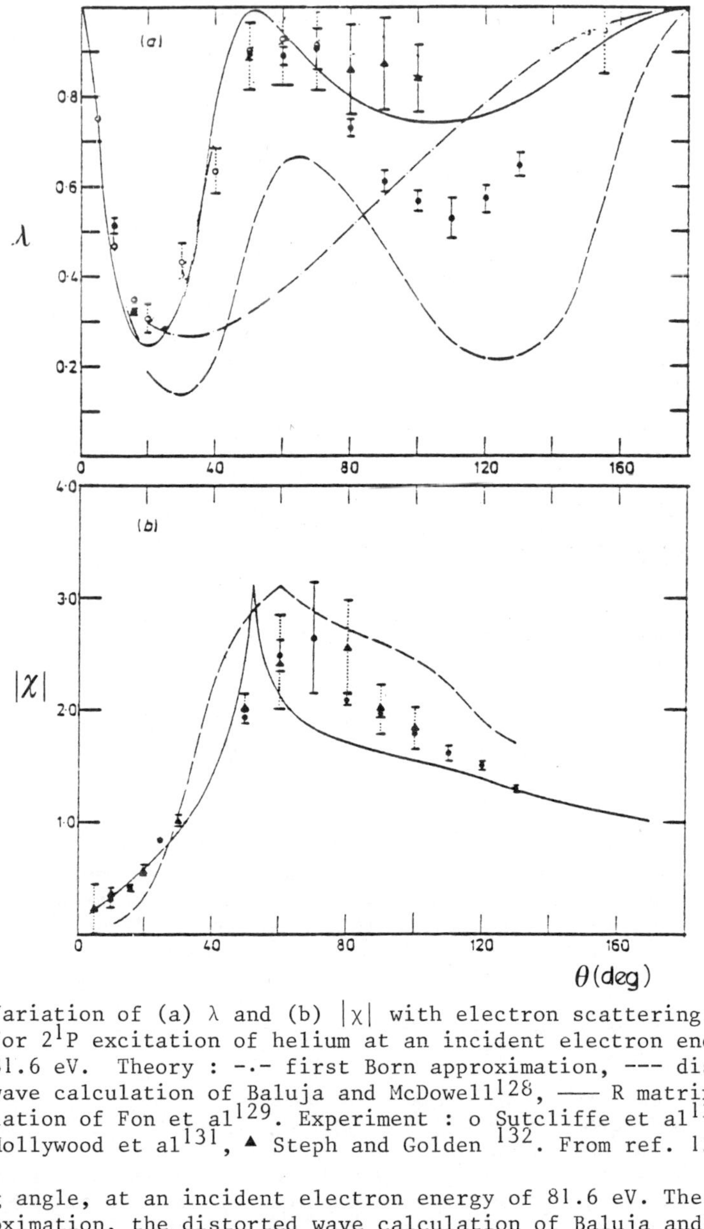

Fig. 13. Variation of (a) λ and (b) |χ| with electron scattering angle for 2^1P excitation of helium at an incident electron energy of 81.6 eV. Theory : -.- first Born approximation, --- distorted wave calculation of Baluja and McDowell[128], —— R matrix calculation of Fon et al[129]. Experiment : o Sutcliffe et al[130], • Hollywood et al[131], ▲ Steph and Golden [132]. From ref. 129.

scattering angle, at an incident electron energy of 81.6 eV. The first Born approximation, the distorted wave calculation of Baluja and McDowell[128] and the five state R-matrix calculation of Fon et al[129] are compared with the experimental data [130-132]. We note in particular that the first Born approximation (which yields the value χ=o) is in very poor agreement with the data, except at small angles. On the other hand, the agreement between the R-matrix calculations[129] and experiment is good, although there are still unexplained discrepancies at large angles.

Instead of characterizing the excited states in terms of excitation amplitudes, as was done in (2.111), one can describe them in terms of state multipoles, which are expectation values of a suitably chosen set of tensor operators. This approach, which exploits the inherent symmetry of the excited states, was suggested by Fano[133]. The state multipoles for n^1P excitations can be written in terms of the excitation amplitudes in the following way[117]

$$\langle T(L)^{+}_{KQ}\rangle = \sum_{MM'} (-1)^{L-M} \langle LM'L-M|KQ\rangle a_{M'} a^{\star}_{M} \qquad (2.117)$$

where $\langle LM'L-M|KQ\rangle$ is a Clebsch-Gordan coefficient, $T(L)^{+}_{KQ}$ is a tensor operator of rank K and component Q, and L=1 in the present case. The number of multipoles required to specify states of definite L is limited by the restrictions satisfied by the Clebsch-Gordan coefficients :

$$K \leqslant 2L \quad , \quad -K \leqslant Q \leqslant K \qquad (2.118)$$

The requirements of hermiticity and reflection invariance impose further restrictions. In particular, Blum[117] has shown that the n^1P states can be described by five multipoles, namely

$$\langle T(1)^{+}_{00}\rangle, \langle T(1)^{+}_{11}\rangle, \langle T(1)^{+}_{20}\rangle, \langle T(1)^{+}_{21}\rangle, \langle T(1)^{+}_{22}\rangle \qquad (2.119)$$

and that only three of them are independent. The state multipoles have simple physical interpretations.For K=0, the monopole $\langle T(1)^{+}_{00}\rangle = \sigma/3$ is a scalar quantity. The K=1 multipole can be related to the expectation value $\langle \underset{\sim}{L}\rangle$ of the atomic orbital angular momentum. That is

$$\langle T(1)^{+}_{11}\rangle = -i\sigma\langle L_y\rangle/2 = -i\sigma[\lambda(1-\lambda)]^{1/2} \sin\chi \qquad (2.120)$$

which shows that only the component $\langle L_y\rangle$ perpendicular to the scattering plane is non-vanishing. The three multipoles with K=2 can be expressed in terms of expectation values of certain products of angular momentum operators[134]; they are related to the alignment of the atomic ensemble, as we shall see below.

The excited state produced by electron impact can also be described in terms of an orientation vector $\underset{\sim}{O}^{col}$, proportional to the average angular momentum of the target atom, and by an alignment tensor A^{col}, the components of which are proportional to the mean values of expressions quadratic in the components of the target angular momentum[135]. The non-vanishing components of both quantities, defined in the "collision frame" are given in the present case (n^1P states) by

$$O^{col}_{1-} = \frac{\langle L_y\rangle}{2} = [\lambda(1-\lambda)]^{1/2} \sin\chi = \frac{i\langle T(1)^{+}_{11}\rangle}{\sigma} \qquad (2.121a)$$

$$A^{col}_{o} = \frac{\langle 3L_z^2-\underset{\sim}{L}^2\rangle}{2} = \frac{1}{2}(1-3\lambda) = \frac{1}{\sigma}\left(\frac{3}{2}\right)^{1/2} \langle T(1)^{+}_{20}\rangle \qquad (2.121b)$$

$$A^{col}_{1+} = \frac{\langle L_xL_z+L_zL_x\rangle}{2} = [\lambda(1-\lambda)]^{2} \cos\chi = \frac{1}{\sigma} 2^{-1/2}\langle T(1)^{+}_{21}\rangle \qquad (2.121c)$$

$$A_{2+}^{col} = \frac{\langle L_x^2 - L_y^2 \rangle}{2} = \lambda - 1 = \frac{1}{\sigma} 2^{-1/2} \langle T(1)_{22}^{+} \rangle \qquad (2.121d)$$

Finally, we remark that the above description of the excited states in the "collision frame" is not unique. In particular, a "natural frame" can be defined, in which the direction perpendicular to the scattering plane is taken as the quantization axis[136]. This frame provides interesting insight into the shape and dynamics of states excited in electron-atom collisions[137].

As a second example of electron-photon coincidence experiments, we shall consider the excitation of the 2p states of atomic hydrogen. Excitation of atomic hydrogen differs from that of helium discussed above in two important ways. Firstly, since atomic hydrogen is a spin -1/2 target, the excitation amplitudes are no longer spin-independent. Secondly, the presence of the fine-structure interaction disturbs the initial alignment and orientation produced in the collision. Assuming that LS coupling holds, excitation of each of the magnetic sublevels $|1M\rangle$ is then described by two independent amplitudes $a_M^{\pm}$ corresponding respectively to singlet (+) and triplet (-) scattering, which are proportional to the singlet and triplet scattering amplitudes defined in (2.23). Because we are considering here experiments in which no spin analysis is made of initial and final particles, the observables are bilinear combinations of the amplitudes averaged over initial spins and summed over final spins. That is

$$\langle a_{M'} \, a_M^{\star} \rangle = \frac{1}{4} a_{M'}^{+} \, a_M^{+\star} + \frac{3}{4} a_{M'}^{-} \, a_M^{-\star} \qquad (2.122)$$

where the symbol $\langle\,\rangle$ denotes the indicated average. Reflection invariance implies that $a_{-1}^{\pm} = -a_1^{\pm}$. Taking into account the hermiticity condition, the density matrix for excitation of hydrogenic p levals can be written in the form

$$\rho = \begin{pmatrix} \sigma_1 & \langle a_1 a_0^{\star} \rangle & \langle a_1 a_{-1}^{\star} \rangle \\ \langle a_1 a_0^{\star} \rangle^{\star} & \sigma_0 & -\langle a_1 a_0^{\star} \rangle \\ \langle a_1 a_{-1}^{\star} \rangle^{\star} & -\langle a_1 a_0^{\star} \rangle^{\star} & \sigma_1 \end{pmatrix} \qquad (2.123)$$

Since $a_{-1}^{\pm} = -a_1^{\pm}$ it follows that $\langle a_1 a_{-1}^{\star} \rangle = -\sigma_1$. Hence the density matrix (2.123) is completely determined by four independent parameters. A set of such independent parameters is given by[138]

$$\sigma = \sigma_0 + 2\sigma_1 \, , \qquad \lambda = \sigma_0/\sigma$$
$$R = \frac{\mathrm{Re}\langle a_1 a_0^{\star} \rangle}{\sigma} \, , \qquad I = \frac{\mathrm{Im}\langle a_1 a_0^{\star} \rangle}{\sigma} \qquad (2.124)$$

In contrast with helium, there is no simple relationship between R and I, so that an additional measurement is needed for the excitation of p states in atomic hydrogen. In practice, angular correlation experiments yield the parameters λ and R while a measurement of the circular polarization of the emitted light gives the quantity I.

In Fig. 14 we show the parameters λ and R for the excitation of the

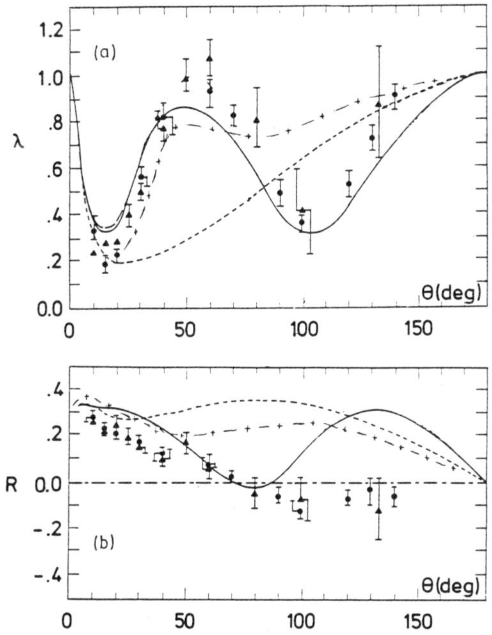

Fig. 14. Variation of (a) λ and (b) R with the electron scattering angle for the excitation of the 2p states of atomic hydrogen at an incident electron energy of 54.4 eV. Theory : --- first Born, Glauber or Wallace approximation, -+- second order potential results of Bransden et al[95], —— UEBS results of Byron, Joachain and Potvliege[79]. Experiments : ▲ Weigold et al[139], ● Williams[80]. From ref. 79.

2p states of atomic hydrogen at an incident electron energy of 54.4 eV. The second order potential results of Bransden et al[95] and the unitarized EBS calculations of Byron, Joachain and Potvliege[79] are compared with the experimental data of Weigold et al[141] and of Williams[80]. The second order distorted wave results of Madison et al[60] are very close to the UEBS results. With regard to the parameter λ the agreement between theory and experiment is satisfactory, but in the case of the parameter R there is a marked disagreement between the theoretical and experimental values at large angles.

The complete determination of the density matrix describing the coherent excitation of the n=2 manifold of atomic hydrogen requires the knowledge of off-diagonal matrix elements characterising the interference between the 2s and 2p states. Back et al[140] have recently reported the first observation of s-p coherence in an electron-photon coincidence experiment. The hydrogen atoms are excited by a 350 eV electron beam in the presence of an electric field of 250 V cm^{-1} which produces a Stark mixing of the 2s and 2p states. The Lyman α radiation emitted perpendicular to the electron beam is observed in coincidence with electrons inelastically scattered with an energy loss of 10.2 eV into a small range of angles near the forward direction. Their experimental values, described in terms of coherence parameters[140] M_1+M_3 and M_2 are compared in Fig. 15 with the UEBS calculations of Potvliege and Joachain[141], at an

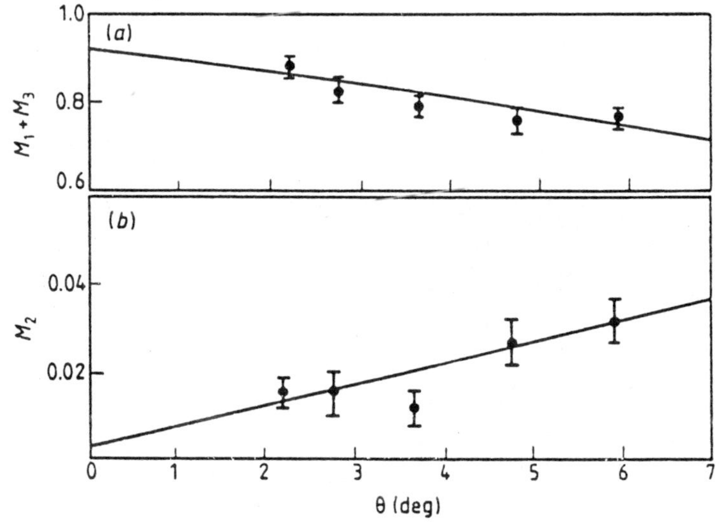

Fig. 15. Variation of (a) M_1+M_2 and (b) M_2 with the electron scattering angle for the excitation of the n=2 manifold of atomic hydrogen at an incident electron energy of 350 eV. Theory : —— UEBS results of Potvliege and Joachain[141]. Experiment : • Back et al[140]. From ref. 141.

incident energy of 350 eV. The agreement is seen to be excellent. Unfortunately, experimental data are only available in the small angle region, where the behaviour of the parameters M_1+M_3 and M_2 is controlled by the first Born 1s-2p amplitudes. It would be very interesting to have experimental values of these parameters outside the small-angle domain, where contributions from higher orders of perturbation theory significantly affect the values of M_1+M_3 and M_2.

3. ELECTRON IMPACT IONIZATION

In this section we shall discuss the process whereby electrons are ejected from the target, giving rise to ionization. We shall limit our discussion to single ionization or (e,2e) reactions of the type

$$e^- + A(i) \rightarrow A^+(f) + 2e^- \qquad (3.1)$$

where A(i) is a neutral atom in state i and $A^+(f)$ is the final ion left in state f. In the course of our analysis we shall also indicate generalizations to the case for which the target A is a positive ion.

3.1. Kinematics

We begin our study by kinematical considerations. We consider an electron of momentum $\mathbf{k}_i$ and energy $E_i=k_i^2/2$ incident on a target atom in the eigenstate $|i>$ corresponding to the eigenenergy w_i. In the final state, two electrons emerge with momenta $\mathbf{k}_A$ and $\mathbf{k}_B$ and corresponding energies $E_A=k_A^2/2$ and $E_B=k_B^2/2$ (see Fig. 16), the remaining ion being left in the eigenstate $|f>$, with eigenenergy w_f. If $\mathbf{Q}$ denotes the recoil momentum of the ion, momentum conservation requires that

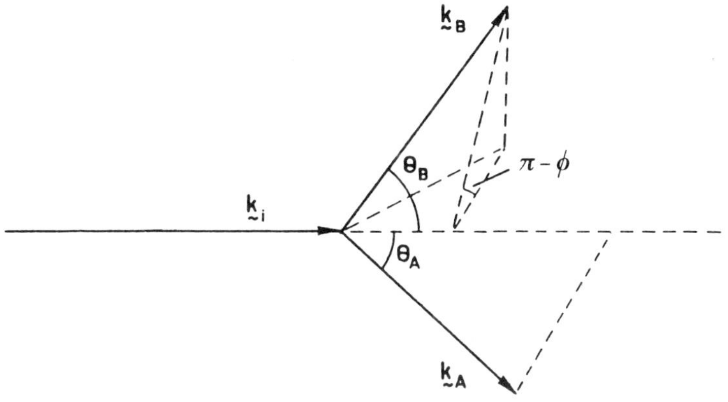

Fig. 16. The kinematics of an (e,2e) reaction. The incident electron momentum is $\mathbf{k}_i$ and the momenta of the outgoing electrons are $\mathbf{k}_A$ and $\mathbf{k}_B$, respectively. Also shown are the angles Θ_A and Θ_B with respect to the incident direction, and the angle $\pi-\phi=\pi-(\phi_A-\phi_B)$ measuring the deviation from the coplanar situation

$$\mathbf{k}_i = \mathbf{k}_A + \mathbf{k}_B + \mathbf{Q} \tag{3.2}$$

The energy conservation condition gives

$$E_i + w_i = E_A + E_B + w_f \tag{3.3}$$

where the recoil energy of the ion, $T = Q^2/2M_{A^+}$ has been neglected since the ion mass M_{A^+} is much larger than unity.

The most detailed information presently available about single ionization reactions of the type (3.1) has been obtained by analyzing triple differential cross sections (TDCS) measured in (e,2e) coincidence experiments. Such experiments were first performed in 1969 by Ehrhardt et al[142] and Amaldi et al[143]. A recent review of (e,2e) experiments has been given by Ehrhardt et al[144]. The TDCS is a measure of the probability that in an (e,2e) reaction an incident electron of momentum $\mathbf{k}_i$ and energy E_i will produce on collision with the target two electrons having energies E_A and E_B and momenta $\mathbf{k}_A$ and $\mathbf{k}_B$, emitted respectively into the solid angles $d\Omega_A$ and $d\Omega_B$ centered about the directions (θ_A,ϕ_A) and (θ_B,ϕ_B). The TDCS is usually denoted by the symbol $d^3\sigma/d\Omega_A\, d\Omega_B\, dE$. For unpolarized incident electrons and targets, which is the case considered here, it is a function of the quantities E_i, E_A (or E_B), θ_A, θ_B and $\phi=\phi_A-\phi_B$. By integrating the TDCS over $d\Omega_A$, $d\Omega_B$ or dE one can form various double and single differential cross sections. Finally, the total ionization cross section is obtained by integrating over all outgoing electron scattering angles and energies, and depends only on E_i, the incident electron energy.

It is useful when studying (e,2e) experiments to distinguish between several kinematical arrangements, since these have important implications on the theoretical analysis of the collision, as we shall see in Section 3.4. In coplanar geometries the momenta $\mathbf{k}_i$, $\mathbf{k}_A$ and $\mathbf{k}_B$ are in the same plane (so that ϕ_A=o and ϕ_B=o or π) while in non-coplanar geometries the momentum $\mathbf{k}_B$ is out of the $(\mathbf{k}_i,\mathbf{k}_A)$ reference plane. Another useful distinction can be made between asymmetric and symmetric geometries. We shall return to this point in Section 3.4, where we analyze (e,2e) reactions with fast incident electrons. Before doing this, however, we shall give a survey of basic ionization theory and discuss the threshold behaviour of ionization cross sections.

3.2. Basic theory

In this section we shall discuss the fundamental aspects of the theory of (e,2e) reactions, following the work of Peterkop[145] and of Rudge and Seaton[146,147]. The key point which distinguishes ionization from elastic scattering or excitation is the fact that in the case of ionization the Coulomb interaction between the two outgoing electrons and between each outgoing electron and the residual ion is present even at large distances. In order to keep calculational details to a minimum, we shall concentrate on the ionization of a hydrogenic system. Since in that case the ion in the final state is just a bare nucleus of charge Z, we may write the energy conservation equation (3.3) in the form

$$E = E_i + w_i = E_A + E_B = \frac{K^2}{2} \tag{3.4}$$

where we have neglected the recoil energy of the nucleus, and we note that $K^2=k_A^2 + k_B^2$. The prototype of this process is of course the ionization of a hydrogen atom, initially in the eigenstate $|i\rangle$ with eigenenergy w_i.

3.2.1. Asymptotic boundary conditions

Denoting by $\mathbf{r}_o$ the coordinates of the incident electron and by $\mathbf{r}_1$ those of the initially bound one, we look for solutions $\Psi_i^{(+)}(\mathbf{r}_o,\mathbf{r}_1)$ of the Schrödinger equation (2.8) corresponding to the initial state $|\mathbf{k}_i,i\rangle$ and to outgoing wave boundary conditions. When the target atom is neutral atomic hydrogen (Z=1) these solutions satisfy the asymptotic boundary condition

$$\psi_i^{(+)}(\underset{\sim}{r}_o,\underset{\sim}{r}_1) \underset{r_o\to\infty}{\to} \psi_i(\underset{\sim}{r}_1)e^{ik_i z_o} + \sum_j \psi_j(\underset{\sim}{r}_1) f_{ji}(\theta_o,\phi_o) \frac{e^{ik_j r_o}}{r_o}$$

$$+ \int_{q_B \leqslant K} \psi^{(-)}_{c,\underset{\sim}{q}_B}(Z=1,\underset{\sim}{r}_1) f_i(q_A\hat{\underset{\sim}{r}}_o,\underset{\sim}{q}_B) \frac{e^{i[q_A r_o+\rho(\underset{\sim}{q}_B,\underset{\sim}{r}_o)]}}{r_o} d\underset{\sim}{q}_B \qquad (3.5a)$$

which replaces (2.10a) and

$$\psi_i^{(+)}(\underset{\sim}{r}_o,\underset{\sim}{r}_1) \underset{r_1\to\infty}{\to} \sum_j \psi_j(\underset{\sim}{r}_o)\, g_{ji}(\theta_1,\phi_1) \frac{e^{ik_j r_1}}{r_1}$$

$$+ \int_{q_B \leqslant K} \psi^{(-)}_{c,\underset{\sim}{q}_B}(Z=1,\underset{\sim}{r}_o) g_i(q_A\hat{\underset{\sim}{r}}_1,\underset{\sim}{q}_B) \frac{e^{i[q_A r_1+\rho(\underset{\sim}{q}_B,\underset{\sim}{r}_1)]}}{r_1} d\underset{\sim}{q}_B \qquad (3.5b)$$

which replaces (2.10b). In these equations $k_j=(k_i^2+2w_i-2w_j)^{1/2}$, $q_A = (K^2-q_B^2)^{1/2}$ and the phases $\rho(\underset{\sim}{q}_B,\underset{\sim}{r}_o)$ and $\rho(\underset{\sim}{q}_A,\underset{\sim}{r}_1)$ have been included since in ionization the Coulomb interaction between the two outgoing electrons and between each electron and the nucleus is present even in the asymptotic region, as we remarked above.

The functions ψ_j and $\psi^{(-)}_{c,\underset{\sim}{q}_B}$ are respectively negative and positive energy solutions of the one-electron Schrödinger equation (2.9). The positive energy solutions are the so-called Coulomb wave functions; at each positive energy there are two linearly independent Coulomb wave functions $\psi^{(+)}_{c,\underset{\sim}{k}}$ and $\psi^{(-)}_{c,\underset{\sim}{k}}$, solutions of (2.9), describing the motion of an electron in a given direction $\underset{\sim}{k}$ in the field of a nucleus of charge Z, and exhibiting either outgoing (+) or incoming (-) spherical wave behaviour. The Coulomb wave functions $\psi^{(\pm)}_{c,\underset{\sim}{k}}$ which we shall use here are defined as follows :

$$\psi^{(\pm)}_{c,\underset{\sim}{k}}(Z,\underset{\sim}{r})=(2\pi)^{-3/2}\exp(\pi Z/2k)\Gamma(1\mp iZ/k)\, e^{i\underset{\sim}{k}\cdot\underset{\sim}{r}}$$

$$\times\ {}_1F_1(\pm iZ/k,1,\pm i(kr \mp \underset{\sim}{k}\cdot\underset{\sim}{r})) \qquad (3.6)$$

They are normalized to a delta functions in momentum space :

$$\langle\psi^{(\pm)}_{c,\underset{\sim}{k}}(Z,\underset{\sim}{r})|\psi^{(\pm)}_{c,\underset{\sim}{k}}(Z,\underset{\sim}{r})\rangle = \delta(\underset{\sim}{k}-\underset{\sim}{k}') \qquad (3.7)$$

It can be shown[2,148] that in order for f_i and g_i in equations (3.5) to have the meaning of direct and exchange ionization amplitudes, respectively, one must use the incoming (-) Coulomb waves in these equations. When the target is an hydrogenic ion one must use in (3.5) the incoming Coulomb wave functions corresponding to charge Z. In addition, one must in that case modify the exponents in (3.5) by including the logarithmic phase factors due to the distortion by the Coulomb field of the target ion.

In order to obtain expressions for the ionization amplitudes, we must consider the asymptotic forms of equations (3.5a) and (3.5b) when both $r_o \to \infty$ and $r_1 \to \infty$. For this purpose, it is convenient to use hyperspherical coordinates.

3.2.2. Hyperspherical coordinates

In the case of two electrons interacting with a nucleus of charge Z the hyperspherical coordinates are defined in terms of the electron spherical polar coordinates (r_o,θ_o,ϕ_o) and (r_1,θ_1,ϕ_1) by

$$R = (r_o^2 + r_1^2)^{1/2} \tag{3.8}$$

and

$$\alpha = \tan^{-1} \frac{r_1}{r_o} \tag{3.9}$$

while the remaining four coordinates are taken to be $(\theta_o,\phi_o,\theta_1,\phi_1)$. The quantity R which describes the "size" of the three-particle system in six-dimensional space is called the hyper-radius and the hyperspherical angle α is a mock angle which characterizes the radial correlation of the pair of electrons. We remark that excitation corresponds to the situation where one electron remains close to the nucleus, so that r_o or r_1 remains small, and α is close to 0 or $\pi/2$. On the other hand, ionization corresponds to the case for which both r_o and r_1 tend to infinity, so that α is close to $\pi/4$. It is also convenient to introduce the angle θ_{01} between the two vectors $\underset{\sim}{r}_o$ and $\underset{\sim}{r}_1$; this angle characterizes the angular correlation of the pair of electrons.

The Schrödinger equation describing the motion of two electrons in the field of a nucleus of charge Z becomes in hyperspherical coordinates

$$\left[\frac{d^2}{dR^2} - \frac{\Lambda^2 + 15/4}{R^2} + \frac{2\zeta(\alpha,\theta_{01})}{R} + 2E \right] (R^{5/2}\Psi) = 0 \tag{3.10}$$

where the factor $R^{5/2}$ has been introduced to remove the first-order derivative with respect to R. In this equation we have defined the effective charge $\zeta(\alpha,\theta_{01})$ via the relation

$$-\frac{Z}{r_o} - \frac{Z}{r_1} + \frac{1}{r_{01}} = -\frac{\zeta}{R} \tag{3.11}$$

so that in hyperspherical coordinates ζ is given by

$$\zeta(\alpha,\theta_{01}) = \frac{Z}{\cos\alpha} + \frac{Z}{\sin\alpha} - \frac{1}{(1-\sin 2\alpha \cos\theta_{01})^{1/2}} \tag{3.12}$$

The operator Λ^2 appearing in (3.10) is the square of the "grand angular momentum" for six dimensions; it is defined by

$$\Lambda^2 = -\frac{1}{\sin^2\alpha\cos^2\alpha}\frac{d}{d\alpha}\left(\sin^2\alpha\cos^2\alpha\frac{d}{d\alpha}\right) + \frac{\ell_o^2}{\cos^2\alpha} + \frac{\ell_1^2}{\sin^2\alpha} \qquad (3.13)$$

where ℓ_o^2 and ℓ_1^2 are squared orbital angular momentum operators for the two electrons, having eigenfunctions $Y_{\ell_o m_o}(\hat{r}_o)$ and $Y_{\ell_1 m_1}(\hat{r}_1)$ and corresponding eigenvalues $\ell_o(\ell_o+1)$ and $\ell_1(\ell_1+1)$. It is readily verified that Λ^2 commutes with ℓ_o^2 and ℓ_1^2 and also with the operators L^2, S^2 and π (where L is the total orbital angular momentum, S the total spin angular momentum and π the parity operator). The eigenfunctions of the operator Λ^2 are called hyperspherical harmonics (or K-harmonics). They satisfy the equation

$$\Lambda^2 - \kappa(\kappa+4)\, Z_{\kappa\gamma}(\Omega) = 0 \qquad (3.14)$$

where κ is a non-negative integer which can be written as

$$\kappa = \ell_o + \ell_1 + 2m \qquad (3.15)$$

and m is a new quantum number associated with the motion in α. The symbol γ in (3.14) represents the remaining quantum numbers required to specify the state, and Ω specifies the angular variables $\alpha, \theta_o, \phi_o, \theta_1$ and ϕ_1.

Looking back at the Schrödinger equation (3.10), we see that the potential energy term factorizes, so that this equation resembles the Schrödinger equation for the radial motion of a particle in a reduced potential $-2\zeta/R$, the centrifugal barrier potential being $(\Lambda^2 + 15/4)/R^2$. However, unlike the corresponding equation for hydrogenic systems, the quantity ζ in (3.10) depends on the angular coordinates α and θ_{01} and does not commute with Λ^2.

The dynamics of the motion of the two electrons depends on the form of the potential function $\zeta(\alpha,\theta_{01})$. In Fig. 17 we show a three dimensional plot[149] of the function $-C(\alpha,\theta_{01}) = -2\zeta(\alpha,\theta_{01})$ for a pair of electrons in the field of H^+ (charge Z = 1), in the range $o \leqslant \alpha \leqslant \pi/2$ and $o \leqslant \theta_{01} \leqslant \pi$. We see that near $\alpha = o$ and $\alpha = \pi/2$ the potential surface exhibits deep valleys corresponding to the electron-nuclear attraction, while at $\alpha = \pi/4$ and $\theta_{01} = 0$ it has a singularity due to the electron-electron repulsion. We also note that there is a saddle point at $\alpha = \pi/4$ and $\theta_{01} = \pi$ corresponding to the situation where the two electrons are equidistant from and on opposite sides of the nucleus. This configuration is important in determining doubly excited states of two-electron systems [149-151]. It also plays a crucial role in the threshold behaviour of the electron-impact ionization cross section, as we shall see in Section 3.3.

3.2.3. Asymptotic form of the wave function when $r_o \to \infty$ and $r_1 \to \infty$

Let us now consider the asymptotic forms (3.5) in the limits $r_o \to \infty$ and $r_1 \to \infty$ simultaneously. Using hyperspherical coordinates and the method of stationary phases, Peterkop[145] and Rudge and Seaton[146] have shown that, for any Z

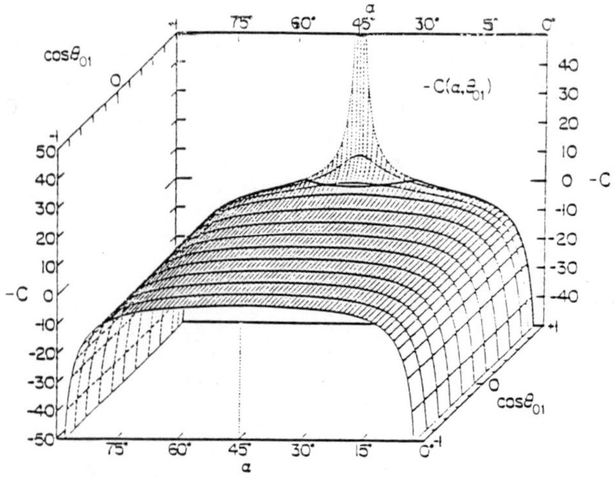

Fig. 17. The two-electron potential function $- C(\alpha,\theta_{01}) = -2\zeta(\alpha,\theta_{01})$ for the case Z=1. Half the potential surface, with θ_{01} running from o to π, is shown. From ref. 149.

$$\Psi_i^{(+)} \underset{R\to\infty}{\to} (-iK)^{3/2}\ f_i(K\hat{\mathbf{r}}_o \cos\alpha,\ K\hat{\mathbf{r}}_1 \sin\alpha)R^{-5/2}\exp[\,iKR+i\gamma_1(K,R,\Omega)] \tag{3.16a}$$

and

$$\Psi_i^{(+)} \underset{R\to\infty}{\to} (-iK)^{3/2}\ g_i(K\hat{\mathbf{r}}_1 \sin\alpha,\ K\hat{\mathbf{r}}_o \cos\alpha)R^{-5/2}\exp\,[\,iKR+i\gamma_2(K,R,\Omega)] \tag{3.16b}$$

The phases γ_1 and γ_2 appearing in the above equations can be determined by solving the Schrödinger equation (3.10) in hyperspherical coordinates in the asymptotic region $R\to\infty$. Here, as in potential scattering, we may neglect the centrifugal term, and the situation becomes identical to that of pure Coulomb scattering[1], giving

$$\Psi_i^{(+)} \underset{R\to\infty}{\to} A(\alpha,\theta_o,\phi_o,\theta_1,\phi_1)R^{-5/2}\exp\,[\,iKR+i\frac{\zeta}{K}\,\ln(2KR)] \tag{3.17}$$

Comparing (3.17) with (3.16), we see that we may write

$$A(\alpha,\theta_o,\phi_o,\theta_1,\phi_1)=(-iK)^{3/2}\ f_i(K\hat{\mathbf{r}}_o \cos\alpha,\ K\hat{\mathbf{r}}_1 \sin\alpha) \tag{3.18a}$$

$$=(-iK)^{3/2}\ g_i(K\hat{\mathbf{r}}_1 \sin\alpha,\ K\hat{\mathbf{r}}_o \cos\alpha) \tag{3.18b}$$

if we choose the phase factors so that

$$\gamma_1=\gamma_2=\frac{\zeta(\alpha,\theta_{01})}{K}\quad \ln(2KR) \tag{3.19}$$

Higher order terms in the asymptotic expansion of $\Psi_i^{(+)}$ have been discussed by Peterkop[145]. From (3.18a) and (3.18b), we also see that

$$g_i(\mathbf{k}_A,\mathbf{k}_B) = f_i(\mathbf{k}_B,\mathbf{k}_A) \tag{3.20}$$

for all $\mathbf{k}_A$, $\mathbf{k}_B$ which satisfy the relation $k_A^2+k_B^2=K^2$. The result (3.20), often called the Peterkop theorem, is in agreement with the fact that the two amplitudes $f_i(\mathbf{k}_B,\mathbf{k}_A)$ and $g_i(\mathbf{k}_A,\mathbf{k}_B)$ describe the same physical process.

3.2.4. Integral expression for the ionization amplitude

In order to obtain an integral expression for the ionization amplitude, from which various approximation schemes can be developed, we consider the expression

$$I = \langle (H-E)\Phi|\Psi_i^{(+)} \rangle \tag{3.21}$$

where $\Psi_i^{(+)}$ is the exact solution of the Schrödinger equation satisfying the asymptotic boundary condition (3.17), and Φ is a non-singular function which essentially describes the final state of the ionization process. In choosing Φ it is useful to arrange matters so that $(H-E)\Phi$ takes on a simple form. A possible choice is

$$\Phi(\mathbf{r}_o,\mathbf{r}_1) = \psi^{(-)}_{c,\mathbf{k}_A}(Z_A,\mathbf{r}_o)\psi^{(-)}_{c,\mathbf{k}_B}(Z_B,\mathbf{r}_1) \tag{3.22}$$

for which we find that

$$(H-E)\Phi = \left(-\frac{Z-Z_A}{r_o} - \frac{Z-Z_B}{r_1} + \frac{1}{r_{01}}\right)\Phi \tag{3.23}$$

and we still have some flexibility through the "effective charges" Z_A and Z_B. We now integrate (3.21) by parts, using hyperspherical coordinates. Setting $k_A=K\cos\beta$ and $k_B=K\sin\beta$, with $0\leqslant\beta\leqslant\pi/2$, and using the method of stationary phases, one finds that the only contribution to the integral (3.21) comes when $\mathbf{r}_o$ lies along $\mathbf{k}_A$ and $\mathbf{r}_1$ lies along $\mathbf{k}_B$ in the asymptotic region. The result of the stationary phase calculation is [145,146]

$$I=-(2\pi)^{-1/2} f_i(\mathbf{k}_A,\mathbf{k}_B)(\cos\beta)^{-2iZ_A/k_A}(\sin\beta)^{-2iZ_B/k_B}$$

$$\times \lim_{R\to\infty} \exp\left[i\left(\frac{Z}{k_A}+\frac{Z}{k_B}-\frac{1}{|\mathbf{k}_A-\mathbf{k}_B|}-\frac{Z_A}{k_A}-\frac{Z_B}{k_B}\right)\ln(2KR)\right] \tag{3.24}$$

In order that this expression should not contain a divergent phase factor it is necessary that

$$\frac{Z_A}{k_A}+\frac{Z_B}{k_B} = \frac{Z}{k_A}+\frac{Z}{k_B}-\frac{1}{|\mathbf{k}_A-\mathbf{k}_B|} \tag{3.25}$$

Therefore the integral expression for the direct ionization amplitude is

$$f_i(\underset{\sim}{k}_A,\underset{\sim}{k}_B)=-(2\pi)^{1/2}\exp\,[i\Delta(\underset{\sim}{k}_A,\underset{\sim}{k}_B)]\langle\Phi|-\frac{Z-Z_A}{r_o}-\frac{Z-Z_B}{r_1}+\frac{1}{r_{01}}|\Psi_i^{(+)}\rangle \tag{3.26}$$

where

$$\Delta(\underset{\sim}{k}_A,\underset{\sim}{k}_B) = \frac{Z_A}{k_A}\,\ell n\,\frac{k_A^2}{k_A^2+k_B^2} + \frac{Z_B}{k_B}\,\ell n\,\frac{k_B^2}{k_A^2+k_B^2} \tag{3.27}$$

and the effective charges Z_A and Z_B are defined by eq. (3.25). We note that since in the asymptotic region $\underset{\sim}{r}_o=\underset{\sim}{k}_A t$ and $\underset{\sim}{r}_1=\underset{\sim}{k}_B t$, where t is the time, eq. (3.25) is equivalent to

$$\frac{Z_A}{r_o}+\frac{Z_B}{r_1} = \frac{Z}{r_o}+\frac{Z}{r_1}-\frac{1}{r_{01}} \tag{3.28}$$

Hence the effective charges Z_A and Z_B which give a uniquely defined phase depend on the relative angles and positions (or momenta) of the two outgoing electrons. We can make this explicit by writing for example $Z_A\equiv Z_A(\underset{\sim}{k}_A,\underset{\sim}{k}_B)$ and $Z_B\equiv Z_B(\underset{\sim}{k}_A,\underset{\sim}{k}_B)$. It is clear from (3.25) that $Z_A(\underset{\sim}{k}_A,\underset{\sim}{k}_B) = Z_B(\underset{\sim}{k}_B,\underset{\sim}{k}_A)$. Thus, with $\Delta(\underset{\sim}{k}_A,\underset{\sim}{k}_B)$ given by (3.27) we have

$$\Delta(\underset{\sim}{k}_A,\underset{\sim}{k}_B) = \Delta(\underset{\sim}{k}_B,\underset{\sim}{k}_A) \tag{3.29}$$

The exchange ionization amplitude g_i may be obtained from the direct amplitude f_i by using the Peterkop theorem (3.20). Taking into account the relation (3.29), it follows that

$$g_i(\underset{\sim}{k}_A,\underset{\sim}{k}_B)=-(2\pi)^{1/2}\exp[\,i\Delta(\underset{\sim}{k}_A,\underset{\sim}{k}_B)]\langle\tilde{\Phi}|-\frac{Z-Z_B}{r_o}-\frac{Z-Z_A}{r_1}+\frac{1}{r_{01}}|\Psi_i^{(+)}\rangle \tag{3.30}$$

where Z_A and Z_B are the same as those used to evaluate the direct amplitude f_i via equation (3.26). The function $\tilde{\Phi}$ in (3.30) is obtained by interchanging $\underset{\sim}{k}_A$ and $\underset{\sim}{k}_B$ and Z_A and Z_B in (3.22), namely

$$\tilde{\Phi}(\underset{\sim}{r}_o,\underset{\sim}{r}_1) = \psi^{(-)}_{c,\underset{\sim}{k}_B}(Z_B,\underset{\sim}{r}_o)\psi^{(-)}_{c,\underset{\sim}{k}_A}(Z_A,\underset{\sim}{r}_1) \tag{3.31}$$

As one would expect this is equivalent to an interchange of $\underset{\sim}{r}_o$ and $\underset{\sim}{r}_1$ in (3.22). We also note that the phase factor $\exp[\,i(\Delta(\underset{\sim}{k}_A,\underset{\sim}{k}_B)]$ is common to the direct amplitude $f_i(\underset{\sim}{k}_A,\underset{\sim}{k}_B)$ and the exchange amplitude $g_i(\underset{\sim}{k}_A,\underset{\sim}{k}_B)$ and therefore will not contribute to any physical process. It should be noted however that in performing approximate calculations of the amplitudes f_i and g_i the Peterkop theorem (3.20) will in general not be satisfied. For this reason, a relative phase $\tau_i(\underset{\sim}{k}_A,\underset{\sim}{k}_B)$ between the amplitudes $f_i(\underset{\sim}{k}_B,\underset{\sim}{k}_A)$ and $g_i(\underset{\sim}{k}_A,\underset{\sim}{k}_B)$ has sometimes been introduced. Various choices for this relative phase have been discussed by Peterkop[145] and Rudge[147].

3.2.5. Cross sections

Ionization cross sections are obtained by taking the ratio of the number of ionization events per unit time and per unit target atom to the incident electron flux. For random electron spin orientations, the triple differential cross section (TDCS) for ionization of a target with one active electron in the initial state $|i\rangle$ is given by

$$\frac{d^3\sigma_i}{d\Omega_A d\Omega_B dE} = \frac{k_A k_B}{k_i}\left[\frac{1}{4}|f_i+g_i|^2 + \frac{3}{4}|f_i-g_i|^2\right] \quad (3.32a)$$

$$= \frac{k_A k_B}{k_i}\left[\frac{1}{4}|\mathcal{f}_i^+|^2 + \frac{3}{4}|\mathcal{f}_i^-|^2\right] \quad (3.32b)$$

where we have written $f_i \equiv f_i(\underset{\sim}{k}_A,\underset{\sim}{k}_B), g_i \equiv g_i(\underset{\sim}{k}_A,\underset{\sim}{k}_B)$ and $\mathcal{f}_i^{\pm} = f_i \pm g_i$

As we mentionned in Section 3.1, various double and single differential cross sections can be formed by integrating the TDCS with respect to $d\Omega_A$, $d\Omega_B$ or dE. By integrating (3.32) over all outgoing electron scattering angles and energies, we obtain the total ionization cross section

$$\sigma_i = \frac{1}{k_i}\int_0^{E/2} dE\; k_A\, k_B \int d\Omega_A \int d\Omega_B \left[\frac{1}{4}|\mathcal{f}_i^+|^2 + \frac{3}{4}|\mathcal{f}_i^-|^2\right] \quad (3.33)$$

where the upper limit of integration over the energy variable is E/2 because the two electrons are indistinguishable.

3.3. Threshold behaviour of the ionization cross sections

The behaviour of ionization cross sections just above threshold has been the subject of many investigations. In particular, threshold laws for ionization, which give the energy dependence of the total ionization cross section near threshold, but not its actual magnitude, have attracted much interest. Since threshold laws only provide limited information, it is not unreasonable to expect that they should be obtainable without knowing the full solution of the many-particle scattering problem. This was emphasized in a fundamental paper by Wigner[152], who pointed out that the derivation of threshold laws does not require a detailed knowledge of the collision dynamics in the "reaction zone", where all the particles are close together. Instead, he showed that threshold laws generally arise from a feature of the escape process, namely lack of kinetic energy for complete escape. Wigner then applied this idea to various cases in which the reaction product consists of two particles escaping from each other.

3.3.1. The Wannier theory of threshold ionization

The extension of Wigner's theory to single ionization of atoms or ions by electrons was carried out by Wannier[153] in the following way.

Firstly, configuration space is divided into three regions, specified in terms of the hyperspherical radius R (see Fig. 18) : the reaction zone ($o \leqslant R \leqslant R_0$), the Coulomb zone ($R_0 < R \leqslant R_1$) and the "free" zone ($R > R_1$).

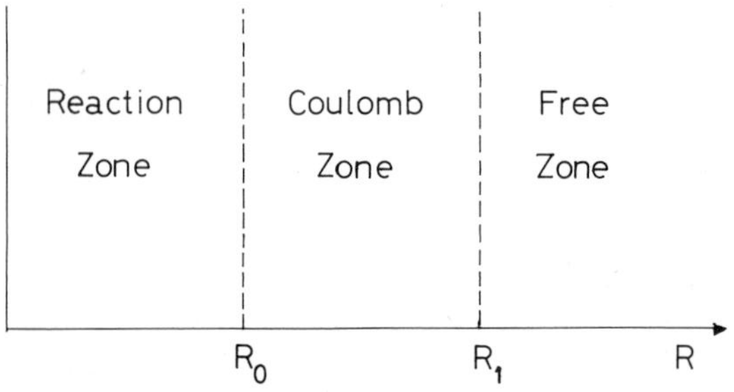

Fig. 18. Partitioning of configuration space in the Wannier theory.

In analogy with Wigner's analysis of single particle threshold escape, Wannier argued that it is not necessary to know the detailed behaviour of the two electrons in the reaction zone. Instead, Wannier assumed that the distribution in phase space of the two electrons is approximately uniform (i.e. quasi-ergodic) when they enter the Coulomb zone. Wannier also assumed that for large enough R_0 the Coulomb potential varies sufficiently slowly for classical mechanics to be valid in the Coulomb zone, even when the total energy $E = E_A+E_B$ tends to zero. This assumption is certainly plausible because in the case of the Coulomb potential the local de Broglie wavelength $\lambda(R)$ is slowly varying for large R, the derivative $d\lambda/dR$ remaining small with respect to unity when R is large. Finally, at very large $R>R_1$ the magnitude of the Coulomb potential energy of the system is less than the combined kinetic energies of the electrons, so that the electrons move essentially freely. As $E\to 0$ the free zone recedes and the Coulomb zone extends to infinity.

The threshold behaviour of the ionization cross section is therefore determined by the motion of the two electrons in the Coulomb zone. By assuming that the distribution in phase space of the two electrons is quasi-ergodic when they enter the Coulomb zone, Wannier could determine their distribution in phase space on leaving that zone, and hence obtain a threshold law for the ionization cross section. Looking back at the potential energy function $-C(\alpha,\theta_{01})$ shown in Fig.17, it is apparent that the deep valleys at $\alpha=o$ and $\alpha=\pi/2$ mean that at E=0 nearly all classical trajectories end up in one or the other valley, corresponding to single escape. Wannier realized that in order to find trajectories leading to double escape, one must look for unstable points of $-C(\alpha,\theta_{01})$. As we have already seen above, there is one such point, namely the "Wannier saddle point" at $\alpha=\pi/4$ and $\theta_{01}=\pi$, such that the two electrons are on opposite sides of the nucleus and equidistant from it. Near this saddle point, the effective charge $\zeta(\alpha,\theta_{01}) = C(\alpha,\theta_{01})/2$ given by (3.12) can be expanded as

$$\zeta(\alpha,\theta_{01})=\xi_o+\frac{1}{2}\zeta_1(\alpha-\frac{\pi}{4})^2 + \frac{1}{8}\zeta_2 (\theta_{01}-\pi)^2 + \dots \tag{3.34a}$$

where

$$\zeta_o = \frac{4Z-1}{\sqrt{2}} \ , \ \zeta_1 = \frac{12Z-1}{\sqrt{2}} \ , \ \zeta_2 = -\frac{1}{\sqrt{2}} \tag{3.34b}$$

so that the motion is stable in θ_{01} but unstable in α at constant R. Clearly, classical trajectories with $\alpha=\pi/4$ lead to double escape since as $R\to\infty$ both r_o and r_1 tend to infinity.

Following Wannier, let us consider the case in which the total orbital angular momentum of the two electrons is equal to zero, L=0. The motion of the electrons can then be described by the three variables R, α and θ_{01}. Let us write

$$\Delta\alpha = \alpha - \frac{\pi}{4} = u_1(R) \ , \ \Delta\theta_{01} = \theta_{01} - \pi = u_2(R) \tag{3.35}$$

and assume that $\Delta\alpha$ and $\Delta\theta_{01}$ are small quantities. Keeping terms of equal order enables one to write the classical equations of motion in the form

$$\frac{d^2R}{dt^2} = -\frac{\zeta_o}{R^2} \tag{3.36a}$$

and

$$\frac{d}{dt}\left(R^2\frac{du_i}{dt}\right) = \frac{\zeta_i u_i}{R} \ , \quad i=1,2 \tag{3.36b}$$

and we note that the equations for u_1 and u_2 are linear and uncoupled. Solving these equations, taking the limit $E\to 0$ and using the quasi-ergodic assumption together with the fact that in the equations determining the trajectories α and θ_{01} depend on E and R only through the product ER, Wannier was able to obtain for the total ionization cross section the threshold law

$$\sigma \sim E^m \tag{3.37a}$$

where

$$m = \frac{1}{4}\left[\left(\frac{100Z-9}{4Z-1}\right)^{1/2} - 1\right] \tag{3.37b}$$

We see that m = 1.127 when Z=1. In the limit $Z\to\infty$ we have $m\to 1$, corresponding to a linear threshold law. This last result is to be expected, since when $Z\to\infty$ the electron-electron interaction can be neglected, and in that instance one can readily show by using the properties of Coulomb wave functions[150] that the threshold law must be linear.

Another important result, obtained from a numerical analysis of classical trajectories by Vinkalns and Gailitis[154], is that the width of the angular distribution in θ_{01} is

$$\Delta\theta_{01} \sim E^{1/4} \tag{3.38}$$

so that the classical differential cross section has a maximum at $\theta_{01}=\pi$, the width of which becomes narrower in proportion to $E^{1/4}$ as E decreases. By using arguments similar to those of Wannier, and by integrating numerically the classical equations of motion for a system of zero angular momentum consisting of an ion and two electrons, Read[155] has obtained the form of the differential cross section (in energy) for near-threshold ionization and also for near-threshold excitation of high-n Rydberg states.

The Wannier result (3.37) has been confirmed by Peterkop[156] and Rau[157] who developed quantum mechanical wave functions in the vicinity of the Wannier saddle point, using the WKB approximation. Experimental evidence[158-161] is also consistent with the Wannier threshold law. In particular, in an experiment on e^--He ionization, Cvejanovic and Read[159] verified the result (3.38) and obtained for the Wannier exponent m the value $m = 1.131 \pm 0.019$, in excellent accord with the theoretical value $m = 1.127$.

3.3.2. Extensions of the Wannier theory

The original classical calculations of Wannier[153] were performed for a system of total angular momentum equal to zero. Likewise, in the semi-classical calculations of Peterkop[156] and Rau[157], only the particular set $^1S^e$ of the quantum numbers ($LS\pi$) of the two electrons were considered. Several authors[162-167] have examined how the Wannier theory can be extended to other ($LS\pi$) values. Klar and Schlecht[163] concluded that all singlet states ($S=0$) have the original Wannier exponent m while all triplet states ($S=1$) have a different, higher exponent m' so that they tend to be suppressed at threshold. This result would lead to a value $A=1$ of the integrated asymmetry parameter [see (2.110)], in contradiction with the experimental evidence[168-170], as seen from Fig. 19. Upon re-examination of this problem, it was found[164-167] that the conclusion of Klar and Schlecht[163] is incorrect. Instead, the threshold behaviour of the ionization cross section is independent of the quantum numbers ($LS\pi$) provided that the Pauli principle does not require the wave function to vanish at the Wannier saddle point. Greene and Rau[164] showed that in all states except $^3S^e$ and $^1P^e$ the wave function can remain finite at the saddle point. Thus except for the $^3S^e$ and $^1P^e$ states, the threshold behaviour is the same for all ($LS\pi$) quantum numbers, in agreement with early expectations[153].

Crothers[171] has recently developed a semi-classical theory of threshold ionization which yields absolute differential and total ionization cross sections. Using the Kohn variational principle, he finds that the total ionization cross section near threshold can be written as

$$\sigma = C\,E^m \tag{3.39}$$

where m is the Wannier exponent (3.35b) and C is a coefficient which he is able to calculate. In particular, for electron-helium ionization he obtains $C = 2.37$ (in au). His theoretical values of σ are in good agreement with the measurements of Pichou et al.[161] in the energy range $0.8\ \text{eV} \leqslant E \leqslant 3.6\ \text{eV}$. The double differential cross sections obtained by Crothers[171] are also in good accord with the data of Pichou et al.[161] and his triple differential cross sections compare favourably with the experimental values of Fournier-Lagarde et al.[172] and Selles et al.[173], obtained by performing (e,2e) coincidence measurements.

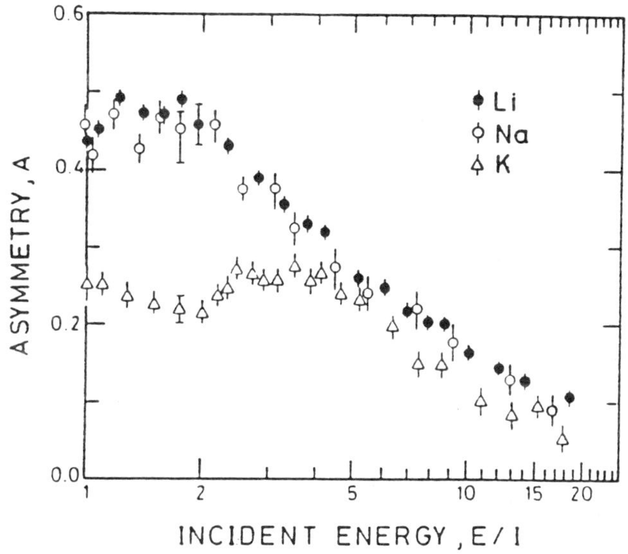

Fig. 19. The integrated asymmetry parameter for ionization of Li, Na and K, as a function of the incident electron energy, in units of the threshold energy I. The experimental data are from Baum et al.[170]

3.3.3. Coulomb-dipole threshold law

In the Wannier theory which we have discussed thus far it is assumed that ionization (double escape) proceeds via the configurations near the Wannier saddle point at $\alpha = \pi/4$ and $\theta_{01} = \pi$. This assumption has been questionned by Temkin[174] who proposed that close to threshold the relevant configurations are those for which the two electrons have unequal energies. The slower electron is then described by a Coulomb wave, while for the case of a residual ion with charge Z=1 the faster electron experiences the attractive long range dipole field due to the ion and the slow electron. This results in a modulated quasi-linear threshold law of the form

$$\sigma \sim E(\ell nE)^{-2} [1 + \sum_L C_L \sin (\alpha_L \ell nE + \mu_L)] \tag{3.40}$$

where C_L, α_L and μ_L are constants. It is worth noting that such an approach gives a linear threshold law, $\sigma \sim E$ for all values of the residual ion charge $Z > 1$. Crothers[171] has estimated the contribution to the ionization cross section arising from the Coulomb-dipole configurations, for the case of electron-hydrogen ionization. He finds that this contribution is only of the order of 5% of that coming from the configurations considered in the Wannier theory.

3.4. Fast (e,2e) reactions

In this section we shall consider (e,2e) experiments in which a "fast" electron is incident on a neutral target atom. We shall focus our attention on coplanar geometries and in particular on two kinematical arrangements known respectively as the Ehrhardt asymmetric geometry and the symmetric geometry.

3.4.1. Theory of coplanar asymmetric (e,2e) reactions

In the Ehrhardt (coplanar) asymmetric geometry, for a given energy E_i of the fast incident electron, a fast ("scattered") electron A is detected in coincidence with a slow ("ejected") electron B. The scattering angle θ_A of the fast electron is fixed and small, while the angle θ_B of the slow electron is varied. It is worth noting that in this geometry the magnitude $\Delta = |\underset{\sim}{k}_i - \underset{\sim}{k}_A|$ of the momentum transfer is small.

A number of measurements of TDCS have been performed in the Ehrhardt asymmetric geometry by several groups; these measurements have been reviewed by Ehrhardt et al.[144]. The basic features emerging from the experiments are illustrated in Fig. 20, which shows the absolute TDCS

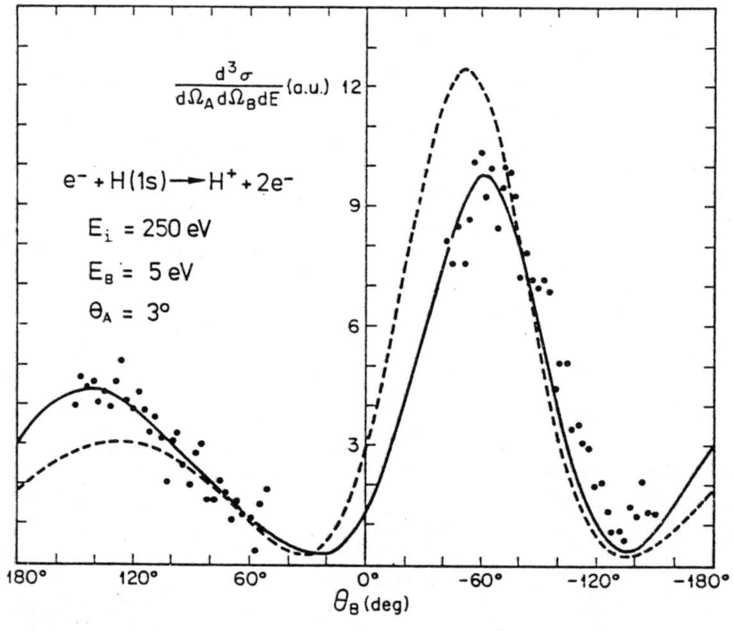

Fig. 20. The TDCS (in a.u.) for the (e,2e) reaction (3.41), for the case E_i=250 eV, E_B=5 eV and θ_A=3°, as a function of θ_B. Theoretical curves : --- first Born approximation, —— EBS results of Byron, Joachain and Piraux[184]. Experimental data : • absolute measurements of Ehrhardt, Knoth, Schlemmer and Jung[175]. From ref.184.

measured by Ehrhardt et al.[175] for the ionization of atomic hydrogen from the ground state

$$e^- + H(1s) \rightarrow H^+ + 2e^- \qquad (3.41)$$

for the case E_i=250 eV, E_B=5 eV and θ_A=3°. As seen from Fig. 20, there is a strong angular correlation between the scattered and ejected electrons, characterized by two peaks : the forward or "binary" peak and the backward or "recoil" peak.

Also shown in Fig. 20 are the values of the first Born TDCS,

$$\frac{d^3\sigma_{B1}}{d\Omega_A d\Omega_B dE} = \frac{k_A k_B}{k_i} |\bar{f}_{B1}|^2 \tag{3.42}$$

where the first Born scattering amplitude is given by

$$\bar{f}_{B1}(\underset{\sim}{k}_A,\underset{\sim}{k}_B) = -(2\pi)^{-1} \langle \exp(i\underset{\sim}{k}_A\cdot\underset{\sim}{r}_o)\psi^{(-)}_{c,\underset{\sim}{k}_B}(\underset{\sim}{r}_1) | \frac{1}{r_{01}} | \exp(i\underset{\sim}{k}_i\cdot\underset{\sim}{r}_o)\psi_{1s}(r_1)\rangle \tag{3.43}$$

In writing down the above expression we have used (3.26) and omitted the subscript i referring to the initial (1s) target state. Furthermore, we have chosen the effective charges to be $Z_A=0$ and $Z_B=1$ since in the Ehrhardt geometry (where $k_A \gg k_B$), equation (3.25) with Z=1 reduces to

$$\frac{Z_A}{k_A} + \frac{Z_B}{k_B} \simeq \frac{1}{k_B} \tag{3.44}$$

According to the first Born approximation (3.42) the maximum of the forward peak should occur in the direction $\hat{\underset{\sim}{\Delta}}$ of the momentum transfer, and that of the backward peak should be in the opposite direction $-\hat{\underset{\sim}{\Delta}}$. However, we see from Fig. 20 that the experimental forward peak is shifted towards larger angles $|\theta_B|$ with respect to the first Born prediction, and that the measured recoil peak is shifted by an even larger amount, also towards larger angles. Moreover, the ratio of the intensity of the forward peak to that of the backward peak is considerably reduced with respect to the first Born prediction.

These features have been illustrated for the (e,2e) reaction (3.41) in atomic hydrogen since in that case the wave function describing the target in the initial (1s) state is known exactly, as is the Coulomb wave function corresponding to the final, unbound target state ($H^+ + e^-$). Moreover, the first Born scattering amplitude (3.43) can be calculated in closed form, so that the comparison with absolute experimental data is unambiguous. In the case of (e,2e) reactions in more complex atoms theoretical difficulties arise in describing accurately the target initial state and especially its final continuum state consisting of an ion and an unbound electron. These problems introduce additional complications in the interpretation of data on TDCS. Nevertheless, the measurements clearly show that at intermediate incident electron energies there are significant departures from the predictions of the first Born approximation.

The experimental work performed since 1969 on coplanar asymmetric (e,2e) reactions was steadily accompanied by theoretical treatments of the problem, including the first Born calculations already mentionned, but also Coulomb-projected Born calculations[176,177] and distorted wave Born approximation (DWBA) treatments[178-181], which are all of first order in the electron-electron interaction. These first order calculations failed to account for the experimental data.

The first theoretical treatment in which all the characteristic features of the measurements were reproduced was a second Born calculation performed by Byron, Joachain and Piraux[182] for the (e,2e) reaction (3.41) in atomic hydrogen, within the framework of the EBS method[55,65]. For an unpolarized e^--H system the TDCS is given by (3.32). According to the EBS method, the direct amplitude is given by eq. (2.91). Now, in the Ehrhardt geometry, where the magnitude Δ of the momentum transfer is small, the third order Glauber term $\bar{f}_{G3}$ is unimportant. Moreover, since one electron emerges with a high velocity and the other with a low velocity, exchange effects are also small. A good approximation to the TDCS in the Ehrhardt geometry is therefore provided by the second Born expression

$$\frac{d^3\sigma_{B2}}{d\Omega_A d\Omega_B dE} = \frac{k_A k_B}{k_i}|\bar{f}_{B1} + \bar{f}_{B2}|^2 \tag{3.45}$$

It is this expression which was evaluated[182], using the closure approximation to obtain $\bar{f}_{B2}$. More recently, Byron, Joachain and Piraux[183,184] have performed full EBS calculations for the reaction (3.41). In addition to the inclusion of the small third order Glauber term $\bar{f}_{G3}$ and the exchange amplitude, the calculation of the second Born term $\bar{f}_{B2}$ was improved by including exactly the contributions of the 1s, 2s and 2p target states. The results of this EBS calculation (which are very close to the second Born values) are shown in Fig. 20 and are seen to be in excellent agreement with the absolute experimental data[175]. Further confirmation of the EBS results has also come recently from the UEBS calculations of Joachain et al.[185], the EBS' calculations of Balyan and Srivastava[186] and the second Born calculation (using pseudo-states) of Curran and Walters[187].

Let us now turn to asymmetric coplanar (e,2e) reactions in other atoms. As an example, we shall consider the (e,2e) reaction in helium

$$e^- + \mathrm{He}(1\,{}^1S) \rightarrow \mathrm{He}^+(1s) + 2e^- \tag{3.46}$$

Second Born calculations for this process have been performed[188,189], using a Hartree-Fock wave function for the helium ground state, and a symmetrized product of the $He^+(1s)$ wave function times a Coulomb wave (orthogonalized to the ground state Hartree-Fock orbital) to describe the final $He^+(1s)+e^-$ continuum state. In Fig. 21 we compare the predictions of the first and second Born approximations[189] with the absolute experimental data of Müller-Fiedler et al.[190] for the case E_i=256 eV, E_B=3 eV and θ_A=4°. In Fig. 22 we display the angular displacement of the recoil peak with respect to the direction $-\hat{\Delta}$ (corresponding to the first Born prediction), as a function of θ_A, for the case E_i=500 eV and E_B=10 eV. Also shown in Fig. 22 are the results of the Coulomb-projected Born exchange approximation[177] which is seen to "rotate" the recoil peak in the wrong direction. It is clear from Figures 21 and 22 that the second Born calculations[188,189] represent a marked improvement over first order results. The remaining discrepancies between theory and experiment should be taken care of mainly by using more elaborate initial and final state helium wave functions in the second Born calculations[192].

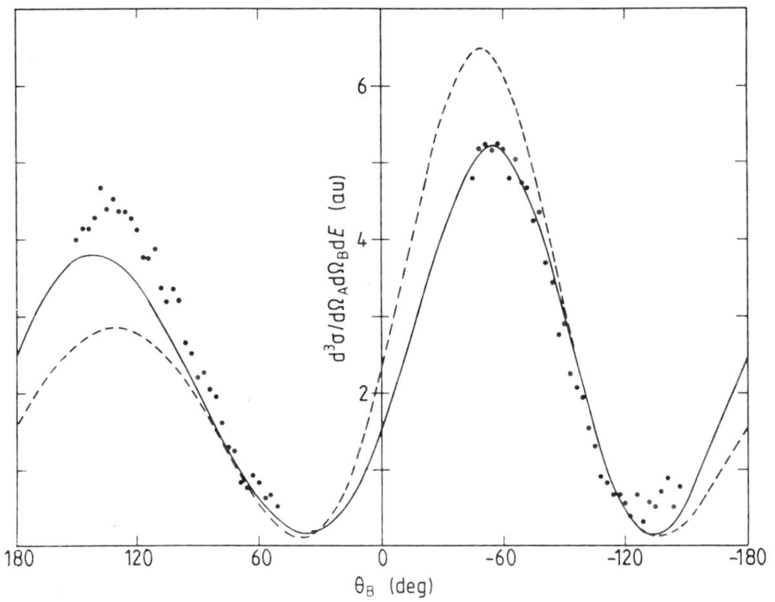

Fig. 21. The TDCS (in a.u.) for the (e,2e) reaction (3.46), for the case E_i=256 eV, E_B=3 eV and θ_A=4°, as a function of θ_B. Theoretical curves : --- first Born approximation, —— second Born results of Byron et al.[189]. Experimental data : • absolute measurements of Müller-Fiedler et al.[190]. From ref. 189.

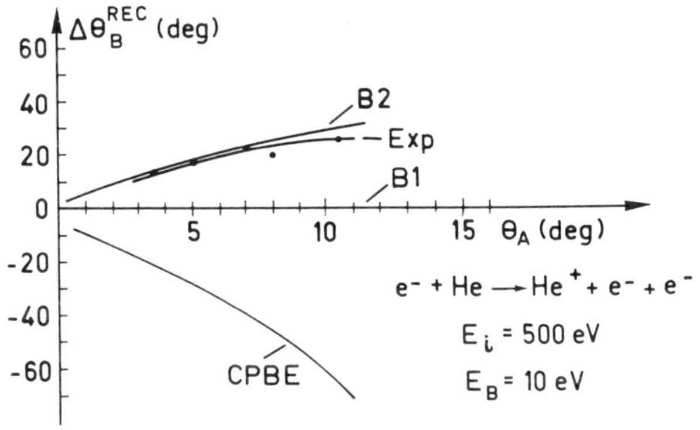

Fig. 22. The angular displacement $\Delta\theta_B^{REL}$ of the recoil peak with respect to the direction $-\hat{\Delta}$ predicted by the first Born approximation, for the (e,2e) reaction (3.46) with E_i=500 eV and E_B=10 eV. The Coulomb-projected Born exchange (CPBE) approximation[177] and the second Born (B2) approximation[189] are compared with the experimental data of Ehrhardt et al.[191]. From ref. 189.

3.4.2. Theory of coplanar symmetric (e,2e) reactions

Symmetric geometries are such that $E_A \simeq E_B$ and $\theta_A \simeq \theta_B$. The first (e,2e) symmetric coincidence experiments were performed by Amaldi et al.[143] and since then a number of measurements of this type have been carried out, which are discussed in the reviews of McCarthy and Weigold[193] and of Giardini-Guidoni et al.[194]. As an example, we shall consider a coplanar, fully symmetric geometry with $E_A=E_B$, $\theta_A=\theta_B$ $(=\theta)$, $\phi_A=o$ and $\phi_B=\pi$. If the incident electron is fast the magnitude of the momentum transfer is given by $\Delta \simeq k_i\ (3/2 - \sqrt{2}\cos\theta)^{1/2}$, so that Δ is never small in this geometry. The recoil momentum of the ion is $Q = |\ 2k_A \cos\theta - k_i\ | \simeq k_i\ |\ \sqrt{2}\cos\theta - 1\ |$.

Two angular regions can therefore be distinguished. For $\theta \lesssim 70°$, Δ is relatively large while Q remains small or moderate. In this case (as well as in non-coplanar symmetric geometries with similar values of Δ and Q), one has impulse-type collisions, from which the electron momentum density distribution of the target atom can be obtained[195]. As an illustration, let us consider the (e,2e) reaction (3.4) in atomic hydrogen, in which the target is initially in the bound state ψ_i. The plane-wave Born (PWB) scattering amplitude is then given by

$$f_{PWB} = f^C_{B1}(\Delta)\ [\tilde{\psi}_i(-\underset{\sim}{Q}) - \tilde{\psi}_i(\underset{\sim}{k}_B)] \tag{3.47}$$

where $\tilde{\psi}_i$ is the Fourier-transform of the wave function ψ_i and $f^C_{B1}=-2/\Delta^2$ is the first Born amplitude corresponding to free Coulomb scattering by the potential $V(r)=1/r$. Now $|\tilde{\psi}_i(\underset{\sim}{p})|$ is small for large values of p, so that the second term on the right of (3.47) can be neglected. Moreover, since Δ is large one can think of the reaction as a "binary" electron-electron collision, with $\underset{\sim}{p}=-\underset{\sim}{Q}$ being the momentum of the struck electron before the collision. We then have

$$\frac{d^3\sigma_{PWB}}{d\Omega_A d\Omega_B dE} \simeq \frac{k_A k_B}{k_i}\ \frac{4}{\Delta^4}\ |\tilde{\psi}_i(\underset{\sim}{p})|^2 \tag{3.48}$$

so that the plane wave Born TDCS is proportional to the momentum density distribution of the target electron[195]. This important property has given rise to the field of (e,2e) spectroscopy[193,194], and for this reason we shall call the angular domain in which Δ is large and Q small or moderate the (e,2e) spectroscopy region. The most elaborate calculations in that region have been performed by using the distorted wave impulse approximation. As an example, we show in Fig. 23 the electron momentum distribution corresponding to the ground state of atomic hydrogen, extracted by Lohmann and Weigold[196] from (e,2e) experiments performed at various energies, and compared with the calculated ground state probability distribution, $|\tilde{\psi}_{1s}(p)|^2 = 8\pi^{-2}(1+p^2)^{-4}$. It is worth noting that in the region for which $\underset{\sim}{Q} \simeq 0$ (corresponding to $\theta \simeq 45°$), the first Born term is of order k_i^{-2} (for large k_i) while the second Born term is of order k_i^{-3} and hence is only a correction to $\bar{f}_{B1}$. The dominance of the first Born approximation for large k_i in this case is in agreement with the experimental results of Van Wingerden et al.[197].

Let us now consider the large angle region ($\theta \gtrsim 70°$) for which both Δ and Q are large. In this region one expects on the basis of the calculations performed for inelastic (excitation) scattering that the second Born term should be very important, and that this term will be dominated by the contributions of the initial and final target states acting as intermediate states. This is confirmed by second Born calculations performed for e^--H(1s) ionization[198]. For large k_i and in the case of a coplanar symmetric geometry at large θ it is found that the first Born term $\bar{f}_{B1}$ is of order k_i^{-6}, while the contribution of the initial (1s) target state to $\bar{f}_{B2}$ is given by

$$\bar{f}_{B2}(1s) = -\frac{8\sqrt{2}}{\pi} k_i^{-5} (1-\sqrt{2}\cos\theta)^{-2} (k_i \cos 2\theta - 2^{3/2} i \cos\theta)^{-1} \quad (3.49)$$

We see from this equation that $\bar{f}_{B2}(1s)$ is in general of order k_i^{-6}, except near $\theta=135°$, where it is of order k_i^{-5}. Thus second order effects due to the term $\bar{f}_{B2}(1s)$ should be very important at large angles, especially near $\theta=135°$.

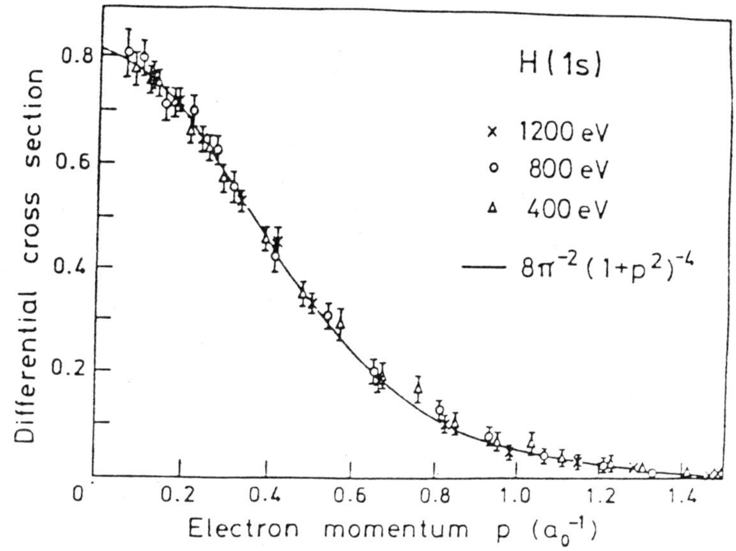

Fig. 23. The electron momentum distribution corresponding to the ground state of atomic hydrogen, as obtained by Lohmann and Weigold[196] at 400 eV (Δ), 800 eV (o) and 1200 eV (x), compared with the square of the calculated ground state probability distribution. From ref. 196.

This analysis is confirmed by an exact (numerical) calculation of the quantity $\bar{f}_{B2}(1s)$. Fig. 24 (a) shows the theoretical TDCS (solid curve), obtained by using $\bar{f}_{B1} + \bar{f}_{B2}$ (1s) as the direct scattering amplitude, and taking into account the fact that the scattering only occurs in the singlet mode. Of particular interest are the dramatic second order effects in the large angle region. Briggs[199] has recently pointed out that the appearance of critical angles (in the present case $\theta \simeq 135°$) is a general feature of three-body atomic collisions, and occurs for break-up as well as capture processes. Such critical angles appear first in double scattering contributions to the scattering amplitude, and their values depend on the masses of the three particles and their mutual relative momenta in the final state. In the present (e,2e) case the critical angle $\theta=135°$ predicted by Byron, Joachain and Piraux[198] corresponds to a situation in which the two outgoing electrons emerge with equal energies at 90° to each other.

A similar striking behaviour of the calculated TDCS, due to second order effects, arises for large angle symmetric (e,2e) reactions in helium (see Fig. 24b) and has recently been observed by Pochat et al.[200]. As seen from Fig. 24(b), the agreement between experiment and theory is considerably better at large angles when the contribution $\bar{f}_{B2}(1^1S)$ of the initial (1^1S) target state to the second Born term $\bar{f}_{B2}$ is included in the scattering amplitude. This agreement could still be improved by including the contribution of the final (continuum) target state to $\bar{f}_{B2}$. Moreover, since the incident electron energy E_i=200 eV is relatively low, the contributions to $\bar{f}_{B2}$ due to other target states acting as intermediate states should also be analyzed, as well as those coming from higher order terms of perturbation theory.

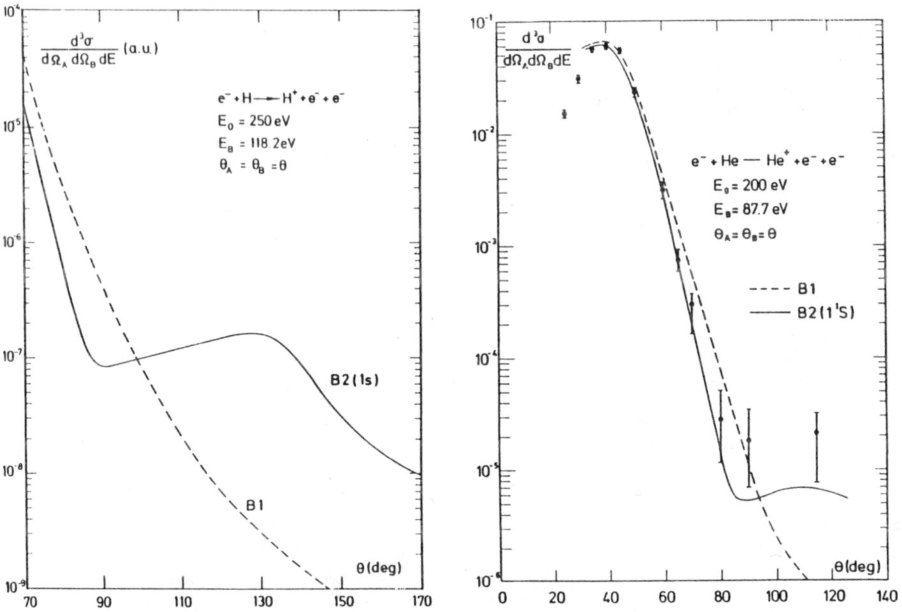

Fig. 24 (a) The TDCS (in a.u.) for the (e,2e) reaction (3.41) in atomic hydrogen, for the case of a coplanar symmetric geometry, with E_i=250 eV, E_A=E_B=118.2 eV, as a function of the scattering angle $\theta=\theta_A=\theta_B$. The full curve, B2(1s), corresponds to a second Born calculation in which $f=\bar{f}_{B1}+\bar{f}_{B2}(1s)$. The broken curve refers to the first Born approximation (B1). From ref. 198.
(b) The TDCS (in a.u.) for the reaction (3.46) in helium, for a coplanar symmetric geometry, with E_i=200 eV and E_A=E_B=87.7 eV, as a function of θ. The full curve, B2(1^1S) corresponds to a second Born calculation in which $f=\bar{f}_{B1}+\bar{f}_{B2}(1^1S)$. The broken curve refers to the first Born approximation (B1). The experimental data (•) are those of Pochat et al.[200]. From ref. 200.

4. PHOTOIONIZATION

In this section we shall study the photoionization of atoms and ions. The basic process which we shall consider is the single photoionization reaction

$$h\nu + A(i) \rightarrow A^+(f) + e^- \quad (4.1)$$

where A(i) is an atom in state i and in the final state we have a residual ion $A^+(f)$ in state f and one free electron. The inverse process of (4.1) is known as radiative recombination. The photoionization reaction

$$h\nu + A^-(i) \rightarrow A(f) + e^- \tag{4.2}$$

involving a negative ion A^- is known as photodetachment.

4.1. Basic theory

Let us consider a photoionization process in which a beam of photons is incident on the target in state i, and the photoelectrons are emitted with momentum $\underset{\sim}{k}$. From energy conservation we have, assuming that the photoelectrons move at non-relativistic velocities

$$h\nu + w_i = \frac{\hbar^2 k^2}{2m} \tag{4.3}$$

where w_i is the eigenenergy of the target in the initial state i, and m denotes the electron mass.

In order to obtain an expression for the ionization cross section, we recall that to first order in the coupling between the N-electron target and the electromagnetic field (described classically by the vector potential $\underset{\sim}{A}$ in the Coulomb gauge, for which $\underset{\sim}{\nabla}.\underset{\sim}{A} = 0$) the interaction Hamiltonian is given (in Gaussian units) by

$$H_{int} = \frac{ie\hbar}{mc} \sum_{j=1}^{N} \underset{\sim}{A}(\underset{\sim}{r}_j,t).\underset{\sim}{\nabla} \tag{4.4}$$

where

$$\underset{\sim}{A}(\underset{\sim}{r}_j,t)=A_o\hat{\underset{\sim}{\varepsilon}} \left\{ \exp[\,i(\underset{\sim}{\kappa}.\underset{\sim}{r}-\omega t+\delta_\omega)] + \text{complex conjugate} \right\} \tag{4.5}$$

Here $\underset{\sim}{\kappa}$ and $\omega = 2\pi\nu$ are respectively the wave vector and the angular frequency of the incident radiation which is assumed to be linearly polarized, $\hat{\underset{\sim}{\varepsilon}}$ being the polarization vector, and δ_ω is a real phase. The first term in (4.5) corresponds to photon absorption and the second one to photon emission.

Using Fermi's Golden Rule, and making the dipole approximation[201] in which $\exp(i\underset{\sim}{\kappa}.\underset{\sim}{r}_j)$ is replaced by unity, one finds that the differential photoionization cross section for a transition i→f is given by [202]

$$\frac{d\sigma_{fi}}{d\Omega} = \frac{4\pi^2 e^2 \hbar^2}{m^2 c\omega} \left|\langle \psi_{fE}^{(-)} | \hat{\underset{\sim}{\varepsilon}}.\underset{\sim}{D}_v | \psi_i \rangle\right|^2 \tag{4.6}$$

where ψ_i is the initial bound state wave function and $\psi_{fE}^{(-)}$ is the final (continuum) wave function satisfying ingoing spherical wave boundary conditions and normalized according to

$$\langle \psi_{fE}^{(-)} | \psi_{f'E'}^{(-)} \rangle = \delta_{ff'} \, \delta(E-E') \tag{4.7}$$

The operator $\underset{\sim}{D}_V$ appearing in (4.6) is the dipole velocity operator

$$\underset{\sim}{D}_V = \sum_{j=1}^{N} \underset{\sim}{\nabla}_j \tag{4.8}$$

so that the expression (4.6) is called the dipole velocity form of the cross section. From the Heisenberg equations of motion, one readily obtains the operator identity

$$\sum_{j=1}^{N} \underset{\sim}{p}_j = (i\hbar)^{-1} \left[\sum_{j=1}^{N} \underset{\sim}{r}_j , H \right] \tag{4.9}$$

where $\underset{\sim}{r}_j$ and $\underset{\sim}{p}_j$ denote respectively the coordinates and momenta of the electrons and H is the target Hamiltonian. Using this identity, it is a simple matter to rewrite the photoionization cross section (4.6) in the dipole length form

$$\frac{d\sigma_{fi}}{d\Omega} = \frac{4\pi^2 e^2 \omega}{c} | \langle \psi_{fE}^{(-)} | \hat{\underset{\sim}{\varepsilon}} \cdot \underset{\sim}{D}_L | \psi_i \rangle |^2 \tag{4.10}$$

where $\underset{\sim}{D}_L$ is the dipole length operator

$$\underset{\sim}{D}_L = \sum_{j=1}^{N} \underset{\sim}{r}_j \tag{4.11}$$

A third expression of the photoionization cross section involving the accelerator form of the dipole operator can also be written down. The above cross section formulae, which have been obtained for linearly polarized photons, must be averaged over the two linearly independent polarization directions in the case of unpolarized incident photons.

If exact wave functions are used the dipole length, dipole velocity and dipole accelerator forms of the cross section are identical. However, if approximate wave functions are used different results will in general be obtained, the magnitude of the difference being often an indication of the accuracy of the approximation. The length form tends to weight the large r part of the wave functions and the velocity form the intermediate part of the wave functions. The acceleration form, which emphasizes the wave functions close to the nucleus usually gives rather poorer results than the two other forms.

It is convenient to rewrite the cross section formulae (4.6) and (4.10) in atomic units. In these units the dipole velocity expression (4.6) becomes

$$\frac{d\sigma_{fi}}{d\Omega} = \frac{4\pi^2\alpha}{E_\gamma} |\langle\psi_{fE}^{(-)}|\hat{\varepsilon}\cdot D_V|\psi_i\rangle|^2 \tag{4.12}$$

while the dipole length form (4.10) reads

$$\frac{d\sigma_{fi}}{d\Omega} = 4\pi^2\alpha E_\gamma|\langle\psi_{fE}^{(-)}|\hat{\varepsilon}\cdot D_L|\psi_i\rangle|^2 \tag{4.13}$$

where $\alpha = e^2/\hbar c$ is the fine-structure constant and E_γ is the photon energy expressed in atomic units.

The differential cross section for photoionization of an unpolarized atom or ion by a polarized photon beam, where the spins of the photoelectron and the residual ion are not observed, is obtained from eqs. (4.12) or (4.13) by summing over the final magnetic quantum numbers and averaging over the initial magnetic quantum numbers. Thus we have for example in the dipole length form

$$\frac{d\sigma_{fi}}{d\Omega} = \frac{4\pi^2\alpha E_\gamma}{(2L_i+1)(2S_i+1)} \sum_{\substack{M_{L_i},M_{S_i} \\ M_{L_f},M_{S_f},m_f}} |\langle\psi_{fE}^{(-)}|\hat{\varepsilon}\cdot D_L|\psi_i\rangle|^2 \tag{4.14}$$

The evaluation of this expression can be carried out by using Racah algebra. It is found that[202]

$$\frac{d\sigma_{fi}}{d\Omega} = \frac{\sigma_{tot}(i\to f)}{4\pi} [1 + \beta P_2(\cos\theta)] \tag{4.15}$$

where $\sigma_{tot}(i\to f)$ is the total photoionization cross section for the transition $i\to j$ and θ is the angle of the direction of the emitted photoelectron relative to the polarization direction of the photon beam. The quantity $\beta\sigma_{tot}(i\to f)$ is called the asymmetry parameter.

If the incident photons are unpolarized, one can obtain the angular distribution from eq. (4.15) by assuming that the incident beam is composed of an incoherent mixture of two beams linearly polarized at right angles. In this case one has[202]

$$\frac{d\sigma_{fi}}{d\Omega} = \frac{\sigma_{tot}(i\to f)}{4\pi} [1 - \frac{\beta}{2} P_2 (\cos\theta)] \tag{4.16}$$

where θ is now the angle of the direction of the emitted photoelectron relative to the incident photon propagation direction. The fact that both cross sections (4.15) and (4.16) must be non-negative at all angles implies that $-1 \leqslant \beta \leqslant 2$. Moreover, if the atom (ion) is initially in an S-state one has $\beta=2$ at all energies. As a result, the differential cross section (4.16) vanishes in the incident photon propagation direction.

4.2. The calculation of photoionization cross sections

It is only for hydrogenic atoms (ions) that photoionization cross sections can be calculated analytically. For example, the total photoionization cross section of an hydrogenic atom (ion) of nuclear charge Z from a state denoted by the principal quantum number n (averaged over ℓ substates) is given by (in a.u.)

$$\sigma_{tot} = \frac{8\pi\alpha}{3^{3/2}} \frac{Z^4}{n^5} E_\gamma^{-3} \, g(E_\gamma, n, Z) \qquad (4.17)$$

where the quantity g is known as the Kramers-Gaunt factor. It is a slowly varying function of the photon energy, falling off like $E_\gamma^{-1/2}$ at high photon energies.

For atoms or ions having two or more electrons, approximations must be used to describe both the initial bound state wave function ψ_i and the final continuum state $\psi_{fE}^{(-)}$. The simplest approximation is based on the assumption that the motion of the active electron, both before and after photoionization can be represented by a model central potential. The calculations based on this single-particle approximation usually agree qualitatively with experiment, but often differ in magnitude, particularly in the vicinity of the first few ionization thresholds. In order to obtain accurate photoionization cross sections, electron correlation effects must be taken into account. A number of theoretical methods have been developed which are capable of achieving this goal. We mention in particular the many-body perturbation theory[204], the random phase approximation[205,206], the multi-channel quantum defect theory[42-44], the R-matrix theory[13,14] and L^2 methods[106,107]. Detailed reviews of the application of these methods to the calculation of photoionization cross sections have been given by Burke[202] and Starace[203].

4.3. Resonances in photoionization

Photoionization cross sections are dominated in certain energy ranges by resonances. At energies close to a resonance the photoionization process can occur either directly

$$h\nu + A(i) \rightarrow A^+(f) + e^- \qquad (4.18)$$

or via an intermediate resonant state $A^\star$

$$h\nu + A(i) \rightarrow A^\star \rightarrow A^+(f) + e^- \qquad (4.19)$$

The interference between these two mechanisms leads to the absorption line profile, which Fano and Cooper[207] have parametrized by the formula

$$\sigma(E) = \sigma_a \frac{(q + \varepsilon)^2}{1 + \varepsilon^2} + \sigma_b \quad (4.20)$$

where σ_a and σ_b are background cross sections which vary slowly with energy, $\varepsilon = (E-E_r)/(\Gamma/2)$, E_r being the resonance energy and Γ the resonance width, and q is the line profile index which defines the line shape.

Early examples of resonances were found by Madden and Codling[208] in the photoionization of helium atoms. Using the continuous light from an electron synchrotron, they observed the resonant process

$$h\nu + He \rightarrow He^{\star}(2snp \pm 2pns\ ^1P) \rightarrow He^+ + e^- \quad (4.21)$$

interfering with the direct process

$$h\nu + He \rightarrow He^+ + e^- \quad (4.22)$$

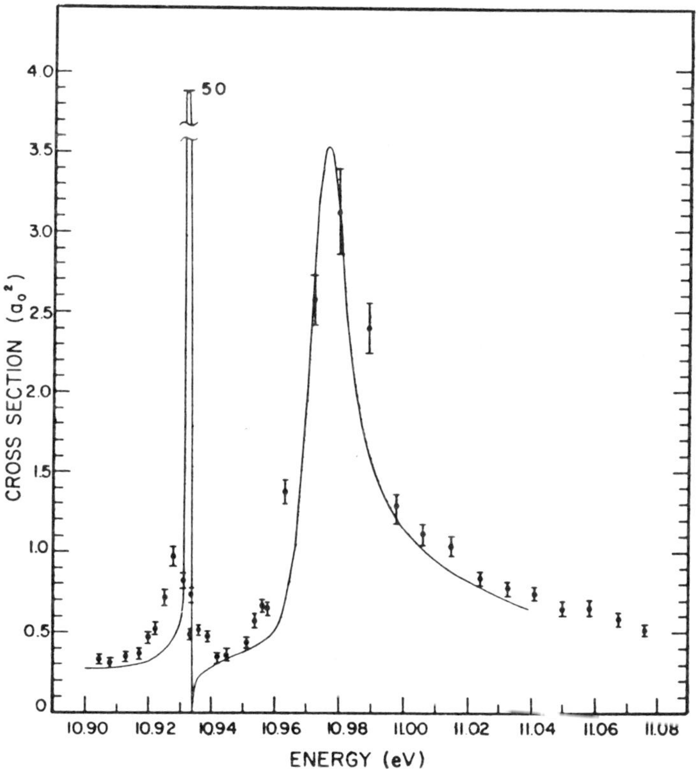

Fig. 25. The photodetachment cross section of H^-, in the energy range between 10.90 eV and 11.08 eV. The solid curve corresponds to the theoretical results of Broad and Reinhardt[211]. The data points are from Bryant et al.[209]. From ref. 209.

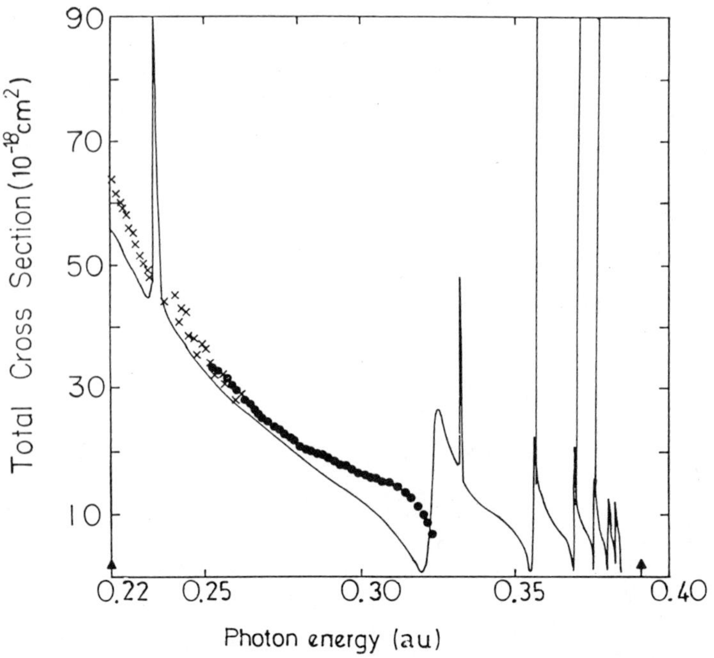

Fig. 26. The total cross section for photoionization of the Al ground state between the $3s^2\ ^1S$ and $3s\ 3p\ ^3P^0$ thresholds. The solid curve corresponds to the R-matrix calculations of Tayal and Burke[214]. The experimental data are : x : measured values of Kohl and Parkinson[215], • measured values of Roig[216]. The arrows on the energy axis show the $2s^2\ ^1S$ and $3s\ 3p\ ^3P^0$ thresholds. From ref. 214.

This gives rise to Rydberg series of Feshbach resonances converging onto the n=2 level of the He^+ ion. Further faint series can also be seen converging onto higher excited levels of He^+.

As another interesting example of resonances, let us consider the photodetachment of the H^- ion ,

$$h\nu + H^- \rightarrow H + e^- \qquad (4.23)$$

Using a colliding beam method in which a laser beam is directed at variable angle across a H^- beam, Bryant et al.[209] have observed a Feshbach and a shape resonance near 11 eV (see Fig. 25) which agree well with the theoretical predictions[210,211]. The shape resonance (2s 2p ^1P) lies just above the n=2 threshold of the H atom and had been previously observed in electron scattering from atomic hydrogen[212,213].

Finally, we show in Fig. 26 the total cross section for photoionization of the ground state of the aluminium atom between the $3s^2\ ^1S$ and $3s\ 3p\ ^3P^0$ thresholds, which also exhibits an important resonance structure. The results of the recent R-matrix calculations of Tayal and Burke[214] are seen to be in good agreement with the experimental data of Kohl and Parkinson[215] and of Roig[216].

REFERENCES

1. C.J. Joachain, Quantum Collision Theory (North Holland, Amsterdam, 3d ed., 1983).
2. P.G. Burke and C.J. Joachain, The Theory of Electron-Atom Collisions (Plenum, New York, to be published).
3. L. Castillejo, I.C. Percival and M.J. Seaton, Proc. Roy. Soc. A 254, 259 (1960).
4. P.G. Burke and H.M. Schey, Phys. Rev. 126, 147 (1962).
5. R. Damburg and E. Karule, Proc. Phys. Soc. 90, 637 (1967).
6. H. Feshbach, Ann. Phys. (N.Y.) 5, 357 (1958); 19, 287 (1962).
7. P.G. Burke and A.J. Taylor, Proc. Phys. Soc. 88, 549 (1966).
8. I.C. Percival and M.J. Seaton, Proc. Camb. Phil. Soc. 53, 654 (1957).
9. T. Kato, Progr. Theor. Phys. 6, 295 (1951); 6, 394 (1951).
10. R.K. Nesbet, Adv. At. Mol. Phys. 13, 315 (1977).
11. J. Callaway, Phys. Rep. 45, 89 (1978).
12. R.K. Nesbet, Variational Methods in Electron-Atom Scattering Theory (Plenum, New York, 1980).
13. P.G. Burke and W.D. Robb, Adv. At. Mol. Phys. 11, 143 (1975).
14. P.G. Burke, Physicalia Magazine 8, 289 (1986); Proceedings of the ICAP-X Conference (Tokyo, 1986), to be published.
15. M.J. Seaton, J. Phys. B 7, 1817 (1974).
16. E.R. Smith and R.J.W. Henry, Phys. Rev. A 7, 1585 (1973); A 8, 572 (1973).
17. P.G. Burke and W. Eissner, in Atoms in Astrophysics, ed. by P.G. Burke, W. Eissner, D.G. Hummer and I.C. Percival (Plenum, New York, 1983), p. 1.
18. L. Hulthén, Kgl. Fysiogr. Sallsk. Lund Förh. 14 (1944); Arkiv. Mat. Ast. Fys. 35 A, 25 (1948).
19. W. Kohn, Phys. Rev. 74, 1763 (1948).
20. C. Schwartz, Ann. Phys. (N.Y.) 16, 36 (1961).
21. S.I. Rubinow, Phys. Rev. 98, 183 (1955).
22. C. Schwartz, Phys. Rev. 124, 1468 (1961).
23. R.L. Armstead, Phys. Rev. 171, 91 (1968).
24. D. Register and R.T. Poe, Phys. Lett. 51 A, 431 (1975).
25. J.F. Williams, J. Phys. B 8, 1683 (1975).
26. R.K. Nesbet, J. Phys. B 12, L243 (1979).
27. D. Andrick and A. Bitsch, J. Phys. B 8, 393 (1975).
28. W.E. Kauppila, T.S. Stein, G. Jesion, M.S. Dababneh and V. Pol, Rev. Sci. Inst. 48, 322 (1977).
29. R.W. Crompton, M.T. Elford and R.L. Jory, Austral. J. Phys. 20, 369 (1967); R.W. Crompton, M.T. Elford and A.G. Robertson, Austral. J. Phys. 23, 667 (1970); H.B. Milloy and R.W. Crompton, Phys. Rev. A 15, 1847 (1977).
30. T.F. O'Malley, P.G. Burke and K.A. Berrington, J. Phys. B 12, 953 (1979).

31. E.P. Wigner, Phys. Rev. 70, 15, 606 (1946); E.P. Wigner and L. Eisenbud, Phys. Rev. 72, 29 (1947).

32. P.G. Burke, A. Hibbert and W.D. Robb, J. Phys. B 4, 153 (1971).

33. P.G. Burke and W.D. Robb, J. Phys. B 5, 44 (1972).

34. C. Bloch, Nucl. Phys. 4, 503 (1957).

35. P.J.A. Buttle, Phys. Rev. 160, 719 (1967).

36. L.C.G. Freitas, K.A. Berrington, P.G. Burke, A. Hibbert, A.E. Kingston and A.L. Sinfailam, J. Phys. B 17, L303 (1984).

37. A.R. Johnston and P.D. Burrow, J. Phys. B 16, 613 (1983).

38. S.J. Buckman, P. Hammond, F.H. Read and G.C. King, J. Phys. B 16, 4039 (1983).

39. P.G. Burke, Adv. At. Mol. Phys. 4, 173 (1968).

40. D.E. Golden, Adv. At. Mol. Phys. 14, 1 (1978).

41. R.J.W. Henry, Phys. Rep. 68, 1 (1981).

42. M.J. Seaton, Mon. Not. Roy. Astron. Soc. 118, 504 (1958); Rep. Progr. Phys. 46, 167 (1983).

43. M. Gailitis, Sov. Phys. JETP 17, 1328 (1963).

44. U. Fano, Phys. Rev. A 2, 353 (1970); A 15, 817 (1977).

45. M. Jones, Phil. Trans. Roy. Soc. A 277, 587 (1975).

46. N.S. Scott and P.G. Burke, J. Phys. B 13, 4299 (1980).

47. N.S. Scott and K.T. Taylor, Comput. Phys. Comm. 25, 347 (1982).

48. D.W. Walker, J. Phys. B 7, 97 (1974).

49. P.H. Norrington and I.P. Grant, J. Phys. B 14, L261 (1981).

50. N.S. Scott, P.G. Burke and K. Bartschat, J. Phys. B 16, L361 (1983).

51. K. Bartschat, N.S. Scott, K. Blum and P.G. Burke, J. Phys. B 17, 269 (1984).

52. J. Kessler, Polarized Electrons (Springer-Verlag, Berlin, 2d ed., 1985).

53. K. Bartschat and P.G. Burke, Comm. At. Mol. Phys. 16, 271 (1985).

54. B.H. Bransden and M.R.C. McDowell, Phys. Rep. 30, 207 (1977); 46, 249 (1978).

55. F.W. Byron, Jr. and C.J. Joachain, Phys. Rep. 34, 233 (1977).

56. H.R.J. Walters, Phys. Rep. 116, 1 (1984).

57. A.R. Holt, J. Phys. B 5, L6 (1972).

58. A.M. Ermolaev and H.R.J. Walters, J. Phys. B 12, L779 (1979); B 13, L473 (1980); H.R.J. Walters, J. Phys. B 14, 3499 (1981).

59. F.W. Byron, Jr. and C.J. Joachain, J. Phys. B 14, 2429 (1981).

60. D.H. Madison, J.A. Hughes and D.S. McGinness, J. Phys. B 18, 2737 (1985).

61. C.J. Joachain, Comm. Atom. Mol. Phys. 6, 69 (1977).

62. R.A. Bonham, J. Chem. Phys. 36, 3260 (1962).

63. V.I. Ochkur, Zh. Eksp. Teor. Fiz. 45, 734 (1963) [Sov. Phys. JETP 18, 503 (1964)].

64. F.W. Byron, Jr., C.J. Joachain and P.M. Potvliege, J. Phys. B 15, 3915 (1982).

65. F.W. Byron, Jr. and C.J. Joachain, Phys. Rev. A 8, 1267 (1973).

66. R.J. Glauber, in Lectures in Theoretical Physics, vol. 1, ed. by W.E. Brittin (Interscience, New York, 1959), p. 315.

67. G. Moliere, Z. Naturforsch. 2 A, 133 (1947).

68. C.J. Joachain and C. Quigg, Rev. Mod. Phys. 46, 279 (1974).

69. D.M. Chase, Phys. Rev. 104, 838 (1956).

70. F.W. Byron, Jr. and C.J. Joachain, J. Phys. B 8, L284 (1975).

71. C.J. Joachain and K.H. Winters, J. Phys. B 10, L727 (1977).

72. G.E. Chamberlain, S.R. Mielczarek and C.E. Kuyatt, Phys. Rev. A 2, 1905 (1970).

73. C.B. Opal and E.C. Beaty, J. Phys. B 5, 627 (1972).

74. H. Suzuki and T. Takayanagi, in Abstracts of Papers of the 8th ICPEAC (Belgrade, 1973),ed. by B.C. Cobic and M.V. Kurepa, p. 286.

75. M.A. Dillon and E.N. Lassettre, J. Chem. Phys. 62, 2373 (1975).

76. S.J. Wallace, Ann. Phys. (N.Y.) 78, 190 (1973).

77. F.W. Byron, Jr., C.J. Joachain and E.H. Mund, Phys. Rev. D 11, 1662 (1975); Phys. Rev. C 20, 2325 (1979).

78. F.W. Byron, Jr., C.J. Joachain and R.M. Potvliege, J. Phys. B 14, L609 (1981).

79. F.W. Byron, Jr., C.J. Joachain and R.M. Potvliege, J. Phys. B 18, 1637 (1985).

80. J.F. Williams, J. Phys. B 14, 1197 (1981).

81. C.J. Joachain, Comm. Atom. Mol. Phys. 6, 87 (1977).

82. F.W. Byron, Jr. and C.J. Joachain, Phys. Lett. A 49, 306 (1974).

83. F.W. Byron, Jr. and C.J. Joachain, Phys. Rev. A 15, 128 (1977).

84. C.J. Joachain, R. Vanderpoorten, K.H. Winters and F.W. Byron, Jr., J. Phys. B 10, 227 (1977).

85. C.J. Joachain, in Electronic and Atomic Collisions, ed. by G. Watel (North Holland, Amsterdam, 1978) p. 71.

86. S. Vucic, R.M. Potvliege and C.J. Joachain, J. Phys. B (to be published).

87. M.H. Mittleman and K.M. Watson, Phys. Rev. 113, 198 (1959); Ann. Phys. (N.Y.) 10, 268 (1960).

88. J.B. Furness and I.E. McCarthy, J. Phys. B 6, 2280 (1973).

89. R. Vanderpoorten, J. Phys. B 8, 926 (1975).

90. M.E. Riley and D.G. Truhlar, J. Chem. Phys. 65, 792 (1976).

91. G.M.A. Hyder, M.S. Dababneh, Y.F. Hsieh, W.E. Kauppila, C.K. Kwan, M. Madhavi-Hezaveh and T.S. Stein, Phys. Rev. Lett. 57, 2252 (1986).

92. C.J. Joachain and R.M. Potvliege, Phys. Rev. A 35, 4873 (1987).

93. C.J. Joachain, in Atomic Physics with Positrons, ed. by J.W. Humberston (Plenum, New York), to be published.

94. B.H. Bransden and J.P. Coleman, J. Phys. B 5, 537 (1972).

95. B.H. Bransden, T. Scott, R. Shingal and R.K. Raychoudhury, J. Phys. B 15, 4605 (1982).

96. I.E. McCarthy and A.T. Stelbovics, Phys. Rev. A 22, 502 (1980); J. Phys. B 16, 1233 (1983).

97. I.E. McCarthy, B.C. Saha and A.T. Stelbovics, J. Phys. B 14, 2871 (1981); B 15, L401 (1982); Phys. Rev. A 25, 268 (1982).

98. D.P. Dewangan and H.R.J. Walters, J. Phys. B 10, 637 (1977).

99. A.E. Kingston and H.R.J. Walters, J. Phys. B 13, 4633 (1980).

100. W.L. Van Wyngaarden and H.R.J. Walters, J. Phys. B 19, 929 (1986); B 19, 1817 (1986); B 19, 1827 (1986).

101. P.G. Burke and T.G. Webb, J. Phys. B 3, L131 (1970).

102. J. Callaway and J.W. Wooten, Phys. Lett. A 45, 85 (1973); Phys. Rev. A 9, 1924 (1974); A 11, 1118 (1975).

103. J. Callaway, M.R.C. McDowell and L.A. Morgan, J. Phys. B 8, 2181 (1976); J. Phys. B 9, 2043 (1976).

104. S.L. Willis and M.R.C. McDowell, J. Phys. B 14, L453 (1981).

105. P.G. Burke, K.A. Berrington and C.V. Sukumar, J. Phys. B 14, 289 (1981).

106. W.P. Reinhardt, Comput. Phys. Commun. 17, 1 (1979).

107. J.T. Broad, in Electron-Atom and Electron-Molecule Collisions, ed. by J. Hinze (Plenum, New York, 1983) p. 91.

108. B.H. Bransden and A.T. Stelbovics, J. Phys. B 17, 1877 (1984).

109. E.J. Heller, W.P. Reinhardt and H.A. Yamani, J. Comput. Phys. 13, 536 (1973).

110. E.J. Heller, T.N. Rescigno and W.P. Reinhardt, Phys. Rev. A 8, 2946 (1973).

111. J.T. Broad, Phys. Rev. A 18, 1012 (1978).

112. J.F. Reading, A.L. Ford, G.V. Swafford and A. Fitchard, Phys. Rev. A 20, 130 (1979).

113. B.H. Bransden and M. Plummer, J. Phys. B 13, 2007 (1986).

114. P.G. Burke, C.J. Noble and P. Scott, Proc. Roy. Soc. (London) A 410, 289 (1987).

115. W.E. Kauppila, T.S. Stein, J.H. Smart, M.S. Dababneh, Y.K. Ho, J.P. Downing and V. Pol, Phys. Rev. A 24, 725 (1981).

116. T.S. Stein and W.E. Kauppila, Adv. At. Mol. Phys. 18, 53 (1982).

117. K. Blum, Density Matrix Theory and Applications (Plenum, New York, 1981).

118. See for example M.S. Lubell, in Coherence and Correlation in Atomic Collisions, ed. by H. Kleinpoppen and J. Williams (Plenum, New York, 1980) p. 663.

119. M.J. Alguard, V.W. Hughes, M.S. Lubell and P.F. Wainwright, Phys. Rev. Lett. 39, 334 (1977).

120. P.F. Wainwright, M.J. Alguard, G. Baum and M.S. Lubell, Rev. Sci. Instr. 44, 571 (1978).

121. G.D. Fletcher, M.J. Alguard, T.J. Gay, V.W. Hughes, C.W. Tu, P.F. Wainwright and M.S. Lubell, Phys. Rev. Lett. 48, 1671 (1982).

122. T.J. Gay, G.D. Fletcher, M.J. Alguard, V.W. Hughes, P.F. Wainwright and M.S. Lubell, Phys. Rev. A 26, 3664 (1982).

123. G.D. Fletcher, M.J. Alguard, T.J. Gay, V.W. Hughes, P.F. Wainwright, M.S. Lubell and W. Raith, Phys. Rev. A 31, 2854 (1985).

124. M.R.C. McDowell, P.W. Edmunds, R.M. Potvliege, C.J. Joachain, R. Shingal and B.H. Bransden, J. Phys. B 17, 3951 (1984).

125. J. Slevin, Rep. Progr. Phys. 47, 461 (1984).

126. M. Eminyan, K. MacAdam, J. Slevin and H. Kleinpoppen, Phys. Rev. Lett. 31, 576 (1973); J. Phys. B 7, 1519 (1974).

127. J. Macek and D.H. Jaecks, Phys. Rev. A 6, 2288 (1971).

128. K.L. Baluja and M.R.C. McDowell, J. Phys. B 12, 835 (1979).

129. W.C. Fon, K.A. Berrington and A.E. Kingston, J. Phys. B 13, 2309 (1980).

130. V.C. Sutcliffe, G.N. Haddad, N.C. Steph and D.E. Golden, Phys. Rev. A 17, 100 (1978).

131. M.T. Hollywood, A. Crowe and J.F. Williams, J. Phys. B 12, 819 (1979).

132. N.C. Steph and D.E. Golden, Phys. Rev. A 21, 759 (1980).

133. U. Fano, Phys. Rev. 90, 577 (1953).

134. K. Blum and H. Kleinpoppen, Phys. Rep. 52, 203 (1979).

135. U. Fano and J. Macek, Rev. Mod. Phys. 45, 553 (1973).

136. H.W. Hermann and I.V. Hertel, J. Phys. B 13, 4285 (1980); Comm. Atom. Mol. Phys. 12, 61 (1982).

137. N. Andersen, I.V. Hertel and H. Kleinpoppen, J. Phys. B 17, L901 (1984).

138. L.A. Morgan and M.R.C. McDowell, J. Phys. B 8, 1073 (1975).

139. E. Weigold, L. Frost and K.J. Nygaard, Phys. Rev. A 21, 1950 (1980).

140. C.G. Back, S. Watkin, M. Eminyan, K. Rubin, J. Slevin and J.M. Woolsey, J. Phys. B 17, 2695 (1984).

141. R.M. Potvliege and C.J. Joachain, J. Phys. B 18, L585 (1985).

142. H. Ehrhardt, M. Schulz, T. Tekaat and K. Willmann, Phys. Rev. Lett. 22, 89 (1969).

143. U. Amaldi, Jr., A. Egidi, R. Marconero and G. Pizzella, Rev. Sci. Instr. 40, 1001 (1969).

144. H. Ehrhardt, K. Jung, G. Knoth and P. Schlemmer, Z. Phys. D 1, 3 (1986).

145. R.K. Peterkop, Theory of Ionization of Atoms by Electron Impact (Colorado Univ. Press, 1977).

146. M.R.H. Rudge and M.J. Seaton, Proc. Phys. Soc. 83, 680 (1964); Proc. Roy. Soc. A 283, 262 (1965).

147. M.R.H. Rudge, Rev. Mod. Phys. 40, 564 (1968).

148. B.H. Bransden, Atomic Collision Theory (Benjamin, Reading, 2d ed., 1983).

149. C.D. Lin, Phys. Rev. A 10, 1986 (1974).

150. A.R.P. Rau, Comm. Atom. Mol. Phys. 14, 285 (1984).

151. A.R.P. Rau, Phys. Rep. 110, 369 (1984).

152. E.P. Wigner, Phys. Rev. 73, 1002 (1948).

153. G.H. Wannier, Phys. Rev. 90, 817 (1953).

154. I. Vinkaln and M. Gailitis, in Abstracts of the 5th ICPEAC (Leningrad, 1967), p. 648.

155. F.H. Read, J. Phys. B 17, 3965 (1984).

156. R.K. Peterkop, J. Phys. B 4, 513 (1971).

157. A.R.P. Rau, Phys. Rev. A 4, 207 (1971).

158. J.W. McGowan and E.M. Clarke, Phys. Rev. 167, 43 (1968).

159. S. Cvejanovic and F.H. Read, J. Phys. B 7, 1841 (1974).

160. D. Spence, Phys. Rev. A 11, 1539 (1975).

161. F. Pichou, A. Huetz, G. Joyez and M. Landau, J. Phys. B 11, 3683 (1978).

162. T.A. Roth, Phys. Rev. A 5, 476 (1972).

163. H. Klar and W. Schlecht, J. Phys. B 9, 1699 (1976).

164. C.H. Greene and A.R.P. Rau, Phys. Rev. Lett. 48, 533 (1982); J. Phys. B 16, 99 (1983).

165. A.D. Stauffer, Phys. Lett. A 91, 114 (1982).

166. R.K. Peterkop, J. Phys. B 16, L587 (1983).

167. J.M. Feagin, J. Phys. B 17, 2433 (1984).

168. D. Hils and H. Kleinpoppen, J. Phys. B 11, L283 (1978).

169. D. Hils, W. Jitschin and H. Kleinpoppen, J. Phys. B 15, 3347 (1982).

170. G. Baum, M. Moede, W. Raith and W. Schröder, Int. Symp. on Polarization and Correlation in Electron-Atom Collisions (Münster, 1983); see also J. Kessler, Comm. Atom. Mol. Phys. 14, 275 (1984).

171. D.S.F. Crothers, J. Phys. B 19, 463 (1986).

172. P. Fournier-Lagarde, J. Mazeau and A. Huetz, J. Phys. B 17, L591 (1984); B 18, 379 (1985).

173. P. Selles, A. Huetz and J. Mazeau, J. Phys. B (to be published).

174. A. Temkin, Phys. Rev. Lett. 49, 365 (1982), Phys. Rev. A 30, 2737 (1984).

175. H. Ehrhardt, G. Knoth, P. Schlemmer and K. Jung, Phys. Lett. A 110, 92 (1985).

176. M. Schulz, J. Phys. B 6, 2580 (1973).

177. S. Geltman and M.B. Hidalgo, J. Phys. B 7, 831 (1974); S. Geltman, J. Phys. B 7, 1994 (1974).

178. K.L. Baluja and H.S. Taylor, J. Phys. B 9, 829 (1976).

179. D.H. Madison, R.V. Calhoun and W.N. Shelton, Phys. Rev. A 16, 552 (1977).

180. B.H. Bransden, J.J. Smith and K.H. Winters, J. Phys. B 11, 3095 (1978); B 12, 1257 (1979); J.J. Smith, K.H. Winters and B.H. Bransden, J. Phys. B 12, 1723 (1979).

181. R.J. Tweed, J. Phys. B 13, 4467 (1980).

182. F.W. Byron, Jr., C.J. Joachain and B. Piraux, J. Phys. B 13, L673 (1980).

183. F.W. Byron, Jr., C.J. Joachain and B. Piraux, Phys. Lett. A 99, 427 (1983); A 102, 289 (1984).

184. F.W. Byron, Jr., C.J. Joachain and B. Piraux, J. Phys. B 18, 3203 (1985).

185. C.J. Joachain, B. Piraux, R.M. Potvliege, F. Furtado and F.W. Byron, Jr., Phys. Lett. A 112, 138 (1985).

186. K.S. Balyan and M.K. Srivastava, Phys. Rev. A 32, 3098 (1985); A 33, 2155 (1986).

187. E. Curran and H.R.J. Walters, J. Phys. B 20, 337 (1987).

188. F.W. Byron, Jr., C.J. Joachain and B. Piraux, J. Phys. B 15, L293 (1982).

189. F.W. Byron, Jr., C.J. Joachain and B. Piraux, J. Phys. B 19, 1201 (1986).

190. R. Müller-Fiedler, P. Schlemmer, K. Jung and H. Ehrhardt, Z. Phys. A 320, 89 (1985).

191. H. Ehrhardt, M. Fischer, K. Jung, F.W. Byron, Jr., C.J. Joachain and B. Piraux, Phys. Rev. Lett. 48, 1807 (1982).

192. F. Furtado and P.F. O'Mahony (to be published).

193. I.E. McCarthy and E. Weigold, Phys. Rep. 27, 275 (1976); E. Weigold and I.E. McCarthy, Adv. Atom. Mol. Phys. 14, 127 (1978), E. Weigold, Comm. Atom. Mol. Phys. 15, 223 (1984).

194. A. Giardini-Guidoni, R. Fantoni, R. Camilloni and G. Stefani, Comm. Atom. Mol. Phys. 10, 107 (1981).

195. A.E. Glassgold and G. Ialongo, Phys. Rev. 175, 151 (1968).

196. B. Lohmann and E. Weigold, Phys. Lett. A 86, 139 (1981).

197. B. Van Wingerden, J.T. Kimman, M. Van Tilburg, E. Weigold, C.J. Joachain, B. Piraux and F.J. de Heer, J. Phys. B 12, L627 (1979); B. Van Wingerden, J.T. Kimman, M. Van Tilburg and F.J. de Heer, J. Phys. B 16, 4203 (1983).

198. F.W. Byron, Jr., C.J. Joachain and B. Piraux, J. Phys. B 16, L769 (1983).

199. J.S. Briggs, J. Phys. B 19, 2703 (1986).

200. A. Pochat, R.J. Tweed, J. Peresse, C.J. Joachain, B. Piraux and F.W. Byron, Jr., J. Phys. B 16, L775 (1983).

201. B.H. Bransden and C.J. Joachain, Physics of Atoms and Molecules (Longman, 1983), Chapter 4.

202. P.G. Burke in Atomic Processes and Applications, ed. by P.G. Burke and B.L. Moiseiwitsch (North Holland, Amsterdam, 1976) p. 199.

203. A.F. Starace, in Encyclopaedia of Physics, Vol. 31, ed. by W. Mehlhorn (Springer-Verlag, Berlin, 1982) p.1.

204. H.P. Kelly, in Photoionization and other Probes of Many-Electron Interactions, ed. by F.J. Wuilleumier (Plenum, New York, 1975) p.75.

205. M. Ya. Amusia and N.A. Cherepkov, Case Stud. At. Phys. 5, 47 (1975).

206. G. Wendin, J. Phys. B 4, 1080 (1971), B 5, 110 (1973), B 6, 42 (1973).

207. U. Fano, Phys. Rev. 124, 1866 (1961); U. Fano and J.W. Cooper, Phys. Rev. A 137, 1364 (1965).

208. R.P. Madden and K. Codling, Phys. Rev. Lett. 10, 516 (1963).

209. H.C. Bryant, B.D. Dieterle, J. Donahue, H. Sharifian, H. Tootoonchi, D.M. Wolfe, P.A.M. Gram and M.A. Yates-Williams, Phys. Rev. Lett. 38, 228 (1977).

210. J. Macek, Proc. Phys. Soc. 92, 365 (1967).

211. J.T. Broad and W.P. Reinhardt, Phys. Rev. A 14, 2159 (1976).

212. J.W. McGowan, J.F. Williams and F.K. Carley, Phys. Rev. 180, 132 (1969).

213. J.F. Williams and B.A. Willis, J. Phys. B 7, L61 (1974).

214. S.S. Tayal and P.G. Burke, J. Phys. B. (to be published).

215. J.L. Kohl and W.H. Parkinson, Astrophys. J. 184, 641 (1973).

216. R.A. Roig, J. Phys. B 8, 2939 (1975).

FREE-FREE TRANSITIONS OF ELECTRON-ATOM SYSTEMS IN INTENSE RADIATION FIELDS

M. Gavrila

FOM-Institute for Atomic and Molecular Physics
Kruislaan 407, 1098 SJ Amsterdam
The Netherlands

1. INTRODUCTION

1.1. Summary

We present a survey of free-free transitions (FFT) in electron-atom collisions occurring in a radiation field. Our emphasis is on theoretical aspects, although we also discuss the existing experimental evidence. The presentation is self-contained and rather detailed (we actually derive most of the basic formulas). The semiclassical treatment is adopted for the interaction of the particles with the radiation field. We envisage intensities ranging from the relatively weak ones of optical sources and CW lasers, to those of the superintense lasers now in operation, yielding values in excess of one atomic unit ($3.5 \cdot 10^{16}$ W/cm^2). Therefore, we consider both the standard perturbation theory, applicable to the former case, and the nonperturbative methods of solution developed for the latter case, which are now in the center of attention. We handle all these by a unified method, based on Floquet theory. This applies to the case of a monochromatic field, the only one we shall discuss in detail.

Since we are interested here primarily in presenting methods of solution, we shall reduce the description of atomic structure to a bare minimum: the atom will be represented by a potential. Whereas in perturbation theory there exist refined FFT computations taking into account details of atomic structure, this is not yet true for the nonperturbative calculations. In the latter case, because of the complexity of the problem, even the potential model has not been fully mastered. It is only peripherally that we shall deal with the role of atomic structure.

The layout of the survey is as follows. The physics of the process and its relevance is considered in Sec. 1.2. We continue by describing the existing experiments (Sec. 1.3). We dwell then on some rather delicate issues required for their interpretation (Sec. 1.4). Chap. 2 is devoted to the general theoretical framework. The electrodynamic gauge problem is considered in Sec. 2.1. The Schrödinger equation for the moving Kramers reference frame is given in Sec. 2.2. By specializing now to a monochromatic field, we can treat the equations of motion by the Floquet method. We outline it first, and then apply it to the forms of interest of the Schrödinger equation (Sec. 2.3). The appropriate boundary condition is discussed in Sec. 2.4. These may be used to cast the Floquet differential system in an integral form, which is a generalized version of the Lippmann-Schwinger equation; FFT amplitudes are thus obtained (Sec. 2.5). Chap. 3 contains the application to FFT of the standard perturbation theory, which yields expansions in powers of the intensity. The basic formulas for the amplitudes are derived and the graphical representation of Feynman is specialized to the present case; Born approximations formulas are also given (Sec. 3.1). In Sec. 3.2 we quote results obtained with the potential description of the electron-atom interaction. Chap. 4 contains the nonperturbative methods. These imply expansions in parameters other than the intensity, or are direct numerical approaches. Sec. 4.1 contains the case of high electron energies and the Born expansion (valid in this case to all orders in the intensity). Sec. 4.2 contains the case of low frequencies, whose leading term is the celebrated Kroll and Watson formula. In Sec. 4.3 we present the opposite case of high frequencies, recently developed, and give an illustrative application. Finally in Sec. 4.4 we discuss other nonperturbative approaches for the potential model, and refer to some of the work done taking into account the detailed atomic structure.

The references we shall be giving are divided into general [G], experimental [E] and theoretical [T]. Among the general ones, we mention here those most relevant for high-intensity FFT: the articles by Rosenberg [G1], and the books by Mittleman [G2] and Faisal [G3]. These contain either alternative points of view on the problem, or aspects not dealt with here.

1.2. Elementary process. Relevance

According to classical electromagnetic theory an electron scattered by an atom (ion) in a field will absorb or emit radiation due to its acceleration. In quantum electrodynamics (QED) this is restated in terms of probabilistic absorption or emission of photons. In fact, in an elementary process, not only one but several photons can be involved, although the corresponding probabilities depend critically on the radiation intensity. In weak fields, of the kind encountered in nature or in conventional optical sources, only one-photon processes have a large enough probability to be detected with present capabilities. For the detection of multiphoton processes, rather intense fields are required, and it was only after the advent of the laser that they could be observed.

Since these radiative collisions involve continuum states of the electron+atom (ion) system, they are called "free-free transitions". We are interested here in *stimulated* FFT, that is in stimulated bremsstrahlung and "inverse bremsstrahlung" (FF photoabsorption); spontaneous bremsstrahlung is entirely negligible in intense fields.

Energy conservation for the elementary process requires that

$$E_f + W_f = E_i + W_i + n\hbar\omega \ . \tag{1.1}$$

We have denoted here by E_α and W_α the initial ($\alpha = i$) and final ($\alpha = f$) energies of the electron and atom respectively, and n is the number of photons involved ($n = 0, \pm1, \pm2, ...$). In Eq.(1.1), n is positive for absorption, negative for emission, and zero for elastic scattering (no net photon exchange). The process may be accompanied by atomic excitation ($W_f > W_i$), or the atom may be left unperturbed ($W_f = W_i$). In the latter case, or when dealing with a structureless target (potential), Eq.(1.1) reduces to

$$E_n = E + n\hbar\omega \ , \qquad (n = 0, \pm1, \pm2, ...) \tag{1.2}$$

with $E_n = E_f$, $E = E_i$.

The physics of the (multiphoton) FFT process is closely related to that of (multiphoton) single electron ionization, since the ionization can be regarded as a "half-collision". However, the fact that for FFT the initial state is free makes the problem harder to handle, both in experiment and theory.

Spontaneous bremsstrahlung has been studied since the early days of atomic physics due to its primary relevance. Interest in stimulated FFT has built up more slowly and originates in astrophysics. It was suggested first by Pannekoek in 1931 [T1] that FFT on H atoms play an important role in the opacity of the atmospheres of the cooler stars (i.e. the attenuation of the radiation flux coming from the core), such as the sun. Although it was later recognized that the dominant absorption mechanism throughout the UV and visible was photodetachment from H^-, it is nevertheless true that FFT are the main absorption cause in the infrared, at wavelengths longer than $\lambda = 1642$ nm corresponding to the photodetachment threshold. This was confirmed by astrophysical observations, after the first successful calculation of (one-photon) FFT on H atoms was carried out by Chandrasekhar and Breen in 1946 [T2]. Among all elements H is the predominant constituent of stellar matter and has attracted most attention, but a number of other elements are also relevant for astrophysical FF absorption (mainly He, and to some extent H_2, C, Ne, Cl). A discussion of the role of FFT in astrophysical problems was given by Myerscough and Peach [G6].

The astrophysical interest has stimulated a great deal of theoretical research on (one-photon) FFT from the relevant atoms, and quite sophisticated computations were carried out. For a review of this work (prior to 1978) see Gavrila [G4], and Gavrila and van der Wiel [G5]. A rather complete bibliography (prior to 1977) was compiled in Ref. [G7].

More generally, FFT are responsible for part of the emission and absorption of hot gases and plasmas. In this case many elementary processes contribute along with FFT to the overall effect, as described in the review article by Biberman and Norman [G8]. For highly ionized plasmas and long wavelengths, FFT involving positive ions are predominant but for smaller fractional ionizations (of about 10^{-3} or less) neutral atom FFT become important. A great deal of experimental work was done on the emission and, to a lesser extent, absorption of hot gases and plasmas under laboratory conditions. A detailed review of the studies concerning radiation from negative atomic ions was given by Popp [G9]. Few experiments, however, have been directed specifically at studying FFT. Their agreement with theory was only fair.

The advent of the laser has been a milestone for the relevance of the FFT process. Even weak lasers for present standards (yielding some 10^4 W/cm^2 in the focal spot at a given frequency) are orders of magnitude more intense than the brightest conventional sources, allowing thus a great enhancement in the FFT yields. From the start much interest has been aroused by the possibility of *laser heating of plasmas*. At normal pressure, a plasma is formed in a gas already at about 10^{11} W/cm^2 in the form of a spark at the focus of the beam ("gas breakdown", see Grey Morgan [G10]). It is generally accepted that once a moderate degree of ionization has been achieved (e.g. by multiphoton ionization of the neutrals), FFT are the main process which heats the plasma in the incipient stage.

In recent times great progress has been achieved towards obtaining very intense laser pulses, of picosecond duration or less. Intensities have been generated in the focal spot in excess of 10^{16} W/cm^2, at selected wavelengths ranging from the IR to the UV, and much higher intensities are expected in the near future. We are thus at the level of the atomic unit of intensity $I_0 = 3.51 \cdot 10^{16}$ W/cm^2, which is defined as the (time averaged) intensity corresponding to an electric field amplitude equal to the electrostatic field on the first Bohr orbit of hydrogen. Such intensities are apt to produce drastic perturbations in the outer shells of atoms. Many electrons will be ripped off by multiphoton ionization, giving rise to highly charged ions. The free electrons of the plasma will, therefore, collide with a statistical distribution of ions of different charges, rather than with neutrals. This in fact enhances the probabilities of FFT, because the atomic field of the ions extends much farther than that of the neutral atom.

However, experiments concerning plasma-radiation yield meagre and uncertain information on the FFT process because of the multitude of the other simultaneous processes. The only way to get direct information on the elementary process is to perform *beam experiments*. In a three beam experiment, an atomic beam is crossed in coincidence by a laser and an electron beam, and the scattered electrons, having undergone FFT, are recorded. Such experiments are quite difficult because of the very low counting rates. They will be described in the following subsection.

1.3. Beam experiments

They have been carried out over the years by two groups: Andrick and collaborators Langhans and Bader, have studied *one-photon FFT* since 1976, [E1]-[E5]; Weingartshofer and collaborators Clarke, Holmes, Jung and Sabbagh, [E6]-[E9], have studied *multiphoton FFT* since 1977, and, more recently, with Wallbank, Connors and Holmes, also one-photon transitions [E10].

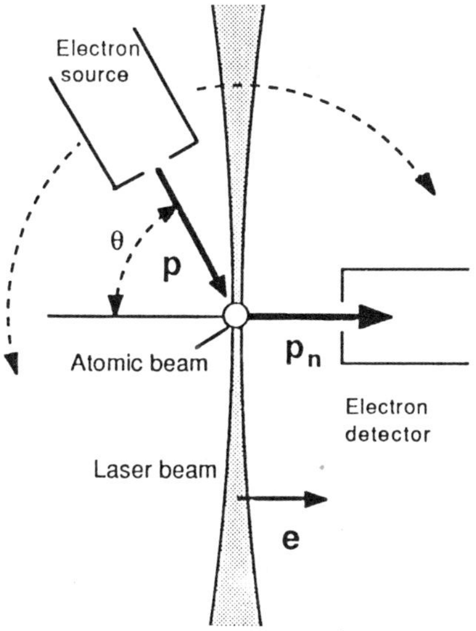

Fig. 1. Scattering geometry used in FFT experiments, e.g. see Ref. [E4]. The scattered electrons (momentum $\mathbf{p}_n$), having absorbed or emitted n photons, are detected in the plane determined by the incident electron beam (momentum **p**) and the laser beam, represented here in the plane of the figure. The atomic beam is perpendicular to the scattering plane.

All experiments were done with a similar scattering geometry and experimental set-up. The scattering geometry is shown in Fig. 1: the atomic beam is perpendicular to the plane defined by the incoming electron beam and the laser beam (which are both represented in the plane of the figure). The (inelastically) scattered electrons are detected in the same plane, at scattering angle θ, by an electron spectrometer, the resolution of which was at best 30-40 meV. The linear polarization of the photon beam was chosen mostly to be parallel to the scattering plane. Energy loss/gain spectra were detected for the final electron with the laser on, and compared to the spectrum with the laser off.

In order to improve on the counting rates, advantage was taken of the fact that (at not too high intensities) the theoretical FFT cross sections are strongly enhanced at low frequencies (see Eqs.(3.34)). Therefore, a CO_2 laser was chosen in all experiments (wavelength $\lambda = 10.6$ μm, photon energy $\hbar\omega = 117$ meV).

The main difference between the approaches of the two groups lies in the kind of laser used and its power. Andrick and collaborators have used a cw CO_2 laser of low intensity, yielding some 10^4 - 10^5 W/cm^2. Under these circumstances only one-photon FFT could be detected. On the other hand, the experimental results could be interpreted quite satisfactorily in terms of the accurate low-frequency cross sections derived from perturbation theory (see Sec.3.2 and references therein). Weingartshofer and collaborators have used a higher intensity pulsed CO_2 laser, yielding some 10^8 W/cm^2 (with the exception of Ref. [E10]). This has enabled the detection of FFT with many photons absorbed or emitted. However, because of the randomness of the multimode operation of the laser, the results were harder to interpret theoretically.

One-photon FFT. The first experiment of the series was carried out by Andrick and Langhans [E1],[E2]. The target atoms were of Ar, the incident electron energy was taken $E_i = 10.6$ eV, and the scattering angle θ fixed at 160° (from Eq.(3.34) it follows that the cross section is enhanced at large momentum transfers which are nearly parallel to the polarization vector, see Fig. 1). With the laser off, only the elastic scattering peak was recorded (see Fig. 2). With the laser on, two relatively quite small shoulders appear, located around the energies $E_i \pm 117$ meV. These represent the electrons having absorbed/emitted one photon. The experiment provided the first direct proof of existence for FFT.

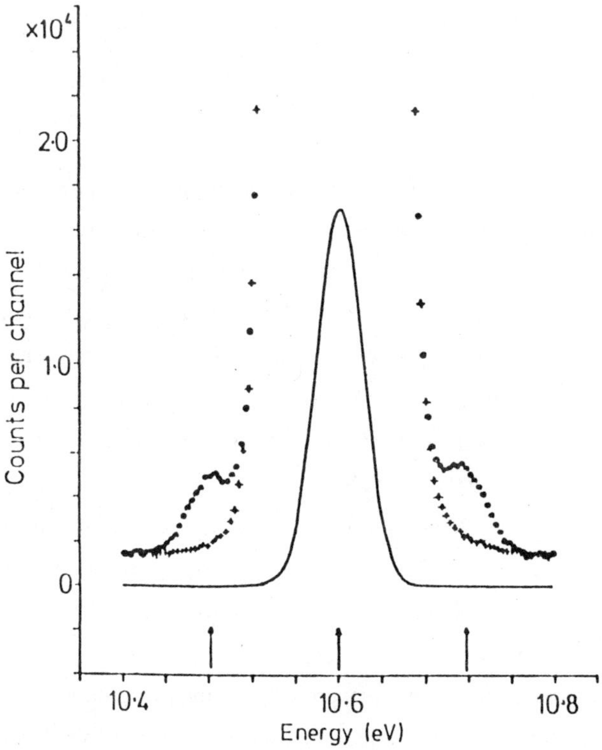

Fig. 2. Experiment of Andrick and Langhans, Ref. [E2]. Energy loss/gain spectrum of electrons scattered at $\theta = 160°$ by Ar in the absence (crosses) and presence (circles) of a CO_2 laser beam (intensity 2.10^3 W/cm^2; photon energy 117 meV). The full curve is a 1/500 reduction of the elastic peak.

In a subsequent experiment, Langhans [E3] has shown the occurence of resonance effects in FF scattering, see Fig. 3. The atoms were again Ar, the scattering angle again $\theta = 160°$, but E_i was varied in the range 11.0 - 11.5 eV, which contains two resonances in elastic scattering. With the laser off, the upper sequence of open circles was recorded for elastic scattering (arbitrary units); note the two dips representing window-type resonances, whose energies we shall denote by E_{R_1} and E_{R_2} ($E_{R_1} < E_{R_2}$). The full curve represents a theoretical evaluation. With the laser on, the final electrons having *lost* the energy of one photon are detected ($E_f = E_i$ - 117 meV), when the incident energy is varied. When E_i is increased, it will pass at one point the first resonant energy E_{R_1} ($E_i = E_{R_1}$), followed afterwards by E_f ($E_f = E_{R_1}$). Now, the low-frequency theory of FFT (see Sec. 3.2 and references therein) indicates that the cross section is expected to resonate when either the initial energy E_i or the final energy $E_f = E_i$ - 117 meV coincides with E_{R_1} or E_{R_2}. Therefore *two* resonances are expected to be detected in the final electron energy spectrum, situated 117 meV apart, for *each* of the elastic scattering resonances E_{R_1} and E_{R_2}. This is clearly borne out by the experimental yields, shown by the lower sequence of crosses in Fig. 2 (arbitrary units). Moreover, using the low-frequency theory of FFT the full lower curve was drawn on the basis of the theoretical upper curve for elastic scattering, indicating good agreement between theory and experiment (up to a scaling factor).

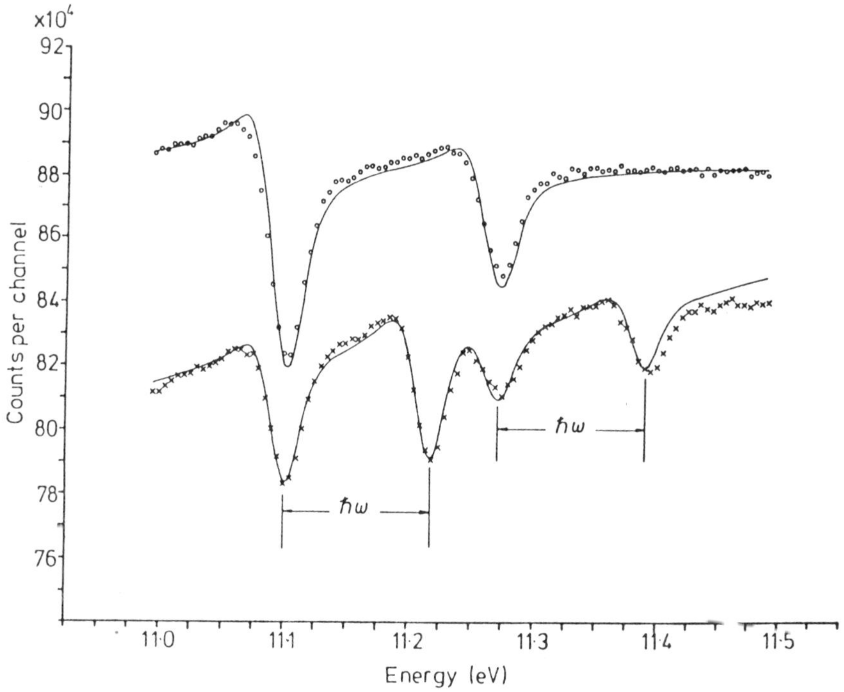

Fig. 3. Experiment of Langhans, Ref. [E3]. Energy dependence of the e-Ar angular differential cross section (arbitrary units) in the resonance region E_i = 11.0 - 11.5 eV, at scattering angle $\theta = 160°$, for: a) Elastic scattering (laser off); upper sequence of open circles: experiment; upper full curve: theory; b) One-photon FF emission of a CO_2 laser photon (laser characteristics as in Fig. 1); lower sequence of crosses: experiment; lower full curve: theory.

In later experiments Andrick and Bader [E4], and then Bader [E5], have extended the Langhans experiment and its interpretation procedure to the case of He and Ne, respectively. Also smaller scattering angles were considered (in the range from 20° to 70°), at which the counting rates are considerably smaller. In general, satisfactory agreement with the low-frequency FFT theory was found (again, up to a scaling factor).

In a recent experiment Wallbank et al. [E10], using Ar atoms as target and electrons of 10.55 eV, have studied one-photon FFT in the intensity regime where the lowest order perturbation theory cross section, proportional to the intensity I (see Sec. 3.2), breaks down. It was demonstrated that above some $2 \cdot 10^6$ W/cm^2 the cross section increases more slowly than linearly with I, in agreement with the trend predicted by the nonperturbative theories developed for low frequencies (see Sec. 4.2).

Multiphoton FFT. Weingartshofer and collaborators, [E6]-[E9], were the first (and only) to demonstrate the existence of multiphoton FFT, by using a pulsed CO_2 laser yielding about 10^8 W/cm^2 in the focal spot. The target atoms were Ar, E_i was first 11.72 eV and then 15.80 eV, the scattering angle θ was 153° and 155°. By upgrading the characteristics of their experimental set-up over the years they could eventually observe as many as 11-photon FF emission and absorption transitions. This is illustrated in Fig. 4 where the open circles indicate the measurements done. Note that because of the long recording times, the measurements were carried out only at final energies E_f differing from E_i by integer or half-integer values of the photon energy. (Even so, some 15 to 24 hours were needed for a complete spectrum.)

As seen in Fig. 4, the relative intensities of two successive FFT peaks (including that for laser modified elastic scattering) are of the same order of magnitude: this shows that we are dealing with a highly nonperturbative regime (compare with Fig. 2, which is typical for the perturbation theory regime). On the other hand, no direct comparison can be made with the prototype of nonperturbative cross sections at low frequencies, the formula of Kroll and Watson (see Sec. 4.2). This is because the latter result was derived for a purely single mode field, whereas the laser used in experiment was in (unknown) multimode operation. (This means that the detector collects signals from FFT events at various space-time points with quite different intensities.) There exists, however, a feature of intense FF scattering at low frequencies which is largely independent of the laser mode operation: the sum rule Eq.(4.12) (see the discussion in Sec. 4.2). It states that the intensities (number of counts) of all FFT peaks should add up to the elastic intensity with the laser off. This could be quantitatively checked. For example, in the case of Fig. 4 the sum of counts for all peaks was 436, whereas the number with the laser off was 431. The improvements of the experimental set-up required to achieve a better comparison with theory were discussed by Weingartshofer and Jung [E9].

The experiments described above were done over a period of more than ten years. A fact which attracts attention is their scarcity. This is even more blatant, when compared to the impressive amount of work done over the same period on the closely related phenomenon of

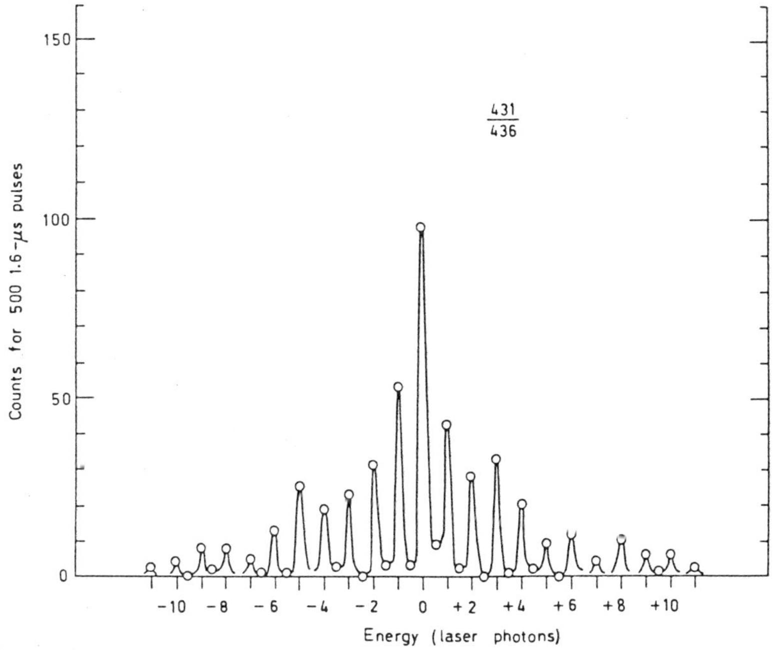

Fig. 4. Experiments of Weingartshofer, Holmes, Sabbagh and Chin, Refs. [E8], [E9]. Energy-loss/gain spectrum of electrons scattered by Ar atoms in a CO_2 laser beam (open circles; the full line was drawn to lead the eye). The abscissa gives the final electron energy in units of the photon energy (117 meV) with the origin fixed on the initial energy, $E_i = 15.80$ eV. Scattering angle $\theta = 155°$, effective laser intensity 10^8 W/cm^2.

multiphoton ionization (for the state of the art in that research and for references, see Refs. [G11] and [G12]). On the other hand, substantially more work was done on the theoretical aspects of multiphoton FFT, as shall become apparent in the following.

1.4. Comments on theoretical premises

In this section we shall discuss a number of theoretical issues with the intention of providing a background for the formal framework to be set up in Chap. 2.

The theoretical description of multiphoton FFT at relatively low intensities can be done (in general) using the formulas of standard *perturbation theory* to lowest nonvanishing order in the interaction of the electron with the radiation field. This is how the one-photon FFT results of astrophysical interest were obtained. The difficulty of such calculations lies in the accurate handling of atomic structure. At high intensities, however, these formulas become insufficient and new, *nonperturbative methods* of solution of the Schrödinger equation need be developed. This represents a considerable theoretical challenge, as one now has to face both the problem of atomic structure and that of the nonperturbative coupling. Besides, to compare with experiment, the structure of the laser pulses should be taken into account. Nevertheless, a wealth of the structure of the laser pulses should be taken into account. Nevertheless, a wealth of theoretical results has been published. On the other hand, in terms of quantitative results, considerably less has been achieved than in the case of weak fields.

There are two possibilities for treating the radiation field and its coupling to the particles. The entirely correct one (according to present day knowledge) is that of QED, which is based on the quantization of the radiation field. Here, the photons are built into the theory from the start, and one can fully account for spontaneous emission phenomena. However, general results show that at the intensities of laser sources, the radiation field can be treated classically (e.g. see I. Bialynicki-Birula and Z. Bialynicka-Birula [T3], Mollow [T4], and also Mittleman [G2] Chap. 1). This yields the *semiclassical interaction theory*. In comparison to QED it has the advantage of a simpler formalism. On the other hand, since photons do not appear automatically in this approach, the results have to be *interpreted* in terms of discrete photon processes. Semiclassical theory being fully justified at the intensities we are interested in, will be adopted in the following.

We shall assume further that we may neglect the spatial dependence of the phases of the electromagnetic waves (i.e. retardation) in the vicinity of the atom, where the absorption and emission processes take place. This is the *dipole approximation*, which is valid if the wavelengths involved are considerably larger than the linear dimensions of the atom. It is satisfied down to soft X rays (unless one is dealing with extremely extended structures, such as Rydberg atoms).

The weak fields appearing in astrophysics, or generated by conventional optical sources, extend over macroscopic space regions and time durations. It is easy to conceive in this case that an electromagnetic plane wave may represent an ideal limiting case. This also applies to continuous wave lasers. For the very intense, short *laser pulses* now used in experiments, the plane wave representation may encounter difficulties. These pulses have had durations ranging from nanoseconds (10^{-9} sec) to picoseconds (10^{-12} sec). Although very short, these durations are quite long (even in the picosecond case) in comparison to optical periods in the UV or IR (e.g. some 10^{-15} sec). One may then still represent the pulses by quasi-monochromatic plane waves, now with slowly ("adiabatically") varying amplitudes. The difficulty with some of the high-intensity pulsed lasers is that many laser modes may be amplified (multimode operation), which introduces the need to account for the light statistics of the field. This may considerably alter single mode results. In the following we shall confine the theory to the case of *single mode radiation,* especially since the technological trend is directed towards obtaining such "clean" pulses and, in many cases, is getting close to achieving them. Occasionally, we shall quote results for multimode operation.

An important theoretical issue is that of the *asymptotic states* of the scattering problem, i.e. the description of the electron entering or leaving the collision. From the point of view of FFT, space can be divided into three regions. We shall enumerate them from the inside accelerated by the atomic forces and the absorption/emission of photons takes place. It has roughly the size of the atom, i.e. linear dimensions of about 10^{-10} m. Region 1 is immersed in the radiation field, extending over a region which coincides with the laser focus in beam experiments (region 2). This may have linear dimensions varying from about 100 μm down to the wavelength of the radiation, say 1 μm. It is orders of magnitude larger than region 1. In region 2 the electron approaching or leaving the atom is only under the influence of the field. Finally, we have the external world (region 3), where no field is assumed to exist, and the electron, shot into the experiment or detected when leaving it, is entirely free.

The asymptotic states may be chosen in two ways. Since the electron is entering the reaction volume (region 1) from the surrounding laser focus (region 2), one may choose as asymptotic states, field-driven free electron states, the so-called "dressed" free particles waves (plane or spherical), see Secs. 2.2 and 2.4. We denote this as *method A*. It can be applied, in principle, to arbitrarily intense fields.

Assuming the collision problem solved in regions 1 and 2, we still have to answer the question of what happens to the electron when passing between the laser focus and the external world. In cases where the field is very intense in the focal region (some 10^{13} W/cm^2 or more), the strong gradient of the electric field amplitude leads to typical "ponderomotive forces", pushing the electron out of the focus. This can severly distort the well defined initial momentum of the electron entering the collision from region 3, and its true angular distribution when emerging from region 2. Experiments proving this point have been carried out by Freeman et al. [E11]. Under these circumstances a detailed knowledge of the shape of the laser pulse is needed in order to be able to extract true angular distributions from experiment.

If the laser pulse is not very intense and has a slowly varying amplitude in space, the ponderemotive forces may be neglected and no difficulties arise when passing between regions 2 and 3. Moreover, the "dressed" free particle waves of region 2 do not differ much from the usual free particle waves existing in region 3. One may then take the latter as asymptotic states and ignore the existence of region 2 altogether (i.e. ignore that the laser focus extends beyond region 1). This we denote as *method B*. Because one now has to assume that the field is sufficiently weak, this method is appropriate for the choice of the asymptotic states in perturbation theory.

2. THEORETICAL FRAMEWORK

2.1. The gauge problem

The general procedure to derive the Schrödinger equation in the presence of an electromagnetic field is to replace the energy-momentum operator of each particle $(i\hbar\,\partial/\partial t, \mathbf{P})$ by $\{i\hbar\,\partial/\partial t - e\phi, (\mathbf{P} - (e/c)\,\mathbf{A})\}$, where $(\phi(\mathbf{r},t),\mathbf{A}(\mathbf{r},t))$ are the electrodynamic potentials describing the field, and e is the algebraic value of the charge. This is the "minimal electromagnetic coupling". In the case of one single particle in a potential V(**r**) we get:

$$[\frac{1}{2m}(\mathbf{P} - \frac{e}{c}\mathbf{A})^2 + V + e\phi]\,\psi = i\hbar\frac{\partial\psi}{\partial t} \quad . \tag{2.1}$$

We are using here Gauss-CGS units. The conserved localization probability density, and its associated probability current density are now

$$\rho = \Psi^*\Psi ,$$

$$\mathbf{j} = \frac{1}{2m}[\Psi^*\mathbf{P}\Psi - \Psi\mathbf{P}\Psi^*] - \frac{e}{mc}\mathbf{A}\Psi^*\Psi \quad . \tag{2.2}$$

The form of the potentials (ϕ,**A**) - the electrodynamic "gauge" - is derived from the fields **E**,**H** through the equations

$$\mathbf{E} = -\nabla\phi - \frac{1}{c}\frac{\partial\mathbf{A}}{\partial t} , \qquad \mathbf{H} = \mathrm{rot}\,\mathbf{A} \quad . \tag{2.3}$$

These define the gauge only up to a *"gauge transformation"*

$$\mathbf{A}' = \mathbf{A} + \nabla f , \qquad \phi' = \phi - \frac{1}{c}\frac{\partial f}{\partial t} , \tag{2.4}$$

depending on an arbitrary real function f(**r**,t).

By appropriately changing the wave function Ψ, the Schrödinger Eq.(2.1) and the probability densities of Eq.(2.2) can be shown to be gauge invariant. This can be proven by recalling that, under the unitary, time dependent transformation

$$\Psi' = U(t)\Psi , \tag{2.5}$$

the Hamiltonian equation

$$H\Psi = i\hbar\frac{\partial\Psi}{\partial t} , \tag{2.6}$$

goes over into

$$H'\Psi' = i\hbar\frac{\partial\Psi'}{\partial t} , \tag{2.7}$$

where the transformed H' is defined as

$$H' = UHU^+ - i\hbar\,U\frac{\partial U^+}{\partial t} \quad . \tag{2.8}$$

By choosing

$$U = \exp[\frac{ie}{\hbar c} f(\mathbf{r},t)] \tag{2.9}$$

it is easily seen that Eqs.(2.1) and (2.2) retain their form. Because of the fundamental character of this transformation, any operator Q corresponding to an observable should be (gauge) invariant under the transformation $Q' = UQU^{+}$. An example is the kinetic momentum operator $\vec{\Pi} = \mathbf{P} - (e/c)\,\mathbf{A}$, corresponding to the classical mv. (N.B. Because of word-processing limitations we shall mark by arrows, instead of boldface, the Greek characters representing vectors.)

By assuming the *dipole approximation* (see Sec. 1.4) for the electromagnetic field (with no sources at finite distance), it is possible to find a gauge in which the electrodynamic potentials depend only on t. Eq.(2.1) can then be simplified by changing to a new gauge in which we have also $\phi' = 0$ (this does not determine fully the gauge, however). The new Hamiltonian is

$$H' = \frac{1}{2m}\,(\mathbf{P} - \frac{e}{c}\,\mathbf{A}')^2 + V \ . \tag{2.10}$$

Since $\mathbf{A}'$ depends only on t, we have div $\mathbf{A}' = 0$ and the $\mathbf{A}'^2$ term in the Schrödinger equation Eqs.(2.7) and (2.10) can be removed by extracting a time dependent phase factor from the wave function Ψ', according to:

$$\Psi'' = \Psi' \exp\,\left(\frac{ie^2}{2mc^2\hbar}\int_0^t \mathbf{A}'^2(t')\;dt'\right) \ . \tag{2.11}$$

The Hamiltonian becomes (by dropping the prime on **A**):

$$H'' = \frac{1}{2m}\,\mathbf{P}^2 - \frac{e}{mc}\,\mathbf{A}\cdot\mathbf{P} + V \tag{2.12}$$

This type of gauge, which couples **A** to **P** in the Hamiltonian Eq.(2.12), will be called the *momentum gauge*.

An alternative possibility for writing Eq.(2.1) (or Eq.(2.10)) in the dipole approximation is to pass to a gauge in which in Eq.(2.4) we take $\mathbf{A}' = 0$. This determines $f = -\mathbf{A}(t)\cdot\mathbf{r}$ up to a constant, and gives $\phi' = -\mathbf{E}\cdot\mathbf{r}$. The U in Eq. (2.9) is now

$$U = \exp\,(-\,\frac{ie}{\hbar c}\,\mathbf{A}\cdot\mathbf{r}) \ . \tag{2.13}$$

and the transformed Hamiltonian becomes

$$H' = \frac{1}{2m}\,\mathbf{P}^2 + V - e\mathbf{E}\cdot\mathbf{r} \ . \tag{2.14}$$

The gauge is now well defined; we shall call it the *length gauge* because it couples **E** to **r**. The associated kinetic momentum operator is $U\vec{\Pi}U^{+} = \mathbf{P}$, and the current is given by Eq.(2.2) with **A** set equal to zero.

2.2. Space-translated Schrödinger equation

Let us first consider the case of a *"free" electron* in a radiation field. By setting V = 0 in the Schrödinger equation corresponding to Eq.(2.12) we get

$$\left(\frac{1}{2m}\,\mathbf{P}^2 - \frac{e}{mc}\,\mathbf{A}\cdot\mathbf{P}\right)\chi = i\hbar\,\frac{\partial\chi}{\partial t} \,. \tag{2.15}$$

This has the exact solution

$$\chi_{\mathbf{p}}(\mathbf{r},t) = N \exp \frac{i}{\hbar}\left\{\mathbf{p}\cdot(\mathbf{r} - \vec{\alpha}(t)) - \frac{p^2}{2m}\,t\right\} , \tag{2.16}$$

where **p** is a constant vector, and

$$\vec{\alpha}(t) = -\frac{e}{mc}\int_0^t \mathbf{A}(t')\,dt' \,. \tag{2.17}$$

$\chi_{\mathbf{p}}$ is an eigenfunction of the canonical momentum, $\mathbf{P}\chi_{\mathbf{p}} = \mathbf{p}\chi_{\mathbf{p}}$. The totality of the $\chi_{\mathbf{p}}$ forms a complete set at all t; in fact (apart for a time dependent phase factor) it is just the Fourier system. The general solution for Eq.(2.15) can thus be written as

$$\chi(\mathbf{r},t) = \int c(\mathbf{p})\,\chi_{\mathbf{p}}\,(\mathbf{r},t)\,d\mathbf{p} \,, \tag{2.18}$$

with constant coefficients c(**p**).

$\chi_{\mathbf{p}}$ will be called a "dressed" plane wave, because it contains the effect of the field on the electron exactly. It is often named "Volkov wave", because it is the nonrelativistic counterpart of an exact relativistic solution derived by Volkov in 1935 [T5] for the Dirac equation.

The significance of the solution Eq.(2.16) can be better understood by adopting a different point of view. Let us first consider a classical electron under the same circumstances. It is acted upon by the Lorentz force **F** = e**E** (**H** = 0 in the dipole approximation). In our restricted gauge ($\phi = 0$) we have $\mathbf{E} = -(1/c)\partial\mathbf{A}/\partial t$, and Newton's equation becomes

$$m\,\frac{d^2\mathbf{r}}{dt^2} = -\frac{e}{c}\,\frac{\partial\mathbf{A}}{\partial t} \,. \tag{2.19}$$

Its solution is

$$\mathbf{r}(t) = -\frac{e}{mc}\int_0^t \mathbf{A}(t')\,dt' + \mathbf{v}_0 t + \mathbf{r}_0 \,. \tag{2.20}$$

By comparing now with Eq.(2.17) we see that $\vec{\alpha}(t)$ describes that part of the classical motion which is due to the field. In practice, because of the oscillatory nature of the electromagnetic field, it has the character of a *quiver motion* (e.g. see Eq.(2.27) below), and this is how we shall be referring to it.

We can now define a frame of reference translated by the vector $\vec{\alpha}(t)$ with respect to the laboratory frame, i.e. which follows the quiver motion of the classical electron. We shall call this the *Kramers frame of reference*. Obviously, Newton's equation Eq.(2.19) becomes in this frame the free particle equation: $m\, d^2\mathbf{r}/dt^2 = 0$.

The Schrödinger equation Eq.(2.15) applies to the laboratory frame of reference. We shall now derive its form for the Kramers frame. Since this is obtained by the *space translation* $\mathbf{r} \to \mathbf{r} + \vec{\alpha}(t)$, the new wave function $\chi'(\mathbf{r},t)$ is related to the laboratory one $\chi(\mathbf{r},t)$ by the unitary operator

$$U = \exp\left(\frac{i}{\hbar}\, \vec{\alpha}(t) \cdot \mathbf{P}\right) , \tag{2.21}$$

acting as

$$\chi'(\mathbf{r},t) = U\chi(\mathbf{r},t) = \chi(\mathbf{r} + \vec{\alpha}(t), t) . \tag{2.22}$$

Using Eqs.(2.5),(2.7),(2.8), Eq.(2.15) becomes

$$\frac{1}{2m}\, \mathbf{P}^2\, \chi' = i\hbar\, \frac{\partial \chi'}{\partial t} . \tag{2.23}$$

This is the usual free particle Schrödinger equation without the field. Its solutions are the usual plane waves, which are connected to the $\chi_{\mathbf{p}}$ of Eq.(2.16) by the transformation Eq.(2.21). We have thus derived the quantum mechanical analogue of the classical case discussed above.

Returning to the general case of a *bound electron*, whenever the binding is weak (i.e. the electron can be considered as nearly free) it is useful to apply the space-translation and go over to the Kramers frame. Starting from Eq.(2.12) and applying the transformation Eq.(2.21), the space-translated Schrödinger equation is obtained in terms of the Hamiltonian:

$$H''' = \frac{1}{2m}\, \mathbf{P}^2 + V(\mathbf{r} + \vec{\alpha}(t)) . \tag{2.24}$$

This contains the potential $V(\mathbf{r} + \vec{\alpha}(t))$, the center of which is subjected to the quiver motion $-\vec{\alpha}(t)$, which is quite natural in view of the fact that it is at rest in the laboratory frame. Obviously, when passing to the Kramers frame, the operators of the observables must be transformed accordingly, e.g. $\mathbf{P}' = U\mathbf{P}U^+ = \mathbf{P}$, $\vec{\Pi}' = U\vec{\Pi}U^+ = \vec{\Pi}$.

The space-translated Schrödinger equation was introduced by Kramers in 1948 [T6]; it was later rediscovered by Henneberger in 1968 [T7], see also Faisal (1972) [T8].

2.3. Monochromatic radiation. The Floquet method.

From now on we shall specialize to the case of monochromatic (single mode) radiation of frequency ω. For simplicity we shall limit ourselves throughout this paper to the case of *linear polarization*. We may then write

$$\mathbf{A}(t) = a\mathbf{e} \cos \omega t \ , \tag{2.25}$$

where $\mathbf{e}$ is the real polarization vector ($\mathbf{e}^2 = 1$; $\mathbf{e}\cdot\mathbf{n} = 0$, with $\mathbf{n}$ the unit vector of the propagation direction).

Eq.(2.25) gives:

$$\mathbf{E} = \frac{a\omega}{c}\, \mathbf{e} \sin \omega t \ , \tag{2.26}$$

and

$$\vec{\alpha}(t) = \alpha_0\, \mathbf{e} \sin \omega t \ , \tag{2.27}$$

where

$$\alpha_0 = -\frac{ea}{mc\omega} \ . \tag{2.27'}$$

α_0 is a length, and for an electron we have in atomic units (Bohr radii)

$$\alpha_0 = \left(\frac{I}{I_0}\right)^{1/2} \left(\frac{2\,Ry}{\omega}\right)^2 \text{a.u.} \ , \tag{2.28}$$

where I is the (time averaged) intensity of the plane wave, the atomic unit of (time averaged) intensity being $I_0 = (8\pi)^{-1}\, m^4\, e^{10}\, c\, \hbar^{-8}$ erg/sec/cm^2 = 3.51×10^{16} W/cm^2. As follows from Sec. 2.2, $\vec{\alpha}(t)$ describes the quiver motion of amplitude $|\alpha_0|$ performed by a classical charge.

We can now insert Eqs.(2.25),(2.26),(2.27) into the Schrödinger equations with the Hamiltonians Eqs.(2.12),(2.14),(2.24), respectively. We thus have three equivalent ways of treating the interaction of the electron with a potential in a monochromatic field, which have been frequently used in practice. The first two (corresponding to Eqs.(2.12) and (2.14)) are attached to the laboratory frame and differ by a gauge transformation, whereas the third is attached to the Kramers reference frame. A common characteristic is that they are partial differential equations with time dependent, *periodic* coefficients. This suggests application of a method of solution developed by Floquet for finite systems of regular differential equations (see [T9]).

We first note that all these cases fall into the form

$$[H + H'(t)]\Psi = i\hbar \frac{\partial \Psi}{\partial t} \ , \tag{2.29}$$

where H and H'(t) are operators in coordinate space, and H'(t) is periodic. (This H'(t) should not be confused with the H' of Sec. 2.1.) Following the *Floquet method*, we seek a solution of the form

$$\Psi = e^{-(i/\hbar)Et} \sum_{n=-\infty}^{+\infty} \Psi_n(\mathbf{r})\, e^{-in\omega t} \ . \tag{2.30}$$

We further expand H'(t) in a Fourier series

$$H'(t) = \sum_{n=-\infty}^{+\infty} H'_n \, e^{-in\omega t} \tag{2.31}$$

Insertion of Eqs.(2.30) and (2.31) into Eq.(2.29) leads to a system of coupled differential equations for the components $\Psi_n(\mathbf{r})$, which we write

$$(E_n - H)\,\Psi_n = \sum_m H'_{n-m}\,\Psi_m \,, \qquad (n = 0, \pm 1, \pm 2, \ldots) \tag{2.32}$$

with E_n defined as in Eq.(1.2). Eq.(2.32) contains the *Floquet system* we want to solve. Let us consider first its specific aspect for the three forms of Schrödinger equation we are interested in.

Momentum gauge . According to Eqs.(2.12) and (2.25), we take

$$H = \frac{1}{2m}\mathbf{P}^2 + V, \qquad H' = -\frac{ea}{mc}\,(\mathbf{e}\cdot\mathbf{P})\cos\omega t \,. \tag{2.33}$$

Hence

$$H'_n = -\frac{ea}{2mc}\,(\mathbf{e}\cdot\mathbf{P})\,(\delta_{n1} + \delta_{n,-1}) \,, \tag{2.34}$$

and Eq.(2.32) becomes

$$(E_n - H)\,\Psi_n = -\frac{ea}{2mc}\,(\mathbf{e}\cdot\mathbf{P})\,(\Psi_{n-1} + \Psi_{n+1}) \,. \tag{2.35}$$

Length gauge. From Eqs.(2.14) and (2.26) we take

$$H = \frac{1}{2m}\,\mathbf{P}^2 + V, \qquad H' = -\frac{ea\omega}{c}\,(\mathbf{e}\cdot\mathbf{r})\sin\omega t \,. \tag{2.36}$$

Consequently

$$H'_n = \frac{ea\omega}{2ic}\,(\mathbf{e}\cdot\mathbf{r})\,(\delta_{n1} - \delta_{n,-1}) \,, \tag{2.37}$$

and Eq.(2.32) becomes

$$(E_n - H)\,\Psi_n = \frac{ea\omega}{2ic}\,(\mathbf{e}\cdot\mathbf{r})\,(\Psi_{n-1} - \Psi_{n+1}) \,. \tag{2.38}$$

Space-translation. Using Eqs.(2.24) and (2.27), we may write

$$H = \frac{1}{2m}\,\mathbf{P}^2 \,, \qquad H' = V(\mathbf{r} + \vec{\alpha}(t)) \tag{2.39}$$

Now, the H'_n coincide with the coefficients V_n of the Fourier expansion of the oscillating potential $V(\mathbf{r} + \vec{\alpha}(t))$, given by

$$V_n(\alpha_0;\mathbf{r}) = \frac{1}{2\pi} \int_0^{2\pi} e^{in\theta}\, V(\mathbf{r} + \vec{\alpha}_0 \sin\theta)\, d\theta \ . \tag{2.40}$$

We are abbreviating here and in the following $\vec{\alpha}_0 = \alpha_0 \mathbf{e}$.

For later use, we give here also the Fourier transform of V_n, defined as

$$\mathcal{V}_n(\alpha_0,\mathbf{k}) = \frac{1}{(2\pi)^3} \int e^{-i\mathbf{kr}}\, V_n(\alpha_0,\mathbf{r})\, d\mathbf{r} \ . \tag{2.41}$$

By inserting Eq.(2.40) into (2.41), changing the variable from $\mathbf{r}$ to $\mathbf{r} + \vec{\alpha}_0 \sin\theta$, and using a well known integral representation for the Bessel function J_n, one finds

$$\mathcal{V}_n(\alpha_0,\mathbf{k}) = J_n(-\vec{\alpha}_0 \cdot \mathbf{k})\, \mathcal{V}(\mathbf{k}) \ , \tag{2.42}$$

where $\mathcal{V}(\mathbf{k})$ is the Fourier transform of the original potential $V(\mathbf{r})$.

Eq.(2.32) reads in the present case

$$(E_n - H)\, \Psi_n = \sum_{m=-\infty}^{+\infty} V_{n-m}\, \Psi_m \ . \tag{2.43}$$

Note that in this equation α_0 and ω appear as the natural independent laser parameters, instead of I and ω; α_0 is contained in the Fourier components V_m, see Eq.(2.40), and ω is contained in the E_n.

2.4. Scattering boundary conditions

In Sec. 1.4. we discussed two possible ways of choosing the boundary conditions for the electron-potential collision problem (methods A and B). Let us now see how these boundary conditions can be expressed analytically.

In the case of *method B* the boundary condition has to account for the existence of a free incoming electron of given energy E and momentum $\mathbf{p}$ represented by a plane wave, and the existence of the infinite set of possible final states with energies E_n (see Eq.(1.2)), into which the electron can be scattered with absorption/emission of photons, represented by spherical outgoing waves. The boundary condition should therefore be of the form

$$\Psi(\mathbf{r},t) \underset{r\to\infty}{\to} \exp\frac{i}{\hbar}(\mathbf{p}\cdot\mathbf{r} - Et) + \sum_{n=n_0}^{\infty} \frac{f_n(\hat{\mathbf{r}})}{r} \exp\frac{i}{\hbar}(p_n r - E_n t) \ . \tag{2.44}$$

We have introduced here the momenta p_n associated to E_n, according to

$$\frac{p_n^2}{2m} = E_n = E + n\hbar\omega \ . \tag{2.45}$$

Note that $n = 0$ represents the purely elastic scattering channel ($E_0 = E$). Whereas n can be arbitrarily large positive, it cannot decrease below a certain integer n_0 ($n_0 \leq 0$), which is the minimal value for which $E_n > 0$. All channels $n \geq n_0$ are open to the FFT process, while those with $n < n_0$ are closed and do not contribute to the flux of outgoing particles. It should be assumed that the wave function Ψ decays exponentially in the closed channels.

In our approach, the f_n in Eq.(2.44) are the transition amplitudes of the electron under the influence of the classical field. These we *interpret* as FFT amplitudes for the absorption ($n > 0$) / emission ($n < 0$) of n photons, or for laser modified elastic scattering ($n = 0$), in agreement with QED.

If we compare the Floquet expansion Eq.(2.30) with the boundary condition Eq.(2.44) we obtain the conditions which should be satisfied by the Floquet components Ψ_n. We first see that in order that the connection be possible the energy parameter E in Eq.(2.30) should be identified with the initial energy. We find

$$\begin{aligned} &\Psi_n(\mathbf{r}) \underset{r\to\infty}{\to} \delta_{n0} \exp\left(\frac{i}{\hbar}\mathbf{p}\cdot\mathbf{r}\right) + (f_n(\hat{\mathbf{r}})/r) \exp\left(\frac{i}{\hbar} p_n r\right) , && n \geq n_0 ; \\ &\Psi_n(\mathbf{r}) \underset{r\to\infty}{\to} 0 \text{ (exponentially)}, && n < n_0 . \end{aligned} \tag{2.46}$$

As discussed in Sec. 1.4 the boundary condition Eq.(2.44) or (2.46) of method B is useful when the radiation field is weak and can be treated as a small perturbation. This boundary condition will be coupled in Sec. 2.5 to the Floquet formulations of the Schrödinger equation, Eqs.(2.35) and (2.38).

Method A of choosing the asymptotic states as dressed free particle states, is useful for intense fields. In this case it is more convenient to work in the Kramers reference frame, by applying the space translation to both Schrödinger equation and the boundary condition of the laboratory frame. The transformed equation is then given in Floquet form by Eq.(2.43). As shown in Sec. 2.2 the dressed plane waves (Volkov waves) of the laboratory frame are transformed into usual plane waves in the Kramers frame; similarly it can be shown that the dressed spherical waves are transformed into ordinary spherical waves (not discussed in Sec. 2.2). Thus the boundary condition for method A, when formulated in the Kramers frame, is expressed in terms of ordinary free particle waves, and should have the same form as in Eqs.(2.44),(2.46). We shall be using it in Eq.(2.59) and in Sec. 4.3 below.

We emphasize, however, that although the boundary conditions used for methods A and B are formally identical, they refer to different reference frames.

As formulated by Eqs.(2.44) or (2.46) the boundary condition applies to short range potentials. For ionic (Coulomb tail) potentials it has to be modified by inclusion of specific logarithmic terms in the phases, as in the case of usual radiationless scattering. (See for example [G15], Sec. XI-7; a specific case will be considered in Sec. 4.3.)

The knowledge of the FFT amplitudes yields the corresponding cross sections, defined as scattering probability divided by the incoming probability current density. By following the

The knowledge of the FFT amplitudes yields the corresponding cross sections, defined as scattering probability divided by the incoming probability current density. By following the usual procedure in scattering theory when several reaction channels are open, one may derive the following FFT differential cross sections:

$$\frac{d\sigma_n}{d\Omega} = \frac{p_n}{p} |f_n(\hat{\mathbf{r}})|^2 \; , \qquad (n \geq n_0) \tag{2.47}$$

where the unit vector $\hat{\mathbf{r}}$ is contained in $d\Omega$. (In the case of the space translated Schrödinger equation, this is the cross section obtained in the Kramers frame; however, translation by the vector $\vec{\alpha}(t)$, Eq.(2.27), does not change directions, so that the cross sections in the laboratory frame are the same.)

The global scattering cross section, irrespective of the number of photons absorbed or emitted, is given by

$$\frac{d\sigma}{d\Omega} = \sum_{n \geq n_0}^{\infty} \frac{d\sigma_n}{d\Omega} \; . \tag{2.48}$$

2.5. **Integral equations. Expressions for FFT amplitudes**

The differential Floquet form of the three Schrödinger equations of interest, Eqs.(2.35),(2.38) and (2.43), can be reexpressed in integral form, by applying the Green's function method.

We shall first recall some general properties of the *Green's function*. (For details see, for example, the books by Messiah ([G15] Chap. XIX), and Joachain ([G16] Chaps. 5,11,14).) The Green's function associated to some Hermitian operator given in coordinate space, $H(\mathbf{r})$, can be defined by the equation

$$[\Omega - H(\mathbf{r})] \, G(\Omega; \mathbf{r}, \mathbf{r}') = \delta(\mathbf{r}-\mathbf{r}') \; , \tag{2.49}$$

with appropriate boundary conditions to insure the unicity of the solution. These require that G vanish for $r \to 0$ and $r \to \infty$. Eq.(2.49) should hold for any complex Ω not belonging to the spectrum (discrete and continuous) of H. It turns out that the Green's function depends in fact on $\hbar k = (2m\Omega)^{1/2}$, and therefore, as function of Ω, is defined on the Riemann surface of $\Omega^{1/2}$. The actual Green's function, as defined by the boundary conditions (more specifically, by its vanishing for $r \to \infty$), corresponds to Ω on the $\mathrm{Im}\, k > 0$ sheet (the "physical" sheet). On this, we shall denote as usual for $E > 0$:

$$G^{(\pm)}(E) = G(E \pm i\varepsilon) \; , \tag{2.50}$$

where $\varepsilon > 0$ is infinitesimal.

The Green's function G associated to $H = (1/2m)\mathbf{P}^2 + V$ satisfies the operator equation

$$G(\Omega) = G_0(\Omega) + G_0(\Omega)\, VG(\Omega) \ , \tag{2.51}$$

where G_0 is the free particle Green's function, associated to $H_0 = (1/2m)\mathbf{P}^2$. As usual, we define

$$G\Psi(\mathbf{r}) = \int G(\Omega; \mathbf{r}, \mathbf{r}')\, \Psi(\mathbf{r}')\, d\mathbf{r}' \ . \tag{2.52}$$

$G_0(\Omega; \mathbf{r}, \mathbf{r}')$ depends in fact only on the difference $\vec{\rho} = \mathbf{r} - \mathbf{r}'$, and has the closed form expression

$$G_0(\Omega;\rho) = -\frac{m}{2\pi\hbar^2}\frac{e^{ik\rho}}{\rho} \ , \qquad (\mathrm{Im}\, k > 0) \ . \tag{2.53}$$

From Eqs.(2.51) and (2.53) the following asymptotic behavior can be derived

$$G^{(\pm)}(E;\mathbf{r},\mathbf{r}') \underset{r\to\infty}{\rightarrow} -\frac{m}{2\pi\hbar^2}\frac{e^{\pm ikr}}{r}\,[\phi^{(\mp)}_{\pm\mathbf{k}}(\mathbf{r}')]^* \ , \tag{2.54}$$

Here $\phi^{(\pm)}_{\mathbf{k}}$ is a scattering solution of the equation $H\phi^{(\pm)}_{\mathbf{k}} = E\phi^{(\pm)}_{\mathbf{k}}$, behaving asymptotically as a plane wave of amplitude 1 and wave vector $\mathbf{k} = k\hat{\mathbf{r}}$, plus outgoing/incoming spherical waves, designated by superscripts (+)/(-).

We can now apply the Green's function method to the three Floquet systems of interest, Eqs.(2.35), (2.38) and (2.43). This consists in treating the coupling terms placed on the right-hand side of the equations, as inhomogeneities. As usual, the desired solution will be obtained by adding a special solution of the inhomogeneous equation to an appropriate solution of the homogeneous equation on the left-hand side, chosen so as to satisfy the boundary condition.

Momentum gauge. By using the Green's function associated to the H of Eq.(2.33), the differential equations Eq.(2.35) may be cast into the system of integral equations

$$\Psi_n = \phi^{(+)}_{\mathbf{k}}\, \delta_{n0} - \frac{ea}{2mc}\, G^{(+)}(E_n)\, (\mathbf{e}\cdot\mathbf{P})\, (\Psi_{n-1} + \Psi_{n+1}) \ . \tag{2.55}$$

The second term of Eq.(2.55) is a solution of the inhomogeneous equation Eq.(2.35), whereas the first is a particular solution of the corresponding homogeneous equation. Using Eq.(2.54) it is easy to check that the boundary condition Eq.(2.46) is satisfied, since $\phi^{(+)}_{\mathbf{k}}$ satisfies by definition Eq.(2.46) for n = 0. At the same time, we find the following expressions for the FFT amplitudes

$$f_n(\hat{\mathbf{r}}) = f_0^{(0)}(\hat{\mathbf{r}})\, \delta_{n0} + \frac{m}{2\pi\hbar^2}\left(\frac{ea}{2mc}\right) < \phi^{(-)}_{\mathbf{k}_n} |(\mathbf{e}\cdot\mathbf{P})|\, \Psi_{n-1} + \Psi_{n+1} >. \tag{2.56}$$

Here $f_0^{(0)}$ is the elastic scattering amplitude in the absence of the field, associated to $\phi^{(+)}_{\mathbf{k}}$, and $\mathbf{k}_n = k_n\hat{\mathbf{r}}$.

Length gauge. With the same Green's function as before (see Eq.(2.36)), we get from Eq.(2.38):

$$\Psi_n = \phi^{(+)}_{\mathbf{k}}\, \delta_{n0} + \frac{ea\omega}{2ic}\, G^{(+)}(E_n)\, (\mathbf{e}\cdot\mathbf{r})\, (\Psi_{n-1} - \Psi_{n+1}) \ , \tag{2.57}$$

$$f_n(\hat{\mathbf{r}}) = f_0^{(0)}(\hat{\mathbf{r}})\,\delta_{n0} - \frac{m}{2\pi\hbar^2}\left(\frac{ea\omega}{2ic}\right) < \phi^{(-)}_{\mathbf{k}_n} |(\mathbf{e}\cdot\mathbf{r})|\, \Psi_{n-1} - \Psi_{n+1} >. \tag{2.58}$$

Space translation. For the space-translated equation, Eq.(2.43), we use the free particle Green's function, associated to H of Eq.(2.39), to find

$$\Psi_n = e^{i\mathbf{k}\mathbf{r}}\,\delta_{n0} + G_0^{(+)}(E_n) \sum_{m=-\infty}^{+\infty} V_{n-m}\,\Psi_m\,, \tag{2.59}$$

By using Eq.(2.54), with $\phi^{(\pm)}_{\mathbf{k}}$ replaced by $e^{i\mathbf{k}\mathbf{r}}$, one easily checks that the boundary condition Eq.(2.46) for the Kramers frame is satisfied, and that

$$f_n(\hat{\mathbf{r}}) = -\frac{m}{2\pi\hbar^2} \sum_{m=-\infty}^{+\infty} < \mathbf{k}_n | V_{n-m} | \Psi_m > . \tag{2.60}$$

We are using here and in the following the ket-notation $|\mathbf{k}>$ for the plane wave of momentum $\hbar\mathbf{k}$ and amplitude 1.

Eqs.(2.56), (2.58) and (2.60) are equivalent formal expressions for the FFT amplitudes. As written, they still contain the unknown Floquet components Ψ_n, and therefore are of little practical value. However, by iteration, i.e. by repeated insertion of the corresponding Ψ_n from Eqs.(2.55),(2.57),(2.59), they can be cast into series of terms containing only the wave functions and Green's functions corresponding to H, assumed to be known. If the successive terms are decreasing sufficiently rapidly, the series may "converge" in some pragmatic sense. This happens indeed for the cases at hand, under specific limitations. Thus, Eq.(2.56) or Eq.(2.58) yield the standard perturbation theory expansion of semiclassical interaction theory in the momentum or length gauges, to be discussed in Chap.3. Eq.(2.60), on the other hand, yields a nonperturbative Born development. (It is nonperturbative in the sense that each term contains the field to all orders.) This, and the nonperturbative expansions for low and high photon frequencies will be considered in Chap. 4.

3. STANDARD PERTURBATION THEORY

3.1. Formalism

It is convenient to write the equations of the momentum gauge Eqs. (2.55), (2.56) and of the length gauge, Eqs.(2.57), (2.58), in the common form:

$$\Psi_n = \phi^{(+)}_{\mathbf{k}}\,\delta_{n0} + C\,G^{(+)}(E_n)\,Q\,(\Psi_{n-1} + \eta\Psi_{n+1})\,, \tag{3.1}$$

$$f_n(\hat{\mathbf{r}}) = f_0^{(0)}\,\delta_{n0} - \frac{m}{2\pi\hbar^2}\,C < \phi^{(-)}_{\mathbf{k}_n} |Q|\, \Psi_{n-1} + \eta\Psi_{n+1} >, \tag{3.2}$$

where for the momentum gauge we need take

$$C = -\frac{ea}{2mc} = \frac{\alpha_0 \omega}{2}, \qquad Q = \mathbf{e}\cdot\mathbf{P}, \qquad \eta = 1, \tag{3.3}$$

and for the length gauge

$$C = \frac{ea\omega}{2ic}, \qquad Q = \mathbf{e}\cdot\mathbf{r}, \qquad \eta = -1. \tag{3.4}$$

By iterating twice we get from Eqs.(3.1), (3.2):

$$\begin{aligned}
f_n(\hat{\mathbf{r}}) = \; & f_0^{(0)}\,\delta_{n0} - \frac{m}{2\pi\hbar^2}\Big[C <\phi^{(-)}_{\mathbf{k}_n}|Q|\,\phi^{(+)}_{\mathbf{k}}> (\delta_{n,1} + \eta\delta_{n,-1}) \\
& + C^2 <\phi^{(-)}_{\mathbf{k}_n}|QG^{(+)}(E_{n-1})\,Q|\,\phi^{(+)}_{\mathbf{k}}> (\delta_{n,2} + \eta\delta_{n,0}) \\
& + C^2 <\phi^{(-)}_{\mathbf{k}_n}|QG^{(+)}(E_{n+1})\,Q|\,\phi^{(+)}_{\mathbf{k}}> (\eta\delta_{n,0} + \delta_{n,-2}) \\
& + C^3 <\phi^{(-)}_{\mathbf{k}_n}|QG^{(+)}(E_{n-1})\,Q\,G^{(+)}(E_{n-2})\,Q|\,\Psi_{n-3} + \eta\Psi_{n-1}> \\
& + C^3 <\phi^{(-)}_{\mathbf{k}_n}|QG^{(+)}(E_{n-1})\,Q\,G^{(+)}(E_{n})\,Q|\,\eta\Psi_{n-1} + \Psi_{n+1}> \\
& + C^3 <\phi^{(-)}_{\mathbf{k}_n}|QG^{(+)}(E_{n+1})\,Q\,G^{(+)}(E_{n})\,Q|\,\eta\Psi_{n-1} + \Psi_{n+1}> \\
& + C^3 <\phi^{(-)}_{\mathbf{k}_n}|QG^{(+)}(E_{n+1})\,Q\,G^{(+)}(E_{n+2})\,Q|\,\Psi_{n+1} + \eta\Psi_{n+3}>\Big],
\end{aligned} \tag{3.5}$$

since $\eta^2 = 1$.

Thus, through order a^3 the nonvanishing FFT amplitudes are:

$$\begin{aligned}
f_0(\hat{\mathbf{r}}) = f_0^{(0)} - \frac{m}{2\pi\hbar^2}\,\eta C^2\,[&<\phi^{(-)}_{\mathbf{k}_0}|QG^{(+)}(E_{-1})\,Q|\,\phi^{(+)}_{\mathbf{k}}> + \\
& + <\phi^{(-)}_{\mathbf{k}_0}|QG^{(+)}(E_{+1})\,Q|\,\phi^{(+)}_{\mathbf{k}}>] + O(a^4),
\end{aligned} \tag{3.6}$$

with the initial ($\mathbf{k}$) and final ($\mathbf{k}_0$) momenta having the same magnitude ($k = k_0$) but different directions;

$$\begin{aligned}
f_{+1}(\hat{\mathbf{r}}) = -\frac{m}{2\pi\hbar^2}\Big[C &<\phi^{(-)}_{\mathbf{k}_1}|Q|\,\phi^{(+)}_{\mathbf{k}}> \\
& + \eta C^3 <\phi^{(-)}_{\mathbf{k}_1}|QG^{(+)}(E_2)\,Q\,G^{(+)}(E_1)\,Q|\,\phi^{(+)}_{\mathbf{k}}> \\
& + \eta C^3 <\phi^{(-)}_{\mathbf{k}_1}|QG^{(+)}(E_0)\,Q\,G^{(+)}(E_1)\,Q|\,\phi^{(+)}_{\mathbf{k}}>
\end{aligned}$$

$$+\eta C^3 <\phi^{(-)}_{\mathbf{k}1}|QG^{(+)}(E_0)\,Q\,G^{(+)}(E_{-1})\,Q|\,\phi^{(+)}_{\mathbf{k}}>] + O(a^5), \tag{3.7}$$

$$f_{+2}(\hat{\mathbf{r}}) = -\frac{m}{2\pi\hbar^2}\; C^2 <\phi^{(-)}_{\mathbf{k}2}|QG^{(+)}(E_1)\,Q|\,\phi^{(+)}_{\mathbf{k}}> + O(a^4)\,, \tag{3.8}$$

$$f_{+3}(\hat{\mathbf{r}}) = -\frac{m}{2\pi\hbar^2}\; C^3 <\phi^{(-)}_{\mathbf{k}3}|QG^{(+)}(E_2)\,Q\,G^{(+)}(E_1)\,Q|\,\phi^{(+)}_{\mathbf{k}}> + O(a^5)\;, \tag{3.9}$$

as well as $f_{-1}(\hat{\mathbf{r}})$, $f_{-2}(\hat{\mathbf{r}})$, $f_{-3}(\hat{\mathbf{r}})$, for which we get similar formulas.

By continuing the iteration in Eq.(3.5), each amplitude f_n may be expressed as a series of terms of increasing order in a (the successive terms actually increase by powers of a^2). This is the standard *perturbation series* of semiclassical interaction theory. At *sufficiently small* a, i.e. at sufficiently small intensity I (as I is proportional to a^2), but for arbitrary ω, the terms of the series decrease in magnitude, and the first nonvanishing one (hopefully amenable to calculation) should give a satisfactory approximation. Nothing is known in general about the convergence of the perturbation series in the strict mathematical sense, although the answer to this question is indeed affirmative in some limiting cases (see Secs. 4.1 and 4.2).

It is easy to show that, to lowest nonvanishing order, the expression for f_n is:

$$f_n(\hat{\mathbf{r}}) = -\frac{m}{2\pi\hbar^2}\; C^{|n|}\begin{pmatrix}1\\ \eta^n\end{pmatrix} \tag{3.10}$$

$$\cdot <\phi^{(-)}_{\mathbf{k}n}|QG^{(+)}(E_{n\mp1})\cdots QG^{(+)}(E_{\pm2})QG^{(+)}(E_{\pm1})Q|\phi^{(+)}_{\mathbf{k}}> + O(a^{|n|+2}),$$

This applies both to absorption and emission. For absorption ($n > 0$) the upper signs in the subscrips of the E_p should be taken, and 1 in the column. For emission ($n < 0$), the lower signs in the subscrips of the E_p are the appropriate ones, and η^n should be taken in the column.

In order to obtain the general expression for f_n one may proceed as follows. $f_n(\hat{\mathbf{r}})$ is made of a sum of matrix elements which we denote by M_n (e.g. the case of f_{+1}, Eq.(3.7)). Each of them has at its right end the ket $|\phi^{(+)}_{\mathbf{k}}>$ of the initial state; this is continued towards the left by a succession of alternating operators Q and $G^{(+)}(E)$, ending with a Q; at its left end stands the bra $<\phi^{(-)}_{\mathbf{k}}|$ of the final state. Each M_n also contains a numerical coefficient which is the product of $(-m/2\pi\hbar^2)$, times a power of C, times 1 or η (recall that $\eta^2 = 1$).

The various M_n differ from each other in two respects:

(1) By the sequence formed by the energies of the initial state, of the (n-1) Green's operators $G^{(+)}(E_j)$ and of the final state, contained in the various M_n. We shall write the sequence of each M_n proceeding from *right to left* (i.e. from the initial towards the final state). Thus in the case of Eq.(3.7), for the second term this is E_0, E_1, E_2, E_1, for the third term E_0, E_1, E_0, E_1, and for the fourth E_0, E_{-1}, E_0, E_1. Each possible sequence will be denoted by a subscript σ added to M_n: $M_{n\sigma}$. The possible energy sequences can be determined by noting that the initial and final energies are fixed (E_0 and E_n) and that two consecutive E_j values

can differ only by one photon energy, i.e. E_j can be followed only by E_{j+1} or E_{j-1}. If E_j is followed by E_{j+1} we shall say that the operator Q appearing in one of the combinations $G^{(+)}(E_{j+1})QG^{(+)}(E_j)$; $G(E_{+1})Q|\phi_{\mathbf{k}}^{(+)}>$; $<\phi_{\mathbf{k}_n}^{(-)}|QG(E_{n-1})$, represents the "absorption" of a photon. If E_j is followed by E_{j-1}, the operator Q appears in one of the combinations $G^{(+)}(E_{j-1})QG^{(+)}(E_j)$; $G^{(+)}(E_{-1})Q|\phi_{\mathbf{k}}^{(+)}>$; $<\phi_{\mathbf{k}_n}^{(-)}|QG^{(+)}(E_{n+1})$, and we shall say that it represents the "emission" of a photon. We thus formally associate to each $M_{n\sigma}$ a sequence of "absorption" and "emission" events. We can now speak of "conservation of energy" at each insertion of an operator Q since, proceeding from right to left, the energy after the insertion of Q is equal to that before it plus/minus the one photon energy "absorbed"/ "emitted".

(2) By the coefficients multiplying their scalar products. Each coefficient contains C at a certain power m, which represents the order with respect to a of the $M_{n\sigma}$ considered. This power is also equal to the number of operators Q contained in the scalar product. It can be equal to $|n|$ or larger than it by an even integer. This can be seen as follows. If $n > 0$, and the total number of "absorptions" in $M_{n\sigma}$ is n+p, (where necessarily p = 0,1,2,...), then

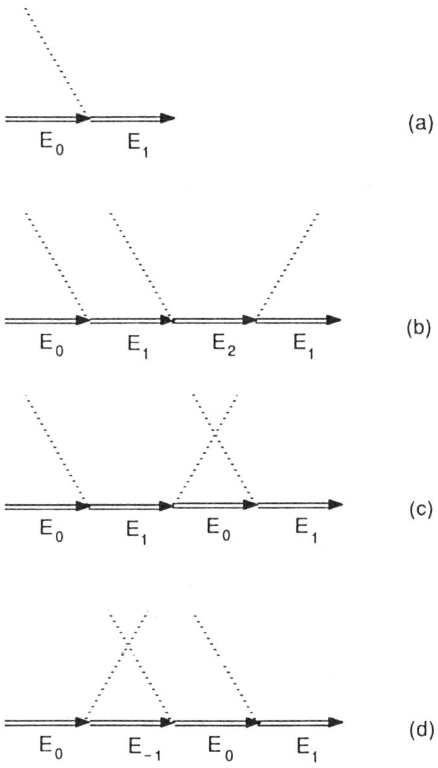

Fig. 5. Feynman diagrams describing the net absorption of one photon through order a^3 (see text). The diagrams correspond to the terms of Eq.(3.7), in the order given there. Diagram (a) represents the absorption to order a; diagrams (b)-(d), to order a^3. Also shown are the energies associated to the electron arrows, in the notation of Eq.(2.45).

the number of "emissions" must equal p so that the net sum of "absorptions" and "emissions" be equal to n. Similarly, if $n < 0$ and $|n|+p$ is the total number of "emissions",the number of "absorptions" must equal p. With this notation the total number of operators Q contained in each $M_{n\sigma}$ is equal to $m = |n|+2p$, which proves the statement. To recall the order in a of each $M_{n\sigma}$, we shall add m as an extra substript: $M_{nm\sigma}$.
Besides C^m, the numerical coefficient of $M_{nm\sigma}$ also contains η^r (with r integer), which is equal to ± 1, depending in general on the sequence σ of energies E_j considered. For the momentum gauge $\eta = 1$, so that $\eta^r = 1$.
For the length gauge $\eta = -1$, and an inspection of the procedure to derive f_n shows that r is equal to the number of "emissions" contained in $M_{nm\sigma}$. Thus, if $n > 0$, $r = p = (m-n)/2$, and if $n < 0$, $r = |n|+p = (m+|n|)/2$. The numerical coefficient of $M_{nm\sigma}$ is thus, finally, $(-m/2\pi\hbar^2)\, C^m\eta^r$.

With the notations introduced above we may write the expression of the amplitude f_n to all order in perturbation theory as:

$$f_n(\hat{\mathbf{r}}) = \sum_{m=|n|}^{\infty} \sum_{\sigma} M_{nm\sigma} , \tag{3.11}$$

where the sum over m should be carried out over the values $m = |n|+2p$, while the sum over σ, over all possible energy sequences.

Each matrix element $M_{nm\sigma}$ can be given a *diagramatic representation*, using a method introduced by Feynman in Q.E.D. These diagrams are useful in keeping track of all contributions to f_n. The incoming/outgoing electron wave functions $\phi_{\mathbf{k}}^{(+)}/\phi_{\mathbf{k}_n}^{(-)}$ (normalized to asymptotic amplitude 1) are represented by double-line arrows entering/leaving the diagram. The initial arrow is connected by a chain of (n-1) double-line arrows to the arrow of the final state (see Fig. 5). Every internal arrow represents one of the Green's functions $G^{(+)}(E_j)$ [defined as in Sec. 2.5] of $M_{nm\sigma}$. To each arrow of the chain one can attach an energy, their sequence being characterized by the index σ introduced above. The connection point of two arrows is called a vertex. To each vertex we attach a dotted "photon" line, indicating that a photon is either "absorbed" or "emitted" at the place where an operator Q is inserted. When the photon is "absorbed", we agree to draw the photon line at an acute angle with the direction of the arrows of the chain, and when "emitted", at an obtuse angle.

The diagram associated to $M_{nm\sigma}$ describes the sequence of "events" happening to the electron which propagates in the matrix element. This description is only figurative and does not indicate how the process evolves in the physical world. The diagram is in one-to-one correspondence with the scalar product contained in $M_{nm\sigma}$ and in order to get its complete expression we need introduce also the numerical coefficient discussed above at point (2). As an example we give in Fig. 5 the Feynman diagrams corresponding to the four matrix elements contained in the amplitude f_{+1} of Eq.(3.7), through order a^3.

The eigenfunctions $\phi_{\mathbf{k}}^{(\pm)}$ and the Green's functions $G^{(+)}(E)$ entering the amplitudes f_n belong to the Hamiltonian H of Eq.(2.33), containing the potential V. If all electron energies involved are high, V may be regarded as a perturbation on the kinetic energy $(1/2m)\mathbf{P}^2$ and we may use their *Born expansions* to express them. These are obtained, by iteration, from Eq.(2.51) and from (see Eqs.(90),(92), (Sec. XIX-13, Ref. [G15]):

$$|\phi_{\mathbf{k}}^{(+)}> = (I + G^{(+)}(E)\,V)\,|\mathbf{k}> \ , \tag{3.12}$$

$$<\phi_{\mathbf{k}}^{(-)}| = <\mathbf{k}_n|(I + VG^{(+)}(E_n)) \ , \tag{3.13}$$

to give

$$G^{(+)}(E) = G_0^{(+)}(E) + G_0^{(+)}(E)\,VG_0^{(+)}(E) + \cdots \ , \tag{3.14}$$

$$|\phi_{\mathbf{k}}^{(+)}> = (I + G_0^{(+)}(E)\,V + G_0^{(+)}(E)\,VG_0^{(+)}(E)\,V + \cdots)\,|\mathbf{k}> \ , \tag{3.15}$$

$$<\phi_{\mathbf{k}_n}^{(-)}| = <\mathbf{k}_n|\,(I + VG_0^{(+)}(E_n) + VG_0^{(+)}(E_n)\,VG_0^{(+)}(E_n) + \cdots) \ . \tag{3.16}$$

Insertion of Eqs.(3.14)-(3.16) into the expression of a $M_{nm\sigma}$ yields a formal expansion in V, which is its Born series. As easily seen, this starts with terms containing the first power of V, because the term independent of V vanishes (for an illustration of the proof in a special case see Eqs.(3.18)-(3.22)). This is quite natural since if V = 0 no FFT can occur.

The Born expansion of $M_{nm\sigma}$ can be represented diagramatically too. We need first introduce some new signs: single-line arrows, to represent the electron wave functions and Green's functions associated to $H = (1/2m)P^2$, and dotted lines ending with a star, to represent the insertion of the potential V within electron single-line arrows. The insertion of a potential line generates a new vertex, but now there is no change of energy along the electron "path" at this vertex. By considering the way in which the various Born terms of $M_{nm\sigma}$ are obtained analytically, one finds that the terms oforder q can be represented diagramatically by inserting in all possible ways q potential lines into the chain of arrows of $M_{nm\sigma}$, and changing all double-line arrows into single-line arrows. To first order in V, this yields in the case of the diagrams (a) and (b) of Fig. 5, the diagrams of Fig. 6.

As an example we consider the first Born approximation (i.e. the first order term in V) of the lowest order perturbation theory result for the n photon absorption (n > 0), Eq.(3.10). We choose in the following the momentum gauge, Eq.(3.3). By inserting Eqs (3.14)-(3.16) into Eq.(3.10), we obtain to first order in the potential:

$$f_n^{B1} = -\frac{m}{2\pi\hbar^2}\,C^n\,(M_{n0} + M_{n1}) + O(a^{n+2}) \ , \tag{3.17}$$

where M_{n0} is the term independent of V and M_{n1} is proportional to V.

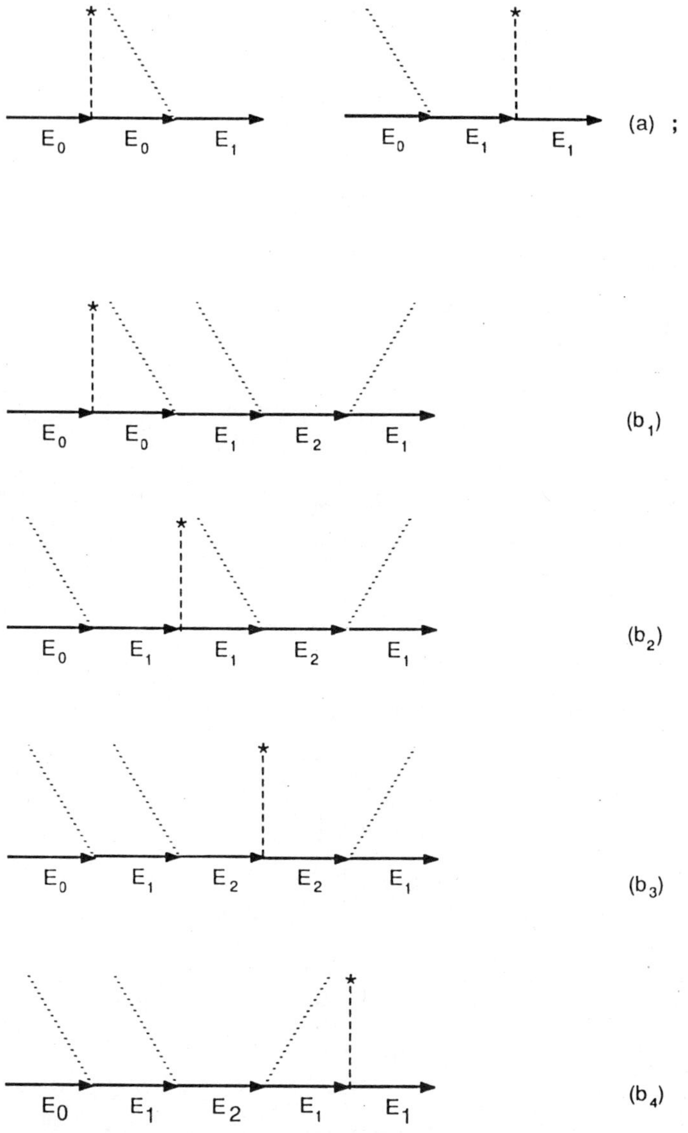

Fig. 6. First Born approximation diagrams corresponding to the Feynman diagrams (a) and (b) of Fig. 5, for net one-photon absorption (see text). Also shown are the energies of the electron arrows.

Thus,

$$M_{n0} = < \mathbf{k}_n |(\mathbf{e}\cdot\mathbf{P})G_0(E_{n-1}) \dots G(E_1)(\mathbf{e}\cdot\mathbf{P})| \mathbf{k} >$$

$$= < (\mathbf{e}\cdot\mathbf{P})\mathbf{k}_n |G_0(E_{n-1})(\mathbf{e}\cdot\mathbf{P}) \dots (\mathbf{e}\cdot\mathbf{P})G(E_1)| (\mathbf{e}\cdot\mathbf{P})\mathbf{k} > . \tag{3.18}$$

We calculate the integral in momentum space by using the Fourier transform of $G_0^{(+)}(E)$, Eq.(2.53):

$$G_0^{(+)}(E; \mathbf{r}\text{-}\mathbf{r}') = (2\pi)^{-3} \iint G_0^{(+)}(E; \mathbf{q},\mathbf{q}')\ e^{i(\mathbf{qr}-\mathbf{q}'\mathbf{r}')}\ d\mathbf{q}d\mathbf{q}' , \tag{3.19}$$

$$G_0^{(+)}(E; \mathbf{q},\mathbf{q}') = \frac{\delta(\mathbf{q}\text{-}\mathbf{q}')}{(E+i\varepsilon) - E_q} , \tag{3.20}$$

where $E_q = q^2/2m$. With our normalization of the plane waves $|\mathbf{k}>$ we have also

$$<\mathbf{k}'|\mathbf{e}\cdot\mathbf{P}|\mathbf{k}> = (2\pi)^3\ (\mathbf{e}\cdot\hbar\mathbf{k})\ \delta(\mathbf{k}\text{-}\mathbf{k}') . \tag{3.21}$$

Inserting these in Eq.(3.18) we obtain

$$M_{n0} = (2\pi)^3\ (\mathbf{e}\cdot\hbar\mathbf{k})^n\ [(E_{n-1} - E_0) \dots (E_1 - E_0)]^{-1}\ \delta(\mathbf{k}_n\text{-}\mathbf{k}) = 0 , \tag{3.22}$$

which vanishes because $\mathbf{k}_n \neq \mathbf{k}$. As noted, this vanishing is common to all matrix elements $M_{nm\sigma}$ when calculated to "zero-th order" in the potential.

M_{n1} can be written as

$$M_{n1} = M_{100} + M_{001} + \sum_{j=1}^{n-1} M_{010}^{(j)} . \tag{3.23}$$

M_{100} and M_{001} contain the first order contributions in V from Eq.(3.16) and Eq.(3.15), respectively:

$$M_{100} = < \mathbf{k}_n |VG_0^{(+)}(E_n)(\mathbf{e}\cdot\mathbf{P})G_0^{(+)}(E_{n-1})\cdots G_0^{(+)}(E_1)(\mathbf{e}\cdot\mathbf{P})| \mathbf{k} > , \tag{3.24}$$

$$M_{001} = < \mathbf{k}_n|(\mathbf{e}\cdot\mathbf{P})G_0^{(+)}(E_{n-1})\cdots G_0^{(+)}(E_1)(\mathbf{e}\cdot\mathbf{P})G_0^{(+)}(E_0)\ V|\mathbf{k} >. \tag{3.25}$$

$M_{010}^{(j)}$ contains the first order contribution of $G^{(+)}(E_j)$:

$$M_{010}^{(j)} = < \mathbf{k}_n |(\mathbf{e}\cdot\mathbf{P})\cdots(\mathbf{e}\cdot\mathbf{P})G_0^{(+)}(E_j)VG_0^{(+)}(E_j)(\mathbf{e}\cdot\mathbf{P}) \cdots (\mathbf{e}\cdot\mathbf{P})| \mathbf{k} >. \tag{3.26}$$

By applying the integration method followed above, we find

$$M_{100} = [(E_n\text{-}E_0)(E_{n-1}\text{-}E_0) \cdots (E_1\text{-}E_0)]^{-1}\ (\mathbf{e}\cdot\hbar\mathbf{k})^n < \mathbf{k}_n |V| \mathbf{k} > , \tag{3.27}$$

$$M_{001} = [(E_{n-1}\text{-}E_n)(E_{n-2}\text{-}E_n) \cdots (E_0\text{-}E_n)]^{-1}\ (\mathbf{e}\cdot\hbar\mathbf{k}_n)^n < \mathbf{k}_n |V| \mathbf{k} > , \tag{3.28}$$

$$M_{010}^{(j)} = [(E_{n-1}\text{-}E_n)\cdots(E_j\text{-}E_n) \cdot (E_j\text{-}E_0)\cdots(E_1\text{-}E_0)]^{-1}\ (\mathbf{e}\cdot\hbar\mathbf{k}_n)^{n-j}\ (\mathbf{e}\cdot\hbar\mathbf{k})^j$$

$$\cdot < \mathbf{k}_n |V| \mathbf{k} > . \tag{3.29}$$

Inserting Eqs.(3.27)-(3.29) into Eq.(3.23), and taking into account that $E_\alpha - E_\beta = (\alpha-\beta)\hbar\omega$, we get

$$M_{n1} = \frac{1}{n!\omega^n}\left[\sum_{j=0}^{n} \frac{n!}{(n-j)!j!} (\mathbf{e}\cdot\mathbf{k})^j (-\mathbf{e}\cdot\mathbf{k}_n)^{n-j}\right] < \mathbf{k}_n |V| \mathbf{k} > . \tag{3.30}$$

The sum in Eq.(3.30) is equal to $(\mathbf{e}\cdot\vec{\Delta}_n)^n$, where

$$\vec{\Delta}_n = \mathbf{k} - \mathbf{k}_n \tag{3.31}$$

is the wave vector transfer. This yields for f_n^{B1} of Eq.(3.17) the simple result:

$$f_n^{B1} = -\frac{(2\pi)^2 m}{\hbar^2} \cdot \frac{(\vec{\alpha}_0\cdot\vec{\Delta}_n)^n}{2^n\, n!} \cdot V(\Delta_n) + O(a^{n+2}), \quad (n > 0) \tag{3.32}$$

with the Fourier transform $V(\Delta_n)$ defined as in Eq.(2.41).

A similar derivation can be given for the case of n-photon emission. The generalization of Eq.(3.32), covering both absorption ($n > 0$) and emission ($n < 0$), can be written

$$f_n^{B1} = -\frac{(2\pi)^2 m}{\hbar^2} \cdot \frac{(\vec{\alpha}_0\cdot\vec{\Delta}_n)^{|n|}}{2^{|n|}(|n|)!} \cdot V(\Delta_n) \begin{pmatrix} 1 \\ (-1)^n \end{pmatrix} + O(a^{|n|+2}), \qquad \begin{matrix} n > 0 \\ n < 0 \end{matrix} \tag{3.33}$$

where in the column we have to take 1 for $n > 0$, and $(-1)^n$ for $n < 0$. The corresponding cross section is

$$\frac{d\sigma_n^{B1}}{d\Omega} = \frac{1}{2^{2|n|}(|n|!)^2} \frac{k_n}{k} (\mathbf{e}\cdot\vec{\Delta}_n a_0)^{2|n|} \cdot \left(\frac{I}{I_0}\right)^{|n|} \left(\frac{2\,Ry}{\hbar\omega}\right)^{4|n|} \frac{d\sigma_{E\ell}^{B1}}{d\Omega} + O(I^{|n|+1}) . \tag{3.34}$$

Here $d\sigma_{E\ell}^{B1}/d\Omega$ is the first Born approximation formula for elastic scattering in the absence of the radiation, at momentum transfer Δ_n, and a_0 is the Bohr radius. (The introduction of $d\sigma_{E\ell}^{B1}/d\Omega$ is possible because it depends only on Δ, and not on the collision energy; recall that in our case we are dealing with two energies, E and E_n, and that $E_n \neq E$ in general.)

Characteristic features of Eq.(3.34) are that it increases rapidly for $\omega \to 0$ (as $\omega^{-4|n|}$ at fixed intensity), and that it increases when Δ_n becomes larger ($k_n - k \leq \Delta_n \leq k_n + k$), or becomes parallel to **e**. These features were taken into account in order to optimize the yields of the FFT experiments (see Sec. 1.3).

3.2. Potential model calculations

We shall now briefly review some of the perturbation theory results obtained by modelling the incoming electron - target atom interaction by a potential. We are thus dealing with a single electron problem, which does not allow for atomic excitation ($W_i = W_f$ in

Eq.(1.1)), but to which the formulas derived in Sec. 3.1 directly apply. More sophisticated (one-photon) FFT calculations, taking into account the internal degrees of freedom of the atom, and capable of describing such fine features as resonances etc., have been discussed in Refs. [G4],[G5].

Most of the calculations done in perturbation theory had in mind either astrophysical applications, or the emission/absorption of hot gases and plasmas, or otherwise referred to bremsstrahlung. Intensity limitations have focused the interest on ***one-photon FFT***. Because the scattering amplitude for stimulated one-photon emission is the same as that for spontaneous bremsstrahlung, many of the results obtained for bremsstrahlung can be taken over, *mutatis mutandis*, for FFT. (The basics on bremsstrahlung are given by Bethe and Salpeter [G13], §§ 76-79; a recent overview is that of Pratt and Feng [G14].)

The relevant scattering amplitude for absorption is given by (see Eq.(3.7)):

$$f_{+1}(\hat{\mathbf{k}}_f,\hat{\mathbf{k}}_i) = -\frac{m}{2\pi\hbar^2} C <\phi^{(-)}_{\mathbf{k}_f}|Q|\,\phi^{(+)}_{\mathbf{k}_i}> , \tag{3.35}$$

whereas the one for emission $f_{-1}(\hat{\mathbf{k}}_f,\hat{\mathbf{k}}_i)$ contains an extra η on the right hand side (see Eq.(3.5)). Here we have explicitly specified the unit vectors $\hat{\mathbf{k}}_i = \hat{\mathbf{k}}$ and $\hat{\mathbf{k}}_f = \hat{\mathbf{k}}_{\pm 1}$ of the initial and final electron momenta the amplitudes depend on. As before, for the momentum gauge we have to choose Eq.(3.3), and for the length gauge Eq.(3.4). Taking into account that $H\phi^{(\pm)}_{\mathbf{k}} = E\phi^{(\pm)}_{\mathbf{k}}$ with the H of Eq.(2.33), and also Eq.(1.2), it can be easily shown that Eq.(3.35) for f_{+1}, and its analogue for f_{-1}, are equivalent to

$$f_{\pm 1}(\hat{\mathbf{k}}_f,\hat{\mathbf{k}}_i) = \pm\frac{i}{4\pi}\cdot\frac{e\,a}{c\omega\hbar^2} <\phi^{(-)}_{\mathbf{k}_f}|\mathbf{e}\cdot\nabla V|\,\phi^{(+)}_{\mathbf{k}_i}> , \tag{3.36}$$

This is the "acceleration formulation" of the amplitudes. The two expressions given by Eqs.(3.35) contain semiconvergent integrals, so that a convergence prescription has to be adopted for their evaluation. The convergence problem does not exist in the case of the acceleration formulation, Eq.(3.36), because the potential V decreases at infinity.

Rotation invariance arguments require that $f_{\pm 1}(\hat{\mathbf{k}}_f,\hat{\mathbf{k}}_i)$ be of the form

$$f_{\pm 1}(\hat{\mathbf{k}}_f,\hat{\mathbf{k}}_i) = A(\mathbf{e}\cdot\hat{\mathbf{k}}_i) + B(\mathbf{e}\cdot\hat{\mathbf{k}}_f) , \tag{3.37}$$

where A en B are invariant amplitudes having the same functional dependence on $\mathbf{k}_i,\mathbf{k}_f$ for emission as for absorption.

The differential scattering cross section is given by Eq.(2.47). When integrated over the directions of the momentum of the final electron $\hat{\mathbf{k}}_f$, this gives the total cross section

$$\sigma_{\pm 1} = \frac{k_f}{k_i}\int |f_{\pm 1}(\hat{\mathbf{k}}_f,\hat{\mathbf{k}}_i)|^2\, d\Omega_{\hat{\mathbf{k}}_f} . \tag{3.38}$$

$\sigma_{\pm 1}$ still depends on the direction of propagation $\hat{\mathbf{k}}_i$ and the polarization **e** of the photon. A quantity of astrophysical and plasma physics interest is the average of $\sigma_{\pm 1}$ over the polarization directions **e** and over the directions $\hat{\mathbf{k}}_i$ of the initial electron:

$$\overline{\sigma}_{\pm 1} = \frac{1}{4\pi} \int \frac{1}{2} \sum_{\mathbf{e}} \sigma_{\pm 1} \, d\Omega_{\mathbf{k}_i} \, . \tag{3.39}$$

$\overline{\sigma}_{\pm 1}$ depends only on the magnitudes k_i and ω (or k_i and k_f). Actually, the quantity of primary interest in this context is the thermal average $\overline{\overline{\sigma}}_{\pm 1}$ of $\overline{\sigma}_{\pm 1}$ over a Maxwell distribution of incident electron velocities. We note that the prevalent definitions in astrophysics for the cross sections Eq.(2.47),(3.38),(3.39) are different from ours (which are the ones used to describe multiphoton FFT). The astrophysical definitions are given in Refs. [T13], [T23].

The case of the *Coulomb potential* ($V = -Ze^2/r$) was considered historically first. In his celebrated paper on bremsstrahlung Sommerfeld derived in 1931 [T10], [T11] (see also Refs. [T12], [T13]) the analytic form of the scattering amplitude Eq.(3.35). This was possible because the continuum wave functions for the Coulomb problem are known exactly (see [G15], Sec. XI-10):

$$\begin{aligned} \langle \mathbf{r} | \phi_i^{(+)} \rangle &= e^{\pi |n_i|/2} \, \Gamma(1+n_i) \, e^{i \mathbf{k}_i \mathbf{r}} \, {}_1F_1(-n_i, 1; i(k_i r - \mathbf{k}_i \cdot \mathbf{r})) \, , \\ \langle \phi_f^{(-)} | \mathbf{r} \rangle &= e^{\pi |n_f|/2} \, \Gamma(1+n_f) \, e^{-i k_f r} \, {}_1F_1(-n_f, 1; i(k_f r + \mathbf{k}_f \cdot \mathbf{r})) \, , \end{aligned} \tag{3.40}$$

where ${}_1F_1(\alpha, \beta; z)$ is the confluent hypergeometric function, and $n = Ze^2 m / i\hbar^2 k$. In order to calculate Eq.(3.35) it is convenient to express the hypergeometric functions by their integral representations. Then, by interchanging the order of integrations in Eq.(3.35), it is possible to carry out the coordinate space integral, and thereafter one of the two parametric integrals. This leaves us with one dimensional integrals which can be identified as integral representations of hypergeometric functions of the Gauss type, ${}_2F_1(a, b, c; z)$. The result for the amplitudes A,B of Eq.(3.37) was written by Véniard [T14] in the symmetric form:

$$\begin{aligned} A &= V \, n_f (1+n_i) \frac{(k_i+k_f)^{n_i+n_f-1}}{(k_i-k_f)^{n_f+1} \, (k_f-k_i)^{n_i+2}} \cdot {}_2F_1(1+n_f, 2+n_i, 2; x) \, , \\ B &= A(i \rightleftarrows f) \, , \end{aligned} \tag{3.41}$$

where

$$V = 2i \, (e\,a/\hbar c) \, k_i k_f \, \exp\frac{\pi}{2}(|n_i|+|n_f|) \cdot \Gamma(1+n_i) \, \Gamma(1+n_f) \, , \tag{3.42}$$

$$x = -\frac{4 k_i k_f}{(k_i-k_f)^2} \sin^2\frac{\theta}{2} \, , \tag{3.43}$$

and θ is the scattering angle.

The first Born approximation of the exact (perturbation theory) result Eqs.(3.37), (3.41)-(3.43) is obtained by retaining the lowest order terms in n (which is proportional to the Ze^2 of the potential). By noting also that

$$_2F_1(1, 2, 2; z) = (1-z)^{-1}$$

one finds

$$f^{B1}_{\pm 1} \cong \mp \frac{Ze^2}{\hbar^2} \frac{e\,a}{\omega c} \frac{(e\cdot\vec{\Delta}_1)}{\Delta_1^2} , \tag{3.44}$$

containing the wave vector transfer Δ_1 defined in Eq.(3.31). This coincides with what one obtains from the Born approximation formula Eq.(3.33), if one takes into account that for a Coulomb potential we have

$$V(\Delta) = - \frac{Ze^2}{2\pi^2} \frac{1}{\Delta^2} . \tag{3.45}$$

By an ingenious method of integration Sommerfeld was able to calculate Eq.(3.39) in closed form (see [T11]). Numerical values of this formula have been tabulated by Grant [T15], and by Karzas and Latter [T16].

We now consider one-photon FFT from atoms, calculated in the one-electron approximation by using a *selfconsistent atomic potential.* Since now the potential and the wave functions are known only numerically, methods of integration different from that of the Coulomb case have to be used. The most direct one is based on a partial wave expansion of the initial and final wave functions $\phi^{(+)}_{\mathbf{k}_i}$, $\phi^{(-)}_{\mathbf{k}_f}$,and yields for f_n of Eqs.(3.35),(3.36), an infinite sum of radial matrix elements (see Johnston [T13]).

A systematic investigation was carried out by Geltman [T17] for a number of neutrals, including the noble gas atoms. He adopted a potential of the form

$$V(r) = V_{HFS}(r) + \frac{1}{r}\,(1 - e^{-r/r_0}) - \frac{\alpha}{2} \frac{(1 - e^{-r/r_p})}{(r^2+r_p^2)^2} . \tag{3.46}$$

Here $V_{HFS}(r)$ is the Hartree-Fock-Slater potential of the neutral atom, the second term was introduced to compensate its undesirable Coulomb tail, and the third term is the polarization potential (α is the atomic polarizability). The free parameters, r_0 and r_p, were adjusted to best fit the known elastic scattering data. The partial wave expansion method was adopted for the calculation. The differential cross section Eq.(2.47) was evaluated for the scattering geometry of Fig. 1, and then the averaged cross sections $\overline{\sigma}_{\pm 1}$ and $\overline{\overline{\sigma}}_{\pm 1}$ were calculated. The errors on the model were estimated at less than 30%.

More recently Coulter [T18] has applied an original method to the calculation of the amplitude Eq.(3.36) and has thus evaluated the cross sections Eqs.(3.38),(3.39) for Ar. The potential was taken of the HF type, with a local approximation for exchange. Coulter et al.

[T19], [T20], have analyzed the polarization of low energy electrons, following one-photon FF absorption on a heavy spinless target (Hg), using the same integration method but a relativistic generalization of Eq.(3.35). A high degree of polarization was found for certain angular configurations. An earlier calculation of the cross section Eq.(3.39) for Ar, done by Ashkin [T21] with the partial wave expansion method, should also be mentioned.

Other works have calculated $\overset{=}{\sigma}_{\pm 1}$, specifically for astrophysical and plasma physics purposes. Lange and Schlüter [T22] have considered the case of the Yukawa (Debye-Hückel) potential, $V(r) = -(e^2/r)\exp(-\mu r)$. Collins and Merts [T23] have carried out a detailed comparison of eight potential models (Yukawa, static, polarization, etc.; with or without local exchange contributions) for a number of neutral atoms. Schlüter [T24] has evaluated $\overset{=}{\sigma}_{\pm 1}$ for the singly charged ions of the rare gases.

A problem which has attracted considerable interest over the years is the *low-frequency limit* of the matrix elements Eqs.(3.35),(3.36) for the case of a short range potential, see [G5], [T13] and references therein. (The low-frequency limit for multiphoton FFT will be discussed in Sec. 4.2.) More recently Rosenberg [G1] [T25] has discussed the extension of these results to the case of a Coulomb potential.

Surprising as it may be (especially when comparing with the situation in the related problem of multiphoton ionization), there was only one attempt to deal in perturbation theory with FFT involving more than one photon. This is the recent work on ***two-photon FFT*** in a Coulomb potential by Gavrila, Maquet and Véniard [T26]. The pertinent amplitudes are given by (see Eqs.(3.5) and (3.8)):

$$f_{\pm 2} = -\frac{m}{2\pi\hbar^2} C^2 < \phi^{(-)}_{\mathbf{k}_f} | Q G^{(+)}(E_{\pm 1}) Q | \phi^{(+)}_{\mathbf{k}_i} > . \tag{3.47}$$

Rotation invariance requires that the result be of the form

$$f_{\pm 2} = P + Q(\mathbf{e}\cdot\hat{\mathbf{k}}_i)^2 + R(\mathbf{e}\cdot\hat{\mathbf{k}}_i)(\mathbf{e}\cdot\hat{\mathbf{k}}_f) + S(\mathbf{e}\cdot\hat{\mathbf{k}}_f)^2 . \tag{3.48}$$

The difficulty with the integration of Eq.(3.47) consists in the presence of the Coulomb Green's function. This, however, is known in closed form as well as in the form of integral representations, both in coordinate and momentum space. The initial and final wave functions are the same as in Eq.(3.40); their Fourier transforms are known in terms of integral representations. In Ref. [T26] the momentum gauge form of the amplitudes was chosen and Eq.(3.35) was integrated in momentum space by using integral representations for all functions involved. By interchanging then the order of integrations, the momentum space integral could be carried out analytically. So could one of the three parameter integrals. The second one was recognized as expressing Gauss functions ${}_2F_1$. The third parametric integral, extending over these Gauss functions, cannot be calculated analytically and requires numerical computation.

(Naturally, the result is one level of complexity higher than that for $f_{\pm 1}$, compare with Eqs.(3.37),(3.41)-(3.43).) Graphs were given for the angular distributions Eq.(2.47), corresponding to various initial scattering configurations (i.e. values of k_i and ω, and the orientation of $\hat{\mathbf{k}}_i$ with respect to **e**). In first Born approximation (to first order in Ze^2) the result reduces to Eqs.(3.33),(3.45) for n = 2, as should be.

4. NONPERTURBATIVE THEORIES

The nonperturbative theories of FFT in high-intensity fields developed in Secs. 4.1, 4.2 and 4.3 cover the cases of high electron energies (Born approximation), low photon frequencies and high photon frequencies, respectively. They can be all obtained from the Schrödinger equation Eq.(2.24) for the Kramers reference frame, or from the equivalent Floquet systems in differential or integral form. In Sec. 4.4 we shall describe other approaches.

4.1. High electron energies

As usually at high electron kinetic energy the potential can be regarded as a small perturbation. It is then convenient to use as a starting point Eq.(2.59), with expression Eq.(2.60) for the FFT amplitudes. By iteration of Eq.(2.60) a formal expansion emerges, whose successive terms contain an increasing number of Fourier components of the oscillating potential, Eq.(2.40). This is the *Born series* for scattering in arbitrarily intense fields:

$$f_n = - \frac{m}{2\pi\hbar^2} \Big[\langle \mathbf{k}_n | V_n | \mathbf{k} \rangle + \tag{4.1}$$

$$+ \sum_{p=1}^{\infty} \sum_{m_1} \cdots \sum_{m_p} \langle \mathbf{k}_n | V_{n-m_1} \, G_0^{(+)} (E_{m_1}) \, V_{m_1-m_2} \cdots G_0^{(+)} (E_{m_p}) \, V_{m_p} | \mathbf{k} \rangle \Big] .$$

It is useful to express it in terms of momentum space integrals, by introducing the Fourier transforms of $V_n(\alpha_0;\mathbf{r})$ and of the Green's function $G_0^{(+)}(E)$, as defined in Eqs.(2.41) and (3.19),(3.20).The result is

$$f_n = - \frac{(2\pi)^2 m}{\hbar^2} \Big[\mathcal{V}_n(\mathbf{k}_n - \mathbf{k}) +$$

$$+ \sum_{p=1}^{\infty} \sum_{m_1} \cdots \sum_{m_p} \int d\mathbf{q}_1 \ldots \int d\mathbf{q}_p \times \mathcal{V}_{n-m_1}(\mathbf{k}_n - \mathbf{q}_1) G_0^{(+)}(E_{m_1}; \mathbf{q}_1) \mathcal{V}_{m_1-m_2}(\mathbf{q}_1 - \mathbf{q}_2)$$

$$\ldots G_0^{(+)} (E_{m_p}; \mathbf{q}_p) \, \mathcal{V}_{m_p}(\mathbf{q}_p - \mathbf{k}) \Big] . \tag{4.2}$$

We have denoted here by $G_0^{(+)}(E;q)$ the function derived from Eq.(3.20):

$$G_0^{(+)}(E;\mathbf{q}) = \frac{2m}{\hbar^2} \, \frac{1}{(k^2 + i\varepsilon) - q^2} . \tag{4.3}$$

The first term of the expansion Eq.(4.2) represents the *first Born approximation* (at arbitrary intensity), originally obtained by Bunkin and Fedorov in 1966 [T27], and rederived in various ways since. In view of Eq.(2.42) this can be written for a central potential ($V(\mathbf{q})$ depends only on $|\mathbf{q}|$) as

$$f_n^{B1} = -\frac{(2\pi)^2 m}{\hbar^2} J_n(\vec{\alpha}_0 \cdot \vec{\Delta}_n)\, V(\Delta_n) \;, \tag{4.4}$$

with the wave vector transfer $\vec{\Delta}_n$ defined as in Eq.(3.31). With Eq.(2.47) the corresponding cross section becomes

$$\frac{d\sigma_n^{B1}}{d\Omega} = \frac{k_n}{k} J_n^2\,(\vec{\alpha}_0 \cdot \vec{\Delta}_n)\, \frac{d\sigma_{E\ell}^{B1}}{d\Omega} \;. \tag{4.5}$$

As in Eq.(3.34), $d\sigma_{E\ell}^{B1}/d\Omega$ is the first Born approximation formula for elastic scattering in the absence of the radiation, at wave vector transfer Δ_n, and the possibility of introducing it can be justified in the same way as there. Eqs.(4.4) and (4.5) display the remarkable feature that their field dependent part factorizes out in the form of a Bessel function.

This Bessel function contains the effect of the field to all orders, since its power series is an expansion in α_0, or a (for their connection recall Eq.(2.27')). Therefore, this expansion should coincide with that obtained for f_n from perturbation theory to all orders in a (see Eq.(3.11)), if in the latter the first Born approximation is made in each of its terms. We shall check the agreement only on the simple case of the lowest order term in α_0 of Eq.(4.4). By using $J_n(x) \cong (x^n/2^n\, n!)$ for $x \to 0$ and $n > 0$, and $J_{-n}(x) = (-1)^n J_n(x)$, one finds from Eq.(4.4) precisely Eq.(3.33), which is indeed the first Born approximation of the n-th (lowest nonvanishing) order perturbation theory result for f_n. One may ascertain that Eq.(4.4) contains indeed the result of the summation of the perturbation theory series for f_n, to first order in V, however. Note that this is a case when the perturbation series is convergent for any (fixed) values of I and ω.

According to Eq.(3.34) the cross section should grow with I as I^n. We know, however, from Eq.(4.5) and the behavior of $J_n^2(x)$ (with $x = \vec{\alpha}_0 \cdot \vec{\Delta}_n$), that this is true only as long as: $x^2 << 2n+2$. Thereafter, the Bessel functions $J_n^2(x)$ start oscillating with an amplitude diminishing slowly to zero. Depending on the value of x it may well happen that for $n_2 > n_1$, $J_{n_2}^2(x) > J_{n_1}^2(x)$, showing that a higher order n_2 cross section has a considerably larger magnitude than one of lower order n_1, contrary to what one should expect from perturbation theory.

Due to its simplicity, Eq.(4.5) can be easily evaluated. For a Coulomb potential ($V = -Ze^2/r$) we just have to insert the Rutherford cross section (which is exact, but also coincides with the first Born approximation):

$$(d\sigma^c/d\Omega) = 4\,(Ze^2 m/\hbar^2)^2\, \Delta^{-4} \;. \tag{4.6}$$

Analytic expressions for the net energy absorption rate by the plasma electrons have been written, mostly assuming a Maxwellian distribution, and then computed for various values of ω and the plasma temperature. For these matters see Geltman [T29], and more recently, Schlessinger and Wright [T31], Bivona, Daniele and Ferrante [T32], and references therein.

4.2. Low frequencies

The case of low frequencies ω has been first considered by Kroll and Watson in 1973 [T33], who have derived the following seminal result for a *short range potential*:

$$f_n^{LF} = J_n(\vec{\alpha}_0 \cdot \vec{\Delta}_n)\, f_{E\ell}(\epsilon_n, \Delta_n)\,(1+O_n) \ . \tag{4.7}$$

Here $f_{E\ell}(\epsilon_n, \Delta_n)$ is the *exact* radiationless (on energy shell) amplitude for elastic scattering from initial wave vector $\bar{\mathbf{k}}$ to the final one $\bar{\mathbf{k}}_n$. These are associated to the physical wave vectors $\mathbf{k}, \mathbf{k}_n$ by

$$\bar{\mathbf{k}} = \mathbf{k} - \mathbf{s}\,, \qquad \bar{\mathbf{k}}_n = \mathbf{k}_n - \mathbf{s}\,; \qquad \mathbf{s} = -(n\omega m/(\mathbf{e}\cdot\vec{\Delta}_n))\,\mathbf{e} \ . \tag{4.8}$$

The wave vector transfer $\vec{\Delta}_n \equiv \bar{\mathbf{k}} - \bar{\mathbf{k}}_n$ is equal to the one defined in Eq.(3.31). It is easily verified that $\bar{\mathbf{k}}$ and $\bar{\mathbf{k}}_n$ both correspond to the same energy, which we have denoted by ϵ_n:

$$\epsilon_n = \frac{\hbar^2 \bar{\mathbf{k}}^2}{2m} = \frac{\hbar^2 \bar{\mathbf{k}}_n^{\,2}}{2m} \ . \tag{4.9}$$

The amplitude $f_{E\ell}(\epsilon_n, \Delta_n)$ is thus completely defined. The relative correction to the Kroll-Watson result has been denoted by O_n in Eq.(4.7).

The cross section corresponding to Eq.(4.7) is

$$\frac{d\sigma_n^{LF}}{d\Omega} = \frac{k_n}{k} J_n^2(\vec{\alpha}_0 \cdot \vec{\Delta}_n) \left(\frac{d\sigma_{E\ell}}{d\Omega}\right)(1 + O'_n) \ , \tag{4.10}$$

where $d\sigma_{E\ell}/d\Omega)$ is the *exact* elastic cross section at energy ϵ_n and momentum transfer Δ_n.

Many proofs have meanwhile been given for the Kroll-Watson formula Eq.(4.7), either from the semiclassical point of view (see Krüger and Jung [T34], Mittleman [T35]) or in QED (see Kelsey and Rosenberg [T36], Bergou and Ehlotzky [T38]). In particular Mittleman [T35] has derived it by using the series of Eq.(4.2) and expanding each of its terms in powers of ω. He found that the corrective terms O_n in Eq.(4.7) were of order ω^2 (and not ω), but that they were no longer expressible in terms of the on - (the energy) - shell amplitude $f_{E\ell}(\epsilon, \Delta)$. A systematic expansion of f_n in ω was recently derived by Krstić and Milosević [T39], with terms which are intricate expressions of off-shell quantities. It appeared thus that the O_n in Eq.(4.7) is in fact of order $(\omega/E)^2$.

were no longer expressible in terms of the on - (the energy) - shell amplitude $f_{E\ell}(\epsilon,\Delta)$. A systematic expansion of f_n in ω was recently derived by Krstić and Milosević [T39], with terms which are intricate expressions of off-shell quantities. It appeared thus that the O_n in Eq.(4.7) is in fact of order $(\omega/E)^2$.

By inserting in Eq.(4.7) the series expansion of the Bessel function $J_n(x)$, one obtains a series expansion in α_0 (or a).This coincides with the (convergent) series expansion of perturbation theory, Eq.(3.11), taken for the low ω limit (recall the similar case of Eq.(4.4)).

An implicit assumption in the derivation of Eq.(4.7) is that the (short range) potential V is such that $f_{E\ell}(\epsilon,\Delta)$ varies smoothly over the energy range in the vicinity of ϵ, in which the FFT occur. If this is not the case, as for example when this energy range contains a resonance of $f_{E\ell}$, Eq.(4.7) has to be modified. Krüger and Jung [T34] have obtained in this case the following formula, valid to lowest order in ω:

$$f_n^{LF} \cong \sum_k J_{n-k}(-\vec{\alpha}_0 \cdot \mathbf{k}_f)\, J_k(\vec{\alpha}_0 \cdot \mathbf{k}_i) \cdot f_{E\ell}(E_k,\Delta_n) \tag{4.11}$$

(This goes over into the lowest order in ω form of Eq.(4.7) if one may assume that, for all relevant values of the summation index k, one can extract the value $f_{E\ell}(E_k,\Delta_n) \cong f_{E\ell}(E,\Delta_n)$ from under the sum; use has then to be made of the sum rule $\sum_k J_{n-k}(x)J_k(y) = J_n(x+y)$.) Jung and Taylor [T40] have discussed how Eq.(4.11) can be applied to resonance scattering in a radiation field. Higher order corrections in ω have been calculated by Mittleman [T41], Milosević and Krstić [T42].

Further extensions of Eq.(4.7) have been made to allow for arbitrary polarization of the plane wave, and for retardation in its propagation (spatial dependence of the phase of the field). For references, see Kelsey and Rosenberg [T36], Rosenberg [T37], Leone, Cavaliere and Ferrante [T43]. A relativistic generalization of Eq.(4.7), using the Volkov solution of the Dirac equation [T5], was given by Kaminski [T44].

To lowest order in ω the Kroll-Watson cross section Eq.(4.10) yields an important sum rule, noted first by Krüger and Jung [T34] (see also [E6]):

$$\sum_n \frac{d\sigma_n^{LF}}{d\Omega} \cong \frac{d\sigma_{E\ell}}{d\Omega} \,. \tag{4.12}$$

In obtaining Eq.(4.12) it was assumed that for all relevant n values under the sum, $k_n \cong k$ and $\epsilon_n \cong E$, and use was made of the sum rule $\Sigma\, J_n^2(x) = 1$. (This assumption may encounter difficulties for extended, though finite range potentials, in forward scattering.) The physical meaning of Eq.(4.12) is that the global cross section for FF scattering in the field, as defined by Eq.(2.48), coincides with the cross section in the absence of the field. The sum rule Eq.(4.12) was generalized by Rosenberg [T45] to include first order corrections in ω and retardation corrections.

The low-frequency and Born approximation results presented above, refer both to the case of a purely single mode field. However, the experiments on multiphoton FFT done so far (Weingartshofer et al., see Sec. 1.2) were carried out with a pulsed laser in multimode operation. This has raised the question of generalizing the previous results to allow for pulse shape and photon statistics. The case of a (quasi) single mode pulse, of average intensity I(t) varying adiabatically on the time scale of the light period involved, was considered by Krüger and Jung [T34]. Assuming that the pulse extends from time t_0 to t_1, they derived the following generalization of Eq.(4.7) for the average of the cross section over the duration of the pulse:

$$\frac{d\bar{\sigma}_n}{d\Omega} \cong \frac{k_n}{k} \left(\frac{1}{t_1 - t_0} \int_{t_1}^{t_0} J_n^2(\vec{\alpha}_0(t)\cdot\vec{\Delta}_n)\, dt \right) \frac{d\sigma_{E\ell}}{d\Omega} \quad . \tag{4.13}$$

Here $\vec{\alpha}_0(t)$ is the value corresponding to I(t) according to Eq.(2.28). They have also shown that assuming reasonable laser pulse shapes, the relative magnitudes of the successive cross sections $(d\bar{\sigma}_n/d\Omega)$ can be entirely different from those expected from the Kroll-Watson formula Eq.(4.10).

The problem of the photon statistics was studied too, under various assumptions (Zoller [T46], Daniele et al. [T47], Curry and Newell [T48], Trombetta et al. [T49], Francken and Joachain [T50]). The result was shown to be quite sensitive to the mode operation of the laser. The conclusion is that in order to be able to compare experimental differential FFT cross sections to theory, a good knowledge of the laser pulse characteristics is required. On the other hand, it turns out that the sum rule Eq.(4.12) is rather insensitive to the latter (e.g. it still holds for Eq.(4.13)) and this is why it could be checked rather well in the experiments of Weingartshofer et al. (see Sec. 1.2).

Other aspects of FFT in intense, low-frequency fields can be found in the review papers by Ehlotzky, Ref. [T51].

4.3. High frequencies

The theory of high-frequency scattering is of more recent date (Gavrila and Kaminski, 1984 [T52]) and was stimulated by the development of very intense high-frequency lasers (e.g. see Rhodes [E12], [E13]). It can be viewed as a counterpart of the low-frequency theory of Kroll and Watson.

Formalism. It is now useful to go back to the space-translated Schrödinger equation, in Floquet form, Eq.(2.43). We rewrite, however, Eq.(2.43) differently, by shifting the term $V_0\Psi_n$ from the right hand side to the left hand side of the equation. With the new definition

$$H = \frac{1}{2m} \mathbf{P}^2 + V_0(\alpha_0;\mathbf{r}) \quad , \tag{4.14}$$

Eq.(2.43) becomes

$$(E_n - H)\,\Psi_n = +\sum_{\substack{m=-\infty \\ m\neq n}}^{+\infty} V_{n-m}\,\Psi_m\,. \tag{4.15}$$

This contains the "dressed potential" $V_0(\alpha_0;\mathbf{r})$ given by Eq.(2.40), so called because it includes the effects of the field on the original $V(\mathbf{r})$ to all orders. From Eq.(2.40), V_0 is simply the time average over a period of the oscillating potential $V(\mathbf{r} + \vec{\alpha}(t))$. The reason for incorporating V_0 in H (as opposed to the partition made in Eq.(2.43), leading to the Born expansion) is that it allows to include the effect of the potential on the asymptotic states of the scattering. In a different context, it allows the description of atomic behavior (structure and ionization) in a high-frequency field (see [T53]).

Solving Eq.(4.15) with the Green's function G associated to Eq.(4.14) yields

$$\Psi_n = u_{\mathbf{k}}^{(\pm)}\,\delta_{n0} + G^{(+)}(E_n) \sum_{\substack{m=-\infty \\ m\neq n}}^{+\infty} V_{n-m}\,\Psi_m\,, \tag{4.16}$$

where we have denoted by $u_{\mathbf{k}}^{(\pm)}(\mathbf{r})$ the scattering solution associated to H:

$$[\frac{1}{2m}\,\mathbf{P}^2 + V_0(\alpha_0,\mathbf{r})]\,u_{\mathbf{k}}^{(\pm)} = E\,u_{\mathbf{k}}^{(\pm)}\,, \tag{4.17}$$

Following the standard procedure described in Sec. 2.5, we can check that Ψ_n satisfies the boundary condition Eq.(2.46) in the Kramers frame, and thereby find the expressions of the FFT amplitudes:

$$f_n = f_0^{(0)}\,\delta_{n0} - \frac{m}{2\pi\hbar^2} \sum_{m\neq n} < u_{\mathbf{k}_n}^{(-)}\,|V_{n-m}|\,\Psi_m > \,, \tag{4.18}$$

where $f_0^{(0)}$ is the elastic amplitude associated to $u_{\mathbf{k}}^{(+)}$. Iterating once we find:

$$f_n = f_0^{(0)}\,\delta_{n0} - \frac{m}{2\pi\hbar^2}[(1-\delta_{n0}) < u_{\mathbf{k}_n}^{(-)}\,|V_n|\,u_{\mathbf{k}}^{(+)} >$$

$$+ \sum_{\substack{m \\ m\neq n}} \sum_{\substack{m' \\ m'\neq m}} < u_{\mathbf{k}}^{(-)}\,|V_{n-m}\,G^{(+)}(E_m)\,V_{m-m'}|\,\Psi_{m'}>]\,. \tag{4.19}$$

Continuing the iteration gives a formal series expansion for f_n. The question arises under which conditions this expansion has decreasing terms ("converges" in some sense), so that the first (tractable) ones represent a good approximation.

A careful analysis ([T52]) shows that the successive terms of the iteration of Eq.(4.17) decrease in magnitude provided that

$$\text{(a)}\quad \omega >> |E_0(\alpha_0)|\;; \qquad \text{(b)}\quad \omega >> E\,, \tag{4.20}$$

where $|E_0(\alpha_0)|$ is the binding energy of the ground state in the potential V_0. The value of α_0 is unrestricted. It can be shown ([T54]) that for small α_0 (i.e. $\alpha_0/a_0 << 1$, where a_0 is the Bohr radius) we are in the realm of perturbation theory (at high frequencies), whereas for $\alpha_0/a_0 \gtrsim 1$

$$f_0^{HF} \cong f_0^{(0)} \; , \tag{4.21}$$

which depends on the laser frequency and intensity only through the parameter α_0, Eq.(2.28). For high, but finite ω, FFT become possible and we find, to lowest nonvanishing order in $1/\omega$, the amplitudes ($n \neq 0$):

$$f_n^{HF} = - \frac{m}{2\pi\hbar^2} < u_{\mathbf{k}_n}^{(-)} |V_n| \, u_{\mathbf{k}}^{(+)} > \; . \tag{4.22}$$

Note that with the conditions Eq.(4.20) satisfied, only absorption ($n > 0$) can occur (emission is excluded).

Modified Coulomb scattering. As an application of the preceding formalism we consider the laser modified Coulomb scattering in the high frequency limit, described by Eqs.(4.17),(4.21). We present here some aspects of the accurate calculation by van de Ree, Kaminski and Gavrila just completed [T56]. Exploratory calculations were done earlier for the Coulomb case by Offerhaus, Kaminski and Gavrila [T57], and for the Yukawa case by Gavrila, Offerhaus and Kaminski [T58].

The dressed potential for the case of linear polarization, Eq.(2.40), can also be written as

$$V_0(\alpha_0,\mathbf{r}) = \int_{-\alpha_0}^{+\alpha_0} V(\mathbf{r}-\xi\mathbf{e}) \, \sigma(\xi) \, d\xi \; , \tag{4.23}$$

where $V(\mathbf{r})$ is the original potential, and

$$\sigma(\xi) = (1/\pi\alpha_0) \, [1 - (\xi/\alpha_0)^2]^{-1/2} \, , \qquad \int_{-\alpha_0}^{+\alpha_0} \sigma(\xi) \, d\xi = 1 \; , \tag{4.24}$$

(Atomic units are used here and in the following.) This shows that $V_0(\alpha_0,\mathbf{r})$ can be envisaged as the potential created by the smearing out of the point-charge of the nucleus, into a linear distribution of density $\sigma(\xi)$, extending from $-\alpha_0$ to $+\alpha_0$ along the direction of $\mathbf{e}$. This interpretation of V_0 is quite natural, since the rapid oscillations of the potential $V(\mathbf{r} + \vec{\alpha}(t))$ in the Schrödinger equation for the Kramers frame of reference, with the Hamiltonian Eq.(2.24), will affect the electron only on the average, and this is equivalent to the static potential Eqs.(4.23),(4.24).

For the Coulomb potential $V(r) = -1/r$, we have:

$$V_0(\alpha_0,\mathbf{r}) = -(2/\pi) \, (r_+ r_-)^{\,1/2} \; K[2^{-1/2} \, (1 \; \hat{\mathbf{r}}_+ \hat{\mathbf{r}}_-)^{1/2}] \; , \tag{4.25}$$

where $\mathbf{r}_\pm = \mathbf{r} \pm \alpha_0 \mathbf{e}$ (the origin of the coordinates is kept at the center of $V(\mathbf{r})$), and $K(x)$ is the complete elliptic integral of the first kind. Eq.(4.25) shows that V_0 has a logarithmic singularity along the line of charges (caused by $K(x)$, for $x \to 1$), and $r^{-1/2}$ type singularities at its endpoints. These are much weaker than the original $1/r$ singularity, so that V_0 is on the whole considerably weaker in the vicinity of the origin than the Coulomb potential. At large distances where $\mathbf{r}_\pm = \mathbf{r} \pm \alpha_0 \mathbf{e}$ (the origin of the coordinates is kept at the center of $V(\mathbf{r})$), and $K(x)$ is the complete elliptic integral of the first kind. Eq.(4.25) shows that V_0 has a logarithmic singularity along the line of charges (caused by $K(x)$, for $x \to 1$), and $r^{-1/2}$ type singularities at its endpoints. These are much weaker than the original $1/r$ singularity, so that V_0 is on the whole considerably weaker in the vicinity of the origin than the Coulomb potential. At large distances ($r >> \alpha_0$), V_0 can be expanded in inverse powers of r, the dominant term being $V_0 \cong V = -1/r$. A graphical representation of V_0 was given in Fig. 1 of Ref. [T52].

$V_0(\alpha_0,\mathbf{r})$ is axially symmetric around $\mathbf{e}$ and has even parity. As a consequence, the azimuthal quantum number ℓ is no longer conserved in the collision, although the magnetic quantum number m associated with the axis $\mathbf{e}$, is. The scattering problem now bears similarity to that for radiationless electron scattering from homonuclear diatomic molecules. Of the many approaches developed in that context we have adapted for our needs here the "one-center expansion" (see Takayanagi [T59]), which analyzes the scattering equation in a spherical harmonics basis. The polar axis of the spherical harmonics was taken along $\mathbf{e}$.

Thus, the scattering solution $u^{(+)}_{\mathbf{k}_i}(\mathbf{r})$ is sought in the form

$$u^{(+)}_{\mathbf{k}_i}(\mathbf{r}) = \frac{2\pi}{kr} \sum_{\ell,\ell',m} i^{\ell'+1} e^{i\sigma_{\ell'}} g^{\ell' m}_{\ell m}(\alpha_0,k;r)\, Y_{\ell m}(\hat{\mathbf{r}})\, Y^*_{\ell' m}(\hat{\mathbf{k}}_i), \tag{4.26}$$

where we are denoting now by $\mathbf{k}_\alpha = k\hat{\mathbf{k}}_\alpha$ ($\alpha = i,f$) the initial and final momenta of the electron, and σ_ℓ are the Coulomb phases. $u^{(+)}_{\mathbf{k}_i}$ should satisfy the same boundary condition as for the usual Coulomb potential (Ref. [G15], Sec. XI-7). This imposes the following asymptotic behavior on the functions $g^{\ell' m}_{\ell m}(r)$:

$$g^{\ell' m}_{\ell m}(r) \underset{r\to\infty}{\to} \delta_{\ell\ell'} \exp\{-i(kr - \gamma \ln 2kr - \frac{\ell\pi}{2} + \sigma_\ell)\}$$

$$- S^m_{\ell\ell'}(\alpha_0,k) \exp\{+i(kr - \gamma \ln 2kr - \frac{\ell\pi}{2} + \sigma_\ell)\}\,. \tag{4.27}$$

Eq.(4.27) contains the S-matrix elements in the spherical harmonics basis $S^m_{\ell\ell'}(\alpha_0,k)$, for given (conserved) quantum number m, and $\gamma = -1/k$.

To determine the S-matrix one has to solve the equations for the functions $g^{\ell' m}_{\ell m}(r)$. These can be obtained by inserting Eq.(4.26) into the integral form of the Schrödinger equation (incorporating the boundary conditions), which results in a system of coupled integral equations. The latter were solved following a method developed by van de Ree [T60]. The S-matrix elements were calculated with a numerical accuracy of 10^{-5} for any given number of coupled ℓ-channels.

As noted, the dressed potential V_0 is a Coulomb potential modified at finite distances. It is then useful to apply the "two-potential" formalism (Ref. [G15], Secs. XI-11 and XIX-11) and write the scattering amplitude in the form of the sum:

$$f_{\mathbf{k}_i}(\alpha_0;\hat{\mathbf{k}}_f) = f_c(\theta) + f'_{\mathbf{k}_i}(\alpha_0;\hat{\mathbf{k}}_f) \ , \tag{4.28}$$

$$f_c(\theta) = \frac{(-\gamma)}{2k\sin^2(\theta/2)} \exp\left[-i\gamma \ln \sin^2(\theta/2) + 2i\sigma_0\right] \ , \tag{4.29}$$

$$f'_{\mathbf{k}_i}(\alpha_0;\hat{\mathbf{k}}_f) = \frac{2\pi}{ik} \sum_{\ell,\ell',m} i^{\ell'-\ell}\, e^{i\sigma_\ell}\, [S^m_{\ell\ell'}(\alpha_0,k) - \delta_{\ell\ell'}]\, e^{i\sigma_{\ell'}}$$

$$\cdot\, Y_{\ell m}(\hat{\mathbf{k}}_f)\, Y^*_{\ell' m}(\hat{\mathbf{k}}_i) \ . \tag{4.30}$$

Here $f_c(\theta)$ is the exact Coulomb scattering amplitude at scattering angle θ, and f' is the extra contribution deriving from the departure of V_0 from the Coulomb form. The scattering amplitude $f_{\mathbf{k}_i}(\hat{\mathbf{k}}_f)$ has a number of symmetry properties, due to the axial symmetry and even parity of V_0 (see [T56]).

The modified elastic scattering cross section is

$$(d\sigma/d\Omega) = |f|^2 = |f_c|^2 + 2\,\mathrm{Re}(f_c^* f') + |f'|^2 \ . \tag{4.31}$$

The first term on the right-hand side is the original Rutherford cross section, $d\sigma^c/d\Omega = |f_c|^2$, Eq.(4.6), and the second represents the interference between the Coulomb and the extra scattering amplitude.

In order to calculate the amplitude Eqs.(4.28)-(4.30) we need choose a set of independent angles to characterize the configuration of the momenta, in which we want to include the scattering angle θ. To this end we shall denote the reference system used above, in Eqs.(4.26)-(4.30), by S, specifying that it has the x axis in the $(\mathbf{e},\mathbf{k}_i)$ plane. We then introduce another reference system S', obtained by a rotation of S by an angle θ_i around the y axis in the direct sense, so that the z axis coincides with $\mathbf{k}_i$. The polar angles of $\mathbf{k}_i$ in S are θ_i, $\phi_i = 0$, and the polar angles of $\mathbf{k}_f$ in S' are θ and ϕ, where θ is the scattering angle; θ_i, ϕ, θ are the angles we shall be using. In terms of these angles the symmetry properties of the amplitude $f(\theta_i;\theta,\phi)$ may be expressed as:

$$f(\theta_i;\theta,\phi) = f(\pi-\theta_i;\theta,\phi), \qquad f(\theta_i,\theta,\phi) = f(\theta_i;\theta,2\pi-\phi) \ , \tag{4.32}$$

These allow to reduce the ranges of the angular variables to $0 \le \theta_i \le \pi/2$, $0 \le \theta \le \pi$ and $0 \le \phi \le \pi$. (Note that for symmetry reasons, at $\theta_i = 0$, f is independent of ϕ for all θ, whereas for geometrical reasons at 0 – 180°, f is independent of ϕ for any θ_i.)

We give here illustrative results for the laser modified scattering by considering $R = (d\sigma/d\Omega)/(d\sigma^c d\Omega)$, the ratio of the cross section Eq.(4.31) to the original Rutherford one Eq.(4.6), as a function of θ at given θ_i and ϕ. We have chosen the case low energy $E = 0.01$ Ry, and the nonperturbative value $\alpha_0 = 2$ a.u. (Fig. 7). The results present some remarkable features:

(1) *Forward scattering.* For $\theta \to 0$, $R(\theta) \to 1$ in all cases. This is due to the fact that $|f_c| \to \infty$, whereas f' stays finite, so that the relative contribution to $R(\theta)$ of the last two terms in Eq.(4.31) tends to zero.

(2) *Coulomb interference oscillations.* At small and for vanishing θ, $R(\theta)$ has an infinite number of oscillations (which cannot be distinguished in Fig. 7). These are due to the interference term in Eq.(4.31) and to the fact that f_c contains the oscillating θ-dependent phase factor, Eq.(4.29).

(3) *Backward scattering.* At large scattering angles, $\theta \geq 90°$, the Coulomb interference pattern is distorted and R may deviate substantially from 1. This is markedly dependent on the configuration of the momenta, in particular at $\theta = 180°$.

(4) *Magic angles.* A striking feature in Fig. 7 is that the $R(\theta)$ curves for *all* ϕ (at given $\theta_i \neq 0$) tend to cross at certain scattering angles θ_m, which we shall call "magic angles". These curve crossings are not exactly pointwise and we have found them in all cases when $\alpha_0 k < 1$. Physically speaking the occurrence of these angles means that, for any $\hat{\mathbf{k}}_i$ (or θ_i), if the scattering angle θ is equal to θ_m the scattering is (approximately) axially invariant around $\hat{\mathbf{k}}_i$.

Thus, laser modified elastic scattering from a Coulomb potential at high frequencies (when multiphoton free-free transitions are quenched) may deviate considerably from the Rutherford formula and shows remarkable interference effects.

4.4. Other approaches

We have discussed so far some of the basic nonperturbative approaches on the case of a potential model for the atom. We shall briefly consider in this section other calculations using the potential model, and the generalizations made to account for detailed atomic structure.

The methods described in Secs. 4.1 - 4.3 produce expansions of the FFT amplitudes f_n in some parameter, different from the intensity. However, only their dominant terms have lent themselves to numerical computation. Thus, for a given potential there is no comprehensive picture of the FFT phenomenon at arbitrary values of ω and I, nor does one know quantitatively the limitations of the previous approximations. The gap is filled to some extent by the discovery of potential models which, although somewhat peculiar, are analytically soluble (up to a point).

Among these *soluble potential models* the best established is the *zero- range potential*, V_0. This can be defined by

$$V_0 \phi(\mathbf{r}) \equiv A \delta(\mathbf{r}) \frac{\partial}{\partial r} (r\phi(\mathbf{r})) , \qquad (4.33)$$

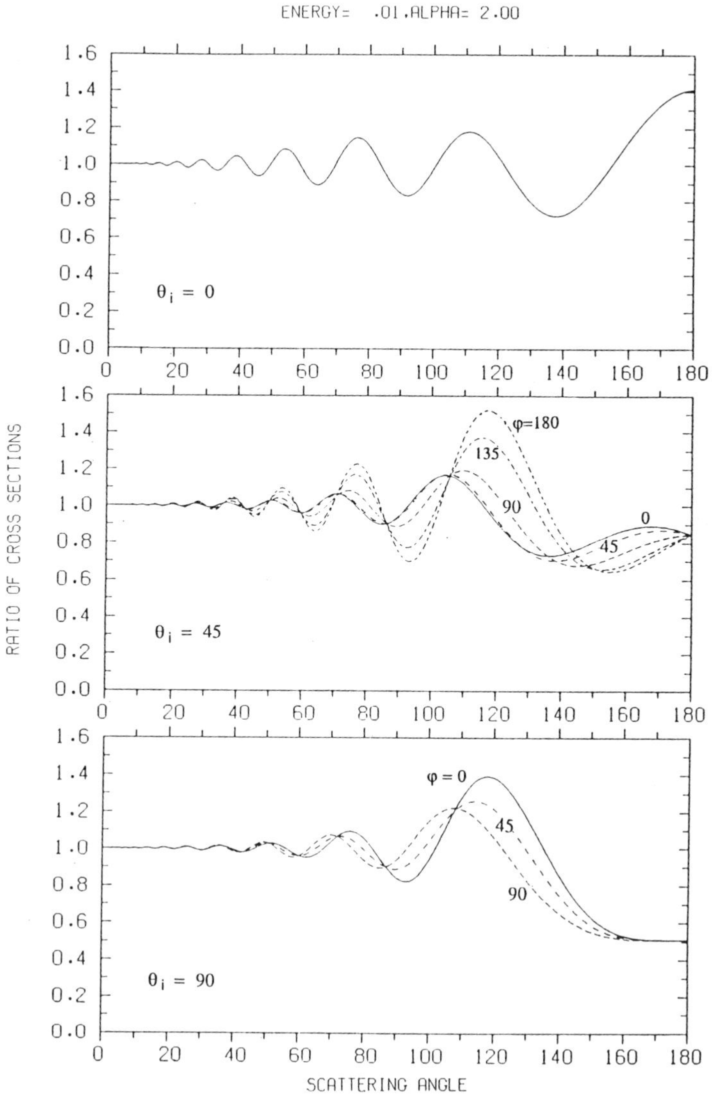

Fig. 7. Ratio R of the elastic scattering cross section for the dressed Coulomb potential, Eq.(4.31), to the Rutherford cross section Eq.(4.6), as function of the scattering angle θ. Values of θ_i and ϕ as indicated. For $\theta_i = 90°$, the R curves for $\phi = 135°$ and 180° coincide with those for $\phi = 45°$ and 0°, respectively (see Ref. [T56]). The electron energy is E = 0.01 Ry and $\alpha_0 = 2$ Bohr radii.

where A is a positive constant. V_0 is a nonlocal potential, capable of supporting one bound *s* state, and can act only on *s* waves in scattering. The coupling of a particle in this potential to a circularly polarized, monochromatic radiation field of arbitrary intensity, leading to multiphoton ionization and FFT, was solved by Berson in 1975 [T61] (the case of multiphoton ionization was considered also in Refs. [T62],[T63]). He derived analytical expressions for the scattering amplitudes f_n and showed that in the appropriate limits they reduce to the forms given in Secs. 4.1, 4.2 (f_n is zero in the high-frequency limit of Sec. 4.3, because of the peculiar form of the potential).

In order to discuss also the high-frequency limit, Kaminski [T64] has generalized V_0 to a finite-range nonlocal potential, which still allows analytic solutions for the f_n. A similar generalization was done by Faisal [T65],[T66].

No numerical investigations have been carried out so far on ordinary potentials of the selfconsistent type, beyond the limiting cases of Secs. 4.1 - 4.3. Various analytical approaches could serve as basis for such computations, in particular the Floquet method in one of the forms discussed in Secs. 2.3, 2.5. *Exploratory computations* have been made only under simplifying assumptions. Thus, Shakeshaft has applied the integral formulation of the Floquet system to study a separable potential projecting on a *s* state [T67], and then to the case of a spherically symmetric Gaussian potential [T68]. In the latter case he has found that the Kroll-Watson approximation extends to rather high frequencies. Bhatt, Piraux and Burnett [T69] have treated by the differential Floquet method (see Eq.(2.43)) a one-dimensional potential model and have discussed the validity of the high-frequency approximation developed in Sec. 4.3.

We now mention some of the works done on ***FFT from atoms***, with the internal structure taken into account. Several channels are now open: the atom can be left in the initial state, can be excited, or ionized. Further, assumptions have to be made as to the extent the atomic structure is perturbed by the radiation field.

In a first group of papers the effort was directed towards developing the *general theory*. Various scattering formalisms were presented by Rosenberg [T70], [T71]. A general framework for computation in terms of Green's functions was presented by Deguchi, Taylor and Yaris [T72], see also [T73]. As the target atom is itself placed in the field, it is undergoing decay by multiphoton ionization; this aspect was taken into account by Fiordilino and Mittleman [T74].

Other papers have addressed the problem of extending to atoms the *low-frequency approximation* of Kroll and Watson. This was done by Mittleman [T75], [T76] for scattering without excitation, and by Banerji and Mittleman [T77] for ionizing collisions on H (for He see Zarcone, Moores and McDowell [T78]). The effects on the scattering of the existence of nearly degenerate atomic levels, or closely spaced resonances were discussed by Rosenberg [T79], [T80]. An extension of the sum rule Eq.(4.12) to atoms was given by Beilin and Zon [T81].

Another case which has attracted interest is that of *radiation of resonant frequency* with respect to an atomic transition. In this case a reasonable approximation is to reduce the atomic description to a two-level model. Work along these lines was done by Gazazian [T82], Gersten and Mittleman [T83], Mittleman [T84], Cavaliere, Ferrante and Leone [T85].

We finally consider the case when the electron energy is large enough so that the *Born approximation* in the interaction of the incoming electron with the target atom is valid, but the frequency may be arbitrary. A first Born calculation was done for H by Prasad and Unnikrishnan [T86] (see also [T87]), using unperturbed atomic eigenfunctions. Francken and Joachain [T88], and Byron, Francken and Joachain [T89] (see also [T90]), have shown that striking effects appear in the scattering without atomic excitation if radiatively perturbed wavefunctions are used (H and He were considered).

We conclude by noting that various other aspects related to the influence of atomic structure can be found in the general references [G1]-[G3]. Further information on FFT from atoms can be found in the Proceedings of the three workshops edited by Rahman, Guidotti and (for the last one also) Allegrini [T91]-[T93].

REFERENCES

General

G1. Rosenberg L., *Theory of Electron-atom Scattering in a Radiation Field*, Adv. At. Molec. Phys. **18**, 1 (1982).

G2. Mittleman M., *Theory of Laser-Atom Interactions* (Plenum Press, 1982).

G3. Faisal F., *Theory of Multiphoton Processes* (Plenum Press, 1987).

G4. Gavrila M., *Free-free Photoabsorption of Electron-atom Systems*, in *Electronic and Atomic Collisions (Proceedings ICPEAC X)*, Ed. Watel G. (North-Holland Publ. Co., 1978), p.165.

G5. Gavrila M. and van der Wiel M., *Free-free Radiative Transitions*, Comments Atom. Mol. Phys. **8**, 1 (1978).

G6. Myerscough V.P. and Peach G., *Atomic Processes in Astrophysical Plasmas*, in *Case Studies in Atomic and Molecular Physics II*, Ed. McDaniel E. and McDowell M. (North-Holland Publ. Co., 1972).

G7. Gallagher J.W., *Bibliography of Free-free Transitions in Atoms and Molecules*, JILA Information Center Rep. **16** (Boulder, Colo. USA, 1977).

G8. Biberman L.M. and Norman G.E., Sov. Phys. Usp. **10**, 52 (1967).

G9. Popp H.P., *The Radiation of Atomic Negative Ions*, Physics Reports **16**, 169 (1975).

G10. Grey Morgan C., *Laser-induced Breakdown of Gases*, Rep. Progr. Phys. **38**, 621 (1975).

G11. Chin S.L. and Lambropoulos P. (Eds.), *Multiphoton Ionization of Atoms* (Academic Press ,1984).

G12. Crance M., *Multiphoton Stripping of Atoms*, Phys. Rep. **144**, 117 (1987).

G13. Bethe H. and Salpeter E., *Quantum Mechanics of One- and Two-Electron Systems* (Springer, 1957).

G14. Pratt R.H. and Feng I.J., *Electron-atom Bremsstrahlung*, in *Atomic Inner-Shell Physics*, Ed. B. Crasemann (Plenum Press, 1985).

G15. Messiah A., *Quantum Mechanics* (North-Holland Publ. Co., 1961).

G16. Joachain C.J., *Quantum Collision Theory* (North-Holland Publ. Co., 1975).

Experimental

E1. Andrick D. and Langhans L., J. Phys. B **9**, L459 (1976).

E2. Andrick D. and Langhans L., J. Phys. B **11**, 2355 (1978).

E3. Langhans L., J. Phys. B **11**, 2361 (1978).

E4. Andrick D. and Bader H., J. Phys. B **17**, 4549 (1984).

E5. Bader H., J. Phys. B **19**, 2177 (1986).

E6. Weingartshofer A., Holmes J., Caudle G., Clarke E. and Krüger H., Phys. Rev. Lett. **39**, 269 (1977).

E7. Weingartshofer A., Clarke E., Holmes J. and Jung C., Phys. Rev. A **19**, 2371 (1979).

E8. Weingartshofer A., Holmes J., Sabbagh J. and Chin S.L., J. Phys. B **16**, 1805 (1983).

E9. Weingartshofer A. and Jung C., in Ref. [G11], p.155.

E10. Wallbank B., Connors V.W., Holmes J.K. and Weingartshofer A., J.Phys. B **20**, L833 (1987).

E11. Freeman R.R., McIlrath T.J., Bucksbaum P.H., Bashkansky M., Phys. Rev. Lett. **57**, 3156 (1986).

E12. Rhodes C.K., Science **229**, 1345 (1985).

E13. Rhodes C.K., Phys. Scripta **T17**, 193 (1987).

Theoretical

T1. Pannekoek A., Mon. Not. R. Astr. Soc. **91**, 162 (1931)

T2. Chandrasekhar S. and Breen G., Astrophys. J. **104**, 430 (1946).

T3. Bialynicki-Birula I. and Bialynicka-Birula Z., Phys. Rev. A **14**, 1101 (1976).

T4. Mollow B.R., Phys. Rev. A **12**, 1919 (1975).

T5. Volkov D.M., Zeits. Physik **94**, 250 (1935).

T6. Kramers H.A., *Collected Scientific Papers* (North-Holland, 1956), p.845.

T7. Henneberger W.C., Phys. Rev. Lett. **21**, 838 (1968).

T8. Faisal F.H.M., J. Phys. B **6**, L89 (1973).

T9. Coddington E.A. and Levinson N., *Theory of Differential Equations* (McGraw-Hill, 1955).

T10. Sommerfeld A., Ann. d. Physik **11**, 257 (1931).

T11. Sommerfeld A., *Atombau und Spektrallinien*, Vol. 2 (F. Vieweg, Braunschweig, 1939), Chap. 7.

T12. Alder K., Bohr A., Huus T., Mottelson B. and Winther A., Revs. Mod. Phys. **28**, 432 (1956).

T13. Johnston, R.R., J. Quant. Spectr. Rad. Transf. **7**, 815 (1967).

T14. Véniard V., Thèse de Doctorat (Chimie Physique, Univ. P. et M. Curie, Paris, Febr. 1986), unpublished.

T15. Grant I.P., Mon. Not. Roy. Astr. Soc. **118**, 241 (1958).

T16. Karzas W.J. and Latter R., Astrophys. J. Suppl. **6**, 167 (1961).

T17. Geltman S., J. Quant. Spectr. Rad. Transf. **13**, 601 (1973).

T18. Coulter P.W., Phys. Rev. A **28**, 2881 (1983).

T19. Coulter P.W., Mian S.N. and Ritchie B., Phys. Rev. A **29**, 509 (1984).

T20. Coulter P.W. and Ritchie B., Phys. Rev. A **24**, 3051 (1981).

T21. Ashkin M., Phys. Rev. **141**, 41 (1966).

T22. Lange R. and Schlüter D., J. Quant. Spectr. Rad. Transf. **33**, 237 (1985).

T23. Collins L.A. and Merts A.L., Los Alamos Report LA-10553 (UC-34A), Jan. 1986.

T24. Schlüter D., Zeits. f. Physik D **6**, 249 (1987).

T25. Rosenberg L., Phys. Rev. A **26**, 132 (1982); **27**, 1879 (1983); **31**, 2180 (1985).

T26. Gavrila M., Maquet A. and Véniard V., Phys. Rev. A **32**, 2537 (1985), and to be submitted for publication.

T27. Bunkin F.V. and Fedorov M.V., Sov. Phys. JETP **22**, 844 (1966).

T28. Brehme H., Phys. Rev. C **3**, 837 (1971).

T29. Geltman S., J. Res. NBS **82**, 173 (1977).

T30. Denisov M.M. and Fedorov M.V., Sov. Phys. JETP **26**, 779 (1968).

T31. Schlessinger L. and Wright J., Phys. Rev. A **20**, 1934 (1979); **22**, 909 (1980).

T32. Bivona S., Daniele R. and Ferrante G., J. Phys. B **15**, 1585 (1982).

T33. Kroll N. and Watson K., Phys. Rev. A **8**, 804 (1973).

T34. Krüger H. and Jung Chr., Phys. Rev. A **17**, 1706 (1978).

T35. Mittleman M.H., Phys. Rev. A **19**, 134 (1979).

T36. Kelsey E.J. and Rosenberg L., Phys. Rev. A **19**, 756 (1979).

T37. Rosenberg L., Phys. Rev. A **20**, 275 (1979).

T38. Bergou J. and Ehlotzky F., Phys. Rev. A **33**, 3054 (1986).

T39. Krstić P.S. and Milosević D.B., J. Phys. B **20**, 3487 (1987).

T40. Jung Chr. and Taylor H., Phys. Rev. A **23**, 1115 (1981).

T41. Mittleman M.H., Phys. Rev. A **20**, 1965 (1979).
T42. Milosević D.B. and Krstić P.S., J. Phys. B **20**, 2843 (1987).
T43. Leone C., Cavaliere P. and Ferrante G., J. Phys. B **17**, 1027 (1984).
T44. Kaminski J.Z., J. Phys. A **18**, 3365 (1985).
T45. Rosenberg L., Phys. Rev. A **20**, 1352 (1979).
T46. Zoller P., J. Phys. B **13**, L249 (1980).
T47. Daniele R., Faisal F.H. and Ferrante G., J. Phys. B **16**, 3831 (1983).
T48. Curry P.J. and Newell W.R., J. Phys. B **17**, 3353 (1984).
T49. Trombetta F., Ferrante G., Wodkiewicz K. and Zoller P., J. Phys. B **18**, 2915 (1985).
T50. Francken P. and Joachain C.J., Europhys. Lett. **3**, 11 (1987).
T51. Ehlotzky F., Can. J. Phys. **59**, 1200 (1981); **63**, 907 (1985).
T52. Gavrila M. and Kaminski J.Z., Phys. Rev. Lett. **52**, 613 (1984), and to be submitted for publication.
T53. Gavrila M., in *Atoms in Unusual Situations*, Ed. J.P. Briand, NATO ASI series B, Vol. 143 (Plenum Press, 1987), p. 225.
T54. Gavrila M. and Kaminski J.Z. (to be published).
T55. Pont M. and Gavrila M., Phys. Lett. A **123**, 469 (1987).
T56. Van de Ree J., Kaminski J.Z. and Gavrila M., Phys. Rev. A **37**, 4536 (1988).
T57. Offerhaus M.J., Kaminski J.Z. and Gavrila M., Phys. Lett. A **112**, 151 (1985).
T58. Gavrila M., Offerhaus M.J. and Kaminski J.Z., Phys. Lett. A **118**, 331 (1986).
T59. Takayanagi K., Progr. Th. Phys. (Suppl.) **40**, 216 (1967).
T60. Van de Ree J., J. Phys. B **15**, 2245 (1982).
T61. Berson I.J., J. Phys. B **8**, 3078 (1975).
T62. Manakov N.L. and Rapoport L.P., Sov. Phys. JETP **42**, 430 (1976).
T63. Muller H.G. and Tip A., Phys. Rev. A **30**, 3039 (1984).
T64. Kaminski J.Z., Phys. Lett. A **120**, 396 (1987).
T65. Faisal F.H.M., Phys. Lett A **119**, 375 (1987).
T66. Faisal F.H.M., Phys. Lett. A **125**, 200 (1987).
T67. Shakeshaft R., Phys. Rev. A **28**, 667 (1983).
T68. Shakeshaft R., Phys. Rev. A **29**, 383 (1984).
T69. Bhatt R., Piraux B. and Burnett K., Phys. Rev. A **37**, 98 (1988).
T70. Rosenberg L., Phys. Rev. A **16**, 1941 (1977).
T71. Rosenberg L., Phys. Rev. A **23**, 2283 (1981).
T72. Deguchi K., Taylor H.S. and Yaris R., J. Phys. B **12**, 613 (1979).
T73. Deguchi K., Chem. Phys. **42**, 201 (1979).
T74. Fiordilino E. and Mittleman M.H., J. Phys. B **16**, 2205 (1983).
T75. Mittleman M.H., Phys. Rev. A **19**, 99 (1979).
T76. Mittleman M.H., Phys. Rev. A **21**, 79 (1980).

T77. Banerji J. and Mittleman M.H., J. Phys. B **14**, 3717 (1981).

T78. Zarcone M., Moores D.L. and McDowell M.R.C., J. Phys. B **16**, L11 (1983).

T79. Rosenberg L., Phys. Rev. A **22**, 2485 (1980).

T80. Rosenberg L., Phys. Rev. A **28**, 2727 (1983).

T81. Beilin E.L. and Zon B.A., J. Phys. B **16**, L159 (1983).

T82. Gazazian A.D., J. Phys. B **9**, 3197 (1976).

T83. Gersten J.I. and Mittleman M.H., Phys. Rev. A **13**, 123 (1976).

T84. Mittleman M.H., Phys. Rev. A **14**, 1338 (1976); **15**, 1355 (1977); **16**, 1549 (1977); **18**, 685 (1978).

T85. Cavaliere P., Ferrante G. and Leone C., J. Phys. B **15**, 475 (1982).

T86. Prasad M. and Unnikrishnan K., J. Phys. B **16**, 3443 (1983).

T87. Unnikrishnan K. and Prasad M., Phys. Rev. A **29**, 3423 (1984).

T88. Francken P. and Joachain C.J., Phys. Rev. A **35**, 1590 (1987).

T89. Byron F.W., Francken P. and Joachain C.J., J. Phys. B **20**, 5487 (1987).

T90. Byron F.W. and Joachain C.J., J. Phys. B **17**, L295 (1984).

T91. Rahman N.K. and Guidotti C. (Eds.), *Photon Assisted Collisions and Related Topics* (Harwood Acad. Publ., London, 1982).

T92. Rahman N.K. and Guidotti C. (Eds.), *Collisions and Half-Collisions with Lasers* (Harwood Acad. Publ., London, 1984).

T93. Rahman N.K., Guidotti C. and Allegrini M. (Eds.), *Photons and Continuum States of Atoms and Molecules* (Springer Proceedings in Physics **16**, 1987).

CORRELATION EXPERIMENTS WITH SECONDARY ELECTRONS

Giovanni Stefani

Istituto Metodologie Avanzate Inorganiche C N R
CP 10, 00016 Monterotondo, Italy

INTRODUCTION

In the final state of an electron induced ionisation event there are at least three unbound charged particles interacting via long range Coulomb forces. Due to the nature of the Coulomb field, they are never independent, not even when the break-up reaction is over and their energies may already be in the continuum. This results in the presence of final state interactions (correlations) whose effect can be of such an extent to determinate a partitioning of energy and angular momentum among the final products, which is noticeably different from the one predicted on the basis of models that don't account for correlations, (i.e. the first order interaction models).

Final state correlations are difficult to deal with in quantum mechanical scattering theories. It is therefore important to gather detailed experimental evidence of their characteristic effects. The probability distribution for the emission of secondary electrons (i.e. others than the primary scattered ones) bears relevant traces of these effects. Among the various experiments, the triple differential cross-section (TDCS) measured by the (e,2e) experiments, was shown to be the most sensitive tool (1).

The (e,2e) technique consists of measuring simultaneously the energy Eo of the incident electron, the energy Ea and Eb of the two final electrons and the probability for these electrons to be emitted into solid angles around the directions (Θa, Φa) and (Θb, Φb). These experiments may then be expected to yield relevant hints for the theory on final state correlations because the kinematics of the ionising collision is fully determined. Furthermore, by tuning the energy (E) and the momentum transferred (K), a large variety of different kinematics can be selected thus allowing for enhancing or dimming the correlation effects.

The coincidence spectrometers used for the (e,2e) experiments are usually crossed beam apparatuses such as the one sketched in Figure 1. It features an electron source that delivers a well collimated beam of monochromatic electrons that crosses an effusive gaseous beam. Two electron spectrometers rotate independently around the scattering center and detect pairs of electrons coincident in time and carefully selected in energy and scattering angle.

The totality of the processes leading to ionisation is represented by the continuum of the Bethe surface (2). Presently most of it has been explored by (e,2e) experiments. This seminar will be confined to those experiments performed at medium and high incident energies (five times the ionisation threshold and higher). Experiments performed at few eV above the ionisation threshold, where correlation effects are dominant, were recently achieved (3).

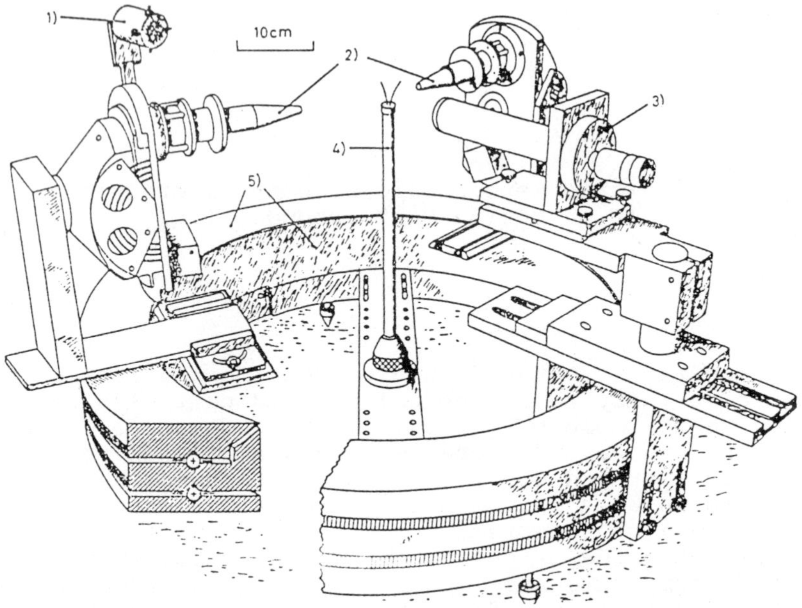

Fig. 1. Schematic view of an (e,2e) spectrometer. The components illustrated include: 1) Faraday cup, 2) electron analysers, 3) electron gun, 4) gaseous beam, 5) independently rotatable turntables. Auxiliary components, electrical connections and electrostatic shields are omitted for graphical clearness.

In the high energy limit, first order interaction models have been extensively applied to describe the ionisation mechanism. First Born models are applied to the "dipolar" regime, characterised by small momentum transfer and highly asymmetric kinematics (1). Departure from First Born theories result from higher order scattering processes, due to short range-like forces and/or long range Coulomb interactions. Several different Impulsive Approximations are used in the "binary" region of the Bethe surface, which in turn is characterised by large momentum transfer and almost symmetric kinematics (4). This regime is dominated by direct knock-out processes and the Plane Wave Impulse Approximation (PWIA) (5) is appropriate to describe the TDCS, provided that Eo both and K are large enough. By reducing either one of them, a simple first order impulsive model becomes inadequate. For both the aforementioned regimes, quite a number of higher order formulations have been put forward, namely the Distorted Wave Impulse Approximation (DWIA) and the Averaged Eikonal Distorted Wave Impulse Approximation (EWIA) (6, 7) for the "impulsive" regime and the Eikonal Born, Coulomb projected Born, Distorted Waves and Second Born approximations for the "dipolar" regime (8). Most of these models, extensively discussed elsewere in this Institute (9), have been successful in describing the main features of the TDCS. Unfortunately, all of them are computationally demanding, thus limiting their application to simple targets. Moreover, they do not allow for an easy tracing of the physical processes beyond first order which are relevant to the dynamics of the collision. It is the aim of this seminar to point out in the simplest possible way the relevance of the Coulomb correlations to the TDCS. Therefore, in addition to the higher order and distorted wave models, an attempt is made to separately take into account the correlations among the final particles, which are due to the long-range part of the Coulomb forces. These correlations in terms of semiclassical correction to the first order interaction models (10, 11). The success of this simple scheme (is discussed in the next paragraphs) suggests that the final state interactions are a prominent feature both in the "dipolar" and "binary" regimes.

VALENCE SHELL IONISATION: SYMMETRIC KINEMATICS

The symmetric kinematics involve even sharing of the energy between the two final electrons and equal scattering angles ($E_a=E_b$, $\Theta_a=\Theta_b$); the TDCS peaks sharply whenever the Bethe ridge conditions are met (2) (i.e. the recoil momentum of the residual ion is negligible). At high incident energy, the small perturbation induced by the incoming electron is sharp and the many-body collisional problem can be reduced to a quasi three-body one where two electrons interact with each other through short range-like Coulomb forces and with the residual ion. The PWIA is the simplest of such models and neglects interactions with the residual ion, thus allowing for factorising the TDCS in two terms, an electron-electron collision term and a structure factor that depends on the unperturbed initial and final target states (6, 12). As the kinetic energy of the free electrons reduces or the ion recoil momentum is increased, this simplified scheme is no longer valid . A simple method to introduce to introduce interactions with the target atom is the Eikonal Wave Impulse Approximation (EWIA) (6, 13), where allowance for the interaction between the unbound electrons and the residual ion is made by a small metric distortion of their wave numbers. Recent reports extended the applicability of this method to the asymmetric kinematics as well (14). In spite of its simplicity the EWIA, especially at intermediate energies and moderate recoil momenta (less than 1au) calculates TDCS which are in good agreement with both the experiments and the DWIA. The DWIA is tho most complete scheme applied to date in describing this kinematical region (15) and it accounts for interactions of the unbound electrons with the target through optical model potentials. The model is in general better than the EWIA and the PWIA. Its superiority is remarkable at high recoil momenta (15). Nevertheless, consistently and for a variety of atoms investigated, all the aforementioned impulsive models fail in accounting for the symmetric TDCSs at the smaller scattering angles ($\Theta < 35^\circ$). In clearly assessing the disagreement between theory and experiment, the capability of measuring absolute TDCS was of the highest importance. All of the experimental findings show that models which include only short-range electron electron and electron ion correlations are not adequate to describe TDCS at the smaller scattering angles. To overcome this dead lock Popov et al. (10, 16) have recently proposed a different model. It stem from the assumption that the impulsive ionisation takes place within a small region of space (short-range like forces), whose radius r0 is expected to be determined by the size of the target atom. Within this region the (e,2e) amplitude can be calculated and factorised as in the PWIA. In the outer region the long range Coulomb interactions of the two outcoming electrons, not accounted by the short-range like forces, are treated by the model of two interacting semiclassical charged particles moving in the field of the residual ion. As a result the free electrons experience a change in the angular

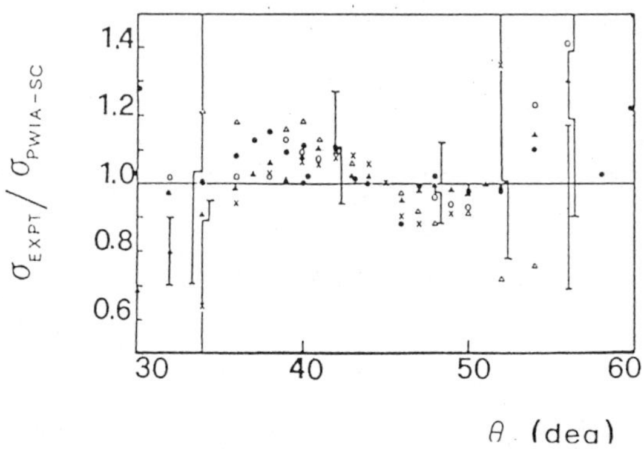

Fig. 2. The ratio between the measured (e,2e) cross section and the theoretical value, calculated in PWIA-SC, is plotted vs the scattering angle ($\Theta=\Theta a=\Theta b$) for symmetric coplanar reactions at 200 (●), 400 (▲), 800 (o), 1600 (△), and 2500 eV (x) incident energy. The experimental data, Ref. 13, have been normalised to the PWIA-SC value at 45°.

momentum and the TDCS is accordingly affected. The model is named the PWIA with semiclassical correction (PWIA-SC) (17). It is to be pointed out that even though the radius r0 is a free parameter, the application of the model to several atoms and upon a variety of symmetric kinematics, has made clear that the best value for the parameter is strictly correlated with the expectation value for the radius of the ionised orbital. Indeed, the value r0=0.7 au obtained for the experiments on He at the incident energy of 424.5 eV, is very close to the mean radius of the He 1s orbital (0.67 au), and was found to be suitable in reproducing all of the other symmetric kinematics investigated (17). This result is summarized in Figure 2 by the ratio of the experimental versus the PWIA-SC TDCS calculated for a variety of scattering angles and for incident energies ranging from 200 to 2500 eV.

Similar agreement is found for the results of (e,2e) experiments done in quite different conditions: the energy sharing (18) and the constant recoil momentum (13) geometries.

In the case of hydrogen (19) the three-body approximation, on which the PWIA-SC is based, is even more realistic and in fact by imposing r0=1 au, the simple PWIA-SC yield a remarkable agreement with the experiments. It is found to be particularly accurate in describing the shape of the TDCS at small scattering angles, where effects of the long-range Coulomb interactions between the final electrons are expected to be larger. None of the other impulsive models is nearly good, as clearly shown by the data reported in Figure 3.

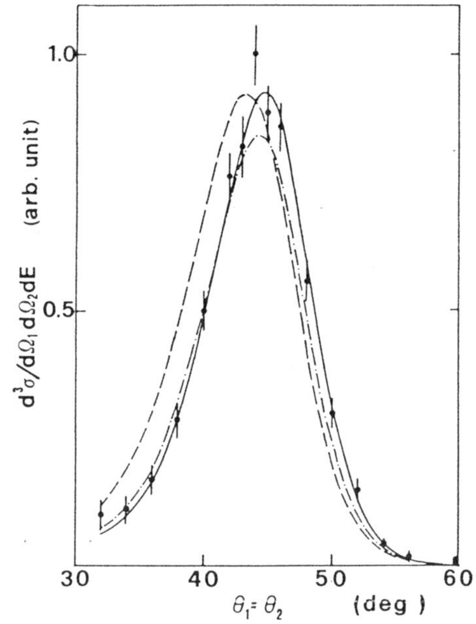

Fig. 3. The H (e,2e) angular correlation measured in coplanar symmetric geometry at 413.6 eV incident energy. The relative experimental data by Weigold et al., Ref. 19, are compared with PWIA (– – – –), EWIA (—·—·—) and the PWIA-SC (——) results. The experimental data have been normalised to best fit the PWIA-SC curve.

It has also to be mentioned that a further relevant issue of these studies concerns the procedure to be used in calculating the half-shell Coulomb T-matrix, i.e. the electron-electron scattering amplitude. A recent investigation indicates (20) the Ford T-matrix as the best one to be used in the factorised approximation of the TDCS (21). Also in developing the PWIA-SC it was found that in order to obtain good results it was crucial to use the Ford T-matrix in describing the break-up reaction in the inner region.

In spite of the success achieved in the symmetric kinematics, the semiclassical approach should give way to more complex models when the momentum transfer is either very small or very large (9).

VALENCE SHELL IONISATION: ASYMMETRC KINEMATICS

The asymmetric kinematics involve uneven sharing of the energy between the two final electrons and different scattering angles ($E_a > E_b$; $\Theta_a < \Theta_b$) and the TDCS peaks in two opposite lobes roughly symmetric around the direction of the momentum transfer. In describing the ionisation upon these kinematics, conventionally termed the "dipolar" regime, the Born approximation represents the traditional starting point. Usually it works in the limit of high incident energy and small momentum transfer. However, quite often the dipolar regime proved difficult to be described by first order Born approximations. Different theoretical approaches ranging from the Eikonal Born to Coulomb projected Born, Distorted waves and Second Born Approximations have been developed (1) and have reached various degrees of accuracy in describing the experimental TDCS (9). Nevertheless, a model that takes into account in a simpler way the physical processes beyond first order that are relevant to the dynamics of the ionisation would be highly desirable from the heuristic view point. Recently Klar et al. (11, 12) developed such a model within a theoretical frame that is similar to the one already discussed in the previous section for the symmetric kinematics. Again the interaction volume is subdivided in two regions. In the inner region, which roughly corresponds to the atomic volume, the ionisation takes place and the scattering is described in terms of first Born, second Born or Glauber series (23). In the simplest case the first Born approximation is used together with a quantum defect wave function for the final ionic state (BA-PCI).

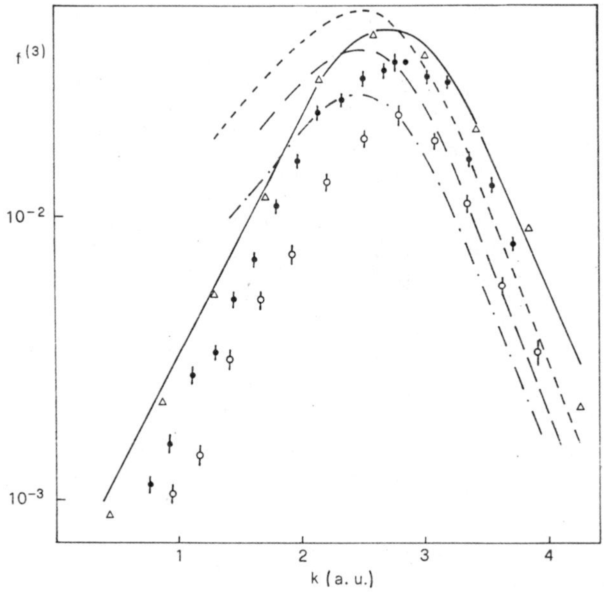

Fig. 4. The quantity $f^{(3)}(K, \Theta_b)$ for the electron impact ionisation of He with E_b=100 eV and K_b parallel to K. The impact energies are: (Δ) 8024 eV, Ref. 27, (●) 1024 eV and (o) 524 eV, Ref. 25. The full line is the BA-OCW, the broken lines are the PWIA at the three impact energies.

In the outer region the long range Coulomb interactions of the two escaping particles in the field of the ion are treated classically. This model predicts an exchange of energy in addition to the the exchange of momentum already accounted for by the PWIA-SC .

Several recent investigations on He (11) have shown that, considering the simplicity of the BA-PCI, the agreement found with the experiments is quite satisfactory. For incident energies larger than 250 eV and for moderate K values (less than 1 au) the experimental TDCS are in good agreement with the BA-PCI predictions. The semiclassical correction on which the BA-PCI is founded, becomes irrelevant either when the incident energy is reduced or when the momentum transfer is increased. In He, for instance, at 150 eV incident energy and at an energy of the ejected electron of 5 eV, when the momentum transfer is roughly 0.8 au, the effect of the long range final state interactions is negligible with respect to the contribution of higher order interactions which instead are well accounted for by the Glauber expansion (23). For even larger momentum transfer, it was recently shown (24, 25) that first Born approximations are inadequate unless the incident energy exceedes 8 KeV. This is clearly summarised in Figure 4 where the Oscillator Strength, differential in the direction of the ejected electron (f(3)(K, Θb)) (1, 2), is reported instead of the more usual TDCS. The quantity f(3) is indeed proportional to the TDCS through kinematical factors. The f(3)'s reported in the figure were measured on He at three different incident energies, for a variety of K values, for a fixed value of the ejected energy (100 eV) and Θb always fixed in the momentum transfer direction (25). The comparison of the experimental f(3)'s with the Born-Orthogonalised Coulomb Wave (BA-OCW) (26, 27) (full line in the figure) demonstrate a difference between experiment and BA-OCW that increases as the incident energy is decreased. The measured f(3) converges towards the first order prediction linearly with the inverse square root of the incident energy (25). Ehrhardt et al. (28) have found that, due to final state long range Coulomb interactions, the binary versus recoil intensity ratio converges linearly with the inverse square root of the incident energy, towards the value predicted by the first Born approximation and reaches this limit for energies larger than 8KeV. These experimental findings confirm a prediction based on the BA-PCI model which is expected to work at low ejected energy (less than 10 eV) and momentum transfer. Even though the scaling law found by this latter investigation is identical to the one observed for the 100 eV ejected energy experiments, the origin seems to be different. Indeed, the magnitude of the effect observed in the experiments at higher ejected energy is too large to be ascribed to final state Coulomb interactions. It is to be noted that the same energy dependence of the correction to first order models is found when the short range electron-electron correlations are accounted for. In the symmetric kinematics, for instance, this effect can be as large as 60% at 600 eV and depends on the incident energy through the relative energy of the final unbound electrons. In Figure 4 the f(3)'s predicted by the impulsive model PWIA are shown by broken curves. These predictions coincide vith the experiments only in the neighborhood of 2.75 a.u. momentum transfer, irrespective of the incident energy. This seemingly fortuitous agreement happens when $K=\sqrt{2Eb}$, which is the Bethe ridge condition, and suggest the Impulse Approximation to be valid only for collisions where the energy and momentum transferred by the incident electron are mostly absorbed by the ejected one (29).

All of the experimental findings so far discussed suggest that both short and long range Coulomb correlations are necessary corrections to first order interaction models at intermediate and low impact energies in order to correctly describe the TDCS.

INNER SHELL IONISATION: ASYMMETRIC KINEMATICS

When the ionisation involves an inner or intermediate orbital the primary ion eventually decays by photon or Auger-electron emission. In this latter case, in the final state of the reaction three free electrons are present in the field of the residual ion and correlations among the secondary electrons should largely influence the behaviour of the TDCS. Coincidence experiments where the Auger electron is detected in coincidence with the primary electron that has suffered a selected energy loss and momentum transfer, were proposed as a tool to investigate the correlation effects among secondary electrons whenever the energy transfer is close to the ionisation threshold (30). In passing it is to be mentioned that identical experiments, hereafter termed by the acronym (e,e'Auger), were also proposed by Berezhko et al. (31) for studying the alignment of the primary ion.

The first successful attempt to accomplish (e,e'Auger) experiments was reported by Sewell and Crowe (32) and in the last few years two other groups have succeeded in similar experiments (33, 34). Even though different energies and kinematics were used in the three experiments, all of them were focussed on the $L_3M^{23}M^{23}(^1S_0)$ Auger transition of Argon. The energy loss and the momentum transfer were comparable, thus constituting an homogeneous body of measurements. The simplest first order interaction model that can be used in interpreting these experiments without taking into account final state correlations is the first Born two-step model (1B2S). The basic assumption are: i) complete independence of the Auger relaxation process from the ionising collision (two step model) and ii) validity of the first Born approximation in describing the primary ionisation collision. The recent investigations in this field have clearly shown that the 1B2S model is inadequate in several kinematics. The continuum correlations between the Auger and the slow emitted electron are responsible for the energy shift and the profile distortion of the Auger line observed by the (e,e'Auger) experiments at few eV above threshold, at an incident energy of 1KeV (32, 35). A similar energy shift of the whole LMM Ar Auger spectrum was found by the (e,e'Auger) experiments performed at 8 KeV incident energy, with an energy loss that is 7 eV (excess energy) larger than the L ionisation threshold and 1.5° scattering angle (34). These kinematics were chosen in order to approach as much as possible the conditions for validity of the first Born approximation. An energy shift of roughly 0.15 eV towards higher energies was detected, which agrees with several predictions (30, 36). Nevertheless, the energy resolution was not sufficient for determining the Auger lineshape, which has been shown to bear further traces of the final state correlations as well as information about interference effects of the Auger decay channel with the direct double ionisation (33). The observed correlation effects among secondary electrons, which have been shown to be relevant for kinematics with low excess energy (34), constitute a violation of the assumption (i), thus preventing the interpretation of the coincident Auger angular distributions in terms of the simple 1B2S approximation.

Coincident Auger angular distributions were measured for several coplanar and non coplanar kinematics. Some results agree with 1B2S predictions (33), while others exhibit substantial violation of the expected symmetry around the direction of the momentum transfer K (32, 37). These works, however, do not allow for assessing whether a more refined description of the unbound electrons wavefunctions (e.g. the distorted-wave Born approximation (38)) would improve the agreement between theory and experiment or the Born approximation itself is to be questioned. To better understand this problem the angular correlations have been separately measured (34) for the two coincident pairs, (e,2e) and (e,e'Auger), of the same process, (i.e. the $Ar(2p)^{-1}$ ionisation) leading to the $L_3M^{23}M^{23}(^1S_0)$ Auger relaxation. The kinematics chosen for measuring these coincident angular distributions are equal to the ones used in measuring the coincident Auger spectrum, i.e. high incident energy, small momentum transfer and two different excess energies (7 and 60 eV). The choice of two different ejection energies was done in order to evidentiate final state correlation effects . The results of the (e,e'Auger) experiment at 1.5° scattering angle and 7eV excess energy are reported in Figure 5a.

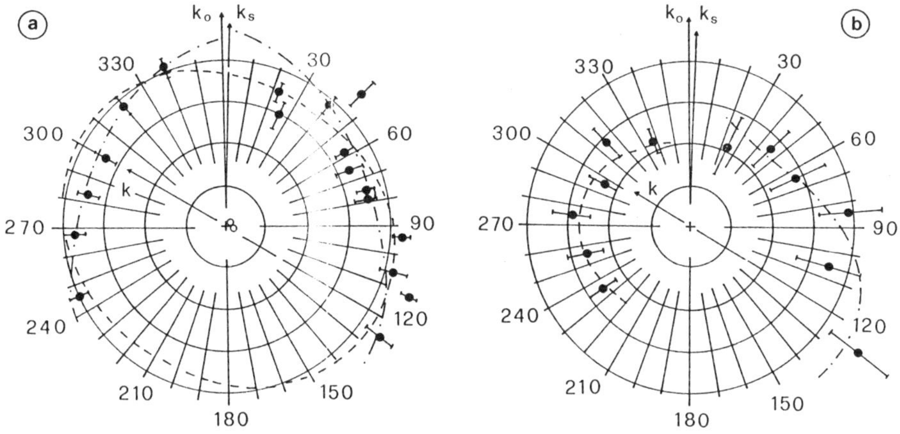

Fig. 5. a) The (e,e'Auger) angular distribution of the $L_3M^{23}M^{23}(^1S_0)$ transition in argon. "Spurious" (e,2e) contributions to the (e,e'Auger) intensity are shown by open circles.
b) The (e,2e) angular distribution "parent" to the Auger distribution in a). The intensity is in arbitrary units.

The momentum transfer is 0.75 au, which is the lowest momentum transfer achieved for this kind of experiments. The broken curve is the 1B2S prediction, while the chain curves are separate best fits to the experiment in the two half planes of the angular correlation. In opposition to the 1B2S theory the Auger angular distribution is not symmetric around the direction of the momentum transfer. Moreover, not even a common symmetry axis exists for the angular distributions in the two opposite half planes. This has been interpreted in terms of final state correlation effects, as it was the shift of the coincident energy spectrum observed under the identical kinematics. Further support to this interpretation is given by the (e,2e) angular distribution reported in Figure 5b. It is relative to the scattered and ejected electrons of the ionisation that creates the initial state for the Auger transition investigated in Figure 5a. The broken and chain curves represent separate best fits to the TDCS. Also the TDCS is not symmetric around the K direction and moreover the recoil lobe is larger than the binary one. These features, already observed in several other (e,2e) experiments, provide evidence for second order interactions in the ionising collision and correlations among the final continuum electrons and the residual ion (1, 7, 39). The companion experiments performed at higher ejected energy (60 eV) show an unquestionable symmetry around the K direction and, limited to the (e,e'Auger) experiment, a fearly good agreement with the 1B2S predictions. Thus a correct interpretation of the experiments at lower excess energy needs the introduction of higher order terms in the description of the ionising collision, together with correlation of the secondary electrons.

CONCLUSIONS

The ionisation TDCS was shown to be a sensitive tool to investigate the Coulomb correlations of the final unbound electrons. To this end the ionisation process has been widely studied by the (e,2e) experiments for large and intermediate electron impact energies and both for valence and inner orbitals. The collisional dynamics have been studied in dipolar ($K \approx 0$) and impulsive ($K=\surd 2Eb$) regimes with an emphasis on the kinematics which are intermediate between these previous two.

For the extreme regimes and for impact energies sufficiently large, i.e. few hundred times the ionisation threshold, Born and Impulse approximations are found adequate in their respective domains.

For impact energies that are within ten and hundred times the ionisation potential substantial disagreement is found between the experiments and first order interaction theories. Within this energy region, both in symmetric and asymmetric kinematics and respectively for small momentum transfer and small ion recoil momentum, the agreement of the experiments with the first order interaction based models is restored by simply including a semiclassical correction that accounts for the long range Coulomb interactions in the final state. A classical treatment of these effects is a reasonable one as they take place in a far zone where the local wave number variations are small. The effects of these long range interactions on the TDCS have been proved to die slowly with the incident energy. For intermediate values of the momentum transfer and for low incident energies, the reaction models that confine the ionisation to a single interaction between the incoming electron and the atom, are insufficient to describe the observed TDCS.

In the case of inner shell ionisation it has been shown that even for impact energies which are 30 times the ionisation threshold and momentum transfer as small as 0.5 au the 1B2S model is not adequate. While the energy shift of the Auger spectrum can be interpreted in terms of exchange of energy due to final state long range Coulomb interactios, the absence of symmetry around the momentum transfer direction exhibited by both the (e,2e) and the (e,e'Auger) angular distributions, is evidence for failure of the first Born approximation itself. Therefore more complex theories including higher-order collisional terms and correlations among secondary electrons should be developed.

REFERENCES

1. H. Ehrhardt, K. Jung, G. Knoth and P. Schlemmer, Z. Phys. D1, 3 (1986)
2. M. Inokuti, Rev. Mod. Phys. 43, 297 (1971)
3. J. Mazeau, A. Huetz and P. Selles, Proceedings XIV ICPEAC in Palo Alto 1985, Invited papers and progress reports, ed. D.C. Lorents, W.E. Meyerhof and J.R. Peterson, p.141 (North-Holland , Amsterdam 1986)
4. G. Stefani in Lecture Notes in Chemistry, Vol.35, ed. F.A. Gianturco and G. Stefani, p.226 (Springer-Verlag,-Berlin,1984)
5. E. Weigold and I.E. McCarthy, Adv. At. Mol. Phys. 14, 127 (1978)
6. I.E. McCarthy and E. Weigold, Phys .Rep. C27, 275 (1976)
7. B. Lohman, I.E. McCarthy, A.T. Stelbovics and E. Weigold, Phys. Rev. A30, 758 (1984)
8. C.J. Joachain and B. Piraux, Comm.At.Mol.Phys. 14, 261 (1986)
9. C.J. Joachain, this volume
10. Yu.V. Popov and J.J. Benayoun, J. Phys.B: At. Mol. Phys. 14, 3513 (1981)
11. H. Klar, A. Franz and H. Tenhagen, Z. Phys. D1, 373 (1986)
12. A. Giardini-Guidoni, R. Fantoni, R. Camilloni and G. Stefani, Comm. At. Mol. Phys. 10, 107 (1981)
13. R. Camilloni, A. Giardini-Guidoni, I.E. McCarthy and G. Stefani, Phys. Rev. A17, 1634 (1978)
14. A. Lahmam-Bennani, Phys. Rev. A29, 962 (1984)
15. A.J. Dixon, I.E. McCarthy, C.J. Noble and E. Waigold, Phys. Rev. A17, 597 (1978)
16. Yu.V. Popov and V.F. Erokin, Phys. Lett. 97A, 280 (1983)
17. L. Avaldi, R. Camilloni, Yu.V. Popov and G. Stefani, Phys. Rev. A33, 851 (1986)
18. G. Stefani and R. Camilloni, J. Phys. B: At. Mol. Phys. 18, 499 (1985)
19. E. Weigold, C.J. Noble, S.T. Hood and I. Fuss, J. Phys. B: At. Mol. Phys. 12, 291 (1979)
20. I.E. McCarthy and M.J. Roberts, J. Phys. B: At. Mol. Phys. 20 L231 (1987)
21. W.F. Ford, Phys. Rev. B133, 1661 (1964)
22. A. Franz and H. Klar, Z. Phys. D1, 33 (1986)
23. H.Klar, A.C. Roy, P. Schlemmer, K. Jung and H. Ehrhardt, J. Phys. B: At. Mol. Phys. 20, 821 (1987)
24. L. Avaldi, R. Camilloni, E. Fainelli, G. Stefani, A. Franz, H. Klar and I.E. McCarthy, J. Phys. B: At. Mol. Phys. 20, 5827 (1987)
25. L. Avaldi, E. Fainelli, A. Lahmam-Bennani and G. Stefani, Abstracts of XV ICPEAC in Brighton 1987, ed.J. Geddes, H.B. Gilbody, A.E. Kingston, C.J. Latimer and H.J.R. Walters, p.256 (North-Holland, Amsterdam 1986)
26. C. Dal Cappello, C. Tavard, A. Lahmam-Bennani and M.C. Dal Cappello, J. Phys. B: At. Mol. Phys. 17, 4557 (1984)

27. A. Lahmam-Bennani, H.F. Wallenstein, C. Dal Cappello, M. Raoult and A. Duguet, J.Phys. B: At Mol.Phys. 16, 2219 (1983)
28. H. Ehrhardt, K. Jung, H. Klar, A. Lahmam-Bennani and P. Schlemmer, J. Phys. B: At. Mol. Phys. 20, L193 (1987)
29. L. Avaldi, R. Camilloni, E. Fainelli and G. Stefani, J. Phys. B: At. Mol. Phys. 20, 4163 (1987)
30. M.J. van der Wiel, G.R. Wight and R.R. Tol, J. Phys. B: At. Mol. Phys. 9, 15 (1976)
31. E.G. Berezhko, N.M. Kabachnik and V.V. Sizov J. Phys. B: At. Mol. Phys. 11, 1819 (1978)
32. E.C. Sewell and A. Crowe, J. Phys. B: At. Mol.Phys. 15, L357 (1982)
33. W. Sandner and M. Vlkel, J. Phys. B: At. Mol. Phys. 17, L597 (1984)
34. G. Stefani, L. Avaldi, A. Lahmam-Bennani and A. Duguet, J. Phys. B: At. Mol. Phys. 19, 3787 (1985)
35. A. Niehaus, J. Phys. B: At. Mol. Phys. 10, 1845 (1977)
36. E.C. Sewell and A. Crowe, J. Phys. B: At. Mol. Phys. 17, L547 (1984)
37. E.C. Sewell and A. Crowe, J. Phys. B: At. Mol. Phys. 17, 2913 (1984)
38. E.G. Berezhko and N.M. Kabachnik, J. Phys. B: At. Mol. Phys. 15, 2075 (1982)
39. A. Lahmam-Bennani, H.F. Wallenstein, A. Duguet and A. Daoud, Phys. Rev. A30, 1511 (1984)

EXPERIMENTS WITH POLARIZED ELECTRONS

Joachim Kessler

Physikalisches Institut
Universität Münster
D 4400 Münster

INTRODUCTION

The focus of this seminar is on the atomic dynamics information which can be extracted from experiments with spin-polarized electron beams. The definition of electron polarization is given in Fig. 1. Since investigations with polarized electrons have been stimulated by advances of the experimental techniques in the past few years, the seminar will start with a brief outline of the state of the experimental art. A breakthrough was the GaAs source of polarized electrons. Polarizations of 35-40% with currents of a few μA are produced with it in several laboratories. The progress with polarized electron sources is, however, not matched by that with polarization analyzers.

As a consequence of the experimental advances, investigations with polarized electrons are now being made in all major fields of physics. Fig. 2 gives a survey which will be discussed in the seminar. For the present summary, I pick out two examples:

1. With the discovery of the Fano effect[2] it became evident that spin polarization of photoelectrons is not an exception, as had been believed

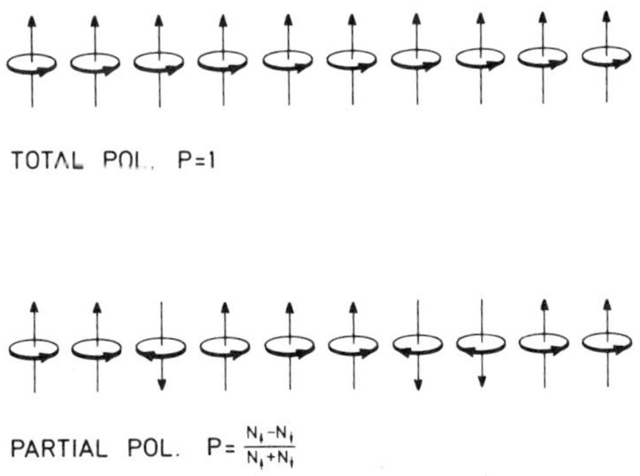

Fig. 1. Definition of the electron polarization P

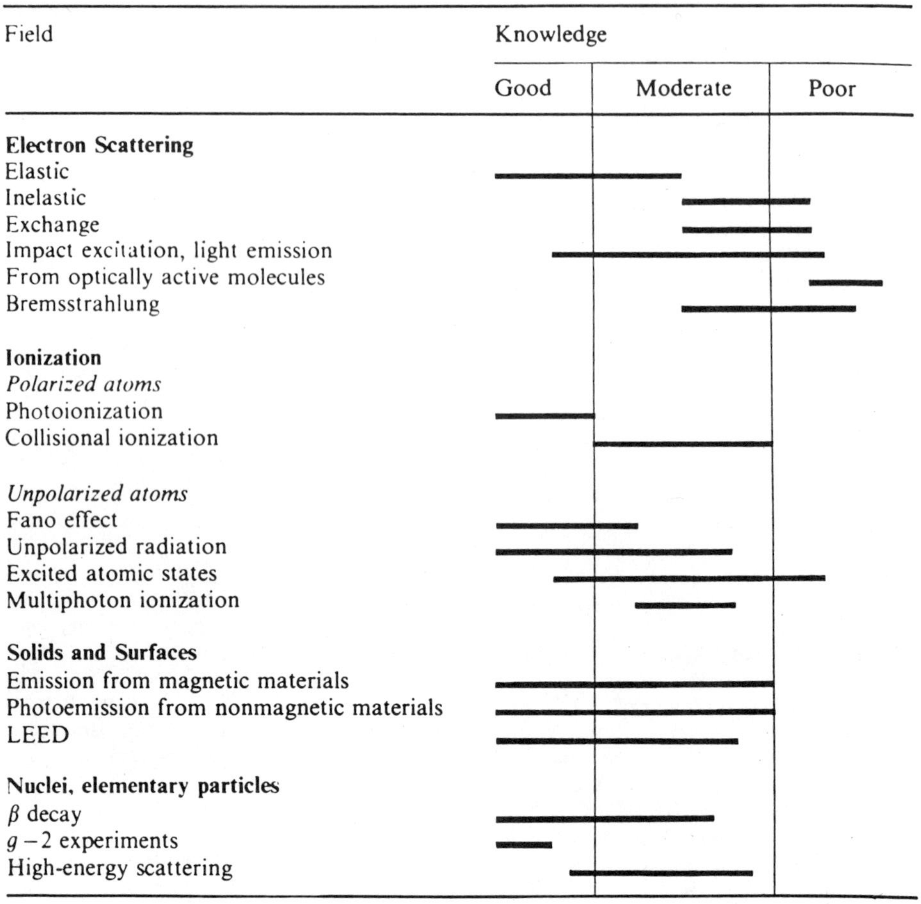
Field
Knowledge
Good
Moderate
Poor
Electron Scattering
Elastic
Inelastic
Exchange
Impact excitation, light emission
From optically active molecules
Bremsstrahlung
Ionization
Polarized atoms
Photoionization
Collisional ionization
Unpolarized atoms
Fano effect
Unpolarized radiation
Excited atomic states
Multiphoton ionization
Solids and Surfaces
Emission from magnetic materials
Photoemission from nonmagnetic materials
LEED
Nuclei, elementary particles
β decay
g − 2 experiments
High-energy scattering

Fig. 2. Present knowledge of electron-polarization effects in various fields of physics[1]

for a long time, but that it is the rule. No matter whether (unpolarized!) atoms are photoionized by circularly polarized, linearly polarized or unpolarized radiation, the photoelectrons are - in general - spin polarized. Their polarization has been utilized to obtain detailed information on the dynamics of the photoionization process[3,4].

2. The majority of polarized-electron studies is now being made in surface and solid state physics. Photoemission, field emission and secondary emission of polarized electrons have opened new dimensions of studying the electronic structure of solids and surfaces. One of the highlights are the impressive images of magnetic domain structures obtained by scanning electron microscopy with polarization analysis[5].

From Fig. 2 it is clear that experiments with polarized electrons cover a wide field. In order to get a deeper understanding we will focus our attention on a specific area: the fundamental process of electron-atom scattering. Polarization studies in electron-atom collisions yield information on spin-dependent interactions such as spin-orbit and exchange interaction. Such information is otherwise hard to obtain because spin-dependent interactions are "weak forces with conspicuous effects" [6]. Because they are weak, they are frequently masked by the much stronger

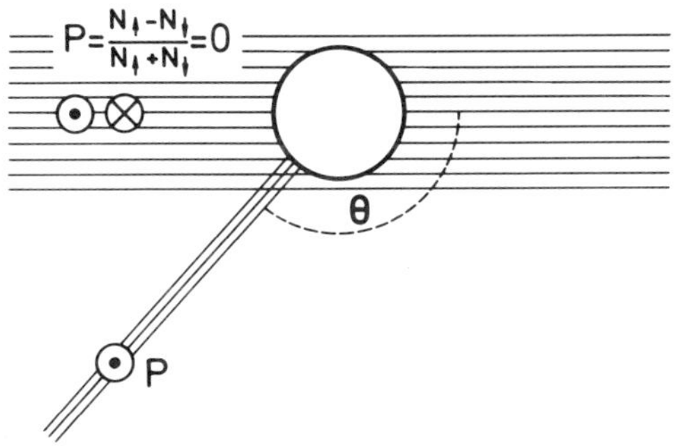

Fig. 3. Polarization caused by scattering

Coulomb interaction. They can, however, be unmasked by polarization experiments with electrons. Let me first try to explain this for the process which, conceptually, is the simplest: elastic electron scattering from spinless atoms, where the spin-dependent effects are caused by spin-orbit interaction[1].

Figs. 3 and 4 illustrate a typical polarization phenomenon, explaining why scattering of an unpolarized electron beam from an unpolarized target engenders a spin polarization of the scattered beam. The unpolarized incident beam can be considered as a mixture of equal numbers of spin up (e↑)

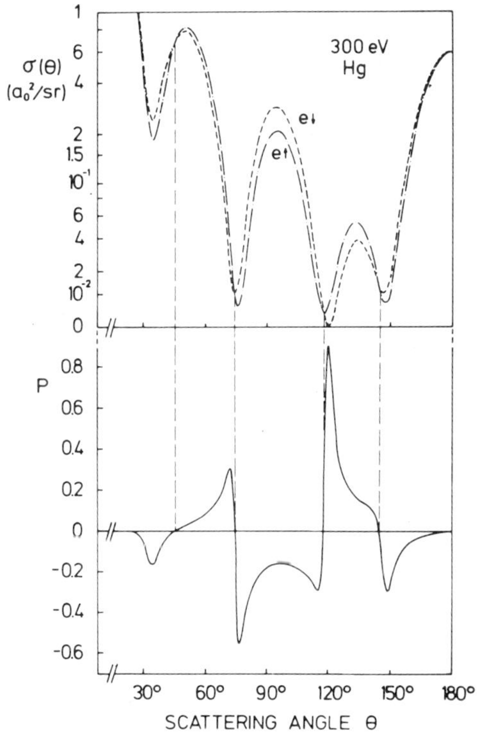

Fig. 4. Construction of the polarization from the cross sections for e↑ and e↓ [1]

and spin down (e↓) electrons. Since, in addition to the Coulomb interaction, the electrons experience a (spin-dependent) spin-orbit interaction, there is a different scattering potential for e↑ and e↓. As a consequence, one has different scattering cross sections for e↑ and e↓. In other words, one finds - in general - different numbers of e↑ and e↓ scattered into a certain direction; i.e., the scattered beam is polarized. The polarization curve can be constructed from the cross sections for e↑ and e↓ as indicated in Fig. 4.

THE "PERFECT" ELASTIC SCATTERING EXPERIMENT

Let me now explain how polarization measurements in conjunction with cross-section measurements enable one to completely understand the elastic scattering process. Elastic scattering from spinless targets is theoretically described by two scattering amplitudes: an amplitude f which is mainly determined by the Coulomb interaction, and an amplitude g which describes the behavior of the electron spins caused by the spin-orbit interaction of the scattered electrons in the atomic field. f and g are complex quantities and may therefore be written in the form

$$f = |f|e^{i\gamma_1}, \quad g = |g|e^{i\gamma_2} .$$

In order to describe the scattering process completely one needs to know the scattering amplitudes completely, i.e., their moduli and their phases as far as possible. The conventional measurements of the cross section

$$\frac{d\sigma}{d\Omega} = |f|^2 + |g|^2 \qquad (1)$$

yield only the sum of the squares of the scattering amplitudes. That is not very much. More information can be obtained by observing the polarization after scattering of an initially unpolarized electron beam. This polarization P is given by a different combination of the scattering amplitudes

$$P = i\,\frac{fg^* - f^*g}{|f|^2 + |g|^2} = -\,\frac{2|f||g|\sin(\gamma_1 - \gamma_2)}{|f|^2 + |g|^2} \qquad (2)$$

so that it yields information that is independent of the information obtained from a cross-section measurement. But these two observables still do not suffice to give all the information hidden in the scattering amplitudes. The missing information can be obtained by scattering a polarized beam and observing the change of the polarization vector caused by the scattering process.

From the change in length of the polarization vector one obtains a new observable T which is given by

$$T = \frac{|f|^2 - |g|^2}{|f|^2 + |g|^2} \qquad (3)$$

and one obtains another observable U which describes the rotation of the polarization vector:

$$U = \frac{fg^* + f^*g}{|f|^2 + |g|^2} = \frac{2|f||g|\cos(\gamma_1 - \gamma_2)}{|f|^2 + |g|^2} \quad . \tag{4}$$

If one has measured the 4 observables, Eqs. (1) to (4), then one has the maximum possible information about the scattering process because one can use these 4 equations to evaluate $|f|$, $|g|$, and $\gamma_1 - \gamma_2$ unambiguously (an absolute determination of the phases from an analysis of the scattered wave is, according to the principles of quantum mechanics, impossible).

Fig. 5 shows that such measurements are no longer science fiction. The polarized electrons come from a GaAs source which is fixed in space. An electrostatic deflection system can be rotated about the target so that the scattering angle θ varies continuously. The rotation does not affect the direction of the initial polarization $\vec{P}_i$ because the electron spins are not affected by the electrostatic deflection fields. The polarization vector $\vec{P}_i$ is always oriented along the axis of observation, which facilitates data evaluation. The transverse components of the final polarization vector $\vec{P}_f$ are determined by the left-right asymmetry measured with the two pairs of counters of the Mott detector. Since such a detector is not sensitive to longitudinal polarization components a Wien filter was introduced which can rotate the longitudinal component by 90° to become transverse, so that it can also be measured.

Some of the experimental and theoretical results for P, T and U will be presented in the seminar. Reasons will be given why the agreement decreases with the electron energy.

An example of an evaluation of the complex scattering amplitudes from the complete set of measured observables, namely $d\sigma/d\Omega$ (absolute), P, T and U, is shown in Fig. 6. When comparing the experimental and theoretical values we must keep in mind that we are considering here the complete set of observables. This is a much more stringent test of the theory than the

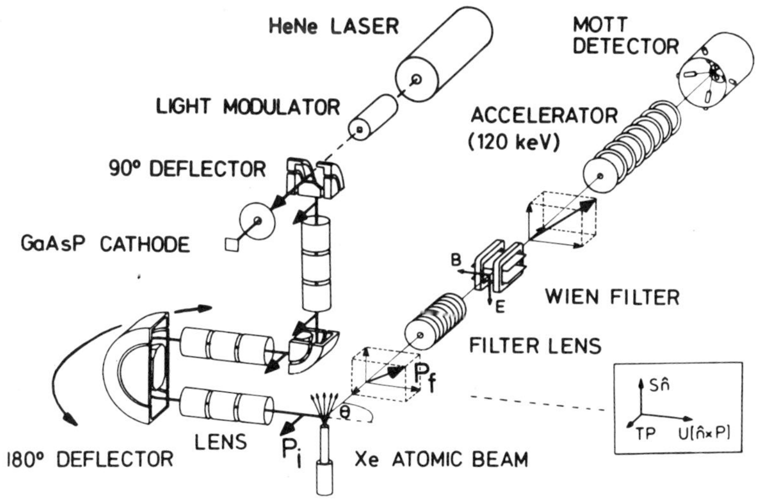

Fig. 5. Measurement of the change of the electron polarization vector[7]

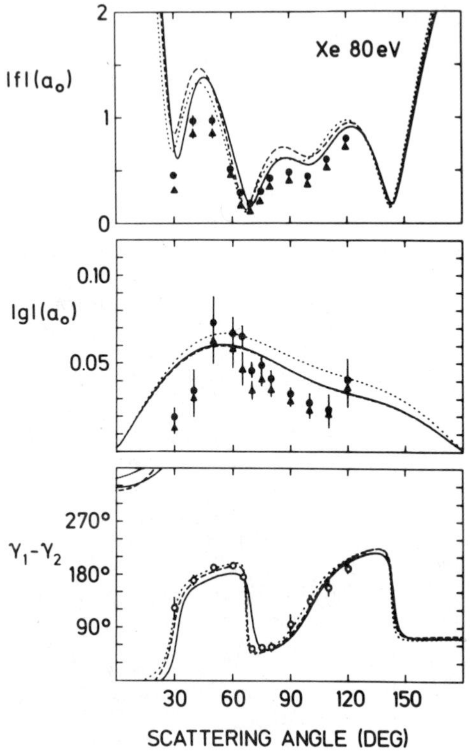

Fig. 6. Evaluation of the scattering amplitudes for Xe at 80 eV. Theory: Haberland et al. —— [8], McEachran and Stauffer .... [9], Awe et al. ---- [10]. Symbols of experimental points see ref. 7

usual one where one is concerned with only one of the observables and does not care about the others. Here we can see whether a theory describes properly the complete scattering process, rather than only a particular aspect of it.

SPIN-DEPENDENT INTERACTIONS IN INELASTIC SCATTERING

This was one of the few examples of a complete or "perfect" scattering experiment (as they are sometimes called in the literature) which is feasible today. Let us now switch to a process which is too involved for such experiments to be practicable, namely inelastic electron scattering.

In inelastic scattering it is not only the state of the electron that is changed, but also the state of the atom, as indicated in Fig. 7. Accordingly, this has to be included in the theory and to be observed in experiment. The excited atomic state usually has several degenerate magnetic sublevels. If one wants to unravel the processes associated with the various transitions in Fig. 7, then one has to do quite a few independent experiments. This is why in inelastic scattering a great number of measurements have to be made before an experiment may be called "perfect". Which observables do we have in this case? First, one can measure the four observables discussed above in elastic scattering. They mainly yield information

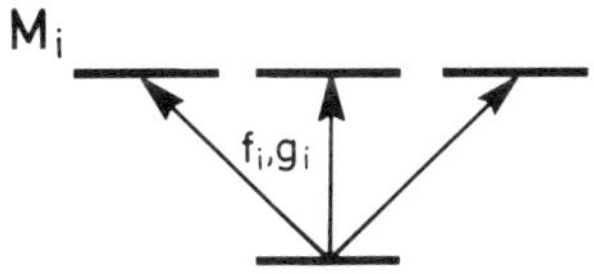

Fig. 7. Excitation of sublevels

about the scattered electrons. One can also study the light emitted by the excited atoms, its intensity distribution and its polarization. This gives a lot of information about the excited atomic state. Another observable is the left-right asymmetry A occurring when polarized electrons are inelastically scattered. In elastic scattering such an asymmetry measurement does not provide new information since there one has $A = P$, i.e., the asymmetry equals the polarization which an unpolarized electron beam obtains by scattering. In inelastic scattering one has, in general, $A \neq P$, so that an asymmetry measurement yields different information than a polarization measurement.

It does not make much sense to discuss here all the inelastic experiments with polarized electrons that have been performed. Instead, it is more rewarding to explain in some detail the virtues of inelastic polarized-electron scattering for a specific example. Let us see what can be learned from light-polarization measurements as they are done in E. Reichert's group at Mainz[11] and in our group at Münster.

Fig. 8 defines the 3 Stokes parameters η_ν one has to measure in order to determine the polarization of a radiation field. I denotes the transmitted intensities for the respective filter orientations indicated.

Fig. 9 shows some results obtained when mercury atoms are excited by polarized electrons and the light emitted along the direction of the polarization vector is observed[12]. The upper panel shows the linear polarization η_1 and the circular polarization η_2 versus the electron energy E for the transition from 6^3P_1 to the ground state in mercury. All the light polarizations are normalized to the electron polarization P. The lower panel shows the circular polarization for the transitions from 7^3S_1 to the fine-structure levels indicated in Fig. 9. There are mainly 3 reasons why such measurements are interesting:

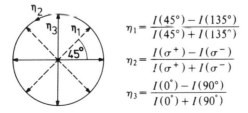

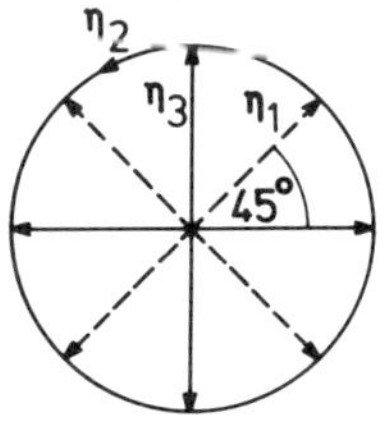

$$\eta_1 = \frac{I(45°) - I(135°)}{I(45°) + I(135°)}$$

$$\eta_2 = \frac{I(\sigma^+) - I(\sigma^-)}{I(\sigma^+) + I(\sigma^-)}$$

$$\eta_3 = \frac{I(0°) - I(90°)}{I(0°) + I(90°)}$$

Fig. 8. Stokes parameters describing the polarization of a radiation field

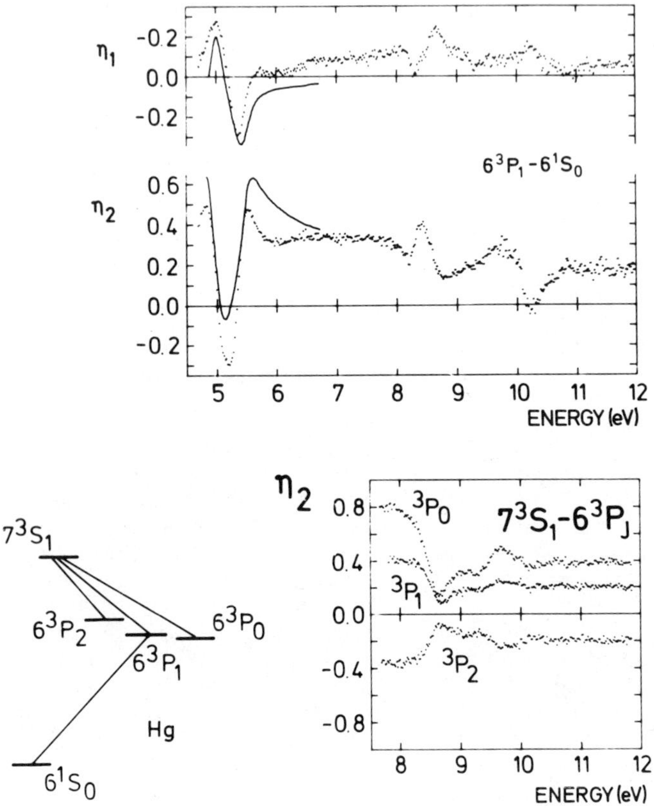

Fig. 9. Light polarization from Hg normalized to incident electron polarization. Dotted curves: experimental[12,13]; full lines: theoretical[14]

First, you see pronounced resonance features in these curves. Generally speaking, short-lived negative ion states cause resonances not only in the cross sections, but also in the polarization of the scattered electrons, in the asymmetries and in the polarization of the emitted light. Such resonances in the polarization effects may be utilized for classification of the compound ion states as has first been shown by the Mainz group[15]. The analysis of the resonances in the light polarization by our theoretical colleagues has decided a controversy about the classification of the compound states of Hg at threshold[12].

Second, the light polarization is of interest because there is a good chance that it can be utilized for measuring the polarization of the electrons by which the radiation is produced. In practice, one has large systematic errors of polarization calibration with the Mott detector. If one wants to use the light polarization, instead, then curves like those of Fig. 9 are not very suitable, because one needs to know the electron energy very precisely. A good candidate is, however, the circular light polarization obtained from helium, as pointed out by Gay[16]. Fig. 10 shows a preliminary experimental result[17]. The light polarization is practically independent of energy, which is very encouraging. Besides it is much easier to predict accurate theoretical values for the light

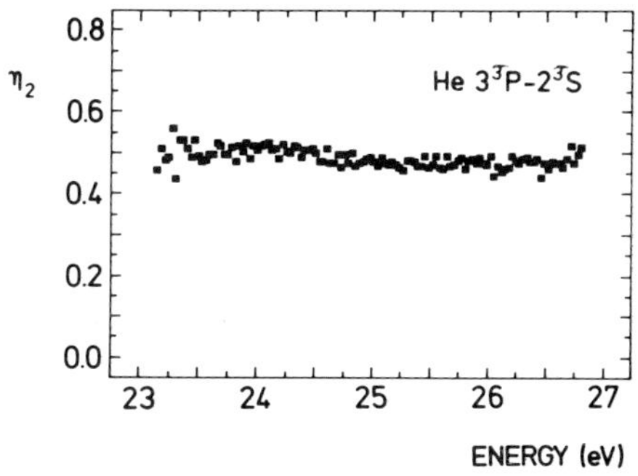

Fig. 10. Circular light polarization from He normalized to incident electron polarization[17]

helium atom than for the heavy mercury atom with its resonances. That is why I consider the He circular light polarization to be a very good candidate for the absolute calibration of electron polarization.

The third and perhaps most important reason for the interest in the light polarization is a close connection between the electron spin and the light polarizations η_1 and η_2: Impact radiation produced by unpolarized electrons in the above mentioned type of experiment cannot have a circular polarization η_2 or a linear polarization described by η_1. This can easily be seen by a mirror reflection of the experiment while recalling the conservation of parity: Suppose you have an unpolarized electron beam propagating along the z axis (Fig. 11). Let us observe the light along the y axis normal to the plane of the diagram. Assume for a moment that the light has circular polarization. If you make a mirror reflection at a plane perpendicular to the plane of the diagram, then the helicity of the light will be reversed since, in the mirror, it has a different sense of rotation. The initial state, however, the unpolarized electron beam, remains unchanged. This means that the same initial state yields different results for this experiment in the laboratory and in the mirror, which is a simple way of saying that parity conservation is violated. Accordingly, our assumption that circularly polarized light can be produced by unpolarized electrons must be wrong.

By the same argument one can see that, with unpolarized electrons, the linear polarization η_1 along the 45° axis must vanish because otherwise it would be reversed in the mirror, where it is in the 135° direction, so that again one and the same initial state would produce different results in the laboratory and in the mirror, which is not allowed. The linear polarization η_3 along the x or z axis behaves, however, quite differently: Polarization along the x direction stays along x in the mirror and polarization along the z direction stays along z. In this case there is no reversal of the polarization and therefore no problem with parity violation, so that η_3 is the only polarization that can be produced by unpolarized electrons in the type of experiment discussed here.

Let me repeat: only if you have polarized electrons can you produce the light polarizations η_1 and η_2, i.e., these observables are closely related to the electron polarization. We can therefore anticipate that they yield direct information about the spin-dependent interaction in electron-impact excitation. This is why not only experimental but also strong theoretical effort has been focused on these observables. A successful start has been made by the Belfast-Münster cooperation[14]. The full

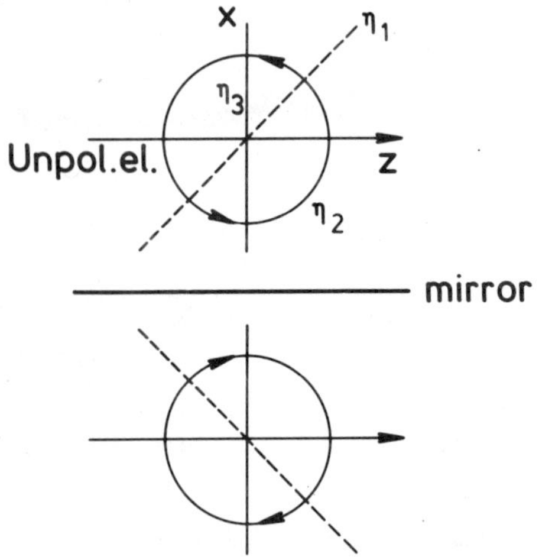

Fig. 11. Light polarization and parity violation

lines in Fig. 9 have been obtained with the R matrix method in conjunction with the close-coupling approximation. Since the computational problems of close coupling increase dramatically as the energy increases, the theoretical results are available only at small energies. Considering the difficulty of such calculations of spin-dependent effects for heavy atoms one can be quite happy about the agreement with the experiment.

Let me mention that our theorists have found a remarkable property of these light polarizations[18]. It turns out that η_1 and η_2 yield different pieces of information on the spin-dependent interactions. η_2 can be different from 0 only if the exchange interaction plays a role in the excitation process, whereas $\eta_1 \neq 0$ can occur only if the spin-orbit interaction plays a role (this will be made plausible in the seminar by a simple model[19]). In other words: one has a situation where certain observables are sensitive to certain atomic interactions. This gives us a chance to disentangle the different spin-dependent interactions. This is a great advantage compared to the usual situation where you have to compare experimental data with the theoretical results as a whole, and if you find deviations between theory and experiment you cannot really tell which of the theoretical assumptions have to be improved or given up. In the investigations discussed here, certain observables convey information on specific interactions so that it is possible to determine which of the specific theoretical assumptions need to be improved and which ones are adequate.

So far I have discussed the following separate aspects of inelastic electron scattering: what happens to the atom in the inelastic process (an answer can be sought by an analysis of the emitted light), and what happens to the electron in the inelastic process? (an answer can be given by measurement of cross sections, electron polarization or asymmetries).

Now we can try to get the whole picture by asking: How is what happens to the atom correlated with what happens to the electron? An answer to this can be found by coincidence experiments. Two such experiments have so far been made where the scattered polarized electrons were observed in coincidence with the photons whose polarization was analyzed[17,20]. By these measurements the experimental averaging over different scattering channels is further reduced.

CONCLUSION

The list of collision experiments with polarized electrons can easily be continued. Much work has been done on scattering of polarized electrons from polarized atoms[21]. Other importantinvestigations were made about polarization phenomena in superelastic scattering[22] and about an interesting polarization mechanism in electron-impact excitation of fine-structure levels[23]. I hope that the seminar will stimulate some of its participants to study also a few of the papers referring to these areas.

REFERENCES

1. J. Kessler, Polarized Electrons (Springer, Berlin 1985)
2. U. Fano, Phys. Rev. 178, 131 (1969)
3. J. Kessler, Comments At. Mol. Phys. 10, 47 (1981)
4. U. Heinzmann, in Electronic and Atomic Collisions, ed. by D.C. Lorents et al. (North Holland, Amsterdam 1986) p. 37
5. R.J. Celotta and D.T. Pierce, Science 234, 333 (1986)
6. U. Fano, Comments At. Mol. Phys. 2, 30 (1970)
7. O. Berger and J. Kessler, J. Phys. B 19, 3539 (1986)
8. R. Haberland, L. Fritsche and J. Noffke, Phys. Rev. A 33, 2305 (1986) and private communication
9. R.P. McEachran and A.D. Stauffer, J. Phys. B 19, 3523 (1986)
10. B. Awe, F. Kemper, F. Rosicky and R. Feder, J. Phys. B 16, 603 (1983)
11. N. Ludwig, A. Bauch, P. Nass, E. Reichert and W. Welker, Z. Phys. D 4, 177 (1986)
12. A. Wolcke, K. Bartschat, K. Blum, H. Borgmann, G.F. Hanne and J. Kessler, J. Phys. B 16, 639 (1983)
13. J. Goeke, G.F. Hanne, J. Kessler and A. Wolcke, Phys. Rev. Lett. 51, 2273 (1983); J. Goeke, thesis (1983)
14. K. Bartschat, N.S. Scott, K. Blum and P.G. Burke, J. Phys. B 17, 269 (1984)
15. A. Albert, C. Christian, T. Heindorff, E. Reichert and S. Schön, J. Phys. B 10, 3733 (1977)
16. T.J. Gay, J. Phys. B 16, L 553 (1983)
17. J. Goeke, Ph.D. thesis 1987, Münster
18. K. Bartschat and K. Blum, Z. Phys. A 304, 85 (1982)
19. J. Kessler, Comments At. Mol. Phys. 17, 15 (1985)
20. A. Wolcke, J. Goeke, G.F. Hanne, J. Kessler, W. Vollmer, K. Bartschat and K. Blum, Phys. Rev. Lett. 52, 1108 (1984)
21. G.D. Fletcher, M.J. Alguard, T.J. Gay, V.W. Hughes, P.F. Wainwright, M.S. Lubell and W. Raith, Phys. Rev. A 31, 2854 (1985); G. Baum, M. Moede, W. Raith and W. Schröder, J. Phys. B 18, 531 (1985); G. Baum, M. Moede, W. Raith and U. Sillmen, Phys. Rev. Lett. 57, 1855 (1986)
22. J.J. McClelland, M.H. Kelley and R.J. Celotta, Phys. Rev. Lett. 55, 688 (1985) and 56, 1362 (1986)
23. G.F. Hanne, Phys. Rep. 95, 95 (1983) and Comments At. Mol. Phys. 14, 163 (1984); H. Borgmann, J. Goeke, G.F. Hanne, J. Kessler and A. Wolcke, J. Phys. B 20, 1619 (1987)

LOW ENERGY ELECTRON-MOLECULE COLLISION EXPERIMENTS

Michel Tronc

Laboratoire de Chimie Physique, Universite Pierre et Marie Curie, 11, rue Pierre et Marie Curie 75231 Paris Cedex 05 France

1. INTRODUCTION

Low energy electron-collision with free molecules is dominated by resonant mechanisms : vibrational excitation, electronic excitation, dissociative attachment. Experimental data have been obtained at a qualitative and sometimes quantitative level for diatomic molecules and for few simple polyatomic molecules, but many measurements have to be done especially for absolute cross sections on a broad energy range. The need for reliable quantitative data is very important for comparison with existing theories , and because these resonant cross sections play a prominent role in many fields including high atmosphere chemistry, plasma physics, gas lasers, catalysis and radiation chemistry.

2. SPECTROMETERS

2.1. Electrostatic and Trochoïdal Electron Spectrometers

Figure 1a shows an electrostatic hemispherical electron spectrometer commonly used in low energy experiments. Electrons emitted by an hairpin filament with a broad energy distribution (0.3 eV) are focussed at the entrance of the hemispherical selector. Monoenergetic electrons (10-30 meV) emerging from the monochromator are accelerated towards the target molecules by a system of field and zoom lenses (Harting and Read 1976). Electrons scattered, at an angle θ , are decelerated and focussed at the entrance of the analyser similar to the monochromator. The scattered electrons analysed in energy are multiplied by a channeltron and counted.

Three modes of operation can be used to obtain :

i) Energy loss spectra at fixed angle and incident electron energy, showing the excited levels of the target,
ii) Differential cross sections of a specific level of the target, by sweeping incident electron energy and keeping constant the energy loss,
iii) Constant residual energy spectra by sweeping both incident electron energy and energy loss to maintain fixed the scattered electron energy.

If high resolution spectra have been obtained by many groups, there is still some difficulty to maintain a constant transmission function for both incident and scattered electrons lenses.

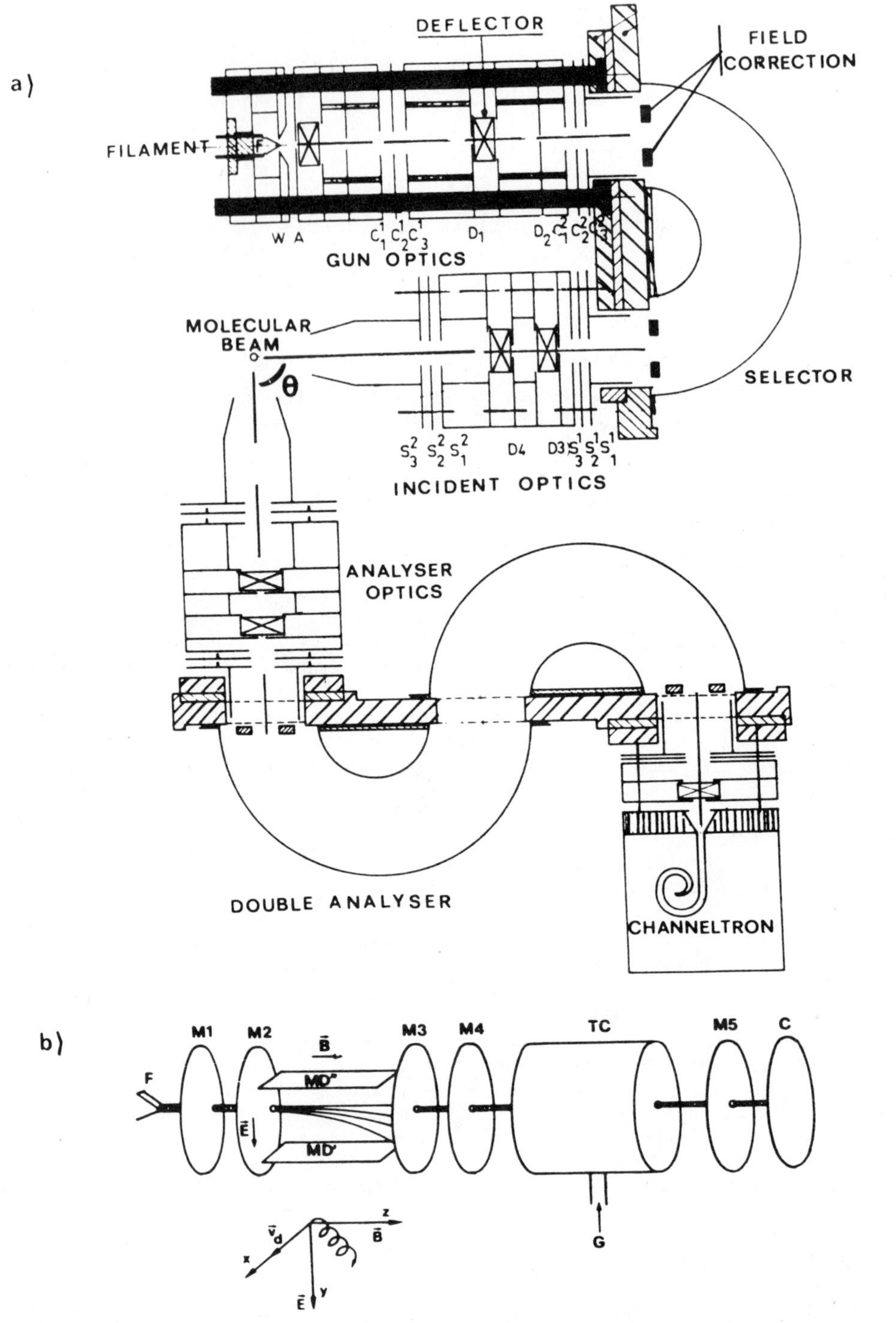

Figure 1 a) Crossed-beams hemispherical electrostatic electron impact spectrometer for low-energy scattering experiments. b) Schematic diagram of a trochoidal electron monochromator: F, filament, M_1-M_5 electrodes, MD deflectors, C collector plate, G gas inlet, TC target chamber.

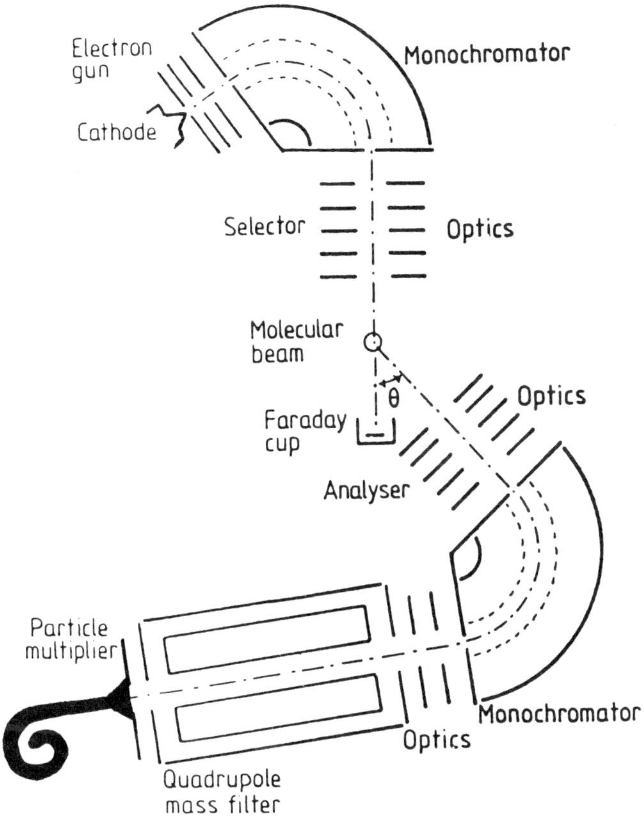

Figure 2. Electron impact spectrometer with quadrupole mass analyser for the study of kinetic energy, angular distribution, and differential cross sections of negative ions. (Le Coat et al. 1982)

This transmission can be controlled by looking at known cross sections in hydrogen or helium for example (Register et al 1980, Srivastava et al 1975), or by looking at near threshold ionisation in helium (Pichou et al 1976). A second difficulty is to maintain a constant collision volume, (defined as the intercept of the molecular beam with incident and scattered electron beams) when the scattering angle is changed, which can be achieved with a geometry where the molecular beam diameter is smaller than electron beams. Improvements in molecular beam definition have been obtained by using multicapillary arrays, by developping supersonic beams giving vibrationally and rotationally cooled molecules by seeded jet technique (Doering 1983) and by pulsed molecular beam for dimers and clusters formation. For determination of absolute cross section, the knowlegde of the density of molecules interacting with electrons is crucial, and progress have been accomplished with a well defined crossed-beam scattering geometry in which target gas densities of the unknown and of a standard gas (N_2, He) are accurately determined by a calibrated mass flowmeter (Srivastava et al 1975).

Multidetection techniques including multielectrode arrays, charge division technique and image scanning technique, have given a striking improvement of more than a factor of 100 in sensitivity over usual channeltron detection technique (Hicks et al 1980), with the possibility to reduce accumulation time and consequently energy shift and drift.

With the difficulty to measure cross sections below 1 or 0.5 eV with conventional electrostatic electron spectrometers, because of the divergence of the incident beam at low energy (Lagrange Helmoltz relation) and because of stray electrostatic and magnetic residual fields especially in the collision region, a trochoïdal monochromator has been developped by Stamatovic and Schulz (1968). This apparatus (Fig. 1b) overcome all the inconvenience of the retarding potential difference method, while maintaining the advantage of axial magnetic field. Two trochoïdal monochromators have been used by Tam and Wong (1979) in a double trochoïdal analyser. Such a system was further developped by Allan (1982) to give a superposition of 0 and 180° differential cross sections as this instrument detects simultaneously forward and backward scattered electrons. It has been shown to give a very high signal to noise ratio with the observation of excitation of high levels up to v = 17 in N_2 through the $^2\Pi_g$ resonance state (Allan 1985) and with the first observation of structure in the energy dependance of vibrational excitation cross section of H_2 ($v \gg 3$) in the $^2\Sigma_u^+$ resonance region (Allan 1985).

2.2. Negative Ion Spectrometers

Very few experimental set up have been devoted to kinetic energy distribution, angular distribution, and differential cross sections of negative ions. The techniques include :

i) Time of flight arrangement for measuring translational energy of negative ions formed by interaction with a pulsed electron beam (Illenberger et al 1979).
ii) Trochoïdal monochromator and a static gas cell, where the negative ions are extracted at 90° of the electron beam and analysed in energy by a 90° cyclindrical condenser (Dressler and Allan 1985);
iii) Crossed-beam experiment using electrostatic filters (Le Coat et al 1982) : both scattered electron and negative ion products of the interaction and dissociation are energy analysed at a selected angle and then electrons and ions are separated by a quadrupole mass filter for which the mass resolution can be adjusted to a given ionic mass (high resolution), or to sort out only the light electrons from all the negative ions having the same translational energy (low resolution). With this apparatus, energy loss spectre give ion kinetic energy spectra and, constant ion energy spectra can be obtained by setting the analyser energy and

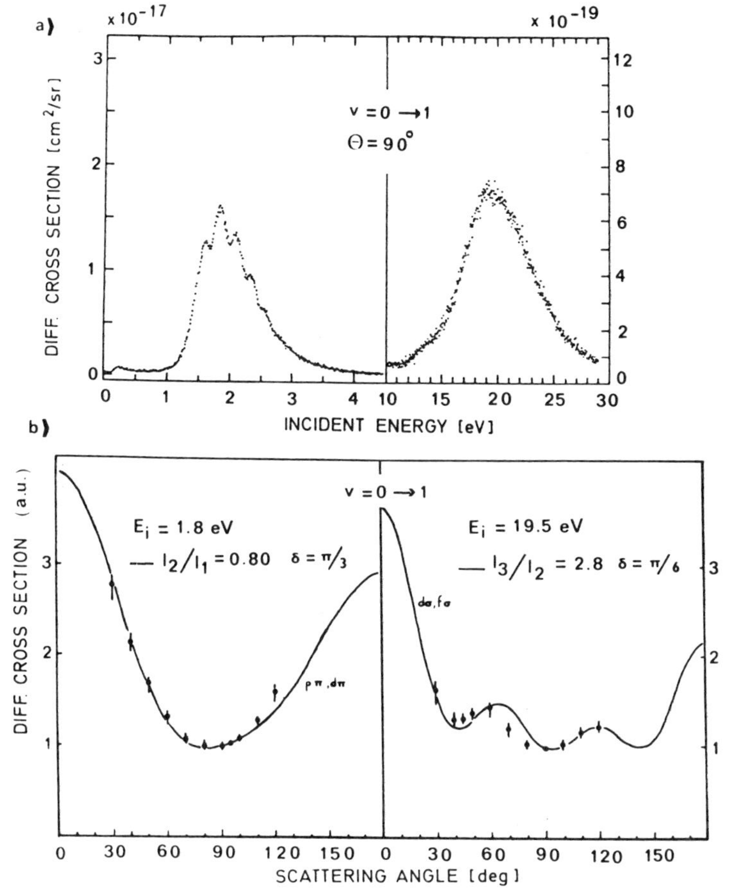

Figure 3. a) Absolute differential cross section for the v=1 level in CO at a 90° scattering angle, showing a low energy Π^* shape resonance around 1.8 eV and a σ^* shape resonance around 19 eV. b) Angular distributions for the v=1 level in CO for the Π^* and σ^* shape resonances.

sweeping the incident electron energy. If the ion energy is fixed at zero, threshold ion spectra are obtained. The energy resolution of this apparatus is limited by thermal broadening which is small at small kinetic energy, but increases rapidly at higher ion energy. The thermal motion of the target molecules in the plane of observation can be reduced from an equivalent temperature of 300 K in a cell to 50 K in an hypodermic needle and to 20 K in a multicapillary arrays.

3. VIBRATIONAL EXCITATION

3.1. Resonant Excitation of CO (0-30 eV)

Vibrational excitation by electron-molecule collision can be very efficient if shape resonance is the dominant interaction mechanism. Absolute differential cross sections of the excited vibrational levels, together with angular distributions are used to caracterise the resonance (energy, width, symmetry, active partial waves).

Figure 3a shows the absolute cross section for the v = 1 level in CO from 0 to 30 eV. At low energy a π^* shape resonance dominates the cross section around 1.8 eV. The oscillatory structure, known as the boomerang effect, results from interferences due to the intermediate lifetime of the resonance state being of the same order as a period of vibration. The angular distribution (fig 3b) can be fitted with admixture of p+d waves (l = 1 + 1 = 2). At higher energy, the cross section, although one order of magnitude smaller is still dominated by a broad shape resonance around 19.5 eV, with trapping of the extra electron in the σ^* antibonding unoccupied molecular orbital. The 5 eV (F W H M) broad peak corresponds to a very short lifetime of 10^{-16}s for the resonance state with no definite vibrational level. The angular distribution is dominated by an f wave (l = 3) with a small contribution from a d wave (l = 2) (figure 3b).

3.2 Near Threshold Excitation in CO

Because of the small permanent dipole moment of the CO molecule, at low impact energy the vibrational excitation cross section proceeds by direct (dipole) excitation. Quantitative measurements are difficult because both incident electron with energy below 1 eV and the scattered electron with an energy even lower are very sensitive to stray electric and magnetic fields.

The Kaiserslautern group has succeed to obtained reliable quantitative absolute cross sections down to threshold and associated angular distributions (Sohn et al 1985, Sohn et al 1986). Figure 4a shows the absolute cross section for the v = 1 level of CO at 90° scattering angle below the $^2\Pi$ resonance and down to threshold, compared to the Born Dipole Approximation (BDA). The discrepancy between experiment and calculation is attributed to the short range potential which is inadequately represented in the BDA by extending the point dipole to the origin. Angular dependance for the fundamental mode of vibration v = 1 at 0.45 eV incident electron energy (figure 4b) is forward peaked and in good agreement with BDA at small angle.

3.3. Selective Excitation in Polyatomic Molecules

Vibrational excitation in polyatomic molecules requires high energy resolution because of the number of vibrational levels, even if resonant excitation can be selective as first shown in Benzene by Wong and Schulz (1975). Moreover various final levels may have different rovibronic symmetry and as such different angular distributions. Rotational branch analysis has been obtained in few molecules like CO_2 (Kochem et al 1985, Antoni et al 1985) and CH_4 (Müller et al 1985).

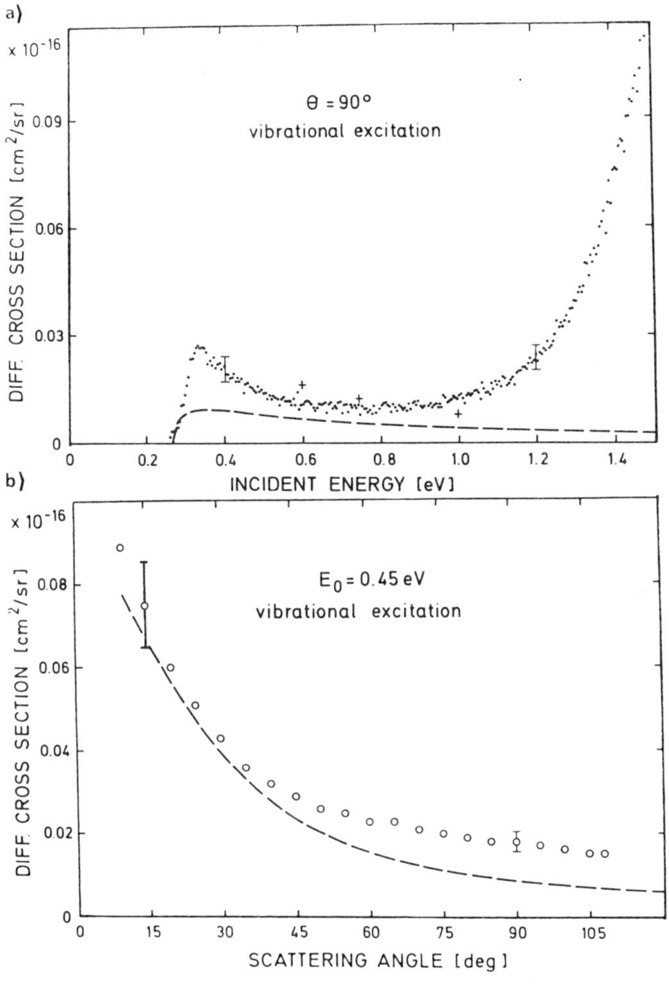

Figure 4. a) Near threshold absolute differential cross section for vibrational excitation in CO. b) Angular distribution. The dots are experimental results and dashed lines are the first Born-Dipole-Approximation. (Sohn et. al. 1985).

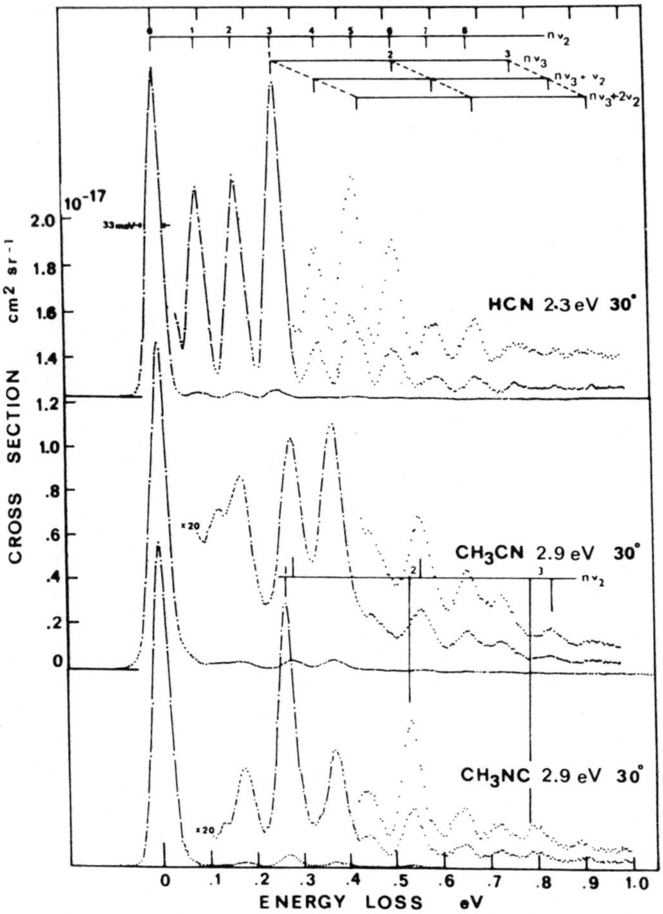

Figure 5. Electron energy loss spectra in HCN, CH3CN and CH3NC in the center of the π^* shape resonance at 2.3, 2.9 and 2.9 eV respectively and for a 30° scattering angle.

Figure 4 shows the vibrational excitation in cyanidric acid HCN, acetonitrile CH_3 CN, and methyl isocyanide CH_3 NC (a stable isommer of acetonitrile). All three molecules have a low energy short lived π^* resonance at 2.3 eV (HCN) and 2.9 eV (CH_3 CN, CH_3 NC). The energy loss spectra are dominated by a series of vibrational peaks associated with the C - N bond stretching and a series of C - N stretching plus one or two quanta of angular deformation. When the symmetry group of the molecule is lowered excitation become more complicated and many overlapping bands are observed. The role of the spatial extend of the antibonding orbital where the extra electron is trapped is shown in CH_3 CN where the π^* is not fully localised between the carbon and nitrogen atom of the cyano group, but has contribution of hydrogen atoms (π^* like CH_2). The different distribution of vibrational levels excited through the resonance in CH_3 CN and CH_3 NC is certainly related to the composition of the π^* CN orbital. (hyperconjugative effect π^*_{CN} - $\pi^*_{CH_2}$).

4. DISSOCIATIVE ATTACHMENT

4.1 Pollution of Electron Spectra by Negative Ions

In electrostatic analysers and electrostatic lenses, particles with the same charge follow identical trajectories and are transmitted equally regardless of their mass : both scattered electrons and negative ions produced by dissociative attachment can participate to the energy loss and excitation spectra (Lassettre 1974). In fact scattered electrons and negative ions cross sections often differ by orders of magnitude: nevertheless negative ions can pollute spectra in low intensity scattered electron regions as first observed by Trajmar and Hall (1974) who show that the strong banded structure observed in electron impact energy loss spectra of H_2O and D_2O in the 4.4 to 6.0 eV with incident electron energy around 6.5 eV were due to H^- (D^-) ions.

Recently negative ions formed with zero on near zero kinetic energy (either because of their heavy mass or because of the dissociation mechanism in which the whole excess energy goes into internal energy of the fragment) have been shown to contribute to threshold peaks in vibrational cross sections (Azria et al 1980).

The opportunity to analyse both electrons and negative ions in electrostatic spectrometers has been used to obtained differential cross sections and kinetic energy distributions of negative ions produced in dissociative attachment processes, by modifying an electron impact spectrometer by incorporating between the analyser and the particle detector, a momentum filter made of a simple coil (Scherman et al 1978) to seperate the light electrons from all negative ions. This system has been successfull for dissociative attachment studies on diatomic molecules (Cadez et al 1973, Tronc et al 1977, Hall et al 1977). Improvement was later obtained by replacing the coil with a quadrupole mass filter allowing observation of mass selected negative ions (Le Coat et al 1982).

4.2 Kinetic Energy Distribution

Kinetic energy distributions of negative ions produced by dissociative attachment in polyatomic molecules brought new insights into the dynamics of the dissociation process (always in competition with autodetachment of the extra electron), and on the potential energy surface of the resonance state (Tronc 1988).

Figure 6a shows the cross section for O^- ions around 8 eV in CO_2.

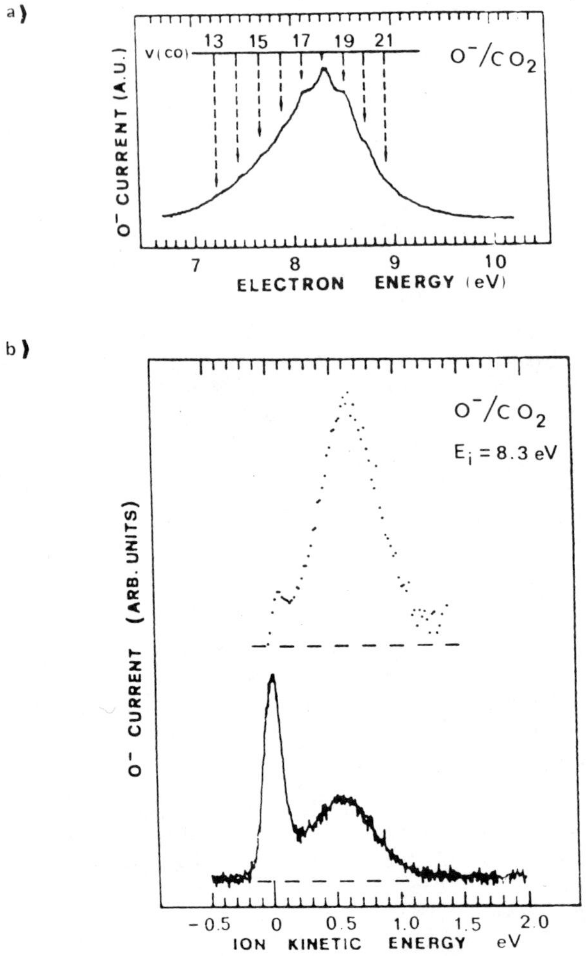

Figure 6. a) Zero kinetic energy O^- ions yield in CO_2 around 8 eV, showing high vibrational excitation of associated CO fragments. b) Kinetic energy distribution at 8.2 eV incident electron energy:the full curve shows the data as recorded, and the dotted curve is qualitatively corrected for low energy transmission effect(Dressler and Allan 1985).

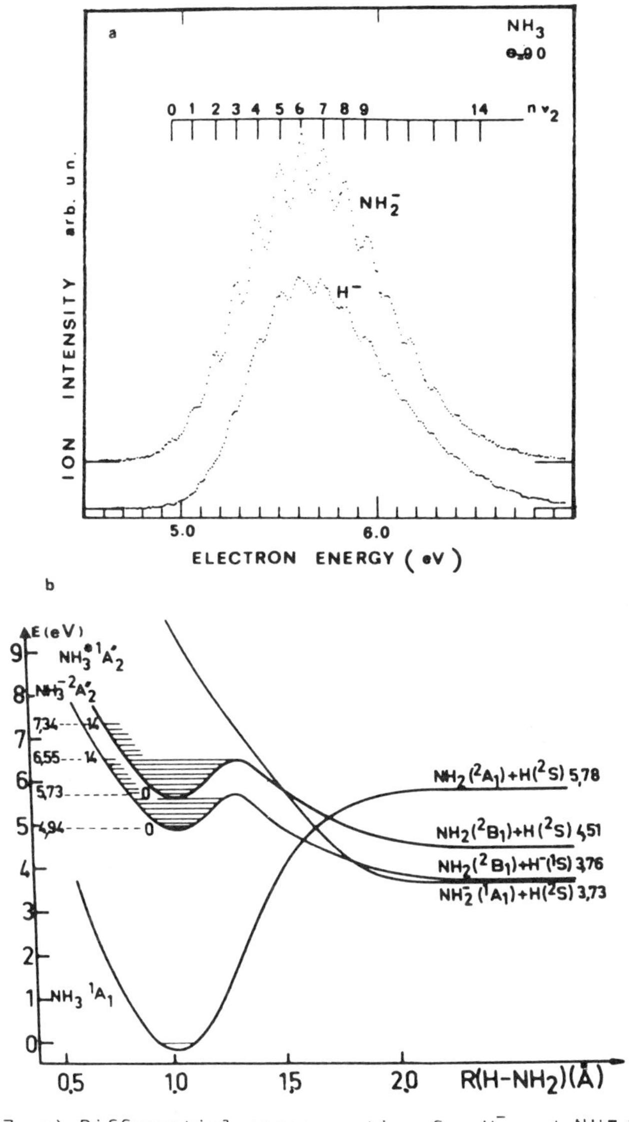

Figure 7. a) Differential cross section for H^- and NH_2^- ions in amonia at 90° scattering angle showing n ν_2 (inversion) vibrational levels of the NH_3^- $^2A_2''$ resonance state. b) Qualitative potential energy curves for the lowest states of NH_3 and NH_3^- in the $H\text{-}NH_2$ coordinate (Tronc et al. 1988).

The weak structures are associated with high vibrational excitation of the CO neutral fragment (Dressler and Allan 1985). The kinetic energy distribution of O^- ions (Figure 6b) is a good illustration of the difficulty to maintain constant lenses transmission for low energy particles. The full line is the measured spectrum, and shows a double maximum with a strong zero energy peak; the dots show the recent transmission corrected kinetic energy distribution of Dressler and Allan (1985).

4.3. H^- and NH_2^- Cross Section in NH_3

In the first published H^- and NH_2^- dissociative attachement results in NH_3, the threshold, the maximum and the shape of the cross sections for H^- and NH_2^- were similar (Compton et al 1969) and it was suggested that both H^- and NH_2^- ions could be produced from the same potential energy surface (Sharp and Dowell 1969). More recently, higher resolution measurements by Stricklet and Burrow (1986) produced a structured total negative ion cross section, explained by predissociation of the quasibound inversion n V_2 levels of the Feshbach NH_3^- $^2A_2''$ resonance state.

Kinetic energy measurements indicate that H^- ions are associated with highly vibrationally and rotationally excited NH_2 radicals, when NH_2^- ions are produced with very low kinetic energy because of their high mass (compared to H atom). Mass analysed angular differential cross sections (figure 7 a) show structures in both $H^- + NH_2$ and $NH_2^- + H$ channels (Tronc 1988). As the $^2A_2''$ NH_3^- state is adiabatically connected to the H- (^1S) + NH_2 ($^2B_1$) limit, structures in the NH_2^- ions yield cannot be explained in a planar dissociation mechanism, because NH_2^- ($^2A_1'$) + H (^2S) must be correlated to a $^2A_1'$ resonance state which does not mix with the $^2A_2''$ resonance state. But if non planar dissociation occurs, as suggested by the kinetic energy distribution, the two A' resonance states (Cs symmetry) can mix strongly at large internuclear separation where they are nearly degenerate (because NH_2 and H have almost the same electron affinity of 0.78 and 0.75 respectively), so that the predissociated bound inversion levels of the Feshbach resonance appear in both the NH_2^- + H channel (adiabatic dissociation) and in the $H^- + NH_2$ channel (diabatic dissociation) (figure 7b).

References

Allan M., 1982, Helv. Chim. Acta, 65 : 2008

Allan M., 1985 a, J.Phys., B.18 : L 451

Allan M., 1985 b, J.Phys., B.18 : 4511

Antoni Th., Jung K, Ehrhardt H, and Chang E.S., 1985, J.Phys., B.19:1377

Azria R., Le Coat Y., Guillotin J.P., 1980, J.Phys., B.13 : L505

Cadez I., Hall R.I., Tronc M., 1973, J.Phys., B.8 : L73

Compton R.N., Stockdale J.A., Reinhardt P.W., 1969, Phys. Rev., 180 : 111

Doering J.P., 1983, J. Chem. Phys., 79 : 2083

Dressler R., Allan M., 1985, Chem. Phys., 92 : 449

Hall R.I., Cadez I., Scherman C., Tronc M., 1977, Phys. Rev. A, 15 : 599

Harting F.H., Read 1976, in "Electrostatic Lenses", Elsevier, N.Y.

Hicks P.J., Daviel S., Wallbanck B. and Comer J., 1980, J.Phys., E : Sci. Instrum., 13 : 713

Illenberger E., Scheuneman H.N., Baungartel H., 1979, Chem. Phys., 37 : 21

Kochem K.H., Sohn W., Hebel H., Jung K., and Ehrhardt H., 1985, J.Phys., B.18 : 4455

Lassettre E.N., Huo W.H., 1974, J. Chem. Phys., 61 : 1703

Le Coat Y., Azria R., and Tronc M., 1982, J.Phys., B15 : 1569

Muller R., Jung K., Kochem K.H., Sohn W., and Ehrhardt H., 1985
J.Phys., B.18 : 3971
Pichou F., Huetz A., Joyez G., Landau M., and Mazeau J., 1976,
J.Phys., B.9 : 933
Register D.F., Trajmar S., and Srivastava S.K., 1980, Phys. Rev. A,
21 : 1134
Scherman C., Cadez I., Delon P., Tronc M., and Hall R.I., 1978,
J.Phys., E : Sci. Instrum. 11 : 746
Sharp T.E., and Dowell J.T., 1969, J. Chem. Phys., 50 : 3024
Sohn W., Kochem K.H., Jung K., Ehrardt H., and Chang E.S., 1985,
J.Phys., B.18 : 2049
Sohn W., Kochem K.H., Schenerlein K.M., Jung K., and Ehrhardt H.,
1986, J.Phys., B.19 : 4017
Srivastava S.K., Chutjian A., and Trajmar S., 1975, J. Chem.
Phys., 63 : 2659
Stamatovic A., and Schulz G.J., 1968, Rev. Sci. Instrum, 39 : 1752
Stricklett K.L., and Burrow P.D., 1986, J.Phys., B.19 : 4241
Tam W.C., and Wong S.F., 1979, Rev. Sci. Instrum., 50 : 302
Trajmar S., and Hall R.I., 1974, J.Phys., B.7 : L 458
Tronc M., Fiquet-Fayard F., Scherman C., and Hall R.I., 1977,
J.Phys., B.10 : 305
Tronc M., 1988, in "Molecules in Physics, Chemistry and Biology",
Vol. 11, Maruani, ed., Kluwer Academic Publishers
Wong S.F., and Schulz G.J., 1975, Phys. Rev. Let., 35 : 1429

DYNAMICS OF MOLECULAR PHOTODISSOCIATION

Reinhard Schinke

Max-Planck-Institut für Strömungsforschung
3400 Göttingen, FRG

INTRODUCTION

IR- and UV-Photodissociation

Photodissociation is the break-up of an initially bound (diatomic or polyatomic) molecule through the absorption of light. If the dissociation energy is small ($\lesssim$ 100meV) as it is usually the case for van der Waals molecules, for example, an infra-red (IR) photon is sufficient to break the weak bond. The energy of the photon is first selectively deposit in an internal mode and then transfered to the van der Waals mode by intramolecular forces. The electronic structure of the parent molecule is not changed by IR-absorption and therefore the dissociation occurs entirely in the ground electronic state. The lifetime of the excited complex is typically longer than a ps and depends sensitively on the strength of the intramolecular forces. This type of (rotational/vibrational) predissociation has been amply studied in the past, both experimentally and theoretically.[1] It is still an active field and many new results are expected in the future.

In this lecture we will exclusively consider the photodissociation through the absorption of an ultra-violet (UV) photon. The parent molecule is electronically excited from the ground state, $\tilde{X}$, to a particular excited state $\tilde{A}$, $\tilde{B}$, etc.. In this case the electronic structure of the molecule is changed. If the corresponding potential energy surface (PES) is anti-bonding, i.e., repulsive along one (or several coordinates) the excited complex will break apart into products, typically within less than 10 fs or so. The UV-photodissociation has been examined for almost all molecules a long time ago.[2,3] Nevertheless, there are many open questions concerning the elementary dynamics of the dissociation process, which can be tackled today with modern experimental and theoretical methods.

In order to be more specific let us consider a triatomic molecule ABC with one dissociation channel A+BC. The appropriate coordinates are the Jacobi-coordinates normally used to describe scattering processes. They are defined in Fig. 1. R is the dissociation coordinate which becomes infinite in the exit channel. The internal motion of the BC product is described by r (vibration) and γ (rotation).

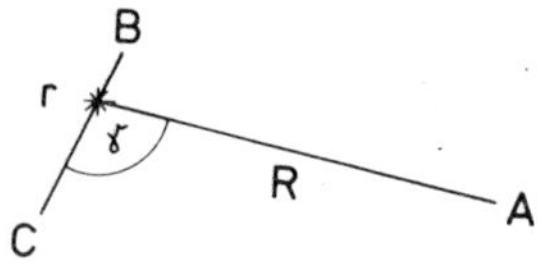

Fig. 1. Jacobi-coordinates for the triatomic ABC molecule.

An UV-photodissociation process is usually envisioned to proceed in two steps according to

$$ABC(\tilde{X}|E_i) + h\nu \xrightarrow{(1)} ABC(\tilde{A})^{*} \xrightarrow{(2)} A+BC(nj). \qquad (1)$$

E_i specifies a particular rotational-vibrational level of the parent molecule within the ground electronic state $\tilde{X}$, $ABC(\tilde{A})$ represents the excited complex within the upper electronic state $\tilde{A}$, and (nj) is a particular vibrational-rotational state of the BC product. Step (1) describes the absorption (i.e., electronical excitation) process and step (2) describes the decay of the excited complex into products A and BC. This separation is well defined only if the lifetime of the excited molecule is long. Nevertheless, it is also a convenient starting point for the discussion of direct processes.

The UV-photodissociation process is schematically illustrated (in one dimension) in Fig. 2, where the PES's for the ground state, $V(\tilde{X})$, and for two excited states, $V(\tilde{A})$ and $V(\tilde{B})$, are plotted vs. dissociation coordinate R. Ψ_{gr} is the bound state wavefunction which describes the nuclear motion in the ground electronic state. If the $ABC(\tilde{X})$ parent molecule is in the lowest vibrational state it is approximately a product of one-dimensional Gaussians in all coordinates. Ψ_{ex} is the continuum wavefunction with energy $E=E_i+h\nu$ which describes the nuclear motion within the excited electronic state. It behaves asymptotically ($R\rightarrow\infty$) as exp(ikR). The proper boundary conditions will be defined below.

Direct and Indirect UV-Photodissociation

If the excited state PES is steeply repulsive along R (case $\tilde{A}$ in Fig. 2) the break-up of the excited complex is fast and direct. The corresponding absorption cross section $\sigma(E)$, shown on the right panel of Fig. 2, is broad and structureless. It "reflects" roughly the R-dependence of the ground state wavefuntion. This is the well known, one-dimensional "reflection principle" characteristic for a direct dissociation process.[4] The width of $\sigma(E)$ is directly related to the steepness of V(A) and therefore related to the "lifetime" of the complex. The $\tilde{X} \rightarrow \tilde{A}$ transition in Fig. 2 is an example of direct photodissociation.

If the excited state PES has a local well along R (case $\tilde{B}$ in Fig. 2) which supports one (or more) quasi-bound levels the lifetime of the excited complex ABC(B) may be relatively long. The barrier may be due to an avoided crossing with a higher state, for example. In this case the cross section exhibits sharp structures near the energies of the quasi-bound states and resembles very much an ordinary bound-to-bound

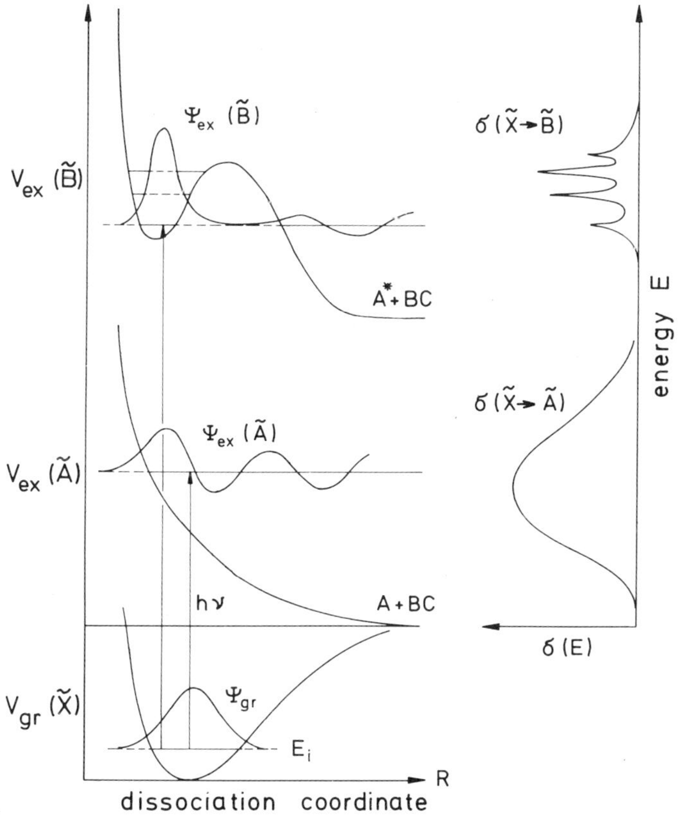

Fig. 2. Schematic illustration of an UV-photodissociation process. $V_{gr}(\tilde{X})$ is the ground state potential and $V_{ex}(\tilde{A})$ and $V_{ex}(\tilde{B})$ are two excited state potentials. The corresponding absorption cross sections are shown vs. energy on the right panel. $\tilde{X} \rightarrow \tilde{A}$ is a direct process while $\tilde{X} \rightarrow \tilde{B}$ represents an indirect process.

absorption spectrum. The decay of the complex is rather slow and may be induced by tunneling through the barrier or by coupling between the internal and the dissociation modes. In this case the two steps in Eq.(1) approximately can be treated separately. The $\tilde{X} \rightarrow \tilde{B}$ transition in Fig. 2 is an example of indirect photodissociation which is somehow similar to an IR-predissociation process described above. The width of the structures of σ(E) are directly related to the lifetime of the complex.

During this lecture we will primarily consider direct photodissociation on steeply repulsive PES's. At the end we will also discuss an example of indirect dissociation. The theoretical treatment in both cases is formally identical. Because of the different time scales, however, the decay mechanisms can be quite different.

Observables

Traditionally the <u>total absorption cross section</u> (absorption spectrum) as a function of the photon energy hν or the photon wavelength λ, $\sigma_{abs}(\lambda)$, has been measured for almost every molecule.[2,3] It is, roughly speaking, the probability for absorbing a photon with wavelength λ, irrespective of the internal states of the products. $\sigma_{abs}(\lambda)$ contains information about the energetical order of the various excited states and the selection rules associated with the electronic transition. The overall shape of the absorption cross section may indicate the general nature of the dissociation process: A broad and structureless spectrum is characteristic for a direct process (reflection principle) whereas sharp and well resolved structures indicate an indirect process (predissociation). Information on the forces between the internal and the translational degrees of freedom are difficult to extract from $\sigma_{abs.}(\lambda)$.

In recent years experimental studies are primarily focussed on the final state distribution of the products.[5-9] This became possible through the availability of powerful light sources (laser) and through the development of efficient detection methods like laser induced fluorescence[10] (LIF) or resonance enhanced multiphoton ionisation[11] (REMPI), for example. In terms of the model ABC problem one can define partial absorption cross sections $\sigma(nj|\lambda)$ which represent the probability for absorbing a photon with wavelength λ and producing BC in a particular vibrational-rotational state (nj). The total absorption cross section is the sum of all partial cross sections, i.e., $\sigma_{abs}(\lambda) = \Sigma_{nj}\ \sigma(nj|\lambda)$.

The main purpose of theoretical studies is to relate specific features of the various cross sections to features of the PES. The state-resolved cross sections certainly contain the most detailed information about the dissociation process, especially the dependence of the excited state PES on the internal degrees of freedom, i.e., r and γ. One of the central points of this lecture will be to demonstrate that under certain conditions the final state distributions can be explained by a generalization of the one-dimensional reflection principle.

Photodissociation and Scattering

In view of Eq.(1) the dissociation of the excited complex ABC^* can be regarded as the second step of a normal A+BC scattering process within the excited electronic state. Therefore it is very plausible to expect certain similarities between a dissociation and a scattering process. For example, we will demonstrate that rotational rainbows, which are well established in direct collisions,[12] are closely related to the rotational reflection principle in direct dissociation.[13,14] The total absorption cross section corresponds to the total cross section in scattering. Likewise, the partial absorption cross sections correspond to the state resolved collisional cross sections.

Nevertheless, differences between scattering and dissociation can be remarkable. They are mainly due to the very different "initial conditions" for both processes. In scattering the initial conditions are prepared for the isolated species A+BC while in dissociation they are defined through the parent molecule ABC in the inner region of the PES when both systems are close together. In scattering experiments with neutral species the energy is typically low ($\lesssim$ 0.1eV or so) while in dissociation the energy is very often of the order of several eV. In scattering many partial waves contribute to the cross section whereas in dissociation very few partial waves contribute because of the $\Delta J=0,\pm1$ selection rule for optical transitions. An important point of this lecture will be to elucidate the similarities and differences between scattering and dissociation. As we will show the theoretical treatment is very similar and standard numerical procedures developed for collisions can be adapted for dissociation with only slight modifications. Also, various dynamical approximations well estabilished in scattering can be readily applied to dissociation.

QUANTAL THEORY OF PHOTODISSOCIATION

Like any "scattering" process photodissociation can be treated either in a time-dependent or in a time-independent formalism. If exact both approaches should obviously give the same results. Although the time-dependent picture resembles more our naive intuition of a dissociation process, most calculations are performed time-independently. The reasons may be the following. First, the exact solution of the time-dependent Schrödinger equation is much more involved than the exact solution of the time-independent Schrödinger equation. Second, there are several efficient numerical methods available to treat collision processes which can be easily adapted for dissociation. Third, photodissociation is in some respects similar to ordinary spectroscopy between two bound states which is usually treated time-independently. Therefore many of the pictures established there may also be helpful to interpret photodissociation processes.

In this lecture we will exclusively present and discuss time-independent quantal calculations. For an overview of the time-dependent quantal formulation and the various approximations evolving from it the reader is refered to the work of Heller and coworkers.[15-18]

The Golden Rule Expression

As usual in photochemical studies it is assumed that the intensity of the photon is weak such that the ligh-matter interaction can be treated in first order perturbation theory. Then the cross section for absorbing a photon with energy $h\nu$ and obtaining products in internal state f is given by Fermi's Golden Rule[19]

$$\sigma_i(f|\nu) \sim \nu|\langle\Psi_{gr}^{(i)}|\mu|\Psi_{ex}^{-(f)}\rangle|^2 \quad . \tag{2}$$

$\Psi_{gr}^{(i)}$ is the nuclear wavefunction for a particular rotational-vibrational level (i) within the ground (gr) electronic state and $\Psi_{ex}^{-(f)}$ is the continuum wavefunction for nuclear motion within the excited electronic state. The supercript -(f) represents a particular choice of boundary conditions which will be defined below. Both wavefunctions are full solutions of Schrödinger's equation for the nuclear Hamiltonian μ is the transition dipole function, i.e., the matrix element of the electronic dipole operator and the corresponding electronic wavefunctions. All quantities in Eq. (2) are multi-dimensional. Constants which do not affect the relative wavelength-dependence of the spectrum and the final state distributions will be omitted throughout this lecture. A rigorous and illuminating derivation of Eq.(2) has been given by Shapiro and Bersohn,[19] for example. The photodissociation cross section is formally equivalent to the absorption cross section for a bound-to-bound transition[20] except that one of the nuclear wavefunctions is a continuum wavefunction.

The formal expression for the dissociation cross section looks rather simple but its evaluation is very difficult in practice. In principle two multi-dimensional wavefunctions have to be calculated exactly. This creates significant difficulties even for the triatomic case if all degrees of freedom have to be included. Two multi-dimensional PES's are required. Normally reasonable information is available for the ground state PES, V_{gr}, from spectroscopic data. Except for a few cases, however, information about the excited state PES, V_{ex}, is usually very insufficient and does not allow a rigorous dynamical study. In addition, the transition dipole function should be known. Fortunately, the transition region is relatively narrow and therefore it seems justified to assume a constant, coordinate-independent transition dipole function. Besides these technical difficulties we found in several applications that the interpretation of the cross sections can be quite complicated because so many ingredients enter the calculation.

Close-Coupling Calculations

The standard procedure to solve the time-independent Schrödinger equation is the expansion of the full wavefunction in terms of product eigenfunctions. The resulting set of coupled equations must be solved numerically. In a beautiful paper Balint-Kurti and Shapiro[21] worked out all the necessary equations for the general ABC → A+BC case in a body-fixed coordinate system including rotation and vibration. The main asset of the theory of Balint-Kurti and Shapiro is that both wavefunctions are calculated in the same set of coordinates. They give explicit expressions for various differential and integral cross sections. In order to illustrate the methodology we consider here the much simpler case of a rigid-rotor neglecting the vibrational degree of freedom of BC. In addition we assume that the total angular momentum J is zero in both electronic states, i.e., we neglect the selection rule $\Delta J=0,\pm 1$ for dipole transitions.

The rigid-rotor Hamiltonian for J=0 is given by[21]

$$H(R,\gamma) = -\frac{1}{2m}\frac{\partial^2}{\partial R^2} + (B_{rot} + \frac{1}{2mR^2})\,\hat{j}^2 + V(R,\gamma) \qquad (3)$$

for the electronic ground state as well as for the excited electronic state. B_{rot} is the rotational constant of the BC rotor, $\hat{j}$ is the molecular rotational angular momentum, and m is the reduced mass of A-BC. Both wavefunctions are expanded in terms of eigenfunctions of $\hat{j}^2$ according to

$$\Psi_{gr}^{(i)}(R,\gamma) = \sum_{j''} \chi_{j'',gr}^{(i)}(R)\, Y_{j'',0}(\gamma,0) \qquad (4a)$$

$$\Psi_{ex}^{-(j)}(R,\gamma) = \sum_{j'} \chi_{j',ex}^{-(j)}(R)\, Y_{j',0}(\gamma,0) \quad , \qquad (4b)$$

where the $Y_{j,0}(\gamma,0)$ are spherical harmonics. For zero total angular momentum the helicity (magnetic quantum number) is restricted to zero, i.e., the motion is confined to a plane.

Insertion of Eq.(4) into Schrödinger's equation and employing the orthonormality of the expansion functions yields the usual set of coupled equations for the radial functions[21]

$$[-\frac{d^2}{dR^2} - k_{j,gr}^2 + \frac{j(j+1)}{R^2}]\,\chi_{j,gr}^{(i)}(R) \; + \qquad (5)$$

$$+\; 2m \sum_{j'} V_{jj'}^{gr}(R)\, \chi_{j',gr}^{(i)}(R) = 0$$

with the definitions

$$k^2_{j,gr} = 2m[E_i - B_{rot}\, j(j+1)] \tag{6}$$

for the wavenumbers and

$$V^{gr}_{jj'}(R) = 2\pi \int_0^\pi d\gamma \, \sin\gamma \; Y_{j,0}(\gamma,0)\; V_{gr}(R,\gamma)\; Y_{j',0}(\gamma,0) \tag{7}$$

for the potential matrix elements. The excited state radial functions fulfill exactly the same equations, however using V_{ex} in Eq. (7).

The coupled equations for the ground state radial functions have to be solved subject to the usual boundary conditions for a bound state, i.e.,

$$\chi^{(i)}_{j,gr}(R) \underset{R\to 0}{\sim} 0 \quad \text{and} \quad \chi^{(i)}_{j,gr}(R) \underset{R\to\infty}{\sim} 0 \ . \tag{8}$$

The coupled equations for the excited state wavefunctions have to be solved subject to the boundary conditions

$$\chi^{-(j)}_{j',ex}(R) \underset{R\to 0}{\sim} 0 \ , \tag{9a}$$

$$\chi^{-(j)}_{j',ex}(R) \underset{R\to\infty}{\sim} k^{-1/2}_{j,ex}\ \delta_{jj'}\ \exp(+ik_{j,ex}R) + k^{-1/2}_{j',ex}\ S_{jj'}\ \exp(-ik_{j',ex}R) \tag{9b}$$

for open channels ($k^2_{j',ex} > 0$) and

$$\chi^{-(j)}_{j',ex}(R) \underset{R\to\infty}{\sim} 0 \tag{9c}$$

for closed channels ($k^2_{j',ex} < 0$). Equation (9) guarantees that the full dissociation wavefunctions have the required asymptotic behaviour

$$\Psi^{-(j)}_{ex}(R,\gamma) \underset{R\to\infty}{\sim} k^{-1/2}_{j,ex}\ \exp(+ik_{j,ex}R)\ Y_{j,0}(\gamma,0) + \sum_{j'} k^{-1/2}_{j',ex}\ S_{jj'} \exp(-ik_{j',ex}R)\ Y_{j',0}(\gamma,0) \ . \tag{10}$$

The minus sign of the superscript indicates that in dissociation the "incoming" solutions of Schrödinger's equation have to be employed rather than the "outgoing" solutions which are more customary in collision problems.[19]

The boundary conditions define for each energy $E=E_i+h\nu > 0$ a whole set of independent and degenerate partial wavefunctions

$$\Psi_{ex}^{-(j=0)}, \Psi_{ex}^{-(j=1)}, \Psi_{ex}^{-(j=2)}, \ldots, \Psi_{ex}^{-(j_{max})}$$

which are distinguished by the single outgoing wave in channel j. Each partial wavefunction is a complete solution of Schrödinger's equation. j_{max} is the highest state which is energetically open at this energy. Each partial cross section is the overlap of Ψ_{gr} with a particular partial wavefunction. Thus, if one wants to calculate all partial cross sections (which is necessary to determine the total cross section) one has to calculate the complete matrix of radial functions for each j and j'. This is different from a scattering calculation where only one row (or one column) of the scattering matrix $S_{jj'}$ is required to calculate all $j_i \rightarrow j_f$ cross sections for a fixed initial state.

Inserting the two expansions 4(a) and 4(b) into the Golden Rule expression Eq. (2) we readily obtain ($\mu=1$)

$$\begin{aligned} \sigma_i(j|\nu) &\sim \nu |\langle \Psi_{gr}^{(i)}(R,\gamma)|\Psi_{ex}^{-(j)}(R,\gamma)\rangle|^2 \\ &\sim \nu |t_i(j|\nu)|^2 \end{aligned} \tag{11}$$

with the definition

$$t_i(j|\nu) = \sum_{j'} \int_0^\infty dR\; \chi_{j',gr}^{(i)}(R)\; \chi_{j',ex}^{-(j)*}(R) \tag{12}$$

for the dissociation amplitude. Deriving Eq. (12) we utilized that both wavefunctions are calculated in the same set of coordinates and expanded in the same set of basis functions. This is of course not essential. The final expression becomes obviously more complex if different coordinates are used for the two electronic states.

The calculation of the cross section is rather straight forward provided the radial wavefunctions are known. In a separate section we will show how they can be obtained directly from an ordinary scattering code. Incidentally we mention that several sophisticated methods have been proposed in the literature to determine dissociation cross sections but avoiding a direct calculation of the wavefunctions.[22-24] In some applications, however, we found it rather worthwhile to interprete the final results directly in terms of the wavefunctions, especially if resonance processes are involved or if their nodal pattern is important.

The time-independent close-coupling (CC) formalism is in principle applicable for all systems regardles which or how many degrees of freedom are involved. In practice such calculations are limited only by the number of states, which unfortunately increases very rapidly for non-zero total angular momenta or for polyatomic molecules. Therefore in most cases it is necessary to introduce approximations. As one example we will briefly discuss the energy sudden approximation (ESA) in the next section which is frequently and very often successfully employed in collision studies. It is easily adapted for dissociation processes.

Energy Sudden Approximation

Within the ESA it is assumed that the internal energy (rotation, vibration,...) is negligibly small compared to the translational energy of the relative motion.[25] In a time-dependent picture this means that the internal motion is slow compared to the translational motion. Thus, the sudden approximation is just the counterpart of the adiabatic approximation. The sudden equations can be obtained by back-transformation of the coupled equations (5) under the assumption that the diagonal term in Eq. (5) is independent of the quantum number j.[26] Alternatively we can start with the exact Hamiltonian and simply replace the angular momentum operator by an effective constant j_{eff}.[27] The necessary condition is that the rotational energy is negligibly small compared to the total energy for all states populated during the collision. The resulting approximate Hamiltonian

$$\tilde{H}(R|\gamma) = -\frac{1}{2m}\frac{d^2}{dR^2} + \left(B_{rot} + \frac{1}{2mR^2}\right) j_{eff}^2 + V(R|\gamma) \tag{13}$$

depends only parametrically on the internal coordinate γ. In practice the effective angular momentum is conveniently set to zero. The famous infinite order sudden approximation (IOSA) would also include the socalled centrifugal sudden approximation (CSA) in which the helicity-changing coupling terms (Coriolis coupling) are neglected.[28] However, this coupling is exactly zero for total angular momentum J=0 and therefore the CSA is not explicitly discussed here.

Approximations to the close-coupling radial functions are readily obtained from

$$\tilde{\chi}^{-(j)}_{j',ex}(R) = \langle Y_{j,0}(\gamma,0)|\tilde{\chi}^{-}_{ex}(R|\gamma)|Y_{j',0}(\gamma,0)\rangle$$

$$= 2\pi \int_0^{\pi} d\gamma \, \sin\gamma \, Y_{j,0}(\gamma,0) \, \tilde{\chi}^{-}_{ex}(R|\gamma) \, Y_{j',0}(\gamma,0) \, , \tag{14}$$

where the one-dimensional, "quasi elastic" wavefunctions $\tilde{\chi}^{-}_{ex}(R|\gamma)$ are solutions of

$$\left[-\frac{d^2}{dR^2} - k^2_{j_{eff},ex} + \frac{j_{eff}(j_{eff}+1)}{R^2} + 2m\, V_{ex}(R|\gamma)\right] \tilde{\chi}^{-}_{ex}(R|\gamma)=0. \tag{15}$$

One can easily check that the approximate radial functions of Eq.(14) solve the set of coupled equations under ESA conditions and fulfill the proper boundary conditions. The one-dimensional wavefunctions depend parametrically on the orientation angle γ.

Insertion of Eq. (14) into Eq. (12) yields for the approximate dissociation amplitude

$$\tilde{t}_i(j|\nu) = 2\pi \int_0^{\pi} d\gamma \, \sin\gamma \, Y_{j,0}(\gamma,0) \, \tilde{t}_i(\gamma|\nu) \tag{16a}$$

$$\tilde{t}_i(\gamma|\nu) = \sum_{j'} Y_{j',0}(\gamma,0) \int_0^{\infty} dR \, \chi^{(i)}_{j',gr}(R) \, \tilde{\chi}^{-*}_{ex}(R|\gamma). \tag{16b}$$

Equation (16) is written in a form which makes closes contact with the ESA transition amplitude in collisions. Please note, that only the excited state wavefunction has been approximated. The ground state wavefunction, which is usually easier to calculate, is incorporated exactly into Eq. (16).

In principle the sudden approximation can be applied to any nuclear degree of freedom. In all cases the methodology is the same. Since the solution of a large set of coupled equations is avoided the required computer times are significantly shorter than for exact CC calculations. In addition, the analytical form of the dissociation amplitude allows a simple semiclassical ($\hbar \to 0$) interpretation of the results.[29] The ESA is often very accurate for rotational excitation provided the overall energy transfer is small. It has been applied in several photodissociation studies. The ESA is naturally less suitable for vibrational excitation where the energy spacing between the levels is much larger.[30]

Numerical Methods

In this section we will very briefly describe the direct calculation of the coupled channel wavefunctions with standard methods. Let us denote the matrix of radial functions and their derivatives by

$$(\underline{\underline{X}})_{j'j} := \chi^{-(j)}_{j',ex}(R) \; , \quad (\underline{\underline{X}})'_{j',j} := \frac{d}{dR} \chi^{-(j)}_{j',ex}(R) \; , \tag{17}$$

where the second (column) index specifies the j-th independent solution defined by the boundary conditions and the first (row) index specifies the expansion channel. Each column of $\underline{\underline{X}}$ represents a complete solution of Schrödinger's equation. The dimension of the matrices, N is determined by the expansion in Eq. (4).

As customary in scattering calculations[31] we first determine N linearly independent solutions by starting the integration at R=0 with initial conditions

$$(\underline{\underline{X}})_{j'j} = 0 \quad \text{and} \quad (\underline{\underline{X}})'_{j'j} = \delta_{j'j} \, \varepsilon \; . \tag{18}$$

Pointwise integration of the CC equations from R=0 up to R_∞, where the potential has decreased to zero, yields the matrices $\underline{\underline{X}}_\infty = \underline{\underline{X}}(R_\infty)$ and $\underline{\underline{X}}'_\infty = \underline{\underline{X}}'(R_\infty)$. They, of course, do not fulfill the required boundary conditions. However, the correct solutions can be easily obtained by taking linear combinations according to

$$\underline{\underline{X}}_\infty \underline{\underline{T}} = \underline{\underline{A}} + \underline{\underline{B}}\,\underline{\underline{S}}$$

$$\underline{\underline{X}}'_\infty \underline{\underline{T}} = \underline{\underline{A}}' + \underline{\underline{B}}'\,\underline{\underline{S}} \quad , \tag{19}$$

where $\underline{\underline{A}}$ and $\underline{\underline{B}}$ are diagonal matrices of the form

$$(\underline{\underline{A}})_{j'j} = k_j^{-1/2} \exp(+ik_j R_\infty)\, \delta_{j'j} \tag{20a}$$

$$(\underline{\underline{B}})_{j'j} = k_j^{-1/2} \exp(-ik_j R_\infty)\, \delta_{j'j} \tag{20b}$$

and $\underline{\underline{A}}'$ and $\underline{\underline{B}}'$ are the corresponding matrices of the derivatives with respect to R. The actually calculated matricess $\underline{\underline{X}}_\infty$ and $\underline{\underline{X}}'_\infty$ are real while all other matrices in Eq. (19) are complex.

The transformation matrix $\underline{\underline{T}}$ and the scattering matrix $\underline{\underline{S}}$ can be easily obtained from Eq. (19) which represents (2N) sets of linear equations each having dimension (2N). For a scattering process all information is contained in the S-matrix. For dissociation, however, one needs the wavefunction in order to make the overlap with the ground state. The wavefunction matrix $\underline{\underline{X}}_i$ at each integration point R_i is obtained by taking linear combinations of the actually calculated matrices according to

$$\underline{\underline{X}}_i = \underline{\underline{X}}_i^{cal}\, \underline{\underline{T}} \quad . \tag{21}$$

The transformation matrix $\underline{\underline{T}}$ is of course the same at each integration point R_i. Since $\underline{\underline{T}}$ is complex (because of the boundary conditions) the correct solution matrix $\underline{\underline{X}}_i$ is also complex.

This procedure requires that the calculated solution matrix $\underline{\underline{X}}_i^{cal}$ is stored at each integration point. Since usually many integration steps are needed these (NxN) matrices must be written on an external file which in turn increases significantly the I/O-time. This is the only disadvantage of the procedure. It has, however, the advantage that the wavefunction can be directly analyzed.

It is well known that the individual vectors of the solution matrix tend to become linearly dependent as the wavefunctions are integrated through the non-classical region, especially if closed channels are involved.[31] This requires an additional <u>stabilization transformation</u> in order to maintain linear independence. A possible transformation is

$$\tilde{\underline{\underline{X}}}_i = \underline{\underline{X}}_i\, \underline{\underline{X}}_i^{-1} = \underline{\underline{1}} \quad , \qquad \tilde{\underline{\underline{X}}}'_i = \underline{\underline{X}}'_i\, \underline{\underline{X}}_i^{-1} \tag{22}$$

which can be readily performed and which makes the calculations very stable. This additional transformation must be accounted for when the final wavefunction is calculated according to Eq. (21).

If one wants to avoid the direct calculation of the wavefunctions one can evaluate the overlap integrals with the ground state,

$$(\underline{\underline{I}})_{j'j} = \int_0^\infty dR\ \chi_{j',gr}(R)\ \chi_{j',ex}^{-(j)*}(R) \tag{23}$$

during the forward integration and finally transform the matrix $\underline{\underline{I}}$ according to Eq. (21). This procedure is certainly more efficient because the storage of many large matrices is avoided and because the (complex) transformation (21) has to be done only once.

CLASSICAL THEORY OF PHOTODISSOCIATION

The quantal theory is exact (within the limit of weak photon intensities) and in principle easy to implement. In practice, however, the calculations are limited to small systems only. Up to about hundred coupled channels are manageable with medium size computers. Another problem is the inherent difficulty to interpret the CC results and to extract the general physics which determines the dissociation process. A CC program is like a "black box" which - loosely speaking - obscures the "mapping" from the initial state to the final state.

Another approach to treat photodissociation processes is ordinary classical mechanics. Classical trajectory calculations have been extensively and in many cases successfully used over the past twenty years or so to examine all kinds of reactive and non-reactive scattering processes.[32] Classical mechanics should be especially suitable for direct (i.e., fast) photodissociation when the "lifetime" of the excited complex is short and when the forces between the various degrees of freedom in the exit channel are strong. Trajectory calculations are in principle easy to implement and also applicable to larger polyatomic systems for which exact quantal calculations are definitely not feasible. Furthermore, it is possible to follow the time evolution of the trajectories which may lead to a simple understanding of the dissociation process. On the other hand, quantum mechanical effects such as interferences or resonances inherently can not be described by classical mechanics.

Cross Section Expressions

Classically a photodissociation process is considered to proceed in two steps: The first step (1) is a vertical Franck-Condon (FC) transition from the ground state to the excited electronic state and the second step (2) is the decay of the excited complex following the classical equations of motion. The electronic transition is assumed to be instantaneous such that the classical coordinates $\underline{q}$ and the corresponding momenta $\underline{p}$ are not changed. Thus, $\underline{q}_0=\underline{q}(t=0)$ and $\underline{p}_0=\underline{p}(t=0)$ as defined by the bounded motion within the ground state constitute the initial conditions for the dissociative trajectory within the excited state. In this sense the nuclear motion on the ground and the excited state PES's are "coupled". Our general intuition of a photodissociation process [Eq. (1)] is primarily based on the classical picture rather than on the formal quantum mechanical theory.

The classical photoabsorption cross section is defined as

$$\sigma_i^{cl}(f|\nu) = \nu \int d\underline{p}_0 \int d\underline{q}_0 \; |\mu(\underline{q}_0)|^2 \; W_i(\underline{q}_0, \underline{p}_0)$$

$$\times \; \delta[H_{ex}(\underline{q}_0, \underline{p}_0) - E] \; \delta[F(q_0, p_0) - f] \tag{24}$$

It is proportional to the modulus square of the transition dipole function. $W_i(\underline{q}_0, \underline{p}_0)$ is the distribution function of the classical coordinates and momenta associated with the nuclear motion of the parent molecule prior to the excitation process. It depends of course on the initial vibrational-rotational state (i). The first delta function selects only those points of the phase-space (i.e., trajectories) which have the specified energy $E=E_i+h\nu$. H_{ex} is the Hamilton function within the excited electronic state. The second delta function selects those trajectories which lead to the specified final product state f, where f represents a set of quantum numbers (n,j,...). $F(\underline{q}_0, p_0)$ is the socalled excitation function depending on the initial variables, $\underline{q}_0$ and $\underline{p}_0$.[33] Examples of F will be presented below. The excitation function is defined by solving Hamilton's equations

$$\dot{\underline{q}} = \frac{\partial H_{ex}}{\partial \underline{p}} \quad , \quad \dot{\underline{p}} = - \frac{\partial H_{ex}}{\partial \underline{q}} \quad , \tag{25}$$

in the limit as t goes to infinity.

The practical calculation of classical cross sections is straight forward. The multi-dimensional integration in Eq. (24) is usually done by Monte-Carlo methods. It is important to stress that the equations of motion must be solved only if the final product state f is resolved. If only the total absorption cross section (i.e., the sum over all product states f) is desired it sufficies to sample the phase-space and to calculate H_{ex} and W_i.

Initial Distribution Functions

So far, the classical theory of photodissociation is very similar to the classical treatment of scattering processes. What remains to be specified is the initial distribution function $W_i(\underline{q}_0, \underline{p}_0)$ for the motion within the ground electronic state. A priori there is no rigorous definition of the distribution function. Both, classical and quantum mechanical choices are in principle possible.

The initial distribution function is very important as we will demonstrate for a simple example. Let us consider a one-dimensional problem with dissociation coordinate R. The excited state potential $V_{ex}(R)$ is considered to be steeply repulsive. For simplicity we further assume that the corresponding initial momentum P_0 is zero. Then the classical absorption cross section is given by (μ=constant)

$$\sigma_i^{cl}(E) = \nu \int dR_0 \; W_i(R_0) \; \delta[V_{ex}(R_0) - E] \tag{26}$$

or alternatively

$$\sigma_i^{cl}(E) = \nu \, W_i\left[R_t(E)\right] \left| \frac{dV_{ex}(R)}{dR} \right|^{-1}_{R=R_t(E)} , \qquad (27)$$

where the properties of the delta function have been used.[34] All quantities in Eq. (27) are evaluated at the classical turning point $R_t(E)$ defined by

$$V^{ex}(R_t) = E \quad . \qquad (28)$$

Equation (27) states that the classical cross section is approximately a reflection of the initial distribution function mediated by $V_{ex}(R)$. This is the well known one-dimensional reflection principle.[4,15] The normalization factor $|dV/dR|^{-1}$ is usually unimportant for repulsive potentials. Any structure in the distribution function (nodal patterns, for example) naturally lead to similar structures in the cross section.

Equation (27) impressively demonstrates that the form of the cross section depends directly on the distribution function W_i. The reflection principle is illustrated in Fig. 3 for a quantal and a classical distribution function. The parent molecule is assumed to be in the ground vibrational state. The quantal distribution is therefore a Gaussian-type function centered around the ground state equilibrium. It tunnels into the non-classical regions. The classical function has singularities at the two classical turning points and it is zero outside the classical region. The corresponding cross sections show the same qualitative behavious as W(R). According to Fig. 3 it is obviously not advisable to use the classical distribution function in photodissociation. The cross section would show spurious features which are in contradiction to all observations. It is possible, however, that the unphysical singularities are partly smeared out if a multi-dimensional phase-space is sampled. In normal scattering the initial conditions are defined for the infinitely separated subsystems. Therefore the proper choice of the initial distribution function is much less important. It is common practice in collision studies to employ the classical distribution function.

Two natural choices for the quantum mechanical distribution function are

$$W_i(R,P) = |\Psi_{gr}^{(i)}(R)|^2 \; |\Phi_{gr}^{(i)}(P)|^2 \qquad (29)$$

and the more sophisticated Wigner-function[15,35,36]

$$W_i(R,P) = \int d\eta \; \Psi_{gr}^{(i)*}(R+\eta) \; \Psi_{gr}^{(i)}(R-\eta) \; \exp(2iP\eta) \quad . \qquad (30)$$

$\Phi_{gr}^{(i)}$ in Eq. (29) is the ground state wavefunction in momentum space. The multi-dimensional extensions are rather obvious. The Wigner function has the advantage to correctly account for the correlation between the coordinates and the corresponding momenta. It has the disadvantage that it can be negative, especially for excited vibrational states. Note, however, that for a harmonic oscillator in the ground vibrational state, both distributions are identical.

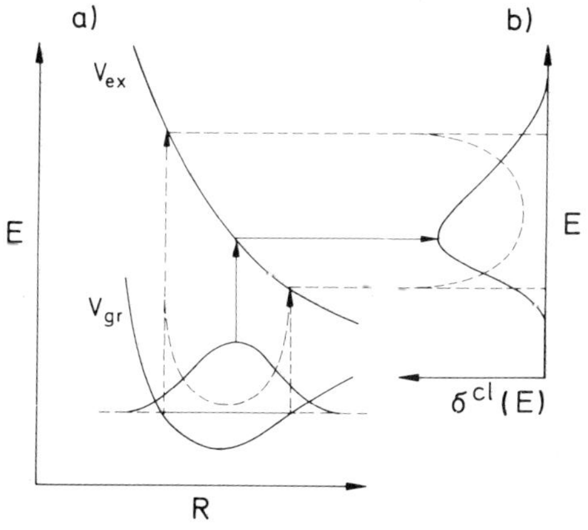

Fig. 3. Schematic illustration of the one-dimensional reflection principle. The solid line in (a) represents a quantum mechanical distribution function and the broken line represents a classical distribution function. The corresponding classical cross sections are shown in (b).

Although classical mechanics has been amply used to study photodissociation processes,[36-41] rigorous tests of its applicability are surprisingly rare. One of the reasons is certainly the lack of exact quantal calculations except for a few cases. In the second part of this lecture we will present several quantal/classical comparisons for various systems and various degrees of freedom. They will intriguingly demonstrate that classical mechanics is indeed very reliable for direct processes and should be used more frequently. For example, the extension of the one-dimensional reflection principle to other degrees of freedom (rotation and vibration) follows immediately from the classical theory.

FINAL STATE INTERACTION AND THE FRANCK-CONDON LIMIT

The two Limiting Cases

The product state distributions following the photodissociation of a polyatomic molecule are determined by the initial preparation of the parent molecule and the subsequent redistribution in the exit channel. According to Eq. (1) the electronical excitation, step (1), prepares the "initial conditions" for the motion in the excited state. The forces between the various degrees of freedom lead to a redistribution in the second step (2) which depends directly on the interaction potential. As in scattering the interaction potential for the A-BC system is defined as

$$V_I(R,r,\gamma) = V_{ex}(R,r,\gamma) - V_{ex}(R=\infty,r,\gamma) \quad . \tag{31}$$

The term final state interaction, which is often used in dissociation studies, simply refers to the forces $\partial V_I/\partial r$ and $\partial V_I/\partial \gamma$.

In order to simplify the discussion let us consider again the two-dimensional case of rotational excitation. We further assume that the ground state wavefunction is separable in the dissociation coordinate R and the orientation angle γ, i.e.,

$$\Psi_{gr}^{(i)}(R,\gamma) = \phi_R^{(i)}(R)\ \phi_\gamma^{(i)}(\gamma) = \phi_R^{(i)}(R) \sum_j a_j^{(i)}\ Y_{j,0}(\gamma,0) \quad (32)$$

and identify the radial functions $\chi_{j,gr}(R)$ of Eq. (4a) with $a_j\ \phi_R(R)$. The a_j are the expansion coefficients of the bending wavefunction in terms of free rigid-rotor wavefunctions. The separability in R and γ is certainly a reasonable ansatz for the lower vibrational states.

If the interaction potential is independent of the orientation angle γ (no final state interaction) the potential coupling matrix in Eq. (7) and therefore the matrix of radial wavefunctions becomes diagonal, i.e.

$$V_{jj'}^{ex}(R) = V_{ex}(R)\ \delta_{jj'}\ , \qquad \chi_{j',ex}^{-(j)}(R) = \chi_{j',ex}^{-(j)}(R)\ \delta_{jj'}. \quad (33)$$

Then the dissociation cross section reduces to the much simpler expression

$$\sigma_i(j|E) = |a_j^{(i)}|^2\ |A^{(i)}(j|E)|^2 \quad (34)$$

$$A^{(i)}(j|E) = \int_0^\infty dR\ \phi_R^{(i)}(R)\ \chi_{j,ex}^{-(j)*}(R). \quad (35)$$

The expansions coefficients a_j depend sensitively on the equilibrium geometry of the parent molecule as it is illustrated in Fig. 4 for a bent molecule (γ_e=104°, H_2O for example) and for a linear molecule (γ_e=180°, CO_2 for example). In both cases we assumed for $\phi_\gamma(\gamma)$ a Gaussian with FWHM of 20° centered at γ_e. Figure 4 clearly shows that an appreciable amount of rotational excitation can already be induced through the motion of the parent molecule without any torque in the exit channel. The expansion coefficients depend also sensitively also on the bending vibrational state and, of course, on the total angular momentum state.

The radial integrals $A^{(i)}(j|E)$ depend directly on the energy E and on the stretching state of the parent molecule. Because of the wavenumbers in Eq. (5) they depend implicitly also on the particular rotational state. The j-dependence of the centrifugal potential is usually small.

If the internal energy, $B_{rot}\ j(j+1)$ in this case, is small compared to the total available energy E, the wavenumbers and therefore the radial wavefunctions become approximately independent of j. The same holds consequently also for the overlap integrals in Eq. (35) and then the final rotational state distribution is simply given by

$$P_i(j) = |a_j^{(i)}|^2 = |\langle\phi_\gamma^{(i)}(\gamma)|\ Y_{j,0}(\gamma,0)\rangle|^2\ . \quad (36)$$

This is the Franck-Condon-(FC-)limit in its simplest form: The (relative) population of rotational states is approximately proportional to the expansion coefficients of the bending part of the ground state wavefunction in terms of free rotor states. The implicit j-dependence of the radial integrals $A(j|E)$ in principle would modify the primitive FC-distribution, however, its influence is usually small for rotations. The total absorption cross section is approximately given by

$$\sigma_i(E) = |A^{(i)}(E)|^2 = | \int_0^\infty dR\ \phi_R(R)\ \chi_{ex}^{-*}(R)|^2 \quad . \tag{37}$$

In analogy to Eq. (36) the primitive FC limit for the vibrational distribution is given by

$$P_i(n) = |a_n^{(i)}|^2 = |\langle\phi_r^{(i)}(r)|\ \phi_n^{BC}(r)\rangle|^2 \quad , \tag{38}$$

where $\phi_r(r)$ is the ground state vibrational wavefunction and $\phi_n^{BC}(r)$ are the asymptotic wavefunctions of the free oscillator. Since the energy spacing between vibrational levels is usually large the radial integrals $A(n|E)$ can not be ignored. They can significantly modify the primitive FC-distribution and also induce a strong energy-dependence. This will be explicitly demonstrated below for a model system.

The FC-limit has been amply discussed in the literature,[42] primarily for rotational excitation.[43,44] The FC expressions become obviously more complicated for dissociation of non-zero total angular momentum states or if the ground state wavefunction is not separable in Jacobi-coordinates. Jacobi-coordinates are most suitable to describe the dissociation step in the exit channel, while normal mode coordinates are usually used for the ground state. If the two wavefunctions are represented in different sets of coordinates all cross section expressions become necessarily more complicated. The work of Freed and coworkers[42,44] is especially devoted to the evaluation of multi-dimensional Franck-Condon factors in the general case.

The coefficient $|a_j|^2$ (or $|a_n|^2$) can be interpreted as the probability that state j (or n) of the free rotor (or oscillator) is prepared in the excited state through the photon absorption [step 1 of Eq. (1)]. If final state interaction is zero these distributions will not be changed during the decay of the excited complex [step 2 of Eq. (1)]. However, if final state interaction is strong, it may completely destroy the initially prepared distribution. The strong coupling limit and its implications will be explicitly discussed for rotation and for vibration in the two following sections.

General Predictions

There are many recent experiments in which the final state distributions of the products are resolved. In most cases it will not be possible to perform reliable dynamical studies because the excited state PES's are usually not known to a reasonable degree of accuracy. Therefore it might be helpful to have a few general rules at hand in order to qualitatively characterize the dissociation process. We will restrict the discussion to rotational distributions for which a vast amount of experimental data has been published in the recent literature.

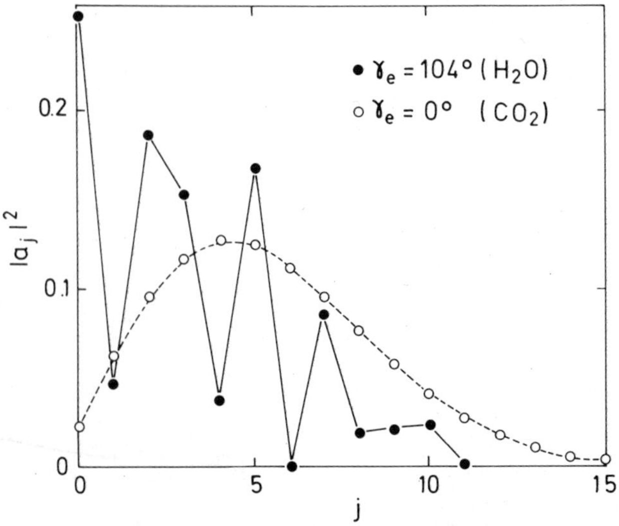

Fig. 4. Expansion coefficients of the ground state bending function $\phi_\gamma(\gamma)$ in terms of free rotor states. The two cases correspond roughly to H_2O and CO_2, respectively.

In the FC-limit we expect

1) that only low rotational states j=0 - 10 or so are significantly populated; higher states are usually not contained in the representation of the ground state wavefunction in terms of free rotor wavefunctions;
2) that the final distribution is almost independent of the photon energy (wavelength);
3) that the final distribution is significantly dependent of the temperature of the parent molecule; this follows from the dependence of the expansion coefficients on the total angular momentum state. The temperature-dependence is usually examined by comparing the dissociation in a room-temperature bulk and in a nozzle beam.

A very nice example of the (rotational) FC-limit is the photodissociation of water in the first absorption band[45] which will be discussed separately below. In this case all the conclusions of above are observed in a series of different measurements including dissociation in the bulk at room-temperature and in a supersonic beam.[46] Very recently Andresen and coworkers[47,48] succeeded to study the dissociation of single rotational states of H_2O and indeed observed a significant J-dependence.

In the strong coupling limit we expect

1) that many rotational states are populated ($j \gtrsim 20$ or so);
2) that the final distribution is energy-dependent;
3) that the final distribution is independent of the temperature of the parent molecule; this follows from the fact that the temperature-dependent "initial" distribution is completely destroyed in the second step of the dissociation process.

There are many examples which clearly show these predictions in the strong coupling limit: ICN,[49] BrCN,[49,50] CH_3O-NO,[51,52] Cl-CN,[53] for example. There are many more. Two other systems will be explicitly discussed in the next section. If the parent molecule dissociates into two diatomic products it is possible that one rotor is determined by the FC-limit while the other is determined by strong coupling. The dissociation of $H_2CO \rightarrow H_2+CO$ is such an example (see below). A general classification is obviously more complicated if final state interaction is weak but not negligible. In that case features of both limits may be observed yielding a more complicated energy- and temperature-dependence.

ROTATIONAL EXCITATION

In this section we will exclusively discuss rotational excitation in direct photodissociation processes and elucidate the general features of the strong coupling limit. First we present exact quantal calculations for the model rigid-rotor A-BC system defined above. These results will be qualitatively explained in terms of simple classical calculations which naturally lead to the rotational reflection principle. This effect, which is characteristic for direct dissociation, subsequently will be analyzed for two realistic systems.

Model Calculations

In order to make the calculations as transparent as possible we assume that the parent wavefunction for the two-dimensional rigid-rotor system in its lowest vibrational state can be written as

$$\Psi_{gr}(R,\gamma) = \phi_R(R)\ \phi_\gamma(\gamma) \tag{39}$$

$$\phi_R(R) = \exp[-\alpha_R(R-R_e)^2] \tag{40a}$$

$$\phi_\gamma(\gamma) = \exp[-\alpha_\gamma(\gamma-\gamma_e)^2]\quad , \tag{40b}$$

where R_e and γ_e specify the equilibrium geometry. This ansatz is based on the assumption that the Hamiltonian is separable and harmonic. This is usually a good approximation for the ground vibrational state. It is, however, not essential for the quantal or the classical calculations and introduced here only for simplicity. In this section we will only consider linear parent molecules, i.e., $\gamma_e=0$.

The excited state potential is parametrized as

$$V_{ex}(R,\gamma) = A\ \exp[-\alpha(R-\bar{R} + f(\gamma) - f(\gamma=0)] \tag{41a}$$

$$f(\gamma) = \varepsilon\ \cos^2\gamma \quad , \tag{41b}$$

where ε controls the coupling between the dissociation coordinate R and the angular coordinate γ. $\varepsilon=0$ corresponds to the FC-limit (no final state interaction). Negative values of ε describe a bent excited state which has a minimum at $\gamma=90°$ and a maximum at $\gamma=0$. These calculation are thus rough models for the dissociation of ICN or OCS, for example. All parameters for the ground and the excited states are given in Refs. 13 and 14. The reduced mass and the rotational constant are appropriate for OC-S.

On the right panel of Fig. 5 we show exact CC rotational distributions for three values of the coupling parameter ε. $\varepsilon=-0.5$Å represents relatively weak coupling and $\varepsilon=-2$Å represents very strong coupling. All

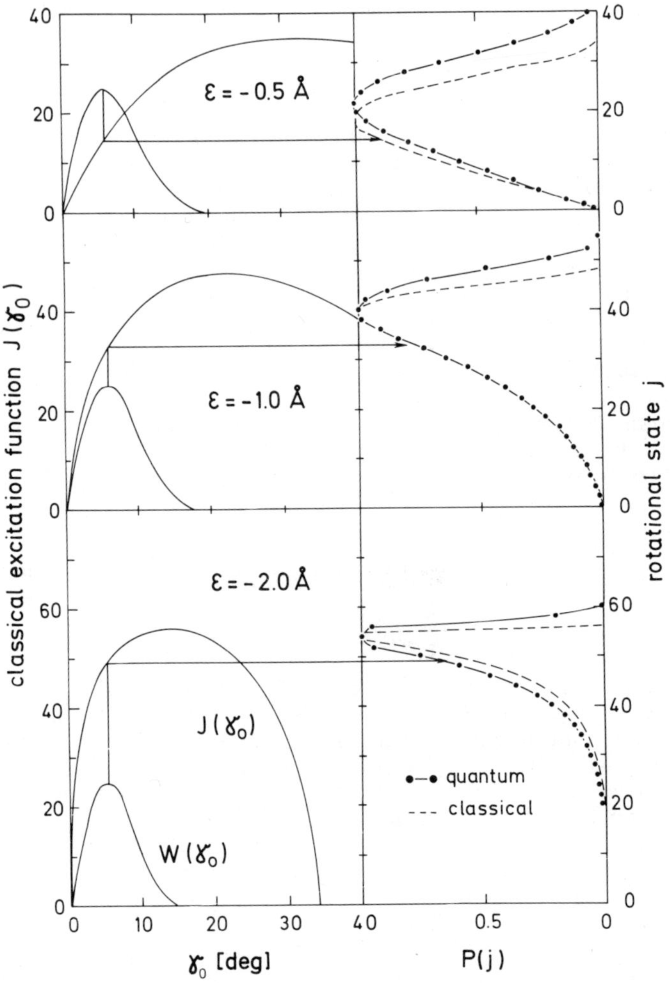

Fig. 5. Left panel: Classical excitation function $J(\gamma_0)$ and weighting function $W(\gamma_0)$ (including $\sin\gamma_0$) vs. initial angle γ_0 for three coupling parameters ε and E=1.3eV. Right panel: Normalized quantal and classical rotational distributions.

distributions have the same qualitative behaviour. They are inverted with a pronounced paucity at low rotational states. In addition, they are very smooth and do not show any finer structures. The peak is considerably shifted as the coupling increases. Many very similar results are reported in Ref. 13 where almost all of the parameters of the model are varied. The corresponding FC-distribution is shown in Fig. 4 ($\gamma_e=0$). It is also smooth and inverted but peaks at much lower j's. Figure 5 nicely illustrates the transition from the FC-limit ($\varepsilon=0$) to the strong coupling limit ($\varepsilon=-1$Å and $\varepsilon=-2$Å).

The exact quantal calculations do not give any hint for the understanding or interpretation of these results. The energy sudden approximation introduced above is unfortunately not applicable in these cases (except for very weak coupling) because the energy transfer during the dissociation process is very efficient. The results of Fig. 5 can be simply interpreted, however, in terms of primitive classical trajectory calculations which also agree almost quantitatively with the exact quantal distributions.

The Rotational Reflection Principle

The classical equations of motion as derived from the Hamiltonian in Eq. (3) are

$$\frac{dR}{dt} = \frac{P}{m} \qquad \frac{dP}{dt} = -\frac{\partial V_{ex}}{\partial R} + \frac{j^2}{mR^3}$$

$$\frac{d\gamma}{dt} = 2j\left(B + \frac{1}{2mR^2}\right) \qquad \frac{dj}{dt} = -\frac{\partial V_{ex}}{\partial \gamma} \quad . \tag{42}$$

In the case of zero total angular momentum j and γ are conjugate action-angle variables. Equation (42) elucidates that the change of j is proportional to the anisotropy of the PES. In the present case we have

$$\frac{dj}{dt} = -2\varepsilon \cos\gamma \sin\gamma \, V_{ex}(R,\gamma) \quad . \tag{43}$$

Following the classical theory as outlined above the classical cross section is defined by

$$\sigma_{cl}(j|E) = \int dR_0 \int dP_0 \int d\gamma_0 \sin\gamma_0 \int dj_0 \, \Psi^2_{gr}(R_0,\gamma_0) \, \Phi^2_{gr}(P_0,j_0)$$
$$\times \, \delta[H_{ex}(R_0,P_0,\gamma_0,j_0)-E] \, \delta[J(R_0,P_0,\gamma_0,j_0)-j], \tag{44}$$

where

$$J(R_0,P_0,\gamma_0,j_0) = \lim_{t\to\infty} j(t|R_0,P_0,\gamma_0,j_0) \tag{45}$$

is the <u>rotational excitation function</u>. It is exclusively determined by the dynamics in the exit channel and depends obviously on all initial variables. In the FC-limit (no final state interaction) it would be identical to the initial angular momentum j_0.

In order to simplify the interpretation we assume that the initial momenta P_0 and j_0 are zero and obtain

$$\sigma_{cl}(j|E) = \int dR_0 \int d\gamma_0 \sin\gamma_0 \, \phi_R^2(R_0) \, \phi_\gamma^2(\gamma_0)$$

$$\times \, \delta[V_{ex}(R_0,\gamma_0)-E] \, \delta[J(R_0,\gamma_0)-j] \quad . \tag{46}$$

The two-dimensional integral can be further reduced if we employ energy conservation and choose R_0 as the classical turning point $R_t(\gamma_0|E)$. This yields

$$\sigma_{cl}(j|E) = \int_0^\pi d\gamma_0 \sin\gamma_0 \, \phi_R^2[R_t(\gamma_0|E)] \, \phi_\gamma^2(\gamma_0) \, \delta[J(\gamma_0)-j] \quad . \tag{47}$$

A normalization constant $|\partial V_{ex}/\partial R|^{-1}$, which is unimportant for steeply repulsive potentials, has been omitted for clarity.

Equation (47) is the desired expression for the classical cross section. The initial angle γ_0 is the only independent variable which completely determines the result of the trajectory. Using the properties of the delta function[34] Eq. (47) can be rewritten in the simpler form

$$\sigma_{cl}(j|E) = \sum_\nu \sin\gamma_{0,\nu} \, W(\gamma_{0,\nu}|E) \left|\frac{dJ}{d\gamma_0}\right|^{-1}_{\gamma_{0,\nu}} \quad . \tag{48}$$

The summation runs over all trajectories $\gamma_{0,\nu}$ which lead to the specified final rotational state, i.e., which are solutions of

$$J(\gamma_{0,\nu}) = j \quad . \tag{49}$$

The weighting function is defined as

$$W(\gamma_0|E) = \phi_R^2[R_t(\gamma_0|E)] \, \phi_\gamma^2(\gamma_0) \quad . \tag{50}$$

If there is only one solution Eq. (49) establishes a unique relation between the initial angular variable γ_0 and the final action variable j. Then the classical rotational distribution is simply a reflection of the ground state wavefunction [primarily its bending part $\phi_\gamma(\gamma)$] onto the momentum axis j. This "mapping" is mediated by the excitation function $J(\gamma_0)$. From analogy with the one-dimensional case, Eq. (27), we call this effect rotational reflection principle. In the one-dimensional case the energy-dependence of the absorption cross section is explained as a mapping of the R-dependence of the ground state wavefunction mediated directly by the excited state PES $V_{ex}(R)$.

On the left pannel of Fig. 5 we show the excitation function $J(\gamma_0)$ and the weighting function $\sin\gamma_0 W(\gamma_0)$ vs γ_0. Because of the $\sin\gamma_0$ factor the weighting function does not peak at $\gamma_0=0$ but is slightly shifted. This extra factor is quite important for linear ground state molecules.[54] The excitation function has the typical shape as usually found in collisions.[12] It is zero at $\gamma_0=0$ because of symmetry, rises to a maximum, and decreases again to zero at $\gamma_0=90°$ (homonuclear molecule). We show only the γ_0-interval where the weighting function is non-zero. The height of the maximum depends sensitively on the coupling strenght ε. Since the weighting function is primarily determined by the bending part of the ground state wavefunction it is almost independent of ε. Note however, that a weak ε-dependence is in principle induced through the γ_0-dependence of the classical turning point $R_t(\gamma_0|E)$.

Figure 5 nicely illustrates the rotational reflection principle. The maximum of P(j) is determined by the excitation function above the peak of $\sin\gamma_0\, W(\gamma_0)$. It shifts to higher values as ε increases. If the maximum of $J(\gamma_0)$ is probed the classical cross section would be infinite because of the normalization factor $|dJ/d\gamma_0|^{-1}$ in Eq. (48). This is the definition of a rotational rainbow which is so prominent in atom-molecule collisions.[12] The exact quantal calculation would yield a broad maximum rather than the unphysical rainbow singularity.

The agreement of the primitive classical results with the exact quantal distributions is asthonishing. The computer time required for the classical calculations is negligibly small compared to CC calculations. Only about twenty trajectories have to be determined. The classical distributions in Fig. 5 are consistently too narrow compared to the exact distributions. This is certainly a consequence of the too simplistic choice of initial conditions $P_0=j_0=0$ used to derive Eq. (46). Varying also the initial momenta generally improves the agreement with the quantal calculations. Figure 6 shows some results for various energies E and fixed coupling ε.[54] In this case the multi-dimensional integral in Eq. (44) is solved by Monte-Carlo methods. We must also note, however, that in some cases the agreement becomes less favourable, especially at low rotational states. The disagreement at low rotational states can be traced back to "untypical" trajectories which have a non-monotonic time-dependence. At present time we do not have a simple explanation of this effect or any recipe for improvement at hand.

At this point it is interesting to compare photodissociation with scattering, i.e., to compare "half" and "full" collisions. The classical excitation functions are in both cases qualitatively similar. In dissociation only a narrow region around the ground state equilibrium is projected out. The initial distribution function acts like a "window" and leads to the rotational reflection maxima. As a consequence, measured distributions yield information only about this restricted region of the excited PES. Since only one single trajectory contributes to the cross section, interference effects are absent and the distributions are smooth functions of j. The classical scattering cross section is formally identical to Eq. (44) except for the distribution function $W(\gamma_0)$. In scattering all angles are equally probable (the "window" is completely open) because the initial conditions are defined for the infinitely separated systems. Therefore rotational rainbows induced by the extrema of $J(\gamma_0)$ are the dominant featurs.[12] In addition, several trajectories lead to the same transition which causes interference effects like supernumerary rainbows, for example. They have indeed been observed experimentally.[55]

Like rotational rainbows in collisions the reflection maxima are direct probes of the anisotropy of the PES. This is most clearly seen if we approximate the excited state PES by a hard potential with angle-dependent contour $\hat{R}(\gamma)$.[56-58] Under energy sudden conditions it can be shown that[59]

$$J_{diss.}(\gamma_0) = \frac{1}{2} J_{scatt.}(\gamma_0) = (2mE)^{1/2} \frac{d\hat{R}(\gamma_0)}{d\gamma_0} , \qquad (51)$$

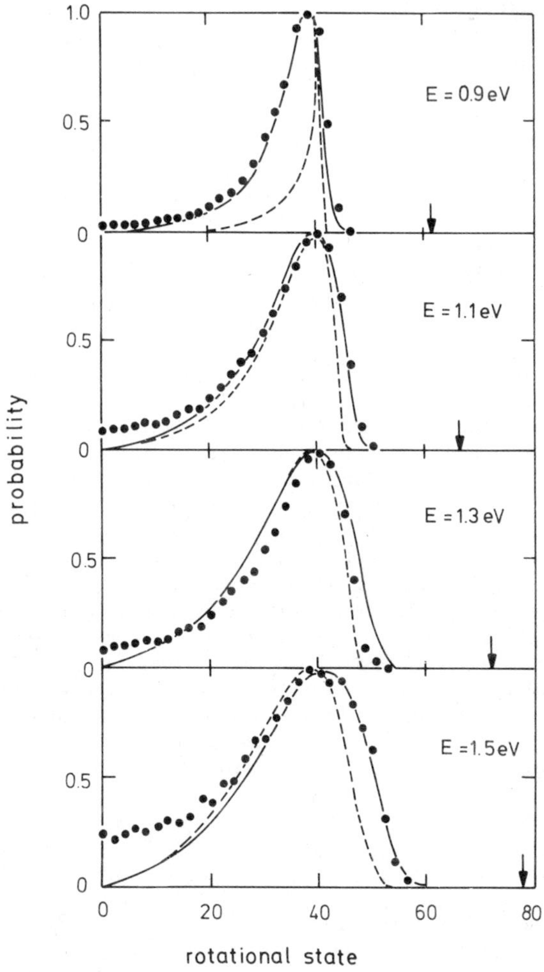

Fig. 6. Rotational state distributions for various energies E. The coupling parameter is ε=-1Å in all cases. The arrows mark the highest state accessible at the respective energy. Solid curve: exact quantal calculations; broken curve: simple classical calculations, Eq. (48); full circles (●●): extended classical calculations, Eq. (44).

where $d\hat{R}/d\gamma$ is the anisotropy of the hard potential. This equation also illustrates very nicely the difference between "full" and "half" collisions. In the first case, the torque acts in the first <u>and</u> in the second step of the collision. According to Eq. (42) we approximately get (for weak coupling)

$$J_{scatt.} \sim - \int_{-\infty}^{t_0} dt\ \partial V_{ex}/\partial \gamma \ - \int_{t_0}^{\infty} dt\ \partial V_{ex}/\partial \gamma \quad , \qquad (52a)$$

where t_0 specifies the classical turning point. In dissociation, however, the dynamics starts at the turning point and

$$J_{diss.} \sim - \int_{t_0}^{\infty} dt \, \partial V_{ex}/\partial \gamma \quad . \tag{52b}$$

There are many experimental distributions which are qualitatively similar to those in Figs. 5 and 6 (for a selection see Fig. 1 of Ref. 13). In the remaining part of this section we will disscuss two examples which clearly show the rotational reflection principle.

Photodissociation of Formaldehyde

The photodissociation of formaldehyde has been extensively studied in the past.[60] Recently Bamford et al.[61] reported CO rotational distributions following the photodissociation of H_2CO, HDCO, and D_2CO at wavelengths of ca. 29500 cm^{-1}. They are shown in Fig. 7. The distributions are highly inverted and roughly symmetric about the peak centre. The maximum shifts to higher states as hydrogen is substituted by HD and D_2, respectively.

The dissociation of formaldehyde with ultraviolet light is on the first sight very different from direct dissociation. According to Bamford et al.[61] formaldehyde is first excited to the S_1-state, is internally converted back to the S_0 electronic ground state, and finally dissociates on the ground state PES into H_2 and CO. The total energy (at 29500 cm^{-1}) is of the order of the barrier height for molecular dissociation. It is assumed that the final state distributions are entirely determined by the dynamics in the repulsive exit valley but not by the highly excited vibrational-rotational motion on the H_2CO side of the barrier. In this sense the dissociation of formaldehyde can be interpreted as direct dissociation starting at the transition state.

The geometry of the transition state (the "reactive window") is known from ab initio calculations[62] which yield an angle of γ_{ts}=32.6°. The PES of the ground electronic state in the exit channel is also known from accurate ab initio calculations and has been used before to model rotational excitation in H_2+CO collisions.[63] Since the energy transfer is relatively small compared to the total available energy we used the energy sudden approximation described above. The semiclassical approximation ($j \gg \hbar$) for the ESA dissociation amplitude in Eq. (16) yields a cross section expression which is formally identical to Eq. (48).[29,59] Details are given in Ref. 64.

The results of these parameter-free calculations[64] are compared in Fig. 7 with the experimental distributions. The agreement is astonishingly good if we consider the simplicity of the model. The peak centers are almost exactly reproduced and show the observed shift for the various hydrogen isotopes. This shift is due to the different reduced masses as predicted by Eq. (51). The agreement is certainly not perfect which is not surprising because of the limitations of the model. The analysis of the excitation function clearly proves that the CO distributions indeed are governed by the rotational reflection principle.[64] Test calculations have shown that the peak centers depend sensitively on the anisotropy of the PES in the region of the transition state. This underlines the accu-

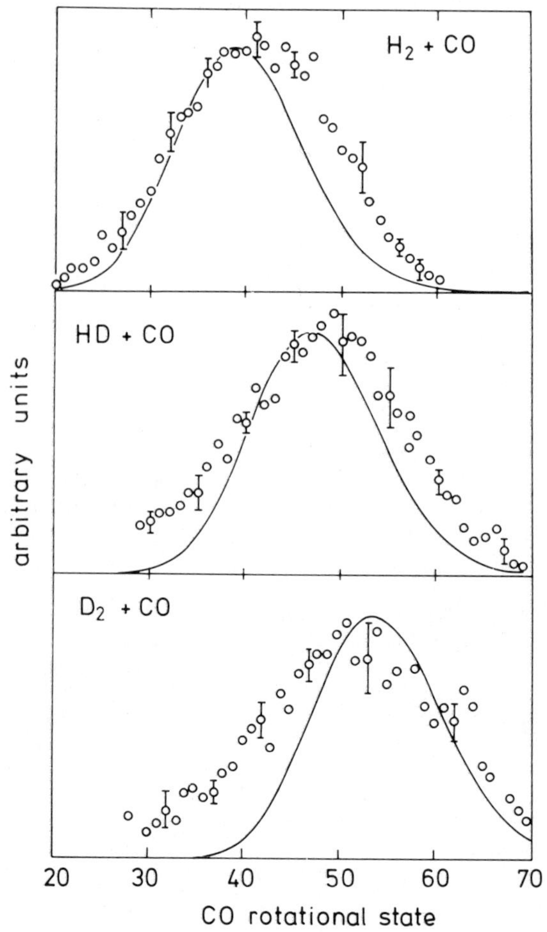

Fig. 7. Experimental[61] (oo) and theoretical[64] (—) CO rotational state distributions following the photodissociation of H_2CO, HDCO, and D_2CO at 29500 cm^{-1}.

racy of the calculated transition state geometry as well as the ab initio PES in the exit channel. Bamford et al.[61] also examined the influence of the total angular momentum J and found a slight broadening of the distribution with increasing J but no detectable effect on the peak center. This demonstrates that strong final state interaction in the exit channel is really the dominant mechanism.

Incidentally we note that the rotational distribution of H_2 has been calculated within the same model.[65] It also agrees very well with recent measurements.[66] However, in this case the anisotropy with respect to the H_2 orientation angle is very weak and therefore it is completely determined by the FC-limit. This result is interesting because it shows that for the same polyatomic molecule one degree of freedom (CO) is determined by strong final state interaction while the other degree of freedom (H_2) is determined by the FC-limit.

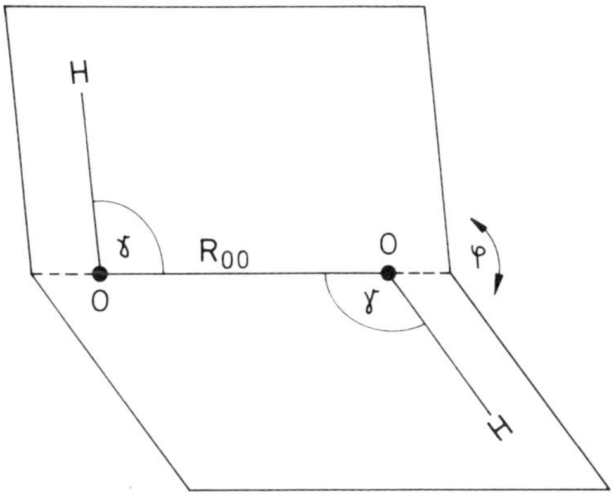

Fig. 8. Coordinates for the dissociation of $H_2O_2 \longrightarrow 2OH$.

Photodissociation of Hydrogen Peroxide

Photodissociation of H_2O_2 at wavelengths above 190nm is currently investigated in great detail by several groups.[67-69] The rotational distributions of $OH(^2\Pi)$ are inverted and the peak center shifts to larger states as the photolysis wavelength is decreased. They resemble very much the rotational reflection principle.

Theoretical models for this tetratomic molecule are certainly much more involved than for the ABC system. Rotational excitation can be due to bending motion associated with the H-O-O angle γ as well as torsional motion associated with the angle φ as schematically illustrated in Fig. 8. The highly sophisticated analysis of Gericke et al.[67] shows that torsional motion is the dominant mechanism and therefore the theory discussed above must be extended.

Recently Meier et al.[70] calculated the corresponding excited state PES, including the O-O distance R_{OO} and the torsional angle φ. All other coordinates are fixed. The PES is steeply repulsive along the dissociation coordinate R_{OO} (direct dissociation) and exhibits a pronounced φ-dependence in accordance with the experimental observation. We used this ab initio PES to perform preliminary dynamical calculations. They certainly can not yield a complete description of this rather complicated dissociation process, however, they nicely manifest the rotational reflection principle for H_2O_2 dissociation.

In our two-dimensional model we include only the coordinates R_{OO} (=R) and the torsional angle φ (Fig. 8). The bending motion is completely ignored and γ is fixed at 90°. The classical Hamilton function for this model is

$$H(R,\varphi,P,j) = \frac{P^2}{2m} + 2B_{OH}\, j^2 + V(R,\varphi) \quad , \qquad (53)$$

where B_{OH} is the rotational constant of OH. The model consists of two OH molecules rotating opposite to each other in a plane perpendicular to the vector which joines the two oxygen atoms. The dissociation dynamics is treated classically and quantum mechanically. The (unnormalized) expansion functions in the quantal treatment are simply the trigonometric functions $\cos(j\varphi)$ and $\sin(j\varphi)$. Only dissociation out of the vibrational ground state is considered here.

On the left panel of Fig. 9 we show the classical excitation function for two energies and the corresponding weighting function. The φ-dependence of the ground state potential is taken from Hunt et al.[71] and the corresponding wavefunction $\phi_\varphi(\varphi)$ is calculated numerically by solving the one-dimensional Schrödinger equation. The equilibrium angle is φ_e=111.5°. The torsional mode is relatively soft and therefore $W(\varphi_0)$ extends over a rather wide range. The right panel of Fig. 9 shows the corresponding classical and quantal rotational state distributions. They are clear manifestations of the rotational reflection principle which in this case is induced by the torsional rather than the bending angle. As observed before in the model ABC problem the classical distributions agree remarkably well with the quantal results. This is important for future, more realistic studies which should also include the bending degree of freedom. An exact quantal treatment is then probably impossible because too many channels must be included.

Also shown in Fig. 9 is the pure FC-distribution as obtained in the limit of zero final state interaction. It is nicely seen how the memory to the initially prepared distribution is successively distroyed as the energy increases. While the true distribution for E=2.5eV is still somehow similar to the FC-limit, the distribution for E=4.25eV is quite different. The experimental distributions at wavelengths of 266nm and 193nm (which roughly correspond to available energies of 2.5 and 4.25eV) are qualitatively similar to the theoretical results in Fig. 9. They peak, however, at j~6 and j~11, respectively, which indicates that our preliminary model is too simplistic. Inclusion of the bending degree of freedom is certainly necessary.[39] Dissociation at 193nm has been studied at room-temperature and in a nozzle beam. The differences are very small which additionally underlines that final state interaction is strong and that the FC-distribution is completely redistributed. Incidentally we note that approximate energy sudden (ES) calculations based on the model Hamiltonian in Eq. (53) are in almost perfect agreement with the measurements. This is of course purely accidental and an artifact due to violation of energy conservation within the ES approximation. Neglecting the energy transfer artificially shifts the peak to higher states.

VIBRATIONAL EXCITATION

The theoretical treatment of vibrational excitation is in most cases simpler than for rotational excitation. Because usually much fewer states are involved exact CC calculations (for the vibrational degree of freedom) are feasible. In this section we first present exact quantal calculations for a model system which again can be simply interpreted in terms of classical trajectories.

Model Calculations

We consider the dissociation of a collinear ABC → A+BC(n) molecule and completely ignore the rotational degree of freedom. The Hamiltonian is given by

$$H(R,r) = -\frac{1}{2m}\frac{\partial^2}{\partial R^2} + h_{vib}(r) + V_I(R,r) \quad , \tag{54}$$

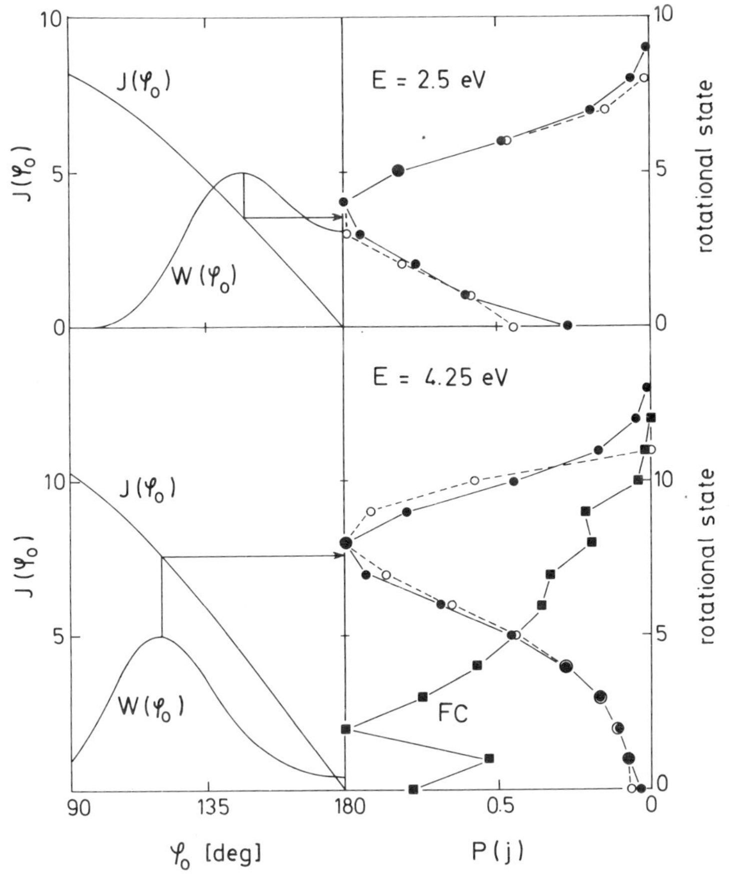

Fig. 9. Left panel: Rotational excitation function $J(\varphi_0)$ and weighting function $W(\varphi_0)$ vs. initial torsional angle φ_0 for the dissociaton of $H_2O_2 \longrightarrow 2OH(j)$. The two energies correspond roughly to dissociation wavelengths of 266nm and 193nm, respectively. Right panel: quantum mechanical (●●) and classical (oo) rotational state distributions of OH. Also shown is the primitive FC-distribution (■■).

where the internal Hamiltonian is defined as

$$h_{vib}(r) = - \frac{1}{2m_{BC}} \frac{\partial^2}{\partial r^2} + v(r) \tag{55}$$

and $V_I(R,r) = V(R,r) - v(r)$ is the interaction potential. Expanding the ground and the excited state wavefunctions leads to the usual set of coupled equations [Eq. (5) without the centrifual term] with wavenumbers $k_n^2 = 2m[E-\varepsilon_n]$ and potential coupling elements

$$V_{nn'}(R) = \int_0^\infty dr\ \phi_n(r)\ V_I(R,r)\ \phi_{n'}(r) \ , \tag{56}$$

where the ϕ_n are the eigenfunctions of $h_{vib}(r)$ with eigenvalues ε_n. If both wavefunctions are expanded in the same set of basis functions (which is not essential for the calculations but convenient for the discussion) the vibrational dissociation amplitude $t(n|E)$ is formally identically to Eq. (12).

In analogy to the rotational problem we represent the ground state wavefunction as

$$\Psi_{gr}(R,r) = \phi_R(R)\ \phi_r(r) \tag{57}$$

$$\phi_R(R) = \exp[-\alpha_R(R-R_e)^2] \tag{58a}$$

$$\phi_r(r) = \exp[-\alpha_r(r-r_e)^2] \quad , \tag{58b}$$

where R_e and r_e specify the equilibrium geometry. The excited state potential is parametrized as

$$V_{ex}(R,r) = A\ \exp[-\alpha(R + f(r)] + 0.5\ k(\bar{r}-r)^2 \tag{59a}$$

$$f(r) = -\varepsilon(r-\bar{r}) \quad , \tag{59b}$$

where ε controls the final state interaction, i.e., the translational-vibrational coupling. ε=0 corresponds to the FC-limit.

We first present simple model calculations for the dissociation of CF_3-I $\rightarrow$ CF_3+I.[30] According to Shapiro and Bersohn[72] and van Veen et al.[73] F_3 is considered as an atom with mass $3m_F$. Only the "umbrella" mode of CF_3 (i.e., the vibration of the C-atom with respect to the plane of the three F-atoms) is taken into account. In this sense the dissociation of CF_3-I is reduced to a collinear-like problem. All parameters are given in Ref. 30. In the quantum study we used a ground state wavefunction which was determined numerically from a non-separable PES.[30] The parameters in Eq. (58) were fit to reproduce this exact wavefunction. Deviations are very small, as it is expected for the lowest vibrational state, and do not detectably affect the comparison between the quantal and the classical results.

Dissociation of CF_3-I was examined experimentally by van Veen et al.[73] using the time-of-flight technique. A large amount of internal energy was observed which was mainly attributed to high excitation of the "umbrella"-mode. This may indicate that the vibrational-translational coupling in the excited state is strong.

On the right panel of Fig. 10 we show exact CC vibrational state distributions for three coupling parameters ε. The general form is similar to the rotational distributions in Fig. 5, for example. They are inverted and structureless. The peak shifts to higher states with increases ε. Since the energy spacing between vibrational levels is significantly larger than between rotational levels much fewer states are populated and the distributions are rather narrow. Also shown in Fig. 10 (lower part) is the primitive FC-distribution, Eq. (38). Since CF_3 is pyramidal both in the ground state and asymptotically (i.e., r_e=0.481Å and $\bar{r}$=0.403Å) the FC-distribution is very narrow. It peaks at n=0 and falls off very rapidly. Therefore, the distributions in Fig. 10 must be

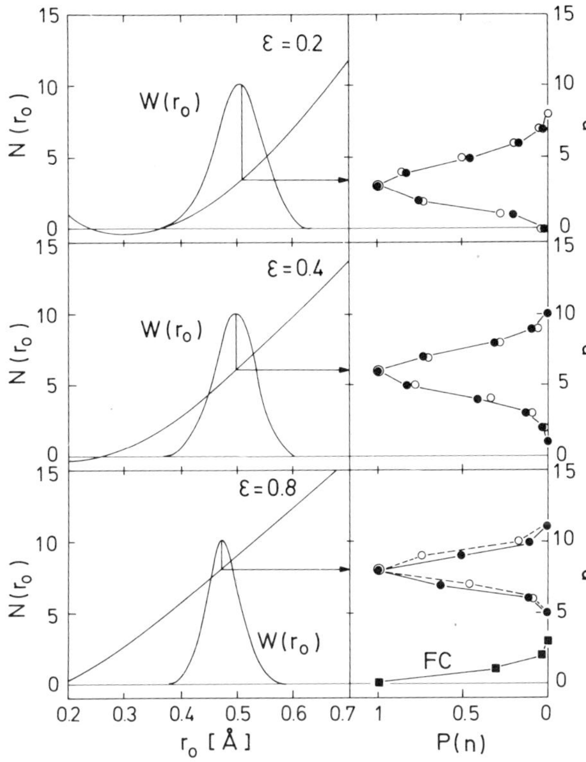

Fig. 10. Left panel: Vibrational excitation function $N(r_0)$ and weighting function $W(r_0)$ vs. initial oscillator coordinate r_0 for three coupling parameters ε. The energy is E=1.88eV. Right panel: Comparison of quantal (●●) and classical (oo) vibrational distributions. Also shown (lower part) is the primitive FC-distribution Eq. (38) (■■).

determined by strong final state interaction, i.e., the second step of the dissociation process. The initial preparation following the electronic excitation is completely destroyed in the exit channel. Figure 11 shows similar results for different energies and fixed coupling parameter ε=0.8.

In Ref. 30 we analyzed these distributions in terms of the energy modified sudden approximation. This analysis was not satisfactory although qualitatively similar distributions were obtained. As for rotations the explanation is extremely simple in terms of classical trajectory calculations.

Vibrational Reflection Principle

The equations of motion following from the Hamiltonian in Eqs. (54) and (55) are

$$\frac{dR}{dt} = \frac{P}{m} \qquad \frac{dP}{dt} = -\frac{\partial V_I}{\partial R}$$

$$\frac{dr}{dt} = \frac{p}{m_{BC}} \qquad \frac{dp}{dt} = -\frac{\partial V_I}{\partial r} - \frac{\partial v}{\partial r} , \qquad (60)$$

where P and p are the linear momenta corresponding to R and r, respectively. As before we choose the initial conditions in the follwoing way: $P_0=p_0=0$, r_0 is changed systematically, and R_0 is chosen as the classical

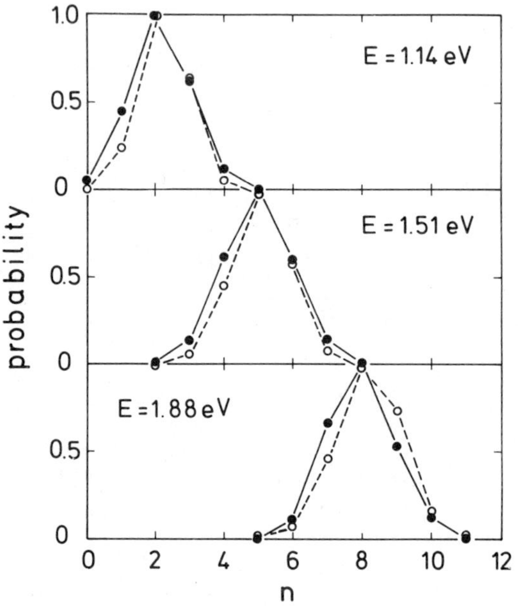

Fig. 11. Quantal (●●) and classical (oo) vibrational distributions for several energies E and coupling strength ε=0.8.

turning point for the particular energy E. The initial oscillator coordinate r_0 is the only independent variable. Solving the equations of motion yields in the limit $t \to \infty$ the vibrational excitation function

$$N(r_0) = \frac{E_{vib}}{\omega} - \frac{1}{2} , \qquad (61)$$

where E_{vib} is the final oscillator energy and ω is the harmonic frequency. The internal energy becomes constant as the interaction potential goes to zero. The extension of Eqs. (55), (60) and (61) for a Morse-oscillator is obvious. N ranges from -1/2 up to $(E/\omega-1/2)$.

In analogy to Eq. (48) the classical cross section is

$$\sigma_{cl}(n|E) = \sum_{\nu} W(r_{0,\nu}|E) \left| \frac{dN}{dr_0} \right|^{-1}_{r_{0,\nu}} , \tag{62}$$

where the summation runs over all trajectories $r_{0,\nu}$ which lead to the specified final vibrational state n, i.e.,

$$N(r_{0,\nu}) = n \quad . \tag{63}$$

The weighting function is given by

$$W(r_0|E) = \phi_R^2(R_0(r_0|E)) \, \phi_r^2(r_0) \; . \tag{64}$$

If there is only one solution Eq.(63) establishes a unique relation between the initial oscillator coordinate r_0 and the final vibrational state n. In that case the final vibrational distribution is a reflection of the ground state wavefunction, especially its vibrational part ϕ_r. We call this effect vibrational reflection principle in total analogy to the rotational reflection principle.

On the left panel of Fig. 10 we show the corresponding vibrational excitation functions $N(r_0)$ and the weighting functions $W(r_0)$. The shape of $N(r_0)$ resembles the harmonic intramolecular potential v(r). It is shifted to the left as the coupling increases. Near the ground state equilibrium the excitation function is monotonic such that only a single solution of Eq. (63) contributes to the cross section. The classical distributions are shown on the right panel and the vibrational reflection principle is clearly manifested. Any further discussion is unnecessary. The vibrational distributions are direct probes of the r-dependence of the interaction potential within the transition region. The classical distributions in Figs. 10 and 11 are again in very good agreement with the quantal results. This is somewhat surprising because only very few vibrational states are populated. The concept of the classical excitation function is borrowed from the classical S-matrix theory of Miller[33] and Marcus[74] for collinear collisions.

Franck-Condon Limit

The vibrational distributions in Figs. 10 and 11 are determined by strong final state interaction. In this section we will discuss in more detail the FC-limit for the vibrational degree of freedom. The reason is that under certain circumstances the FC-limit yields distributions which are qualitatively similar to those in Figs. 10 and 11. This can complicate the analysis of measured distributions, because the strength of the final state interaction (i.e., the excited state PES) is in most cases, completely unknown.

The calculations of the foregoing section were performed to roughly model the dissociation of CF_3-I. The bound as well as the free CF_3-radicals are pyramidal, i.e., the respective equilibrium distances (r_e and r) are approximately equal. The primitive FC-distribution given in Eq.(38) is therefore narrow with a peak at n=0. This situation is completely different for the dissociation of CH_3-I. CH_3 is planar in the parent molecule but it is pyramidal asymptotically. Therefore the FC-distribution is quite broad and inverted.

In order to demonstrate the differences we compare in Fig. 12 final vibrational distributions for "pyramidal CF_3" (i.e., $\bar{r}$=0.403Å in Eq. (59), as before) and for "planar CF_3" (i.e., $\bar{r}$=0). All other parameters are the same as in the previous calculations. The final state interaction is zero in both cases (ε=0). The distributions for "pyramidal CF_3" are very narrow and peak at n=0. They are almost independent of the energy. The distributions for "planar CF_3" are much broader, inverted and show a significant energy-dependence.

The energy-dependence of the distributions in Fig. 12 can be easily explained in terms of Eqs. (34) and (35).[42] Because of the relatively large spacing between the channel energies $E_n=E-\varepsilon_n$ the radial integrals A(n|E) of Eq. (35) are strongly dependent on n and on the energy E. This is schematically illustrated in Fig. 13 where we show the (diagonal) radial functions for several final states n. The overlap with the ground state wavefunction $\phi_R(R)$ is mainly determined by the first maximum beyond the classical turning point $R_t(n|E)$, which is of course dependent on n as well as E. In Fig. 13 the overlap is most favourable for n=1 and decreases rapidly with increasing n. For a higher energy it would be largest for n=2 and so on. Combined with the pure FC-factor $|a_n|^2$ this leads to the behaviour observed in Fig. 12. In the case of "pyramidal CF_3" the pure FC-distribution is mainly restricted to n=0 and so the energy-dependence of the radial term becomes unimportant. For a more detailed discussion see Freed and Band.[42]

The results for "planar CF_3" in Fig. 12 are qualitatively similar to the distributions in Fig. 11, although the dynamical mechanism is completely different in both cases. In Fig. 12 the distributions are determined by the first step of the photodissociation process. Due to the FC-principle a relatively high vibrational state n_0 is prepared at the turning point (t=0). Because the interaction potential is elastic (ε=0, no final state interaction) n(t) does not change during dissociation. In the second case (Fig. 11) a low vibrational state, $n_0 \sim 0$, is prepared at the turning point. Because the interaction potential is inelastic ($\varepsilon \neq 0$, strong final state interaction) the oscillator is excited during the dissociation step, i.e., n(t) changes significantly along the trajectory. The extent of the excitation depends on the coupling strength.

In both cases the distributions are determined by a "mapping" of the nuclear ground state wavefunction onto the "momentum" axis n. This is of course not unexpected because the dissociation cross section is given as a matrix element which explicitly contains the parent wavefunction. In the FC-limit this "mapping" is via Eq. (38), possibly modified by the radial overlap integral (35). This type of "mapping" has been discussed by Shapiro[75] and by Child and Shapiro,[76] for example. Child and Shapiro[76] explicitly ignored final state interaction and evaluated the two-dimensional FC-integral by semiclassical methods. Shapiro[75] studied the "mapping" for the model CH_3-I system.[72] Although final state interaction is formally included in his model, it is relatively weak and does not significantly influence the results. A structure in the ground state wavefunction due to exciation of the R- or the r-mode, for example, leads naturally to similar structures in the energy- or the state-dependence of the cross section. This has been nicely demonstrated by Shapiro.[75] One could describe this process as "Franck-Condon mapping". In the strong coupling limit the "mapping" of the ground state wavefunction is via Eqs. (62) and (63). Final state interaction, which determines the classical excitation function, is the dominant factor.

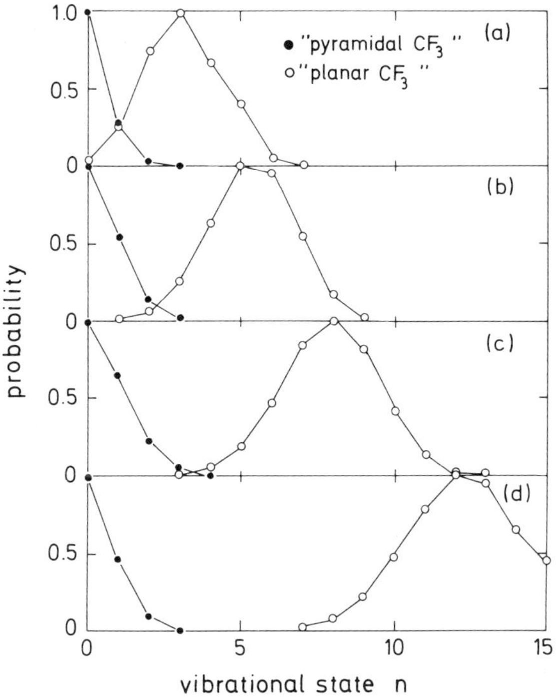

Fig. 12. Quantum mechanical vibrational distributions for "pyramidal CF_3" and "planar CF_3". For definitions see the text. The final state interaction is zero in all cases (ε=0). The energies are (a) E=1.142eV, (b) E=1.482eV, (c)E=1.801eV, and (d) E=2.311eV.

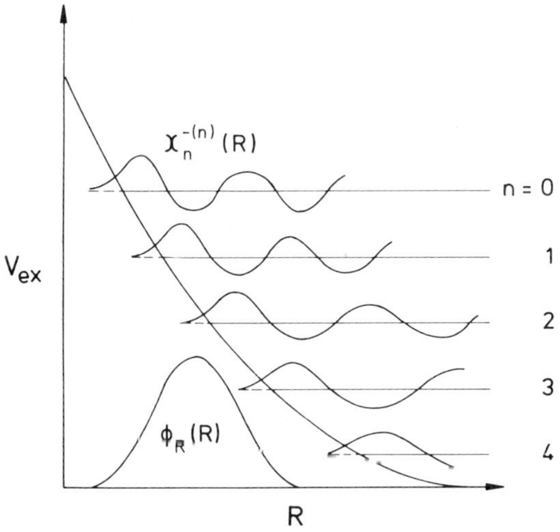

Fig. 13. Schematic explanation of the state- and energy-dependence of the radial overlap integrals A(n|E) of Eq. (35). The horizontal lines are the channel energies $E-\varepsilon_n$.

Usually only few vibrational states or populated for product molecules like OH, CO, NO, for example. This can make the interpretation of measured distributions quite cumbersome, because it is a prior not possible to decide whether final state interaction or a significant FC-preparation is dominant. If the final distribution is broad and significantly energy-dependent, although the equilibrium separations for the bound and the free product molecules are roughly equal, then final state interaction must be strong. The dissociation of H_2O in the first absorption band is a good example for this situation (see the next section). If the final distribution is narrow and peaks at n=0 and if it does not show a significant energy-dependence, final state interaction is probably very weak. The dissociation of H_2O_2 in the first band is a good example for this behaviour.[67-69]

PHOTODISSOCIATION OF WATER IN THE FIRST ABSORPTION BAND

In order to perform reliable dynamical calculations for a photodissociation process it is essential to know, at least qualitatively, the corresponding excited state PES. Unfortunately, this information is in almost all cases not available. The ground state PES is usually sufficiently known from spectroscopic studies which yield the equilibrium geometry and the vibrational force field. The lack of good excited state PES's is the main reason why accurate ab initio studies for photofragmentation processes are very rare. Such calculations are further complicated if two (or even more) excited states are ènergetically close together. Then the absorption cross section may reflect a superposition of several spectra and furthermore non-adiabatic coupling between the various states may also be important.

In the next two sections of this lecture we will discuss in more detail two systems, first the dissociation of water and second the dissociation of methylnitrite. In both cases dynamical calculations are performed using ab initio PES's. The dissociation of water is an example for a direct process while the dissociation of methylnitrite is an example for an indirect process. Both fragmentations evolve on a single excited state PES, which is well separated from other states. The theoretical results are all compared to experimental data.

The photodissociation of water is one of the best studied molecular processes. The absorption spectra of H_2O and D_2O in the wavelength region $\lambda \gtrsim$ 120nm consist of two broad, well separated bands with maxima at ~165nm and ~128nm.[77] The corresponding electronic transitions are $\tilde{X}(^1A_1) \longrightarrow \tilde{A}(^1B_1)$ and $\tilde{X}(^1A_1) \longrightarrow \tilde{B}(^1A_1)$, respectively.[2,3] At wavelengths below ~125nm excitation into the next higher $\tilde{C}$ state is also possible which leads to a progression of sharp and pronounced structures superimposed on the second band.

In this lecture we discuss the photodissociation of water in the first band, i.e.,

$$H_2O(\tilde{X}^1A_1,i) + h\nu \longrightarrow H_2O(\tilde{A}^1B_1) \longrightarrow H(^2S) + OH(^2\Pi,f). \qquad (65)$$

Dissociation in the second band has been studied some time ago by Segev and Shapiro[78] and very recently by Dunne et al.[79] and by Weide and Schinke.[80] In Eq. (65) (i) represents a particular vibrational-rotational state $(\nu_1\nu_2\nu_3|JK_aK_c)$. Since water is an asymmetric top each rotational level J is split into (2J+1) non-degenerate sub-levels specified by K_a and K_c. Likewise (f) represents a particular vibrational-rotational state (n,j) of OH. Since OH is a $^2\Pi$ molecule each rotational level is split into two multiplet states, $^2\Pi_{1/2}$ and $^2\Pi_{3/2}$, respectively. Moreover, each multiplet is split into two Λ-doublet states. Thus, each rotational level j consists of four electronic sub-states. This will be important for the subsequent comparison with true state-to-state distributions.

Potential Energy Surface

The $\tilde{A}$ state PES of H_2O has been calculated by Staemmler and Palma[81] with quantum chemical methods including configuration interaction. The two OH distances R_{OH} and the HOH bending angle are varied separately. The ca. 250 ab inito points are fit to an analytical expression which is used without any modification in all of our calculations.[82,83]

Figure 14 shows a contour plot of $V_{ex}(\tilde{A})$ for fixed bending angle $\gamma=104°$. The potential is symmetric in the two OH distances and has two equivalent dissociation channels. It has a barrier of ~2eV for the exchange of the two H-atoms at $R_{OH} \sim 1Å$. The asymptotic potential of the free OH-radical is significantly perturbed as the other H-atom approaches. This indicates strong vibrational-translational coupling and predicts efficient vibrational excitation. Typical classical trajectories (Fig. 14) start in the FC-region near the line of symmetry and move towards the saddle point. The simultaneous streching of both oscillators just after the transition is responsible for the pronounced vibrational excitation to be discussed later. The $\tilde{A}$ state PES is steeply repulsive at short distances (except for the barrier region) and therefore the dissociation of H_2O in the first band is a direct process. The absorption spectrum is broad and (almost) structureless.[77]

The extent of rotational excitation is determined by the angular variation of the excited state PES which is shown in Fig. 15 for several H-OH distances and fixed OH separation. $V_{ex}(\tilde{A})$ is globally highly anisotropic and rotational excitation would be very efficient in an ordinary "full" collision.[84] It is, however, locally isotropic around the ground state equilibrium at $\gamma_e \sim 104°$. $V_{gr}(\tilde{X})$ and $V_{ex}(\tilde{A})$ have roughly the same binding angles. According to the rotational reflection principle we therefore do not expect a significant redistribution of the initial FC-distribution in the dissociation step. The corresponding excitation function $J(\gamma)$, which is approximately proportional to the anisotropy of V_{ex},

$$J(\gamma) \sim \frac{\partial V_{ex}(\tilde{A})}{\partial \gamma} \quad , \qquad (66)$$

is zero near $\gamma_e \sim 104°$. For this reason rotational excitation in the dissociation of water in the first band is an excellent example for the pure FC-limit. The initially prepared rotational distribution (approximately given by Fig. 4) is not destroyed in the exit channel. This is true only for dissociation of the ground ($\nu_2=0$) bending state. The wavefunction for excited bending states ($\nu_2>0$) extends over a wider angular interval and thus the more anisotropic parts of $V_{ex}(\tilde{A})$ are also sampled. Then the final state interaction becomes obviously more important.[84]

Factorization of State-to-State Cross Sections

Several important conclusions follow immediately from the above discussion of the $\tilde{A}$ state PES. The rotational degree of freedom can be described within the FC-limit which makes large scale close-coupling calculations for the rotational degree of freedom unnecessary. As a direct consequence only very low OH rotational states are populated although energies of several eV are available for distribution. This was

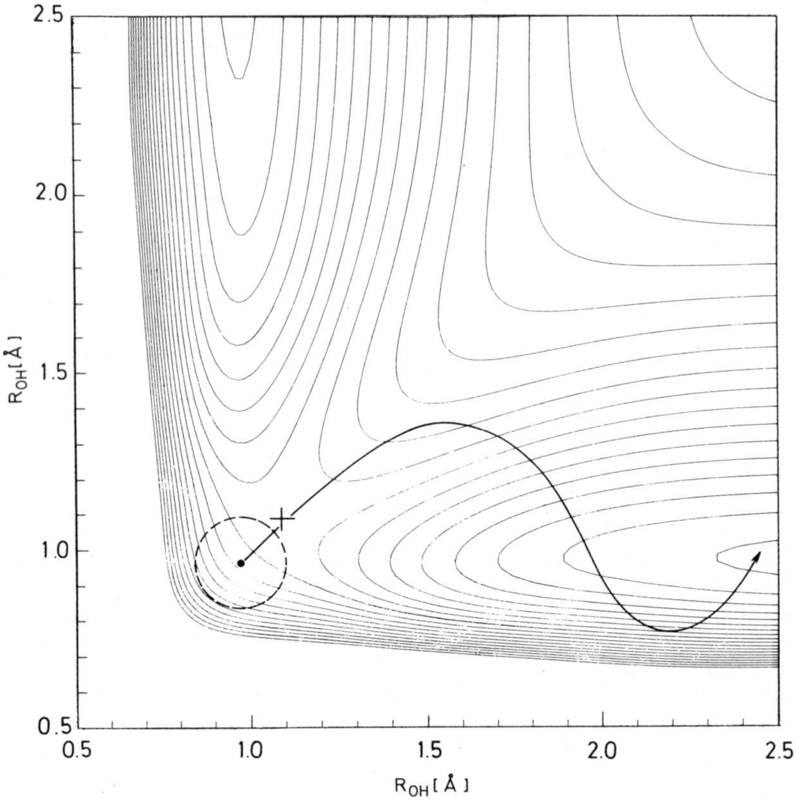

Fig. 14. Contour plot of the $\tilde{A}$ state PES of H_2O for fixed HOH bending angle γ=104°. The highest energy is E=0 (corresponding to three infinitely separated ground state atoms) and the spacing is 0.25eV. The cross marks the saddle point and the dashed circle indicates the width of the ground state wavefunction. Schematically shown is a typical trajectory.

indeed observed experimentally some time ago.[46,85] Since the rotational energy is extremely small ($E_{rot}/E_{avai.} \approx$ 0.01-0.02) it is sufficient to use the simpler FC expression in Eq. (36) rather than the expression in Eq. (34).

The completely resolved photodissociation cross section can be approximately written in the following factorized form,

$$\sigma(\nu_1\nu_2\nu_3, JK_aK_c \longrightarrow nj|E)$$
$$= \sigma_{abs}(E|\nu_1\nu_3)\ P_{vib}(n|\nu_1\nu_3|E)\ P_{rot}(j|\nu_2, JK_aK_c), \tag{67}$$

where j represents the angular momentum of OH including the electronic degrees of freedom (multiplets and Λ-doublets). The total absorption cross section σ_{abs} (i.e., the sum over all internal degrees of freedom) and the vibrational probability P_{vib} depend both on the two stretch quantum numbers (ν_1, ν_3) and the energy E (or wavelength λ). They do not depend on the bending state ν_2 or the initial rotational state. In contrast, the rotational probability P_{rot} depends on ν_2 and JK_aK_c but it does not depend on the energy (FC-limit) or the two stretch quantum numbers. Moreover, it is independent of the vibrational OH level as observed experimentally[46] and later confirmed theoretically.[86] The separability of the detailed cross section follows from the (approximate) factorization of the ground state wavefunction according to

$$\Psi_{gr}(R,r,\gamma|\nu_1\nu_2\nu_3,JK_aK_c) \sim \phi_R(R|\nu_1)\ \phi_r(r|\nu_3)\ \phi_\gamma(\gamma|\nu_2,JK_aK_c) \quad (68)$$

and from the lack of final state interaction for the rotational degree of freedom in the exit channel.

Calculations

The rotational distribution can be calculated within the FC-limit in analogy to Eq. (36). Since these calculations are rather straight forward it is possible to include also the electronical degree of freedom of OH. This has been worked out especially for water by Balint-Kurti.[87] The resulting expression is very lengthy and not repeated here. It contains a large number of Clebsch-Gordon coefficients and the angular part of the ground state wavefunction $\phi_\gamma(\gamma|\nu_2=0,JK_aK_c)$ for the specific rotational state JK_aK_c. The latter is determined within the close-coupling formalism as described above. The analytical potential of Sorbie and Murrell[88] is employed for $V_{gr}(\tilde{X})$.

The evaluation of the absorption spectrum and the vibrational probability requires full dynamical calculations. The rotational (bending) degree of freedom is treated within the energy sudden approximation (ESA). This is perfectly justified for this process because the rotational energy is extremely small compared to the total available energy.[84] Within the ESA the three-dimensional problem is reduced to a two-dimensional (collinear-like) problem which depends parametrically on the orientation angle. The partial cross section for vibrational excitation, summed over all rotational states, is given by

$$\sigma(n|E) = \int_0^\pi d\gamma\ \sin\gamma\ \sigma(n|E|\gamma) \quad . \quad (69)$$

The γ-dependent cross sections are evaluated from the two-dimensional version of the Golden Rule expression (2).

These two-dimensional calculations (for fixed bending angle γ) are more complicated than described above because of the two dissociation channels. The continuum wavefunction is a "reactive" wavefunction which spreads over both channels. An expansion in terms of the asymptotic OH states is therefore impossible when the energy is equal or larger than the barrier height. We have solved this problem by employing Delves's coordinates which are very successively used to treat collinear reactive scattering.[89] In two dimensions they are simply the polar coordinates in the plane of mass-scaled Jacobi-coordinates. All details are fully described in Ref. 83.

The $\tilde{X}$-$\tilde{A}$ transition dipole function $\mu(R,r,\gamma)$ has been calculated with SCF electronic wavefunctions and has been included in our study. Test calculations have shown that it is indeed not very important as one usually assumes. However, ignoring it would certainly render the excellent agreement with the measured absorption cross sections.

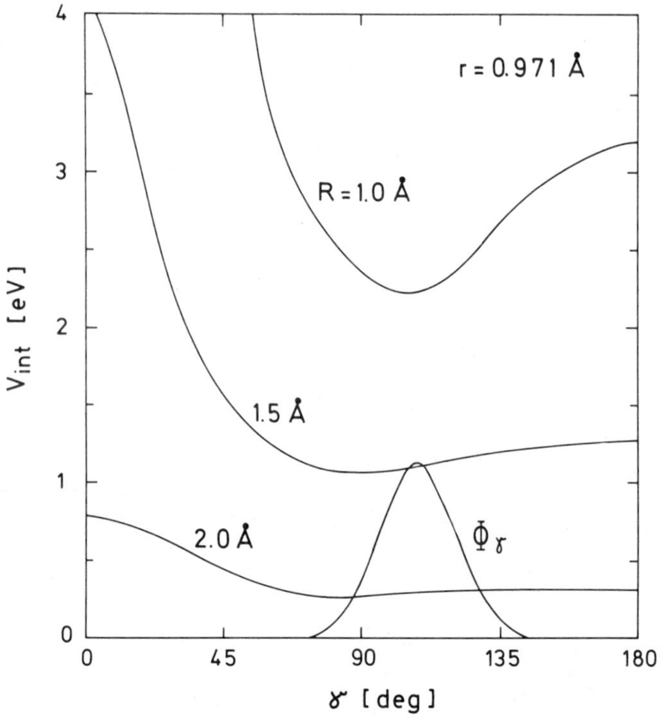

Fig. 15. Angular dependence of the interaction potential of the $\tilde{A}$ state of H_2O for several H-OH distances R and fixed OH separation r. Φ_γ is the bending part of the ground electronic state.

In conclusion, our study includes a calculated PES for the $\tilde{A}$ state and a calculated $\tilde{X}$-$\tilde{A}$ transition dipole function. The dynamical treatment is based on the ESA for the rotational degree of freedom which is nearly exact for this dissociation system. The vibrational degree of freedom is treated exactly using Delves's coordinates. The calculations do not include any adjustable parameter.

Rotational Distributions

Very recently Andresen and coworkers[47,48,90] measured - for the first time- true state-to-state product rotational distributions for a direct dissociation process. This was achived by first exciting a particular rotational-vibrational level of $H_2O(\tilde{X})$ with an IR-photon. Experimental details are given in Ref. 48. These measurements are a big step towards the detailed understanding of photodissociation processes, because thermal averaging over many (bulk) or at least several (beam) initial states is avoided. Many of the conclusions will be important for other systems as well. The data are state-specific on the level of six quantum numbers: Three quantum numbers, J, K_a, and K_c specify the parent state and three quantum numbers, j, $^2\Pi_{1\pm1/2}$, $\Lambda_\pm$, specify the final product state.

Figure 16 depicts three representative examples. Shown are rotational distributions for the $^2\Pi_{3/2}$-multiplet and the lower Λ-doublet state. Because of the electronic spin j is half-integer. The initial total angular momentum is J=4 in all cases but the projection quantum numbers are different. More examples are given in Refs. 47 and 48. In all cases

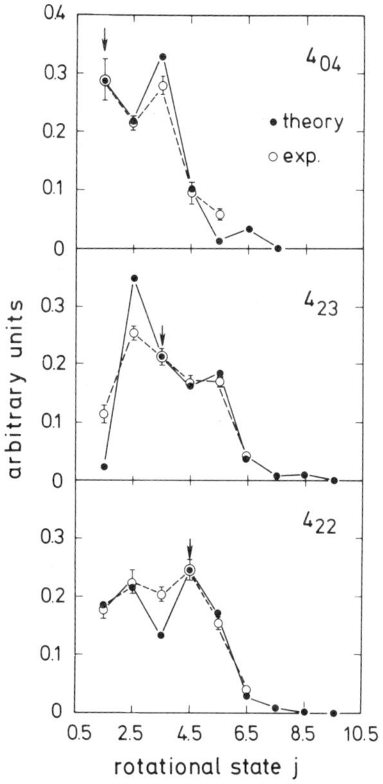

Fig. 16. Rotational distributions of OH following the dissociation of single rotational states JK_aK_c as indicated. The distributions are for the lower Λ-doublet of the $^2\Pi_{3/2}$-multiplet. Normalization between experiment and theory is made separately for each initial state indicated by the arrows.

only very few rotational states of OH are populated as expected in the FC-limit. The theoretical results, which are obtained from Balint-Kurti's FC-theory including the electronic structure of OH, agree perfectly with the measurements. All details are nicely reproduced. This is a comparison on the level of six quantum numbers.

The distributions exhibit a pronounced oscillatory behaviour which has not been observed before. It is nicely reproduced by the FC-calculations. The phase of the oscillations depends sensitively on the initial state of H_2O and the electronic state of OH. The oscillations are a trivial reminiscence of the even-odd propensity observed in Fig. 4. They would be most pronounced for a ground state equilibrium of $\gamma=90°$, i.e., when the bending wave function is symmetric around 90°. Likewise they would vanish for a linear molecule ($\gamma_e=180°$ in Fig. 1). The oscillations are significantly damped, however, by the electronic degrees of freedom of OH, as it was observed before in collisions with open-shell molecules.[91,92] If the comparison between experiment and theory is made on such a detailed level it is absolutely necessary to properly include the electronic structure.

The oscillations are rapidly washed out if the distributions are averaged over a few initial states, for example, as it is normally done even in a molecular beam, not to mention a bulk experiment. The averaged distributions are structureless Boltzmann-distributions and can be characterized by one parameter, the rotational "temperature" T(OH).[46] Since the rotational distributions depend on the initial state of H_2O it is not surprising to find that T(OH) varies with $T(H_2O)$. Within the FC-limit the rotational motion of the parent molecule is completely "transfered" to the products and therefore the final distribution becomes "hotter" as $T(H_2O)$ increases. Consequently, T(OH) increases with $T(H_2O)$ as it was observed experimentally[46] and similarly for H_2S.[93] Since we are now able to calculate rotational distributions for each initial H_2O state it is possible to perform the Boltzmann-average and thus to determine distributions for each H_2O-temperature. This has been done and satisfactory agreement with the bulk data has been obtained.[48] There are many more interesting aspects concerning the rotational and electronical excitation of OH, for example, the inversion of Λ-doublet states. For more details the reader is referred to the original publications and a recent review.[45]

Absorption Cross Sections for H_2O and D_2O

In the following discussion the various cross sections represent sums over all OH rotational states. In all cases dissociation is from the $\nu_1=\nu_2=\nu_3=0$, J=0 ground state. Figure 17 compares the λ-dependence of the calculated and the measured[94] total absorption cross sections for H_2O. As expected for a direct dissociation process the spectrum is broad and does not show distinct structures, except for the very weak undulations superimposed on the background. Theory and measurement are normalized to each other at the maximum. The agreement is excellent. The peak center at ~165nm, the width of the spectrum, and even the slight structures are nicely reproduced by our ab initio, completely parameter-free calculation. The theoretical spectrum is slightly shifted to the blue side.

The measurements are performed at room-temperature and more than hundred initial rotational states contribute to the spectrum. In order to make the comparison with experiment most rigorous we thermally averaged the calculated cross section in an approximate way as described in Ref. 83. The final results for H_2O and D_2O are shown in Fig. 18. The agreement with the experimental spectra speaks for itself. The onset of the absorption spectrum and the peak center depend directly on the barrier height of the $\tilde{A}$ state PES. The width of the spectra is determined by the slope of the potential within the FC-region. Thus, the excellent agreement in Fig. 18 underlines the high accuracy of the calculated $\tilde{A}$ state PES.

In Ref. 94 the slight undulations superimposed on the broad background are attributed to selective absorption of the bending mode. Our calculations prove unambigiously that this interpretation is wrong.[83] Selective energy transfer (i.e., resonant absorption) into the bending mode is impossible if this mode is treated within the energy sudden approximation (ESA). However, our calculations are performed within the ESA and moreover, the γ-dependent partial cross sections $\sigma(\lambda|\gamma)$ already exhibit these weak undulations. They are in fact partly smeared out by the incoherent sum of cross sections according to Eq. (69).

Figure 17 clearly shows that the "progression" in the total cross section arises from the summation of several vibrational cross sections $\sigma(n|\lambda)$. Each partial cross section exhibits one modulation. Since they are successively shifted towards higher energies (smaller wavelenghts) this structure appears as a "progression" in the total cross section. The shift of the partial cross sections is roughly given by the energy spacing of the symmetric stretch mode on top of the barrier of the $\tilde{A}$ state PES.[83,95] The modulation of the individual partial cross sections

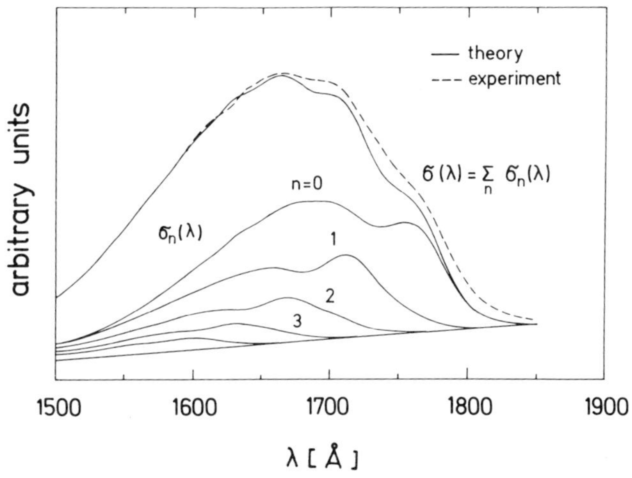

Fig. 17. Comparison of the experimental[94] and the theoretical [82,83] total absorption cross sections vs. wavelenght (linear scale). The experimental base-line is added to the theoretical results. The measurement is performed at room-temperature while the calculation is for the vibrational-rotational ground state. Also shown are the partial absorption cross sections $\sigma_n(\lambda)$ for dissociation into specific vibrational states n of OH. The total cross section is the sum of all partial cross sections.

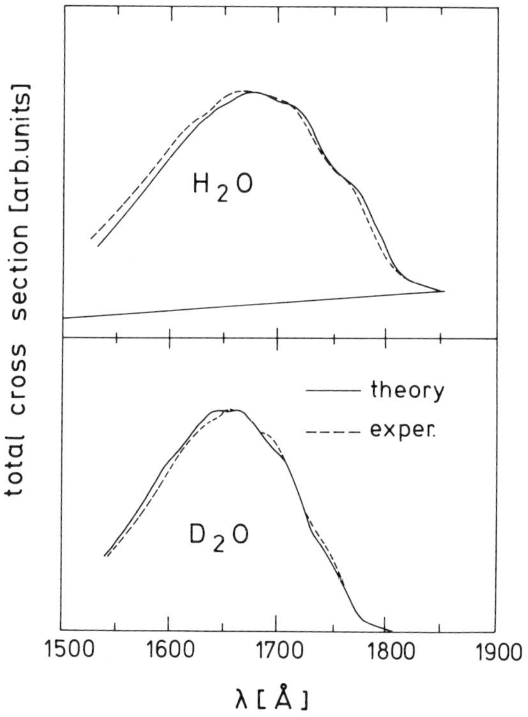

Fig. 18. Comparison of experimental[94] and thermally-averaged (300K) theoretical[83] absorption spectra (linear scale) for H_2O and D_2O. The experimental base line for H_2O is added to the calculated spectrum. It is essentially zero for D_2O.

can be explained in terms of a slight shoulder of the adiabatic potential curves (in polar coordinates) and the one-dimensional reflection principle. Because of this shoulder the normalization factor $|dV(R)/dR|^{-1}$ in Eq.(27) leads to a modulation superimposed on the broad background. A more detailed discussion can be found in Ref. 83.

Vibrational Distributions

Figure 17 shows also the individual vibrational cross sections $\sigma(n|\lambda)$. The vibrational probability for given wavelength is defined as

$$P_{vib}(n|\lambda) = \sigma(n|\lambda) / \sum_n \sigma(n|\lambda) \quad . \tag{70}$$

It is the probability that a photon of wavelength λ is absorbed and that OH is produced in state n. Although the ground level n=0 is the dominant channel over the entire spectrum, the distribution becomes significantly "hotter" as λ decreases.

Figure 19 shows the vibrational distribution vs. n for several wavelengths. It is strongly energy-dependent. At smaller wavelengths vibrational states up to n=6 are significantly populated. Comparison with experiment is possible only for λ=157nm and the three lowest states.[46] The experimental ratio n=0:1:2 at 157nm is 1:1:0.58 compared to 1:0.81:0.53 as obtained from our calculation. Note, that the experimental value for n=2 was later corrected for predissociation in the upper electronic state of OH.[45] What one really would like to have is a comparison of the complete distribution at several wavelengths. This would most rigorously test the accuracy of the PES and of our dynamical calculations. Experiments in this direction are in progress.

The OH equilibrium separation in the water molecule is 0.957Å[88] compared to 0.971Å asymptotically. Therfore, the pure FC-distribution peaks at n=0 and is very narrow. According to the above discussion this implies that the high degree of vibrational excitation is induced by strong final state interaction. This is not surprising for a symmetric triatomic molecule with two dissociation channels. The same behaviour was observed before in model calculations for the collinear CO_2 system.[24,96] The free OH oscillator is drastically perturbed as the other H-atom approaches, i.e., one bond is broken and a new bond is formed.

Photodissociation of HOD

The photodissociation of the non-symmetric isotope variation HOD has not yet been studied experimentally. It is apparently very difficult to produce pure HOD and therefore it is challenging for us to make theoretical predictions prior to any experiment. The calculations become also more difficult because the symmetry of the two dissociation channels is lifted.[97]

Figure 20 shows the branching ratio σ(H+OD)/σ(D+OH) for the production of OD and OH, respectively, as a function of energy E.[97] The bending angle is fixed at γ=104°. Averaging over γ according to the sudden prescription does not change the overall picture. Dissociation into the H+OD channel is significantly more probable over the entire energy regime. The slight undulations are due to structures of the corresponding cross sections which have the same origin as discussed before for H_2O and D_2O.

Also shown in Fig. 20 are simple classical results using the approximate Hamiltonian

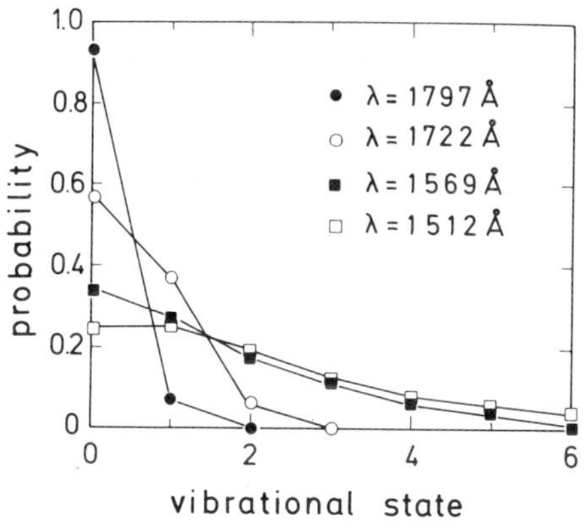

Fig. 19. Vibrational state distributions following the dissociation of H_2O in the first absorption band.

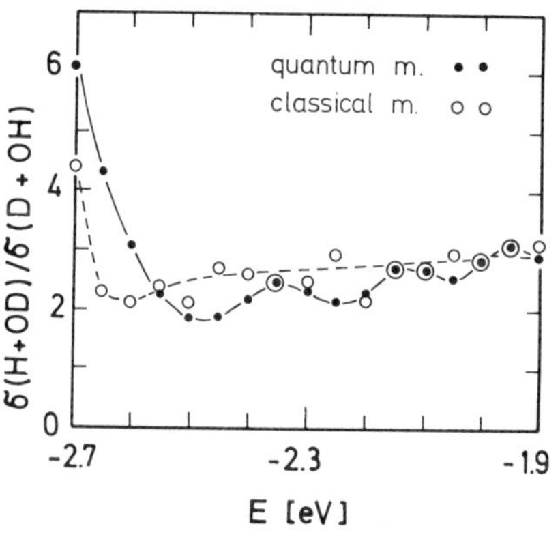

Fig. 20. Comparison of the quantum mechanical and the classical branching ratio for the dissociation of HOD. The bending angle is fixed at 104°. Energy normalization is such that three separated ground state atoms correspond to E=0.

$$H = \frac{P_H^2}{2m_H} + \frac{P_D^2}{2m_D} + V_{ex}(R_H, R_D | \gamma = 104°) \quad , \tag{71}$$

where R_H is the OH separation and R_D is the OD distance. In Eq.(71) we assumed that the oxygen atom is infinitely heavy in consistency with the quantal calculation. P_H and P_D are the corresponding momenta. The classical cross sections are obtained by sampling the four-dimensional phase-space and using the distribution function of Eq. (29). The classical branching ratio is in very good agreement with the quantal curve. The structures are due to an insufficient number of trajectories.

Within the classical theory the branching ratio is determined by two factors, the weighting through the initial wavefunction and the dynamics in the excited state. The ground state wavefunction is approximately given by a product of two Gaussians

$$\Psi_{gr}(R_H,R_D) = \phi_H(R_H)\ \phi_D(R_D) \quad . \tag{72}$$

Because the mass of hydrogen is half the mass of deuterium $\phi_H(R_H)$ is significantly broader than $\phi_D(R_D)$. This means that on the average more trajectories start within the H+OD channel. In addition, the acceleration at or just beyond the starting point of the trajectory is larger for hydrogen than for deuterium for the same argument. Therefore the H-atom leaves the FC-zone faster than the D-atom which consistently explains the observed branching ratio. The dissociation of HOD is a further example of the applicability and usefulness of ordinary classical mechanics. More results for the vibrational distributions of OH and OD and for the dissociation of vibrationally excited HOD are discussed in Ref. 97.

PHOTODISSOCIATION OF METHYLNITRITE

The photodissociation of methylnitrite in the region of 300-400nm ($S_0 \rightarrow S_1$) has been studied by several experimental groups in recent years.[51,52,98,99] The absorption spectrum[100] consists of a well defined vibrational progression which is attributed to selective absorption by the N=O chromophore within the S_1 excited state. The vibrational distribution of the NO products is narrow, inverted and systematically wavelenght-dependent.[51] The rotational distribution of NO is highly inverted and symmetric around the peak center at $j{\sim}35$.[52] It seems to be a beautiful example of the rotational reflection principle.

Theoretical studies to interprete the wealth of experimental data started only very recently, on a relatively low level of sophistication. Nonella and Huber[101] calculated the S_1 excited state PES including two coordinates, the O-N distance (R) and the N=O separation (r). All other coordinates are fixed. We used this two-dimensional PES in a collinear-like study and calculated the total absorption cross section and the final NO vibrational distribution.[102] Although the calculations are rather limited they give relatively good agreement with most of the exerimental data. Moreover, we are certain that they provide a realistic description of the overall dissociation process.

Potential Energy Surface

Figure 21 shows a contour plot of the calculated S_1-PES. Details are given in Ref. 101. Two features, which make this system so interesting for dynamical theory, are immediately apparent. First, the NO equilibrium distance and the NO frequency change significantly as the CN_3O-radical

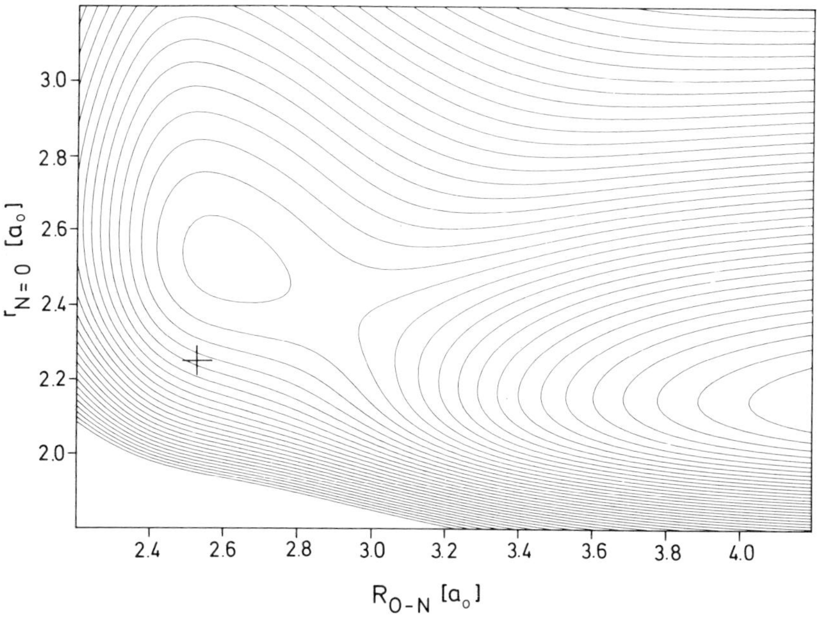

Fig. 21. Contour plot of the two-dimensional S_1-PES (in eV) of methylnitrite. The cross marks the equilibrium of the electronic ground state.

approaches. Second, it has a local well along the dissociation coordinate which is probably due to an avoided crossing with a higher state. As it will be seen below, this well is deep enough to support two quasi-bound states. Therefore, we expect sharp resonance structures superimposed on a broad background as schematically illustrated in Fig. 2. The $S_0 \rightarrow S_1$ photodissociation of methylnitrite is a beautiful example of indirect dissociation or vibrational-predissociation within the excited electronic state.

For interpretation purposes it is useful to define the (corrected) adiabatic potential curves

$$U_n(R) = \varepsilon_n(R) - \frac{1}{2m} < \phi_n(r|R) \mid \frac{\partial^2}{\partial R^2} \phi_n(r|R)> \quad , \tag{73}$$

where the $\varepsilon_n(R)$ and $\phi_n(r|R)$ are determined by the one-dimensional Schrödinger equation for the vibrational degree of freedom and fixed dissociation coordinate R,

$$[- \frac{1}{2m_{BC}} \frac{d^2}{dr^2} + V_{ex}(R|r) - \varepsilon_n(R)] \phi_n(r|R) = 0 \quad . \tag{74}$$

Asymptotically ($R \rightarrow \infty$) the adiabatic energies $\varepsilon_n(R)$ become the energy levels of the free NO oscillator. Figure 22 shows these potential curves $U_n(R)$ vs. dissociation coordinate in the region of the local well. The energy is measured with respect to the bottom of the asymptotic NO oscillator. Because of the bond-dilution the energy spacing within the well region is significantly smaller than for the free NO molecule.

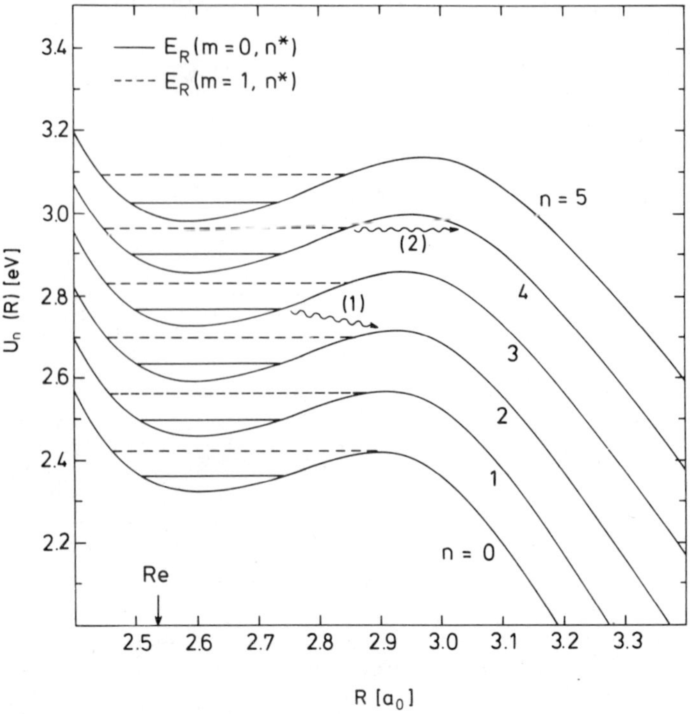

Fig. 22. Corrected adiabatic potential curves $U_n(R)$, Eq. (73), within the transition region. The horizontal lines indicate the resonance energies $E_R(m,n^*)$ as obtained from the exact CC calculations. The arrows indicate schematically the two possible decay mechanisms, i.e., vibrational-predissociation (1) and tunneling (2). R_e is the ground state equilibrium.

The energies of the two-dimensional quasi-bound (i.e., resonance) states can be approximately written as

$$E_R(m,n^*) = \varepsilon_m^{(n)} + \varepsilon_{n^*} \quad , \tag{75}$$

where ε_{n^*} is the local NO vibrational energy within the well. The $\varepsilon_m^{(n)}$ are the one-dimensional quasi-bound (i.e., resonance) energies associated with the radial motion on the adiabatic potential curve $U_n(R)$. The corresponding wavefunctions are

$$\Psi_{m,n^*}^{(n)}(R,r) = \phi_m^{(n)}(R)\ \phi_{n^*}(r) \quad , \tag{76}$$

where $\phi_m^{(n)}(R)$ is a one-dimensional radial function in state m defined through $U_n(R)$ and $\phi_{n^*}(r)$ is the vibrational wavefunction in state n. Resonances in the various scattering or dissociation cross sections are expected whenever the total energy matches one of the resonance energies.

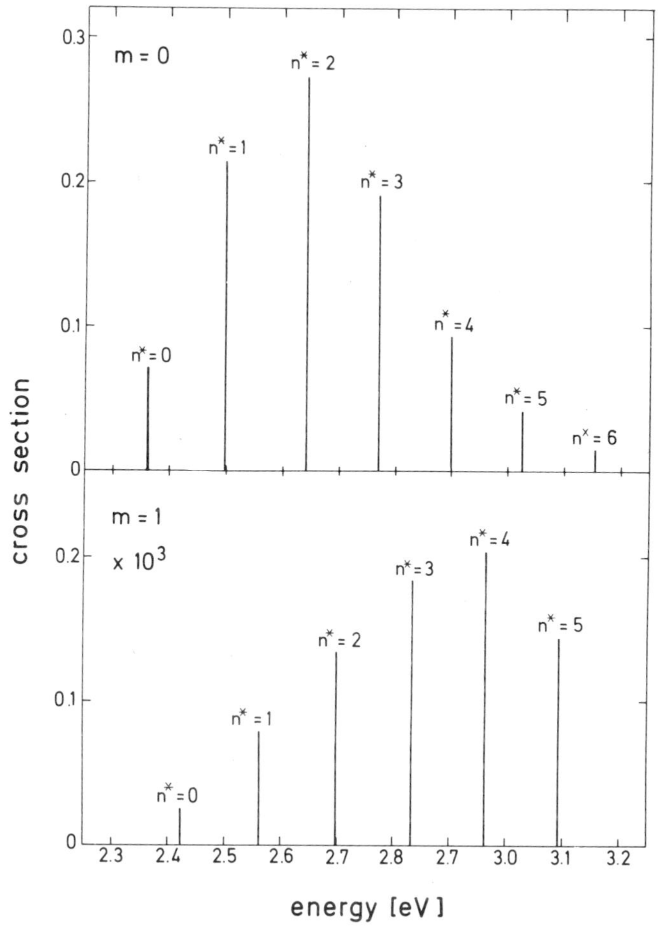

Fig. 23. Calculated absorption spectrum (arbitrary units) vs. energy. The assigment with quantum numbers is described in the text. m represents excitation of the O-N bond and n^* describes excitation of the N=O bond. The peaks of the m=1 band are multiplied by 10^3.

Absorption spectrum

The following calculations are performed exactly within the close-coupling formalism. Because the oscillator potential changes significantly with R we found it necessary to use an adiabatic basis as defined by Eq. (74). Figure 23 shows the calculated absorption spectrum over a relatively wide energy range. It consists of two progressions, m=0 and m=1, of very narrow resonance lines. The background is roughly five orders of magnitude smaller than the m=0 peaks. The intensities of the m=1 progression are about three orders of magnitude smaller than the intensities of the first series.

The line shape of the individual m=0 absorption profiles is analyzed on the left panel of Fig. 24 on a much finer energy grid than in Fig. 23. Plotted is the total absorption cross section $\sigma_{abs.}(E)$ vs. detuning (E_R-E). All profiles are very accurately represented by a simple Lorentzian[103]

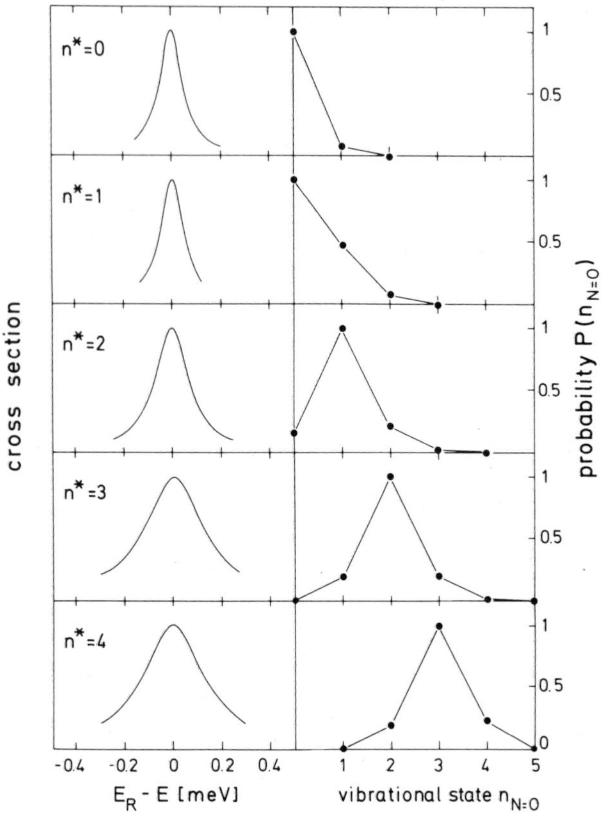

Fig. 24. Left panel: Total absorption spectra (linear scale) vs. detuning (E_R-E) in the vicinity of the (m=0,n*) resonances. The maxima are set to the same values. Right panel: Corresponding vibrational distributions of NO products at the peak maxima.

$$\sigma(E) \sim \left[(E_R-E)^2 + (\Gamma/2)\right]^{-1} \qquad (77)$$

where Γ is the full width at half of the maximum (FWHM). The width of the m=0 progression increases from 0.12meV for n^*=0 to 0.31meV for n =4. The m=1 resonances are much broader and their widths decreases with n^*.[102]

The exact resonance energies $E_R(m,n^*)$ as obtained from the CC calculations are indicated in Fig. 22. It is obvious that our tentative assingment in Fig. 23 completely agrees with the quasi-bound spectrum discussed above in the adiabatic approximation. This was further substantiated by an analysis of the continuum wavefunctions at resonance. They indeed show the behaviour of a quasi-bound state as approximately described by Eq. (76). Quantum number m counts the nodes (within the well region) along the dissociation coordinate and n^* counts the nodes along the internal coordinate.

The photodissociation of methylnitrite is a nice example of an indirect process. It proceeds as

$$\begin{aligned} CH_3ONO(S_0) + h\nu &\longrightarrow CH_3ONO(S_1|m,n^*) \\ &\longrightarrow CH_3O + NO(n) \quad . \end{aligned} \qquad (78)$$

The first step is a resonant transition in which a quasi-stable level (m,n^*) within the S_1 electronic state is excited. In the second step the excited complex decays into produces NO(n) and CH_3O. The decay of the complex can be induced either by non-adiabatic (translational-vibrational) coupling or by tunneling through the barrier. This is schematically indicated by the arrows in Fig. 22. In the first case the break-up is initiated by de-excitation from n^* to (n^*-1) and adiabatic dissociation along the (n^*-1) potential. The NO products are then populated mainly in the (n^*-1) level. The vibrational distributions on the right panel of Fig. 24 clearly manifest that this kind of vibrational-predissociation is the dominant decay mechanism for the m=0 mainfold. In the second case the break-up is initiated by tunneling through the barrier and adiabatic dissociation along the n^* potential curve. NO products are then primarily populated in level n^*. The vibrational distributions for the m=1 progression show exactly this type of behaviour and therefore indicate that tunneling is the main dissociation pathway.[102]

Comparison with Experiment

The measured absorption spectrum shows a well resolved progression of band structures which is identified with the N=O stretching vibration (ν_3-mode) in the $S_1(n\pi)$ state. Excitations from S_0 to the vibronic states $\nu_3=1$ to $\nu_3=6$ are clearly observed with the maximum at $\nu_3=3$. The spacing is about 125meV and the average width is about 60meV.[98,99] In our notation the ν_3-quantum number corresponds to n^* and the O-N vibration corresponds to m.

The calculated spectrum for m=0 as depicted in Fig. 23 agrees qualitatively well with the measured spectrum, except for the width. The intensities for the second, m=1, band are three orders of magnitude smaller and therefore unimportant for the comparison with experiment. The energy spacing of the calculated spectrum ranges from 137meV to 127meV and is thus in reasonable accord with experiment. This underlines the quality of the calculated S_1-PES with respect to the N=O degree of freedom within the FC-region.

The theoretical widths of the resonances are too small by about two orders of magnitude. This is not surprising in view of the simplicity and the limitations of our model. First, incoherent broadening due to thermal averaging is not included. Second, only two degrees of freedom are incorporated. Inclusion of other modes would certainly give rise to coherent broadening due to an increased coupling. The inverted NO rotational state distributions, for example, indicate a strong coupling to the O-N=O bending motion.

The main asset of our calculation is the good agreement found for the final vibrational distributions of NO after excitation into the peaks of the absorption spectrum, associated with the transitions $S_0(n^*=0) \longrightarrow S_1(n^*=1,2,3,4)$. The experimental[51] and calculated (m=0) distributions are compared in Fig. 25. The general trend that NO is produced primarily in state n^*-1 if state n^* was excited is nicely seen both experimentally and theoretically. This manifests that vibrational-predissociation is indeed the true decay mechanism of the excited S_1-complex. The experimental distributions are generally broader. However, one should not forget that these measurements are very difficult and that the uncertainties are probably quite substantial.

Although the present calculations are rather limited they explain some of the experimental findings and unambigously clarify the general dissociation mechanism. In future studies we will test the applicability of classical mechnics. The resonance structures of the absorption cross section certainly can not be described classically. However, it may be possible that the general behaviour of the vibrational distributions can be satisfactorily reproduced. If so, it will be worthwile to include also the rotational degree of freedom, which would be impossible in the CC formalism. Calculations similar to those described in this lecture have also been started for the dissociation of HO-NO which has been examined experimentally some time ago.[104] Preliminary results are quite encouraging.

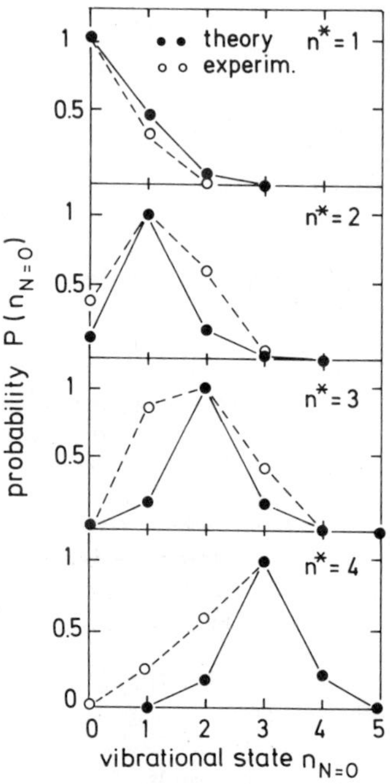

Fig. 25. Comparison of calculated (●●) and experimental[51] (○○) NO vibrational state distributions following excitation of the m=0, $n^{*}(\nu_3)$=1,2,3, and 4 levels within the S_1-state. All distributions are normalized to unity at the maxima.

CONCLUSIONS

At the end of this lecture we will shortly summarize those points which we find most interesting and of general importance.

1. Photodissociation is the second half of a full collision process. The classical equations of motion are the same and the quantum mechanical wavefunctions are identical. Therefore it is not surprising to find many similarities. For example, the classical excitation functions (for rotation and for vibration) are qualitatively very similar. On the other

hand, the "initial conditions" are defined in a completely different manner which leads to some pronounced differences. In scattering the "initial conditions" are defined for the separated systems and the full phase-space is sampled. In dissociation "initial conditions" are defined for the "united molecule" and a narrow region of the phase-space is probed. The nuclear wavefunction of the parent molecule acts like a multi-dimensional "window". For example, H-OH scattering on the $\tilde{A}$ state PES would yield very large rotational energy transfer. The photodissociation process evolving on exactly the same potential surface, however, yields negligible rotational energy.

2. The restricted "initial conditions" defined for the united molecule lead naturally to the multi-dimensional reflection principle: The energy-dependence of the absorption spectrum is a reflection of the R-dependence of the parent wavefunction and of the excited state potential (one-dimensional reflection principle). The j-dependence of the product distribution is a reflection of the angular-dependence of the parent wavefunction and of the rotational excitation function (rotational reflection principle). The n-dependence of the product distribution is a reflection of the r-dependence of the parent wavefunction and of the vibrational excitation function (vibrational reflection principle). In all cases the final distributions (E,j,n,...) are direct probes of the respective degree of freedom of $V_{ex}(R,\gamma,r,...)$ above the ground state equilibrium (R_{eq}, γ_{eq}, r_{eq},...).

3. The reflection principle is a characteristic feature of direct photodissociation processes. The distribution of initial coordinates ($R_0,\gamma_0,r_0,...$) is directly mapped onto the final "momenta" (E,j,n,...). This mapping is obviously smeared out or completely destroyed in indirect photodissociation processes, when the trajectories live for a long time. Then the memory of the initial state is lost. However, recent experiments seem to indicate that a reflection principle, in a modified version of course, may be operative even in predissociation processes. This topic deserves certainly future investigations.

4. Ordinary classical mechanics seems to be a very reliable and powerful tool to study direct photodissociation processes. In all cases where we could compare with exact quantum mechnical calculations we observed excellent agreement. Classical calculations are easy to implement and to interprete, they can be straight forwardly extended to more complicated systems, and they are much cheaper than comparable quantal calculations. On the other hand, they can not describe quantal featurs like resonances which might be important for many dissociation processes.

5. In order to perform realistic dynamical calculations it is essential to know the corresponding PES's, at least qualitatively. It is not so important to have a PES which is correct on an absolute scale but it should describe the qualitative features correctly. Because of avoided crossings, for example, excited state PES's can have a complicated appearance. Such potential features may lead to particular structures of the cross sections and therefore it is essential to know the potential at least qualitatively. It is very dangerous to propose simple models without knowing the excited PES. The conclusions can be wrong and misleading.

6. The photodissociation of water in the first absorption band is completely understood on the basis of reliable quantum chemical and dynamical calculations. All measured data are satisfactorily reproduced in all details. This includes the total absorption cross sections for H_2O and D_2O, the vibrational distributions, and the rotational distributions for dissociation of single initial states of H_2O, the dissociation of

HOD, etc. We are not aware of any other system for which such detailed investigations have been performed. We should not forget to mention, however, that this success became possible due to some fortunate circumstances. The rotational degree of freedom can be described within the simple FC-limit and does not have to be treated dynamically. As a consequence it is possible to include the electronic structure of OH and to use the IOS approximation in the dynamical calculations for the other degrees of freedom. This may be very different for other systems.

ACKNOWLEDGEMENTS

I am very grateful to my colleagues and coworkers Dr. P. Andresen, Dr. V. Engel, S. Hennig, K. Weide and A. Untch who contributed significantly to my understanding of photodissociation processes. Their contributions are "reflected" at several places of this lecture.

REFERENCES

1. K.C. Janda, in: "Photodissociation and photoionization," K.P. Lawley, ed., Wiley, New York (1985).
2. H. Okabe, "Photochemistry of small molecules," Wiley, New York (1978).
3. M.B. Robin, "Higher excited states," Academic Press, New York (1974).
4. G. Herzberg, "Molecular spectra and molecular structure I. Spectra of diatomic molecules," Van Nostrand, Princeton (1950); "Molecular spectra and molecular structure III. Electronic spectra of polyatomic molecules," Van Nostrand, Princeton (1966).
5. S.R. Leone, Adv. Chem. Phys. 50: 255 (1982).
6. J.P. Simons, J. Phys. Chem. 88: 1287 (1984).
7. R. Bersohn, J. Phys. Chem. 88: 5145 (1984).
8. W.M. Jackson and H. Okabe, in: "Advances in photochemistry," Vol. 13, D.H. Volman, K. Gollnick, and G.S. Hammond, eds., Wiley, New York (1986).
9. S. Buelow, M. Noble, G. Radharkrishnan, H. Reisler, C. Wittig, and G. Hancock, J. Phys. Chem. 90: 1015 (1986).
10. U. Hefter and K. Bergmann, in: "Atomic and molecular beam methods," G. Scoles, ed., Oxford University Press, Oxford (1986).
11. M.N.R. Ashfold, Mol. Phys. 58:1 (1980); D.H. Parker, in "Ultrasensitive laser spectroscopy," D.S. Klinger, ed., Academic Press, New York (1983).
12. R. Schinke and J.M. Bowman, in: "Molecular collision dynamics," J.M. Bowman, ed., Springer, Heidelberg (1983).
13. R. Schinke, J. Chem. Phys. 85: 5049 (1986).
14. R. Schinke and V. Engel, Faraday Discuss. Chem. Soc. 82: paper 11 (1986).
15. E.J. Heller, J. Chem. Phys. 68: 2066 (1978).
16. E.J. Heller, J. Chem. Phys. 68: 3897 (1978).
17. K.C. Kulander and E.J. Heller, J. Chem. Phys. 69: 2439 (1978).
18. S.Y. Lee and E.J. Heller, J. Chem. Phys. 76: 3035 (1982).
19. M. Shapiro and R. Bersohn, Ann. Rev. Phys. Chem. 33: 409 (1982).
20. M. Weissbluth, "Atoms and molecules," Academic Press, New York (1978).
21. G.G. Balint-Kurti and M. Shapiro, Chem. Phys. 61: 137 (1981); 72: 456 (1982).
22. M. Shapiro, J. Chem. Phys. 56: 2582 (1972).
23. M. Shapiro and G.G. Balint-Kurti, in: "Photodissociation and Photoionization," K.P. Lawley, ed., Wiley, New York (1985).
24. K.C. Kulander and J.C. Light, J. Chem. Phys. 73: 4337 (1980).

25. F.A. Gianturco, "The transfer of molecular energy by collisions", Springer, Heidelberg (1979).
26. D. Secrest, J. Chem. Phys. 62: 710 (1975).
27. G.A. Parker and R.T. Pack, J. Chem. Phys. 68: 1585 (1978).
28. P. McGuire, Chem. Phys. Lett. 23: 575 (1973); P. McGuire and D.J. Kouri, J. Chem. Phys. 60: 2488 (1974); R.T. Pack, J. Chem. Phys. 60: 633 (1974)
29. R. Schinke, J. Phys. Chem. 90: 1742 (1986).
30. S. Hennig, V. Engel, and R. Schinke, J. Chem. Phys. 84: 5444 (1986).
31. L.D. Thomas, M.H. Alexander, B.R. Johnson, W.A. Lester, Jr., J.C. Light, K.D. McLenithan, G.A. Parker, M.J. Redmond, T.G. Schmalz, D. Secrest, and R.B. Walker, J. Comput. Phys. 41: 407 (1981).
32. R.N. Porter and L.M. Raff, in: "Modern theoretical chemistry," W.H. Miller, ed., Plenum, New York (1976);
G.C. Schatz, in: "Molecular collision dynamics," J.M. Bowman, ed., Springer, Heidelberg (1983).
33. W.H. Miller, Adv. Chem. Phys. 25: 69 (1974), 30: 77 (1975).
34. A. Messiah, "Quantum mechanics," North Holland, Amsterdam (1972)
35. E. Wigner, Phys. Rev. 40: 749 (1932).
36. S. Goursaud, M. Sizun, and F. Fiquet-Fayard, J. Chem. Phys. 65: 5453 (1976).
37. M.D. Pattengill, Chem. Phys. 68: 73 (1982); Chem. Phys. 78: 229 (1983); Chem. Phys. Lett. 104: 462 (1984); Chem. Phys. Lett. 105: 651 (1984); Chem. Phys. 87: 419 (1984).
38. E.M. Goldfield, P.L. Houston, and G.S. Ezra, J. Chem. Phys. 84: 3120 (1986).
39. R. Bersohn and M. Shapiro, J. Chem. Phys. 85: 1396 (1986).
40. M.G. Sheppard and R.B. Walker, J. Chem. Phys. 78: 7191 (1983).
41. N.E. Henriksen, Chem. Phys. Lett. 121: 139 (1985).
42. K.F. Freed and Y.B. Band, in "Excited states," Vol. 3, E.C. Lim, ed., Academic Press, New York (1978).
43. J.A. Beswick and W.M. Gelbart, J. Phys. Chem. 84: 3148 (1980).
44. M.D. Morse, K.F. Freed, and Y.B. Band, J. Chem. Phys. 70: 3604 (1979); M.D. Morse and K.F. Freed, Chem. Phys. Lett. 74: 49 (1980): J. Chem. Phys. 74: 4395 (1981); J. Chem. Phys. 78: 6045 (1983).
45. P. Andresen and R. Schinke, in "Molecular photodissociation dynamics,"J.E. Baggott and M.N.R. Ashfold, eds., Roy. Soc. Chem. (1987).
46. P. Andresen, G.S. Ondrey, B. Titze, and E.W. Rothe, J. Chem. Phys. 80: 2548 (1984).
47. R. Schinke, V. Engel, P. Andresen, D. Häusler, and G.G. Balint-Kurti, Phys. Rev. Lett. 55: 1180 (1985).
48. D. Häusler, P. Andresen, and R. Schinke, J. Chem. Phys., in press (1987).
49. W.J. Marinelli, N. Sivakumar, and P.L. Houston, J. Phys. Chem. 88: 6658 (1984); I. Nadler, D. Mahgerefteh, H. Reisler, and C. Wittig, J. Chem. Phys. 82: 3885 (1985); W.H. Fisher, R. Eng, T. Carrington, C.H. Dugan, S.V. Filseth and C.M. Sadowski, Chem. Phys. 89: 457 (1984).
50. J.A. Russel, I.A. McLaren, W.A. Jackson, and J.B. Halpern, J. Phys. Chem., to be published.
51. O. Benoist d'Azy, F. Lahmani, C. Lardeux, and D. Solgadi, Chem. Phys. 94: 247 (1985).
52. U. Brühlmann, M. Dubs, and J.R. Huber, J. Chem. Phys. 86: 1249 (1987).
53. J.B. Halpern and W.M. Jackson, J. Phys. Chem. 86: 3528 (1982).
54. R. Schinke, J. Phys. Chem., in press.
55. E. Gottwald, K. Bergmann, and R. Schinke, J. Chem. Phys. 86: 2685 (1987).
56. W. Schepper, U. Ross, and D. Beck, Z. Phys. A 290: 131 (1979).
57. S. Bonanac, Phys. Rev. A 22: 2617 (1980).
58. J. Korsch and R. Schinke, J. Chem. Phys. 75: 3850 (1981).

59. R. Schinke and V. Engel, J. Chem. Phys. 83: 5068 (1985).
60. C.B. Moore and J.C. Weishaar, Ann. Rev. Phys. Chem. 34: 31 (1983).
61. D.J. Bamford, S.V. Filseth, M.F. Foltz, J.W. Hepburn, and C.B. Moore, J. Chem. Phys. 82: 3032 (1985).
62. J.D. Goddard and H.F. Schaefer III, J. Chem. Phys. 70: 5117 (1979).
63. R. Schinke, H. Meyer, U. Buck, and G.H.F. Diercksen, J. Chem. Phys. 80: 5518 (1984).
64. R. Schinke, Chem. Phys. Lett. 120: 129 (1985).
65. R. Schinke, J. Chem. Phys. 84: 1487 (1986).
66. D. Debarre, M. Lefebvre, M. Pealat, J.P.E. Taran, D.J. Bamford, and C.B. Moore, J. Chem. Phys. 83: 4476 (1985).
67. A.U. Grunewald, K.H. Gericke, and F.J. Comes, Chem. Phys. Lett. 132: 121 (1986); K.H. Gericke, S. Klee, and F.J. Comes, J. Chem. Phys. 85: 4463 (1986).
68. M.P. Docker, A. Hodgson, and J.P. Simons, Faraday Discuss. Chem. Soc. 82: paper 10 (1986).
69. A. Jacobs, M. Wahl, R. Walter, and J. Wolfrum, Applied Physics B, 42: 173 (1987).
70. U. Meier, V. Staemmler, and J. Wasilewski, to be published.
71. R.H. Hunt, R.A. Leacock, C.W. Peters, and K.T. Hecht, J. Chem. Phys 42: 1931 (1965).
72. M. Shapiro and R. Bersohn, J. Chem. Phys. 73: 3810 (1980).
73. G.N.A. van Veen, T. Baller, A.E. de Vries, and M. Shapiro, Chem. Phys. 93: 277 (1985).
74. R.A. Marcus, J. Chem. Phys. 54: 3965 (1971), 56: 311 (1972); 59: 5135 (1973).
75. M. Shapiro, Chem. Phys. Lett. 81: 521 (1981).
76. M.S. Child and M. Shapiro, Mol. Phys. 48: 111 (1983).
77. K. Watanabe and M. Zelikoff, J. Opt. Soc. Am. 43: 753 (1953).
78. E. Segev and M. Shapiro, J. Chem. Phys. 73: 2001 (1980); 77:5604 (1982).
79. L.J. Dunne, H. Guo, and J.N. Murrell, Mol. Phys., in press.
80. K. Weide and R. Schinke, J. Chem. Phys., in press (1987).
81. V. Staemmler and A. Palma, Chem. Phys. 93: 63 (1985).
82. V. Engel, R. Schinke, and V. Staemmler, Chem. Phys. Lett. 130: 413 (1980).
83. V. Engel, R. Schinke, and V. Staemmler, J. Chem. Phys., in press (1987).
84. R. Schinke, V. Engel, and V. Staemmler, J. Chem. Phys. 83: 4522 (1985).
85. P. Andresen, G.S. Ondrey, and B. Titze, Phys. Rev. Lett. 50: 486 (1983).
86. R. Schinke, V. Engel, and V. Staemmler, Chem. Phys. Lett. 116: 165 (1985).
87. G.G. Balint-Kurti, J. Chem. Phys. 84: 4443 (1986).
88. K.S. Sorbie and J.N. Murrell, Mol. Phys. 29: 1387 (1975); 31: 905 (1976).
89. J. Manz, Comments At. Mol. Phys. 17: 91 (1985).
90. P. Andresen, V. Beushausen, D. Häusler, and H.W. Lülf, J. Chem. Phys. 83: 1429 (1985).
91. T. Orlikowski and M.H. Alexander, J. Chem. Phys. 79: 6006 (1983).
92. H. Joswig, P. Andresen, and R. Schinke, J. Chem. Phys. 85: 1904 (1986).
93. W.G. Hawkins and P.L. Houston, J. Chem. Phys. 76: 729 (1982).
94. H.T. Wang, W.S. Felps, and S.P. McGlynn, J. Chem. Phys. 67: 2614 (1977).
95. R.T. Pack, J. Chem. Phys. 65: 4765 (1976).
96. K.C. Kulander and J.C. Light, J. Chem. Phys. 85: 1938 (1986).
97. V. Engel and R. Schinke, to be published.

98. F. Lahmani, C. Lardeux and D. Solgadi, Chem. Phys. Lett. 102: 523 (1983).
99. B.A. Keller, P. Felder, and J.R. Huber, J. Phys. Chem. 91: 1114 (1987).
100. P. Tarte, J. Chem. Phys. 20: 1570 (1952).
101. M. Nonella and J.R. Huber, Chem. Phys. Lett. 131: 376 (1986).
102. S. Hennig, V. Engel, R. Schinke, M. Nonella, and J.R. Huber, J. Chem. Phys. 87: 3522 (1987).
103. J.R. Taylor, "Scattering Theory," Wiley, New York (1972).
104. R. Vasudev, R.N. Zare, and R.N. Dixon, J. Chem. Phys. 80: 4863 (1983).

PART II

NON-ADIABATIC SCATTERING PROCESSES

NON-ADIABATIC ATOM-ATOM COLLISION

B. H. Bransden
Department of Physics, University of Durham
Durham DH1 3LE, England

ABSTRACT

Theoretical models to describe ion-atom collisions at intermediate and high energies are discussed. First the two centre expansion method is introduced and the problems posed by correlation in many body systems are outlined. This is followed by a treatment of perturbation and distorted wave methods paying particular attention to the proper Coulomb boundary conditions to be satisfied. The many applications of the theory to individual systems are not described in detail but a few illustrative examples are given for each theoretical model.

1. INTRODUCTION

At the Advanced Study Institute held at Cortona in 1980, McCarroll[1] described the theory of ion-atom collisions, concentrating particularly on the low energy or near-adiabatic region. To avoid repetition, the present lectures will deal particularly with collisions at intermediate and high energies which are completely non-adiabatic in character. In this context, intermediate and high energies are those for which the velocity of the incident ion is comparable or greater than the Bohr velocity of a target electron, respectively. In these energy regimes, the wave length associated with the heavy particle motion is very small compared with the extent of the interaction region, and classical conditions apply. Furthermore, the heavy particle scattering is concentrated in a very narrow cone about the forward direction, so that the relative motion of the heavy particles can be represented by a straight line trajectory traversed at a constant velocity v. We have

$$\underline{R}(t) = \underline{b} + \underline{v}t \qquad 1.1$$

where $\underline{R}$ is the internuclear distance and $\underline{b}$ is a two-dimensional impact parameter vector. The model in which the electronic motion is described by quantum mechanics, but the heavy particle motion is determined by (1.1), is termed the impact parameter approximation. It has been shown,[1,2] that total cross sections calculated within this model are the same to order (m_e/m_p), Where m_e and m_p are the electron and proton masses, as those determined from a fully quantal theory. This is also true for differential cross sections found by using a Bessel function transform. An alternative approach, which has been rather successful in predicting cross sections at intermediate energies is the classical trajectory Monte Carlo method (CTMC) based on a classical three body model in which even the electronic motion is determined by classical mechanics, but we shall not discuss this model in these lectures. Some account of this and other classical models can be found in a review by Bransden and Janev.[3]

For simplicity, the theory will be developed for a one-electron system, for which typical reactions are

$$B + (A+e^-) \rightarrow B + (A+e^-) \qquad (a)$$
$$\rightarrow B + A + e^- \qquad (b)$$

where A and B represent ionic cores of charge Z_A and Z_B respectively in atomic units (which will be used throughout). Reactions (a) and (b) represent excitation and ionisation, while (c) represents charge transfer in which electron is captured by the incident ion B. At intermediate energies the cross sections for all three types of reaction are usually comparable, but at high energies the charge exchange cross section decreases rapidly (at first like v^{-12} and ultimately like v^{-11}) compared with those for excitation or ionisation (v^{-2} or $(\log v)v^{-2}$).

The time-dependent Schrödinger equation which determines the electronic wave function is

$$\left[-\frac{1}{2}\nabla^2_{r_A} + V_A(r_A) + V_B(r_B) + W_{AB}(R) - i\frac{\partial}{\partial t}\right]\Psi(\underline{r}_A,t) = 0 \qquad 1.3$$

where (see Fig. 1) $\underline{r}_A$ and $\underline{r}_B$ are the position vectors of the electron

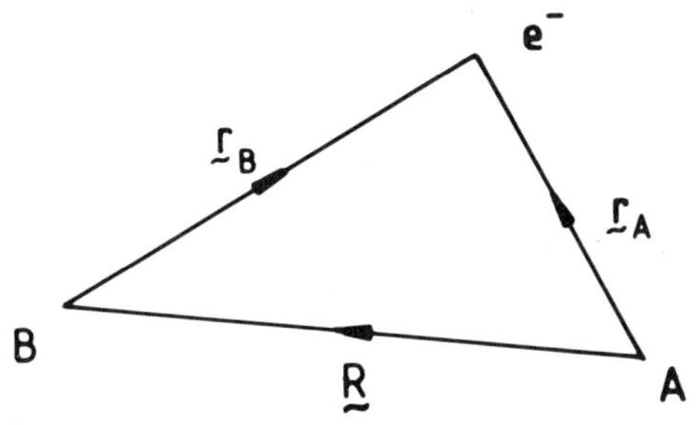

Fig. 1 The coordinate system

with respect to A and B, W_{AB} is the effective potential between the ionic cores A and B, and V_A, V_B are the effective potentials between the electron and A and B respectively. The origin of coordinates will be taken at A, and the derivative $\partial/\partial t$ is performed with the independent variable $\underline{r}_A$ held fixed. The variable $\underline{r}_B$ depends both on $\underline{r}_A$ and $\underline{t}$, since $\underline{r}_B = -\underline{R}(t) + \underline{r}_A$ and the time dependence of $\underline{R}$ is given by (1.1). As discussed in detail by McCarroll,[1] or by Bransden,[2] equivalent results can be obtained by any choice of origin along the level AB and the theory is translationally invariant.

Since the relative motion of A and B is predetermined through (1.1), the probability of an electronic transition cannot depend on the inter-ionic potential W_{AB} and, indeed, this can be removed by the phase transformation $\Psi(\underline{r}_A, t) = \psi(\underline{r}_A, t) \exp\left[-i \int_0^t W_{AB}(R)\, dt\right]$, where $\psi(\underline{r}_A, t)$ satisfies

$$\left[-\frac{1}{2}\nabla^2_{r_A} + V_A(r_A) + V_B(r_B) - i\frac{\partial}{\partial t}\right]\psi(r_A, t) = 0 \qquad 1.4$$

It should be noted, however, that it is necessary to take into account W_{AB} when angular distributions are calculated, as we shall discuss briefly below.

When A and B are widely separated and $|t| \to \infty$, the unperturbed solutions of (1.4) are

$$F_n^{\pm}(\underline{r}_A, t) = \phi_n^A(\underline{r}_A) \exp\left[-i\epsilon_n^A t \mp i\sigma(\pm t)\right]$$

$$G_j^{\pm}(\underline{r}_A, t) = \phi_j^B(\underline{r}_B) \exp\left[-i\epsilon_j^B t + i(\underline{v}.\underline{r}_A - \tfrac{1}{2}v^2 t) \mp i\eta(\pm t)\right] . \qquad 1.5$$

The functions $\phi_n^A(\underline{r}_A)$, $\phi_j^B(\underline{r}_B)$ are the wave functions of the isolated atoms ($A+e^-$) and ($B+e^-$) with corresponding eigenenergies $\epsilon_n^A, \epsilon_j^B$. The <u>translational factor</u> $\exp i(\underline{v}.\underline{r}_A - \frac{1}{2}v^2 t)$ allows for the fact that the electron attached to centre B is travelling with a velocity $\underline{v}$ with respect to the chosen origin situated at A. Finally the phases $\sigma(t) = (Z_B/v)\log(R-vt)$ and $\eta(t) = (Z_A/v)\log(R-vt)$ take into account the long range overall Coulomb interactions in the arrangements B + ($A+e^-$) and ($B+e^-$) + A respectively.[4,5] The channel functions $F_n^{\pm}$ and $G_j^{\pm}$ satisfy the equations

$$\left[-\frac{1}{2}\nabla^2_{r_A} + V_A - \frac{Z_B}{R} - i\frac{\partial}{\partial t}\right]F_n^{\pm}(\underline{r}_A, t) = 0 \qquad (a)$$

$$\left[-\frac{1}{2}\nabla^2_{r_A} + V_B - \frac{Z_A}{R} - i\frac{\partial}{\partial t}\right]G_j^{\pm}(\underline{r}_A, t) = 0 \qquad (b) \qquad 1.6$$

Let us consider the case in which before the collision the ion ($A+e^-$) was in the state labelled i. If $\psi_i^+(\underline{r}_A, t)$ is the corresponding solution of (1.4), the boundary condition satisfied by ψ^+ is

$$\ell t_{t\to-\infty}\ \psi_i^+ = F_i^+ \qquad 1.7$$

The probability amplitude $A_{ni}(\underline{b})$ for finding atom $(A+e^-)$ to be in an excited (or ionised) level n after the collision is then given by

$$A_{ni}(\underline{b}) = \ell t_{t\to\infty}\ \langle F_n^-|\psi_i^+\rangle \qquad 1.8a$$

where, explicitly,

$$\langle F_n^-|\psi_i^+\rangle = \int d\underline{r}_A\ F_n^{-*}(\underline{r}_A,t)\ \psi_i^+(\underline{r}_A,t) \qquad 1.8b$$

The corresponding cross section is found by integrating $|A_{ni}(\underline{b})|^2$ over all impact parameters

$$\sigma_{ni}^A(b) = \int d^2\underline{b}\ |A_{ni}(\underline{b})|^2 \qquad 1.9$$

Correspondingly, the probability amplitude for finding the electron captured into a level j of the ion $(B+e^-)$ is

$$C_{ji}(\underline{b}) = \ell t_{t\to\infty}\ \langle G_j^-|\psi_i^+\rangle \qquad 1.10$$

and the charge exchange cross section is

$$\sigma_{ji}^B(\underline{b}) = \int d^2\underline{b}\ |C_{ji}(\underline{b})|^2 \qquad 1.11$$

The corresponding angular distributions can be obtained by constructing the wave function of the three body system. Taking the relative motion of A and B to be described by the plane wave $\exp(i\underline{K}.\underline{R})$ where $K = \mu v$ and μ is the reduced mass, the total wave function of the system becomes

$$\Phi_i^+(\underline{r}_A,\underline{R}) = \exp(i\underline{K}.\underline{R})\Psi_i^+(\underline{r}_A,Z) \qquad 1.12$$

where $\underline{R}$ has components b_x, b_y and $Z = vt$. This wave function can now be employed in the quantum mechanical integral expression for the scattering amplitude. Following McCarroll and Salin,[6] it is found in the small angle approximation that the scattering amplitude for excitation can be expressed in terms of $A_{ni}(\underline{b})$ as

$$f_{ni}(\theta) = -\frac{i\mu v}{2\pi}\int d\underline{b}\, \exp\left[-i\underline{\eta}.\underline{b} - \frac{i}{v}\int_{-\infty}^{\infty} W_{AB}(R')dZ'\right]A_{ni}(\underline{b}) \tag{1.13}$$

where $R'^2 = b^2 + Z'^2$. A similar expression for the differential cross section for charge exchange can be obtained in terms of $C_{ji}(\underline{b})$. The vector $\underline{\eta}$ is in the direction of the X-axis and is of magnitude $\mu v \sin\theta$. Details of the derivation were discussed in McCarroll's lectures and will not be repeated here.

2. COUPLED CHANNEL MODELS

2.1 Pseudostate Basis Sets

A straightforward method of obtaining approximate solutions of the impact parameter equation (1.4) is to expand $\psi(\underline{r}_A, t)$ in a finite basis set. Such a set can be chosen in various ways. For example, at low energies where the collision is quasi-adiabatic an expansion in molecular orbitals is appropriate and this has been discussed by McCarroll.[1] In contrast, as high energies if we wish to calculate ionisation or excitation cross sections, the charge exchange channels can be neglected and an expansion in orbitals centred on the target A can be used. A finite discrete basis of functions $\bar{\phi}_n(\underline{r}_A)$, called psuedostate functions, can be obtained by diagonalising the target Hamiltonian in terms of a set of Slater orbitals, or some other suitable set of functions,

$$\langle\bar{\phi}_n(\underline{r}_A)|H_A|\bar{\phi}_m(\underline{r}_A)\rangle = \bar{\epsilon}_n^A\,\delta_{nm} \qquad n,m = 1,2,\ldots N \tag{2.1}$$

where

$$H_A = -\frac{1}{2}\nabla^2_{r_A} + V_A(\underline{r}_A) \tag{2.2}$$

If the basis is well chosen, the lowest lying members of the set $\bar{\phi}_m$ and the energies $\bar{\epsilon}^A$ will provide good approximations to the corresponding exact eigenfunctions ϕ_m^A and energies ϵ_m^A. Indeed for hydrogenic systems with a suitable Slater basis, this corespondance can be made exact for any desired number of low lying states. Of the remaining members of the set, some will be associated with negative values of the energies $\bar{\epsilon}_m^A$ and will represent the remaining bound states of $(A+e^-)$ while others will be associated with positive discrete energies $\bar{\epsilon}_m^A$ and these provide a discrete representation of the continuum states of $(A+e^-)$ which can not be employed directly. In fact it can be shown that over a limited range of the variable r_A, a positive energy pseudo-state differs from the continuum state $\phi^A(E,\underline{r}_A)$ at the same energy only by a normalisation factor[7,8]

$$\phi^A(\epsilon,\underline{r}_A)\Big|_{\epsilon=\bar{\epsilon}_m^A} \simeq N\,\bar{\phi}_m^A(r_A) \qquad 2.3$$

It can also be shown that integrals over the continuum can be represented as sums over the discrete pseudostates. For example if $f(H_A)$ is a non-singular function of H_A then the spectral representation of $f(H_A)$ is

$$\int_0^\infty dE\,|\phi_E^A\rangle\, f(E)\langle\phi_E^A| \simeq \sum_n |\bar{\phi}_n^A\rangle f(\bar{\epsilon}_n^A)\,|\bar{\phi}_n^A\rangle \qquad 2.4$$

where for simplicity the sum over the negative energy states of H_A has been omitted. The sum over n is in fact an integration rule[9] for integrals over E and in certain cases these integration rules can be shown to be of Gaussian type.[10]

In the finite basis spanned by the pseudostates, ψ_i^+ can be expressed as

$$\psi_i^+(\underline{r}_A,t) = \sum_{n=1}^{N} \bar{\phi}_n(r_A)\exp(-i\bar{\epsilon}_n^A t)a_{ni}(b,t) \qquad 2.5$$

The amplitudes $a_{ni}(bt)$ satisfy coupled equations found by inserting (2.5) into the Schrödinger equation (1.4) and projecting with each of the functions $\bar{\phi}_n(r_A)\exp(-i\bar{\epsilon}_n^A t)$ in turn. We find

$$i\,\dot{a}_{ni}(b,t) \;=\; \sum_m \langle\bar{\phi}_n^A|V_B|\bar{\phi}_m^A\rangle \exp[i(\bar{\epsilon}_n^A-\bar{\epsilon}_m^A)t]\; a_{mi}(\underline{b},t) \qquad 2.6$$

These equation can be integrated starting from an initial condition at $t = -t_o$, where t_o is some large value chosen so that for $|t| = t_o$ the coupling terms in (2.6) can be considered to be negligible to some final time $t = +t_o$.

$$a_{ni}(\underline{b},-t_o) \;=\; \delta_{ni} \qquad 2.7$$

It will be noticed that in forming equation (2.6), the Coulomb phase factors which occur in the exact asymptotic expressions (1.5) have been omitted. Since we are integrating over a finite (but large) range of t, the only effect of this omission is that the final values of the amplitude $a_n(\underline{b},+t_o)$ differ from the probability amplitudes $A_{ni}(\underline{b})$ defined by (1.8) by a phase factor (varying with t_o), which is of no significance when calculating total cross sections through (1.9)

$$|A_{ni}(\underline{b})| \;\simeq\; |a_{ni}(\underline{b},t_o)| \qquad 2.8$$

By ensuring that the initial and final states of the atom $(A+e^-)$ are well represented by the basis, the excitation cross sections can be calculated directly. This is not the case for ionisation. Since probability is conserved within the discrete basis by the coupled channel approximation, that is

$$\sum_n |A_{ni}(\underline{b})|^2 \;=\; 1 \qquad 2.9$$

a simple prescription for the ionisation cross section is to subtract from the total cross section the cross sections for the excitation of all the negative energy channels. This method can be improved by calculating the overlap of the discrete basis with the true continuum states and applying a correction factor.

At sufficiently high energies, the first Born approximation is expected to be accurate, in which case $a_m(\underline{b},t)$ on the right hand side of (2.6) can be replaced by the unperturbed value of δ_{mi}. At lower energies the single centre expansion method produces accurate cross sections for single particle excitation or ionisation, down to energies simple example of the n=2 excitation of atomic hydrogen by proton impact,[11] the first Born approximation is accurate above an impact energy of ~ 300 keV, and the single centre expansion model begins to fail below 60 keV when charge exchange becomes important. For systems in which the target electron is more strongly bound, for instance in the K shell excitation of heavier atoms, the energy at which the Born approximation is valid becomes larger in proportion to the larger binding energy. No <u>finite</u> discrete single centre basis can represent the rearranged system $(B+e^-) + A$ as $t \to \infty$, and hence this model is bound to fail at the lower energies. However when charge exchange is important, an expansion about the centre B in terms of pseudostates $\bar{\phi}_n^B$ which result from a diagonalisation of the Hamiltonian H_B

$$H_B = -\frac{1}{2}\nabla^2_{r_B} + V_B , \qquad 2.10$$

can be added to the expansion about the centre A. These additional terms allow charge exchange probabilities to be calculated and also represent the influence of the charge exchange channels on excitation or ionisation. The augmented expansion is

$$\psi_i^+(\underline{r}_A, t) = \sum_{n=1}^{M} \chi_n(r_A, t)\, a_{ni}(\underline{b}, t) \qquad 2.11$$

where

$$\chi_n(\underline{r}_A, t) = \begin{cases} \bar{\phi}_n^A(\underline{r}_A)\exp(-i\bar{\epsilon}_n^A t) & n \leq N \\ \bar{\phi}_{n-N}^B(\underline{r}_B)\exp\left[-i\bar{\epsilon}_{n-N}^B t + i(\underline{v}\cdot\underline{r}_A - \tfrac{1}{2}v^2 t)\right] & n \geq N \end{cases}$$

2.12

The amplitudes $a_n(\underline{b}, t)$ for $1 \leq n \leq N$ represent direct excitation and ionisation, while those for $N < n \leq M$ represent capture into discrete states, as well as ionisation in which the motion of the ejected electron is centred on the ionic core B. This latter process, called charge exchange into the continuum, contributes to the ionisation process and cannot in principle be distinguished from 'direct' ionisation, however at high energies the concept has been useful in explaining the spectrum of electrons ejected in the forward direction, particularly at high energies.

By inserting (2.10) into the Schrödinger equation (1.4) and

projecting with the functions χ_n, the amplitudes $a_n(\underline{b},t)$ are found to satisfy

$$\sum_{n=1}^{M} [N_{jn}\, \dot{a}_{ni}(\underline{b},t) - M_{jn}\, a_{ni}(\underline{b},t)] = 0 \qquad 2.13$$

where N_{jn} is an 'overlap' matrix element

$$N_{jn} = \langle \chi_j | \chi_n \rangle \qquad 2.14$$

and M_{jn} is a 'potential' matrix element

$$M_{jn} = \langle \chi_j | -\frac{1}{2}\nabla^2_{r_A} + V_A + V_B - i\frac{\partial}{\partial t} | \chi_n \rangle \qquad 2.15$$

The boundary condition is again expressed by (2.7) and the probabilities for excitation or charge exchange are approximated by

$$\begin{aligned} |A_{ni}(\underline{b})| &\simeq |a_{ni}(\underline{b},t_o)| \quad n \leq N \\ |C_{ni}(\underline{b})| &\simeq |a_{ni}(\underline{b},t_o)| \quad n > N \end{aligned} \qquad 2.16$$

Direct ionisation or charge exchange into the continuum can be calculated from the probabilities for exciting the positive energy pseudostates, as discussed above.

In general, to obtain converged solutions for the most important reaction channels the basis set may have to be rather large, perhaps containing upwards of sixty pseudostates distributed between the two centres. This leads to a major computing task, the exchange matrix elements being particularly lengthy to evaluate,[12] and it follows that it is important to use simplified methods, if possible.

One such method has been developed by Reading et al.[13] for use in those cases in which it is desired to compute excitation or ionisation cross sections in circumstances in which the charge exchange channels are less important. In this method, termed the one and a half centred expansion (OHCE) a large basis about the target A is employed, but only one or a few terms are retained in the expansion about B and, in addition, for these terms the probability amplitude $a_{ni}(\underline{b},t)$ are given a predetermined form:

$$a_n(\underline{b},t) = a_{ni}(\underline{b},\infty)\, Q_n(t) \qquad n > N \qquad 2.17$$

where $Q_n(t) \to 0$ as $t \to -\infty$, $Q_n(t) \to 1$ as $t \to \infty$. The functions $Q_n(t)$ are chosen to increase from 0 to 1 rapidly near the point of closest approach at t = 0. The differential equations satisfied by the amplitude $a_{ni}(\underline{b},t)$, $n \leq N$ are now coupled to a few algebraic equations for the constant amplitudes $a_{ni}(\underline{b},\infty)$ for $n > N$. Only a few of the tine consuming exchange matrix elements need to be computed and a very large basis set can be employed about the target. This model is particularly effective at the higher energies for which charge exchange is improbable, but naturally it cannot be expected to be accurate in the low energy region.

2.2 Alternative Basis Sets

Pseudostates formed by diagonalising the target or projectile Hamiltonians are convenient, but are not the only choice of basis by any means. Alternatives are to employ sets of Gaussian functions[14] centred on A and B which have the advantage that the potential and overlap matrix elements can be evaluated analytically. The chief disadvantage is that it often requires a rather large number of Gaussian functions to represent the initial and final channel wave functions, so that usually many more coupled equations must be solved. In principle any complete set of functions can be used to represent ψ_i^+ approximately. One set which has proved useful is the Sturmian set originally used for ion-atom collisions by Gallaher and Wilets and more recently exploited by Winter.[16] In this case the expansion functions $\bar{\phi}_n^A$ or $\bar{\phi}_n^B$ are represented as

$$\bar{\phi}_n = r^{-1}\, S_{n\ell}(r)\, Y_{\ell m}(\theta,\phi) \qquad 2.18$$

where the radial functions $S_{n\ell}$ satisfy

$$\left[-\frac{1}{2}\frac{d^2}{dr^2} + \frac{\ell(\ell+1)}{r^2} - \frac{\alpha_{n\ell}}{r}\right] S_{n\ell}(r) = E_1\, S_{n\ell}(r) \qquad 2.19$$

In (2.19) E_1 is a fixed parameter and $\alpha_{n\ell}$ acts as an eigenvalue, for the boundary condition $S_{n\ell}(o) = 0$, $S_{n\ell}(r) \to 0 \quad r \to \infty$. The attraction of this approach is that the $S_{n\ell}(r)$ form a complete set which is entirely discrete with no continuum. The only disadvantage is that a fairly large number of terms may be necessary to represent the actual physical channel functions so that it can be difficult to extract the physical amplitudes.

It should be noticed that further flexibility can be introduced into the approximate wave function, by adding to the two centre expansion terms centred on some other point, for example the centre of charge.[17,18] This procedure can be helpful in obtaining improved ionisation cross sections, and may be more effective than adding additional terms to the expansions about A and B.

2.3 Illustrative Examples

Out of the considerable number of coupled channel calculations, which have been performed, some results for the two systems p + Li^{2+} and p + He^{+} will be chosen for illustrative purposes. In both systems the interaction potentials are purely Coulombic so that there are no

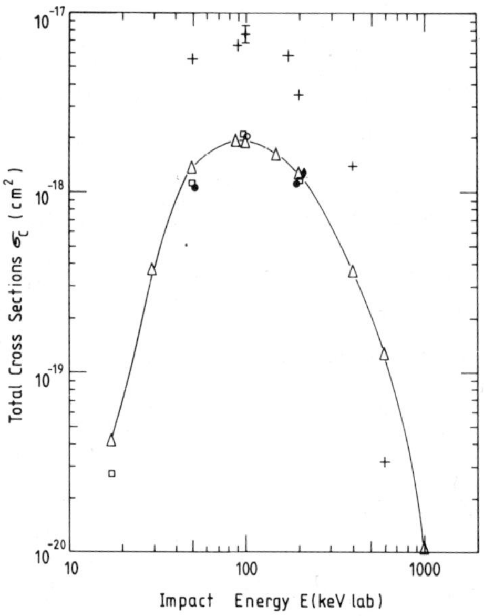

Fig. 2 Theoretical total cross sections for electron capture by protons from Li^{2+}. Δ 32 pseudostate two centre basis.[19] O OHCE model with large target centred basis.[20] □ Up to 38 Sturmian two centre basis.[16] ♦ Continuum distorted wave cross section (see Section 3). + Classical CTMC cross sections.[19]

differences between various calculations resulting from the introduction of differing effective potentials. In Fig. 2, various coupled channel results are shown for electron capture by protons from Li^{2+}. Good agreement is found between the two centre expansions based on pseudostates and on Sturmians, and also between those results and those of the OHCE model. This is a very favourable case for the OHCE model since capture is predominantly into a single state - the ground state of H. Also included are the classical Monte-Carlo cross sections, which

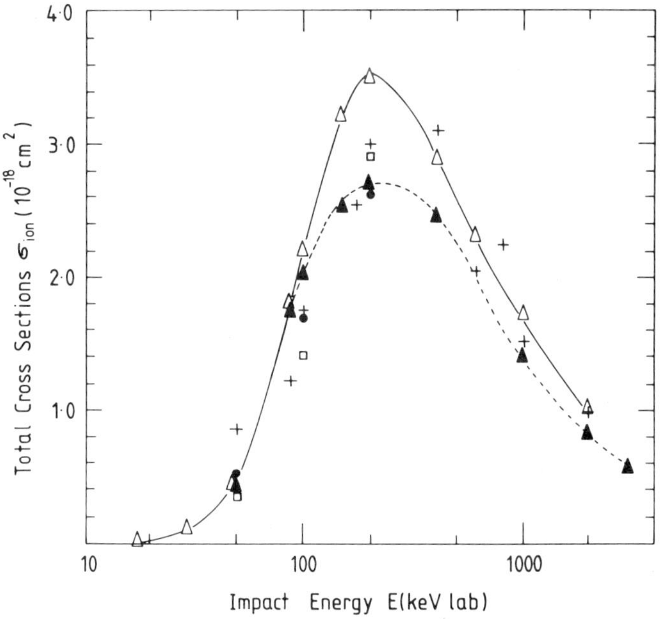

Fig. 3. Theoretical cross sections for the ionisation of Li^{2+} by proton impact. Notation as in Fig. 2 together with Δ 59 pseudostate centre basis.

for this system are rather large compared with those of the coupled channel model. Fig. 3 shows the coresponding cross section for ionisation. In this case tho classical results are closer to the quantal results, presumably because the classical representation of the ionised state is more accurate than a classical representation of the bound final state.

Figs. 4 and 5 show corresponding capture and ionisation cross sections for the p + He^{+} case, for which experimental data is also

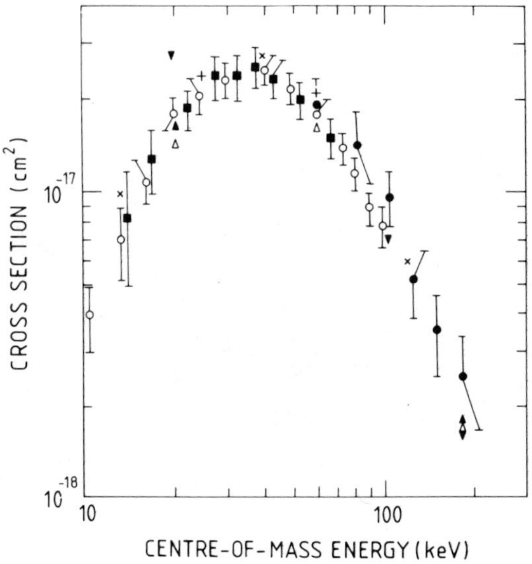

Fig. 4. Total cross sections for electron capture by protons from He^+. Theoretical cross sections: Δ,▲ Different sets of 23 two centred pseudostates.[21] ▽ OHCE model with large target centred basis.[20] □,+ 35 and 19-24 two centred Sturmian basis respectively.[16] Other coupled pseudostate calculations by Ftitsch and Lin[22] agree with the experimental data and are not shown for clarity. Experimental : ● Angel et al.[23] ■ Peart et al.[24] ○ Rinn et al.[25] At 60 kev, the pseudostate result ▲ and the OHCE result ▼ coincide with the Sturmian value + and have been omitted for clarity.

available. The agreement between the theoretical results and between theory and experiment is again good, with the exception that the OHCE model cross sections are too large at the lowest energy (20 kev). However as we have already stated, the OHCE model would not be expected to be accurate at the lower energies, so this is not a serious defect.

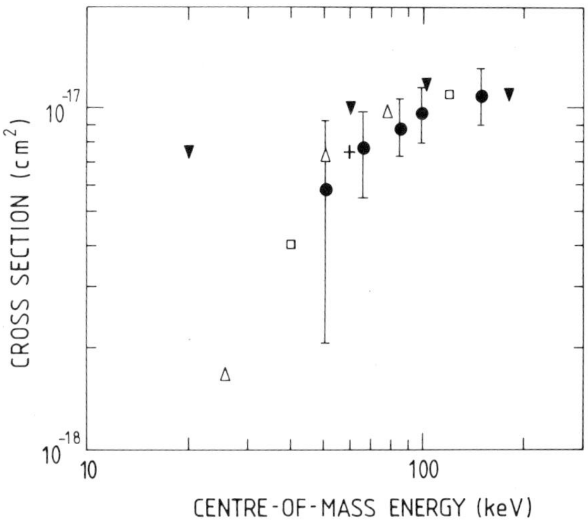

Fig. 5. Cross sections for ionisation of He^+ by proton impact. Notation as in Fig. 4 with the exception Δ theoretical results with a pseudostate basis calculated by Fritsch and Lin.[26] ⧳ Experimental data of Watts et al.[27]

The general conclusion, based on many coupled channel calculations of single electron systems (or of systems with a single active electron) is that over a very wide range of impact energies extending from ~ 1 kev per amu to several MeV per amu the coupled channel method can produce reliable cross sections for excitation and charge exchange and (with less accuracy) for ionisation. However, depending on the system the basis sets may have to be quite large with up to ~ 100 terms for convergence.

2.4 Optical Potentials

Let us consider again the case of excitation under conditions in which the charge exchange channels can be omitted. Expanding the wave function $\psi(\underline{r}_A, t)$ in the complete set of target eigenfunctions ϕ_n^A, the coupled equations analogous to (2.6) for the expansion coefficients $a_{ni}(\underline{b}, t)$ are

$$i\, \dot{a}_{ni}(\underline{b}, t) = \sum_m M_{nm}(t)\, a_{mi}(\underline{b}, t) \qquad 2.20$$

where

$$M_{nm}(t) = \langle \phi_n^A | V_B(t) | \phi_m^A \rangle \exp\left[i(\epsilon_n^A - \epsilon_m^A)t\right] \qquad 2.21$$

This is an infinite set of equations which must be approximated in practice. The idea of the optical potential method is to form a <u>finite</u> set of equations connecting a group of channels which contain the initial and final states of interest and others coupled strongly with those channels. Let this particular group of channels be labelled by the index n with $1 \leq n \leq N$ and introduce a projection operator P so that

$$\begin{aligned} P\, a_{ni} &= a_{ni} \quad n \leq N\ ; & P\, M_{nm} &= M_{nm} \quad n < m \\ &= 0 \quad n > N & &= 0 \quad n > m \end{aligned} \qquad 2.22$$

Then if Q = 1-P, the set of equations (2.20) can be written using a matrix notation as

$$iP\dot{\underline{a}} = P\underline{M}P\underline{a} + P\underline{M}Q\underline{a} \qquad 2.23a$$

$$iQ\dot{\underline{a}} = Q\underline{M}Q\underline{a} + Q\underline{M}P\underline{a} \qquad 2.23b$$

Substituting the solution of (2.23b) into (2.23a), we find a finite set of coupled equations for the amplitudes a_n, $n \leq N$

$$i\, P\dot{\underline{a}} = \left[P\underline{M}P + P\underline{M}Q \frac{1}{Q[i \frac{\partial}{\partial t} - M]Q} Q\underline{M}P \right] \underline{a} \qquad 2.24$$

The second term on the right hand side is a matrix optical potential. The optical potential contains the whole complexity of the problem and, of course, must be approximated in practice. By expanding $\left[Q\left(i \frac{\partial}{\partial t} - \underline{M}\right)Q\right]^{-1}$ in series of powers of $\underline{M}$, we find a sum of time ordered products

$$P\underline{M}Q \frac{1}{Q[i \frac{\partial}{\partial t} - \underline{M}]Q} Q\underline{M}P = -iP\underline{M}Q\, \theta(t-t')\underline{a}\underline{M}P$$

$$+ (-i)^2 P\underline{M}Q\, \theta(t-t')Q\underline{M}\underline{Q}\, \theta(t'-t'')Q\underline{M}P$$

$$+ \ldots \qquad . \qquad 2.25$$

If the leading term is retained, we obtain the second-order model of Bransden and Coleman.[28] Explicitly we find

$$i\, \dot{a}_{ni}(t) = \sum_{m=1}^{N} \left[M_{nm}(t) + \sum_{q>N} \int_{-\infty}^{t} dt'\, M_{nq}(t)\, M_{qm}(t') \right] a_{mi}(t')$$

$$n = 1,2,\ldots N \qquad 2.26$$

The sum over intermediate states can in principle be achieved numerically, or approximated by using a pseudo-state expansion. An alternative approach is to use a closure approximation. In one version of this approach, the intermediate energies ϵ_q^A for $q > N$ are replaced by some fixed 'average' value $\bar{\epsilon}$. We then have

$$\sum_{q>N} M_{nq}(t)\, M_{qm}(t') = \sum_{q>N} \langle \phi_n^A | V_B(t) | \phi_q^A \rangle \langle \phi_q^A | V_B(t') | \phi_m^A \rangle \; x$$

$$x \exp i \left[(\bar{\epsilon} - \epsilon_m^A) t' + (\epsilon_n^A - \bar{\epsilon}) t \right] \qquad 2.27$$

Setting $\sum\limits_{q>N} = \sum\limits_{(\text{all } q)} - \sum\limits_{q=1}^{N}$

the second finite sum can be evaluated explicitly, while the first sum can be evaluated using the closure property of the functions ϕ^A, since

$$\sum_{(\text{all } q)} \langle\phi_n^A|V_B(t)|\phi_q^A\rangle\langle\phi_q^A|V_B(t')|\phi_m^A\rangle = \langle\phi_n^A|V_B(t)V_B(t')|\phi_m^A\rangle \qquad 2.28$$

This approach has been used to discuss excitation of atomic hydrogen by proton (and electron) impact[29], using the simplest approximation in which just the initial and final stats (N=2) of these states together with two others (N=4) are treated explicitly in P space. Some other systems have also been treated in this approximation.[30] The results were encouraging, but large scale applications have not, as yet, been carried out.

2.5 Many Electron Systems - Correlation

A large number of reactions can be described accurately in the one-active electron approximation defined by eq. (1.4) in which the active electron moves in the effective potentials V_A and V_B. The coupled channel method provides accurate cross sections in these cases provided a basis set of adequate size is chosen. Similar calculations are feasible for several electron systems in which the electronic wave function is fully antisymmetrised providing only single particle excitations are important. For example, calculations have been reported for p + He,[31] Li + He^+,[32] p + H^-,[33] H + H[34], He^+ + He^+,[35] in which two-electron wave functions are used. The chief technical difficulty in these cases is the evaluation of the two-electron exchange integral describing exchange between electrons attached to different ionic centres. The corresponding integrals in the molecular orbit expansion with simplified translational factors are easier to compute, and systems of several electrons have frequently been investigated. Another approach which has been explored is to employ the time dependent Hartree-Fock approximation[36] in which the system is described at each time t by a single Slater determinant. However, major difficulties arise for systems in which correlation is very important. To describe a fully correlated two electron system by a pseudostate expansion requiring the excitation of both electrons to be taken into account would require a double pseudostate expansion of the order of $10^2 \times 10^2 = 10^4$ terms, which is impractical. Some years ago Reading et al.[37] introduced a new method to overcome this difficulty which they termed the Forced Impulse Method (FIM) which has proved very successful when applied to the highly correlated problem of determining the double ionisation cross section for proton impact on helium.[38]

To introduce the Forced Impulse Method, consider an N electron atom centred on a nucleus A which is excited by the impact of a proton B. The standard impact parameter approximation is used in which the projectile B moves along a straight line trajectory at constant velocity. In a zero-order approximation, all the N electrons can be taken to move in the same local potential U(r), which, for example, can be taken to be a local approximation to the Hartree Fock potential. The single electron wave functions $\psi(\underline{r},t)$ satisfy the Schrödinger equation (cf. 1.4)

$$\left[-\frac{1}{2}\nabla_r^2 - \frac{Z_B}{|\underline{R}-\underline{r}|} + U(r)\right]\psi(\underline{r},t) = i\frac{\partial}{\partial t}\psi(\underline{r},t) \qquad 2.29$$

where the projectile has been taken to a fully stripped ion of charge Z_B. By diagonalising the Hamiltonian $\left[-\frac{1}{2}\nabla_r^2 + U(r)\right]$ a finite discrete set of pseudostates $\bar{\phi}_i^A$ can be introduced, i = 1,2,...M, and the one-electron problem represented by (2.29) can be solved, as explained in Section 2.1, to obtain the expansion coefficients $a_{ni}(\underline{b},t)$ satisfying the boundary conditions (2.5). Since the basis set is orthonormal (charge exchange which employs a non-orthogonal basis is not being considered). The coefficients $a_{ni}(\underline{b},t)$ form a M x M dimensional unitary matrix for each t. In fact we can identify this matrix with the evolution matrix $U_{ni}(t,-t_o)$ which transforms the wave function at time $-t_o$ into that at time t. We have

$$U_{ni}(t,-t_o) \equiv a_{ni}(\underline{b},t) \qquad 2.30$$

where the dependence of U_{ni} on $\underline{b}$ is not displayed explicitly. The evolution matrix has the group property

$$U_{ni}(t,t) = \delta_{ni}\,, \qquad U_{ni}(t,t') = \sum_j U_{nj}(t,t'')\,U_{ji}(t'',t') \qquad 2.31$$

The N-electron target wavefunctions $\Phi_\alpha(\underline{r}_1,\ldots\underline{r}_N)$ in the common potential approximation satisfy the Schrödinger equation

$$\sum_{i=1}^{N}\left[-\frac{1}{2}\nabla^2_{r_i} + U(r_i)\right]\Phi_\alpha(\underline{r}_1..\underline{r}_N) = \epsilon_\alpha\,\Phi_\alpha(\underline{r}_1..\underline{r}_N) \tag{2.32}$$

It is clear that the functions Φ_α are the N dimensional Slater determinants that can be formed from the 2M spin-orbitals $\bar{\phi}_n(\underline{r}_i)\alpha_i$ and $\bar{\phi}_n(\underline{r}_i)\beta_i$ (n = 1,2,...M; i = 1,2,...N). The N-electron wavefunction $\Psi_\alpha(t)$ perturbed by the Coulomb interaction with the projectile B, satisfies

$$\sum_{i=1}^{N}\left[-\frac{1}{2}\nabla^2_{r_i} + U(r_i) - \frac{Z_B}{|\underline{R}(t)-\underline{r}_i|}\right]\Psi_\alpha(t) = i\,\frac{\partial\Psi_\alpha(t)}{\partial t} \tag{2.33}$$

and can be expanded in the basis set Φ_α:

$$\Psi_\alpha(t) = \sum_{a'} U^N_{\alpha'\alpha}(t,-t_o)\Phi_{\alpha'}(\underline{r}_1,\underline{r}_2,...\underline{r}_N)\exp(-i\,\epsilon_\alpha t) \tag{2.34}$$

with the boundary conditions

$$U^N_{\alpha'\alpha}(-t_o,-t_o) = \delta_{\alpha'\alpha} \tag{2.35}$$

Since the Hamiltonian is completely separable, the evolution matrix is the product of the single electron U-matrices:

$$U^N_{\alpha'\alpha}(t,-t_o) = U_{ik}(t,-t_o) \times U_{jn}(t,-t_o) \tag{2.36}$$

where $\alpha' \equiv ij$; $\alpha \equiv kn$. This solves the uncorrelated common potential problem completely. The true target Hamiltonian differs from the model single particle Hamiltonian H_{sp} by the potential term V where

$$V(\underline{r}_1,\underline{r}_2,...\underline{r}_N) = H_A - H_{sp}$$

$$= \sum_{i=1}^{N}\left[\frac{Z_A}{r_i} - U(r_i)\right] + \sum_{i,j}\frac{1}{|\underline{r}_i-\underline{r}_j|} \tag{2.37}$$

Approximations to the true target eigenfunctions $\chi(\underline{r}_1,\underline{r}_2,\ldots\underline{r}_N)$ which satisfy

$$H_A\chi_\beta(\underline{r}_1,\underline{r}_2,\ldots\underline{r}_N) = E_\beta\chi_\beta(\underline{r}_1,\underline{r}_2,\ldots\underline{r}_N) \tag{2.38}$$

can be found by diagonalizing H in the discrete basis of uncorrelated functions $\Phi_\alpha(\underline{r}_1,\underline{r}_2,\ldots\underline{r}_N)$. This determines the coefficients $a_{\beta\alpha}(-t_o)$, where

$$\chi_\beta(\underline{r}_1,\underline{r}_2,\ldots\underline{r}_N) = \sum_\alpha a_{\beta\alpha}(-t_o)\Phi_\alpha(\underline{r}_1,\underline{r}_2,\ldots\underline{r}_N) \tag{2.39}$$

The wave function $\Psi^c_\beta(t)$ for the system composed of the projectile B interacting with the atom A satisfies

$$\left[H_A + \sum_{i=1}^{N} \frac{Z_B}{|R(t)-\underline{r}_i|}\right]\Psi^c_\beta(t) = i\frac{\partial}{\partial t}\Psi^c_\beta(t) \tag{2.40}$$

with the boundary condition

$$\Psi^c_\beta(-t_o) = \chi_\beta \exp(+iE_\beta t_o) \tag{2.41}$$

The wave function $\Psi^c_\beta(t)$ at any time t can always be expressed as a superposition of the uncorrelated functions $\Psi_\alpha(t)$:

$$\begin{aligned}\Psi^c_\beta(t) &= \sum_\alpha a_{\beta\alpha}(t)\exp[i(\epsilon_\alpha-E_\beta)t]\Psi_\alpha(t)\\ &= \sum_\alpha a_{\beta\alpha}(t)\left[\sum_{\alpha'} U^N_{\alpha'\alpha}(t,-t_o)\exp[i(\epsilon_\alpha-\epsilon_{\alpha'}-E_\beta)t]\Phi_{\alpha'}(\underline{r}_i-\underline{r}_N)\right]\end{aligned} \tag{2.42}$$

The idea of the impulse approximation is that if a projectile is rapid, the target electrons do not have time to interact with one another during the collision. The impulse approximation can be forced to be accurate by using it only over a small time interval. In particular if the time interval from $-t_o$ to t is sufficiently small, the system must evolve without the correlating interaction V acting, so that the basic approximation can be made

$$a_{\beta\alpha}(t) \simeq a_{\beta\alpha}(-t_o) \qquad 2.43$$

The evolution matrix $U^c_{\beta'\beta}$ between fully correlated states satisfies

$$\begin{aligned} U^c_{\beta'\beta}(t,-t_o) &= \langle \chi_{\beta'} | \Psi^c_{\beta}(t) \rangle \exp(i\, E_{\beta'} t) \\ &= \delta_{\beta\beta'} + \int_{-t_o}^{t} \frac{\partial}{\partial t} \left[\langle \chi_{\beta'} | \Psi^c_{\beta}(t) \rangle \exp(iE_{\beta'} t) \right] \end{aligned} \qquad 2.44$$

Substituting the expression (2.42) into (2.44) and employing (2.41) together with (2.43) the forced impulse approximation to $U^c_{\beta'\beta}$ becomes

$$U^c_{\beta'\beta}(t,-t_o) = \delta_{\beta\beta'} + \sum_{\alpha,\alpha'} a_{\beta\alpha}(-t_o)\, a^*_{\beta'\alpha'}(-t_o) \int_{-t_o}^{t} \frac{\partial}{\partial t} \left[U^N_{\alpha'\alpha}(t,-t_o) e^{i\Lambda t} \right] \qquad 2.45$$

where $\Lambda = \epsilon_{\alpha} - \epsilon_{\alpha'} + E_{\beta'} - E_{\beta}$.

The approximation represented by (2.45) can be applied over any time interval $t_1 < t < t_2$ provided $(t_2 - t_1)$ is small enough for the impulse approximation to be sufficiently accurate. In their application to the single and double ionisation of helium, Reading and Ford split the total time interval from $-t_o$ to t_o into two, and used the group property to write

$$U^c_{\beta'\beta}(t_o,-t_o) = \sum_{\beta''} U^c_{\beta'\beta''}(t_o,0)\, U^c_{\beta''\beta}(0,-t_o) \qquad 2.46$$

The system is correlated at the times $-t_o$, 0 and t_o.

Although this represents a formidable numerical calculation, unfortunately it is not the end of the affair. The calculation of the small cross section for double ionisation in the presence of a large single ionisation cross section is difficult because the positive energy two particle pseudostates cannot be assigned to one channel or the other. To overcome this Reading and Ford constructed a complete set of single and double ionisation channel functions at each two particle pseudo-state energy. Each of these pseudostates was then re-expressed in terms of the channel functions which enables a definite division of the wave function into single and double ionised channels to be made.

In their numerical calculations for $p^{\pm}$ and α-particles ionising helium, Reading and Ford used 36 s and p single electron pseudostate orbitals which lead to 423 two electron correlated states. The calculated ratio R for double to single ionisation showed the characteristics of the experiment, but were some 35% smaller than the observed values. The particular case of α + He is shown in Fig. 6, scaled by 35% to agree with experimental high energy values. It should be noted that the independent particle model, making no allowance for correlation, fails to agree with the experimental data at both high and low energies.

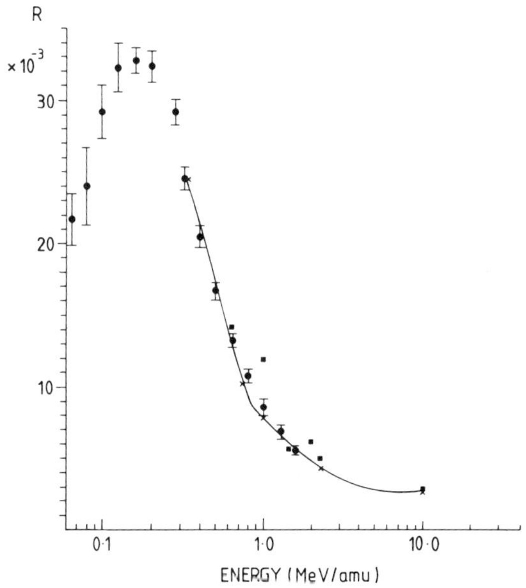

Fig. 6 The ratio R between double and single ionisation for α + He collisions. The solid curve represents the calculations of Reading and Ford[38] scaled upwards by 35% to agree with the high energy data. Experimental data: □ Knudsen et al.[39] ⊥ Shah and Gilbody.[20]

3. HIGH ENERGY MODELS

3.1 Perturbation Series

Excitation and ionisation. At high energies it is natural to look for a Born series from which approximations to these probability amplitudes $A_{ni}(\underline{b})$ or $C_{ji}(\underline{b})$ (see 1.8 and 1.10) can be obtained. This is straightforward in the case of excitation or ionisation provided there is no need to represent the charge exchange channels explicitly, which is possible at high energies. In this case, for the reaction $B + (A+e^-) \rightarrow B + (A+e^-)$ there is an unambiguous separation between the Hamiltonian of the target H_A and the perturbation $(V_B + W_{AB})$. If we consider the case in which the target atom A is overall neutral $(Z_A = 1)$, then the boundary conditions on Ψ are

$$\ell t_{t\to-\infty} \Psi_i^+(\underline{r}_A, t) = \phi_i^A(\underline{r}_A)\exp(-i\epsilon_i^A t) \equiv \Phi_i^A(\underline{r}_A, t) \qquad 3.1$$

$$\ell t_{t\to+\infty} \Psi_j^-(\underline{r}_A, t) = \phi_j^A(\underline{r}_A)\exp(-i\epsilon_j^A, t) \equiv \Phi_j^A(\underline{r}_A, t) \qquad 3.2$$

The solutions $\psi_j^-(\underline{r}_A, t)$ and $\psi_j^+(\underline{r}_A, t)$ are related by $\psi_j^-(\underline{v}, \underline{r}_A, t) = \psi_j^{+*}(-\underline{v}, \underline{r}_A, t)$ where the dependence on $\underline{v}$ has been shown explicitly. The Born series for $\Psi_{i,j}^{\pm}$ are easily found to be

$$\Psi_i^+(\underline{r}_A, t) = \left[1 + \mathcal{G}_A^+(V_B+W_{AB}) + \mathcal{G}_A^+(V_A+W_{AB})\mathcal{G}_A^+(V_B+W_{AB})\ldots\right]\Phi_i(\underline{r}_A, t)$$

3.3

$$\Psi_j^-(\underline{r}_A, t) = \left[1 + \mathcal{G}_A^-(V_B+W_{AB}) + \mathcal{G}_A^-(V_B+W_{AB})\mathcal{G}_A^-(V_B+W_{AB})\ldots\right]\Phi_j(\underline{r}_A, t)$$

3.4

where

$$\mathcal{G}_A^+(\underline{r}_A, t; \underline{r}_A', t') = -i\,\theta(t-t') \sum_n \phi_n^A(\underline{r}_A)\, \phi_n^{A*}(\underline{r}_A')\exp[i\epsilon_n^A(t'-t)]$$

3.5

and

$$\mathscr{G}_A^-(\underline{r}_A,t;\underline{r}_A',t') = +i\,\theta(t'-t)\sum_n \phi_n^A(\underline{r}_A)\,\phi_n^{A*}(\underline{r}_A')\exp[i\epsilon_n^A(t'-t)] \tag{3.6}$$

The sums in (3.5) and (3.6) over n run over all the states of H_A, both discrete and continuous. The probability amplitude for exciting the atoms $(A+e^-)$ to a level j from an initial level i is

$$\bar{A}_{ji}(\underline{b}) = \lim_{t\to\infty} \langle\Phi_j^A|\Psi_i^+\rangle = -i\int_{-\infty}^{\infty} \langle\Phi_j^A|(V_B+W_{AB})|\Psi_i^+\rangle dt \qquad i \neq j \tag{3.7}$$

In the first Born approximation, we then find from (3.3) and (3.7), that

$$\bar{A}_{ji}(\underline{b}) = -i\int_{-\infty}^{\infty} \langle\phi_j^A|V_B|\phi_i^A\rangle \exp\left[i(\epsilon_j^A-\epsilon_i^A)t\right]dt \qquad i \neq j \tag{3.8}$$

where we have used the orthogonality of ϕ_j^A and ϕ_i^A to elliminate the term in W_{AB}. Had we chosen to elliminate W_{AB} from the outset by a phase transformation and to work from (1.4), the solution ψ_i^+ satisfying the boundary condition (1.7) would be given by the Coulomb-Born series

$$\psi_i^+(r_{\underline{A}},t) = \left[1 + g_A^+\left[V_B + \frac{Z_B}{R}\right] + g_A^+\left[V_B + \frac{Z_B}{R}\right]g_A^+\left[V_B + \frac{Z_B}{R}\right] + \ldots\right]F_i^+(r_{\underline{A}},t) \tag{3.9}$$

where the Coulomb-Born Green's function is

$$g_A^+(r_{\underline{A}},\underline{t};r_{\underline{A}}',t') = -i\theta(t-t')\sum_n F_n^-(r_{\underline{A}},t)F_n^{-*}(r_{\underline{A}}',t') \tag{3.10}$$

and $F_n^{\pm}$ is defined by (1.5).

The probability amplitude $A_{ji}(\underline{b})$ defined by (1.8), is, in the first order approximation,

$$A_{ji}(\underline{b}) = -i\int_{-\infty}^{\infty} \langle F_j^-|V_B + \frac{Z_B}{R}|F_i^+\rangle\, dt$$

$$= -i \int_{-\infty}^{\infty} \langle \phi_j^A | V_B | \phi_i^A \rangle \exp\left[i(\epsilon_j^A - \epsilon_i^B)t - i\{\sigma(t)+\sigma(-t)\}\right] dt \qquad 3.11$$

From the definition of $\sigma(t)$ we see that

$$\sigma(-t) + \sigma(t) = \frac{Z_B}{v} \{\log (R+rt) + \log (R-vt)\}$$

$$= \frac{Z_B}{v} \log (R^2-v^2t^2) = 2 \frac{Z_B}{v} \log(b) \qquad 3.12$$

so that

$$\exp[-i\{\sigma(-t) + \sigma(t)\}] = b^{-(2iZ_B/v)} \qquad 3.13$$

Comparing (3.11) and (3.8) and using (3.13), we find

$$A_{ji}(\underline{b}) = b^{-(2iZ_B/v)} \bar{A}_{ji}(b) \qquad 3.14$$

and

$$|A_{ji}(b)| = |\bar{A}_{ji}(b)| \qquad 3.15$$

It follows that in the impact parameter model the first order approximation to the total cross section for excitation is the same whether the Coulomb boundary conditions are satisfied, or not. We emphasise this point because the corresponding result is not correct for the case of charge exchange, as we shall see shortly.

For the excitation of optically allowed transitions, the first Born approximation becomes accurate when the relative velocity of the colliding ions exceeds the Bohr velocity of the target electron by a factor of 4 or 5, while for optically forbidden transitions the final Born approximation is not always accurate until much higher impact velocities.[2,41] The second Born approximation does not always represent an improvement and indeed can provide a <u>less</u> accurate cross section than the first Born approximation. This is well illustrated in some results of Bransden, Dewangan and Noble[42] who examined higher Born approximations to the cross section for excitation of the n=2 levels of hydrogen by proton impact. This was achieved by expanding the Green's function g_A in a pseudo-state basis, chosen so that the exact numerical solution of the coupled channel model using the same basis provided good agreement with the experimental data. It was also shown[43] that the exact second Born approximation was well represented by the pseudostate expansion of g_A. The results are shown in Table 1, where it is seen clearly that the second Born approximation is less accurate than the first over the energy range considered. Calculation of higher order terms indicates that the Born series for excitation converges, but slowly, agreement with the exact coupled channel result being obtained when 4 or 5 terms are retained in the Born series.

Charge exchange. The most obvious and important difference between excitation and charge exchange[44] is that the perturbing potential V_B which appears in the expansion (3.3) of the wave function is also responsible for binding the electron in the atom $(B+e^-)$, and it follows that there is no clear cut distinction between binding potentials, which must be treated to all orders, and perturbing potentials. The same is, of course, true if we expand the wave functions in terms of V_A. For example, if we ask for a solution $\bar{\psi}_j^-(r_A,t)$ of 1.4 satisfying the boundary condition

Table 1 Cross sections for 2s and 2p excitation of atomic hydrogen by proton impact (units of πa_0^2)

E(keV)	75		105		145		200	
	2s	2p	2s	2p	2s	2p	2s	2p
C	0.19	1.01	0.13	0.93	0.09	0.81	0.06	0.68
B1	0.13	1.26	0.08	1.06	0.07	0.88	0.05	0.73
B2	0.37	1.37	0.20	1.12	0.14	0.92	0.09	0.74

C - results of a coupled channel calculation[43] agreeing closely with experiment.

B1,B2 - First and second Born approximations.

$$\ell t_{t\to\infty} \bar{\psi}_j^-(\underline{r}_A, t) = G_j^-(\underline{r}_A, t) \qquad 3.16$$

we obtain the series

$$\bar{\bar{\psi}}_j^-(\underline{r}_A, t) = \left[1 + g_B^-\left[V_A + \frac{Z_A}{R}\right] + g_B^-\left[V_A + \frac{Z_A}{R}\right] g_B^-\left[V_A + \frac{Z_A}{r}\right] + \ldots\right] \bar{\bar{\psi}}_j^-(\underline{r}_A', t') \qquad 3.17$$

in which powers of the potential V_A appear. The Green's function g_B^- is defined analogously to (3.10)

$$g_B^-(r_{\underline{A}}, t;\ r_{\underline{A}}', t') = +i\ \theta(t'-t) \sum_n G_n^+(r_{\underline{A}}, \underline{t}) G_n^{+*}(r_{\underline{A}}', t') \qquad 3.18$$

where the functions G_n^+ are defined by (1.5). The potential V_A, however, creates the initial bound state $\phi_i(\underline{r}_A)$, in which V_A acts to all orders. Another difficulty is that, if we are interested in excitation, the most important intermediate states in g_A^+ given by 3.10 are discrete states, and while the continuum is important, the results are not extremely sensitive to the continuum contribution which can be approximated in various ways. Since $\{\phi_n(\underline{r}_A)\}$ is a complete set, the charge exchange channels are represented among the intermediate states in g_A^+. However since the overlap $\langle G_j^{\pm}|F_n^{\pm}\rangle$ vanishes as $|t| \to \infty$ if j and n represent discrete states, the intemediate states concerned can only come from the sum over the continuum states in $\{\phi_n(\underline{r}_A)\}$ and any approximation may lead to a severe loss of information about the charge exchange channel. On the positive side, it should be noted that the problem we are trying to solve is not the complete three-body problem because the motion of the heavy particles A and B is predetermined. It follows that some of the difficulties with 'disconnected diagrams', which make the expansions of the Green's function in true three body systems diverge,[45] do not occur, although convergence is not guaranteed.

The probability amplitudes $C_{ji}(\underline{b})$ for charge exchange can be expressed in series form as follows.

$$C_{ji}(b) = \langle \bar{\bar{\psi}}_j^-(t)|\psi_i^+(t)\rangle = \underset{t\to\infty}{\ell t} \langle G_j^-(t)|\psi_-^+(t)\rangle \quad \text{(a)}$$

3.19

$$= \underset{t\to-\infty}{\ell t} \langle \bar{\bar{\psi}}_j^-(t)|F_i^+(t)\rangle \quad \text{(b)}$$

Since

$$\underset{t\to-\infty}{\ell t} \langle G_j^-(t)|\psi_i^+(t)\rangle = \underset{t\to-\infty}{\ell t} \langle G_j^-(t)|F_i^+(t)\rangle = 0 \qquad 3.20$$

we can write

$$C_{ji}(\underline{b}) = \int_{-\infty}^{\infty} \frac{\partial}{\partial t} \langle G_j^-(t)|\psi_i^+(t)\rangle dt$$

$$= -i \int_{-\infty}^{\infty} \langle G_j^-(t)|V_A + \frac{Z_A}{R}|\psi_i^+(t)\rangle dt \qquad 3.21$$

where we have used the fact that ψ_i^+ satisfies (1.4) and $G_j^-(t)$ satisfies (1.6b).

Similarly using (3.19b) we find the alternative form

$$C_{ji}(\underline{b}) = -i \int_{-\infty}^{\infty} \langle \bar{\bar{\psi}}_j^-(t)|V_B + \frac{Z_B}{R}|F_i^+(t)\rangle dt \qquad 3.22$$

Combining (3.9) with (3.21) and (3.17) with (3.22), we find two series expansions for the charge exchange probability amplitude $C_{ji}(\underline{b})$: the 'post' form ${}^{A}C_{ji}(\underline{b})$

$$C_{ji}(\underline{b}) = {}^{A}C_{ji}(\underline{b}) = -i \int_{-\infty}^{\infty} dt \, \langle G_j^-(t)| \left[V_A + \frac{Z_A}{R}\right] + \left[V_A + \frac{Z_A}{R}\right] g_A^+ \left[V_B + \frac{Z_B}{R}\right] +$$

$$+ \left[V_A + \frac{Z_A}{R}\right] g_A^+ \left[V_B + \frac{Z_B}{R}\right] g_A^+ \left[V_B + \frac{Z_B}{R}\right] + \ldots |F_i^+(t)\rangle$$

3.23

and the 'prior' form

$$C_{ji}(\underline{b}) = {}^{B}C_{ji}(\underline{b}) = -i \int_{-\infty}^{\infty} dt \, \langle G_j^-(t) | \left[V_B + \frac{Z_B}{R}\right] + \left[V_A + \frac{Z_A}{R}\right] g_B^+ \left[V_B + \frac{Z_B}{R}\right] +$$

$$+ \left[V_A + \frac{Z_A}{R}\right] g_B^+ \left[V_A + \frac{Z_A}{R}\right] g_B^+ \left[V_B + \frac{Z_B}{R}\right] + \ldots \; | F_i^+(t)\rangle$$

3.24

where we have used the result:

$$g_B^+ = g_B^{-\dagger}$$

The Nth order perturbation approximations to ${}^{A}C_{ji}$ and ${}^{B}C_{ji}$ are

$${}^{A}C_{ji}^{(N)}(b) = -i \int_{-\infty}^{\infty} \langle G_j^-(t) | \left[V_A + \frac{Z_A}{R}\right] \sum_{n=1}^{N} \left[g_A^+ \left[V_B + \frac{Z_B}{R}R\right]\right]^{n-1} | F_i^+(t)\rangle dt$$

3.25

and

$${}^{B}C_{i}^{(N)}(b) = -i \int_{-\infty}^{\infty} \langle G_j^-(t) | \sum_{n=1}^{N} \left[\left[V_A + \frac{Z_A}{R}\right] g_B^+\right]^{n-1} \left[V_B + \frac{Z_B}{R}\right] | F_i^+(t)\rangle dt$$

3.26

respectively.

Provided the internal wavefunctions ϕ_i^A, ϕ_j^B are exact it is clear that ${}^{A}C_{ji}^{(1)}(\underline{b}) = {}^{B}C_{j}^{(1)}(\underline{b})$ so there is no 'post-prior discrepancy' for the first order approximation. This is not the case for higher order approximations, and the question arises as to which series should be preferred. Clearly for nearly symmetrical collisions, in the sense that $\left[V_A + \frac{Z_A}{R}\right]$ and $\left[V_B + \frac{Z_B}{R}\right]$ are potentials of similar magnitude (an example is the symmetrical system $p + H \rightarrow H + p$), both series will converge equally well, or equally badly. On the other hand as pointed out by Macek and Shakeshaft[46] (who were discussing a slightly different series), if one of the perturbations is much stronger than the other, for example if $V_A \gg V_B$, then the sequence of approximations ${}^{A}C_{ji}^{(1)}$, ${}^{A}C_{ji}^{(2)}$, ..., should converge faster than the sequence ${}^{B}C_{ji}^{(1)}$, ${}^{B}C_{ji}^{(2)}$ ..., because ${}^{A}C_{ji}$ represents an expansion in powers of the weaker interaction, the stronger interaction being summed to all orders in the Green's function g_A^+. Likewise if $V_B >> V_A$, the approximations ${}^BC_{ji}^{(N)}$ are to be preferred. For situations in which $V_A \approx V_B$, some more symmetrical set of approximations is preferable,[47] and this can be most easily achieved by distorted wave expansions as we shall discuss later.

The series (3.23) and (3.24) are examples of Coulomb Born series in which Coulomb waves replace plane waves, in this case in order to satisfy the proper boundary conditions of the Schrödinger equation (1.4). As the term Coulomb-Born approximation has been used in a different sense by a number of authors,[48] we shall follow Dewangan and Eichler and call these series (3.22) and (3.24) boundary corrected Born series. Although Coulomb boundary conditions that should be used in solving (1.4), or the corresponding Schrödinger equation in the wave formulation, were pointed out by Cheshire[4] and by Belkic et al.,[5] the practical importance of this was, perhaps, not generally recognised until Dewangan and Eichler[49] identified a serious flaw in the perturbation series using plane waves. The history of the development of perturbation methods is interesting and instructive. In 1928, Oppenheimer[50] and, in 1930, Brinkman and Kramers[51] looked for a first order perturbation expression for charge exchange amplitudes. Making use of the fact that the probability amplitude for an electronic reaction is independent of W_{AB} (provided small terms of the order of the ratio of the electronic mass to the proton mass are neglected), the perturbing potential could be taken to be V_A or, alternatively, V_B. Since charge exchange only takes place when the atomic wave functions centred on A and B overlap to an appreciable extent, and this overlap is only large for small distances of separation, it seemed that the Coulomb boundary conditions which are consistent with (1.4) were unimportant and that a suitable first order approximation, usually called the OBK approximation, would be

$$C_{ji}^{(1)} \approx -i \int_{-\infty}^{\infty} \langle \tilde{G}_j(t) | V_{A,B} | \tilde{F}_i(t) \rangle dt \qquad 3.27$$

where $\tilde{G}_j$ and $\tilde{F}_i$ differ from $G_j^{\pm}$ and $F_i^{\pm}$ given by (1.5) by the omission of the Coulomb phases $\exp[\pm i\sigma(\mp t)]$ and $\exp[\pm i\eta(\mp t)]$.

$$\tilde{F}_n(\underline{r}_A,t) = \phi_n^A(\underline{r}_A)\exp(-i\epsilon_n^A t)$$

$$\tilde{G}_j(\underline{r}_A,t) = \phi_j^B(\underline{r}_B)\exp[-i\epsilon_j^B t + i(\underline{v}.\underline{r}_A - \tfrac{1}{2}v^2 t)] \qquad 3.28$$

When it became possible, much later in the nineteen fifties, to compare the Brinkman-Kramers cross section σ_{BK} obtained using (3.27) with experiment, it became apparent that this was a very poor approximation, exceeding the experimental data in the case of the symmetrical reaction p + H → H + p by factors of five or more at energies above 50 keV. In 1952 Bates and Dalgarno[52] and independently Jackson and Schiff[53] suggested that, in p + H reactions, the internuclear potential should be retained in the first order matrix element, since in that case the signs of V_A and W_{AB} are opposite and errors in the matrix elements of V_A and W_{AB} should tend to cancel. In fact it turned out that the first order Jackson and Schiff approximation using $(V_A + W_{AB})$ as the perturbing potential provided a rather accurate total cross section at high energies E > 100 kev for the reaction p + H → H + p. Subsequently it has been shown that the first order Born approximation retaining W_{AB} fails to agree with experiment for other systems. In any case, the very fact that the matrix element of W_{AB} turns out to be large suggested that, in general, higher order terms were necessary, since W_{AB} does not contribute in the exact theory. The fact that the first order Born approximation for the total cross section in the case of p + H → H + p agrees with the experimental data can be explained by noting that if $Z_A = Z_B$, and <u>only in this case</u>, we have $\sigma(\pm t) = \eta(\pm t)$ and it follows that

$$^{A}C_{ji}^{(1)}(\underline{b}) = -i\, b^{-(2iZ_A/v)} \int_{-\infty}^{\infty} \langle\tilde{G}_j(t)|V_A + \frac{Z_A}{R}|\tilde{F}_i(t)\rangle dt$$

then if $Z_A = Z_B = 1$

$$^{A}C_{ji}^{(1)}(\underline{b}) = b^{-2iZ_A/v}\, C_{ji}^{B}(\underline{b}) \qquad 3.29$$

where C_{ji}^{B} is the first order Born approximation amplitude. Since C_{ji}^{B} and $^{A}C_{ji}^{(1)}(b)$ only differ by a phase factor, the total cross sections obtained are identical.

Although, as we shall see, second order terms are certainly required to explain charge exchange at very high energies, at intermediate energies (for example from 100 kev/amu to 10 MeV/amu) it is entirely possible that the first order boundary corrected amplitude represented by ${}^{A}C_{ji}^{(1)}$ (or ${}^{B}C_{ji}^{(1)}$) may agree with the data. To test this hypothesis, several calculations have been carried out recently, for a number of systems.[54,55] In general rather good agreement has been found with the experiments at intermediate energies. For example, in the work of Belkic et al.,[55] cross sections have been computed in the first order approximation for capture by a range of fully stripped ions (H^+ to C^{6+}) from H(1s). In each case the calculated cross section agrees with the experimental data for impact energies above ~ 30 kev/amu. (Recollect that for an atomic hydrogen state target, the impact velocity equals the Bohr velocity of the target electron at 25 kev/amu.) In Fig. 7, these results are illustrated for the particular case of a ${}^{7}Li^{3+}$ projectile. The poor agreement found when the first order Born (i.e. Bates-Dalgarno, Jackson-Schiff) approximation is employed is apparent, and as expected

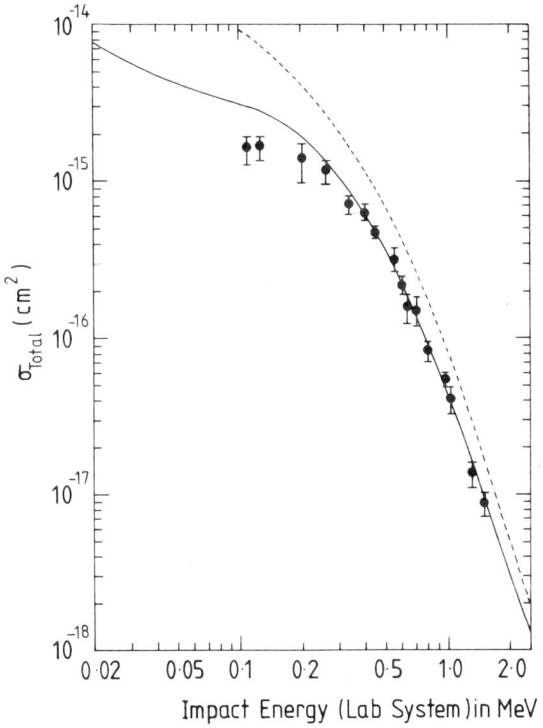

Fig. 7. Total cross section for electron capture by ${}^{7}Li^{3+}$ from H(1s).
——— Boundary corrected Born approximation.[55]
- - - First order Jackson-Schiff Born approximation.[55]
● ● Experimental data.[93]

this disagreement increases with increasing projectile charge.

Second order Born approximation. We have already noted that for asymmetrical collisions in which $V_A >> V_B$, the expansion (3.23) is likely to be more rapidly convergent than the form (3.24). If we include only the linear term in the expansion in powers of the weaker potential, we obtain

$$A_{C_{ji}^{(2)}}(\underline{b}) = -i \int_{-\infty}^{\infty} \langle G_j^-(t)| \left[V_A + \frac{Z_A}{R}\right] + \left[V_A + \frac{Z_A}{R}\right]$$

$$+ \left[V_A + \frac{Z_A}{R}\right] g_A^+ \left[V_B + \frac{Z_B}{R}\right] |F_i^+(t)\rangle \qquad 3.30$$

Similarly if $V_B >> V_A$, and retaining the linear term in V_A we find

$$B_{C_{ji}^{(2)}}(\underline{b}) = -i \int_{-\infty}^{\infty} \langle G_j^-(t)| \left[V_B + \frac{Z_B}{R}\right] + \left[V_A + \frac{Z_A}{R}\right] g_B^+ \left[V_B + \frac{Z_B}{R}\right] |F_i^+(t)\rangle$$

3.31

These are both forms of second-Born approximation and have not yet been evaluated, although it is known (as we shall see in the next section), that at sufficiently high energies the second order terms are significant.

Before the importance of the correct boundary conditions was fully appreciated, the second order Born approximation (for $V_A > V_B$) was taken to be

$$A_{\tilde{C}_{ji}^{(2)}}(\underline{b}) = -i \int_{-\infty}^{\infty} \langle \tilde{G}_j(t)| V_A + V_A \, \tilde{g}_A^+ \, V_B | \tilde{F}_i(t)\rangle dt \qquad 3.32$$

where $\tilde{g}_A$ is the Green's function (cf. 3.10)

$$\tilde{g}_A(\underline{r}_A, t; \underline{r}_A', t') = -i \, \theta(t-t') \sum_n \tilde{F}_n(\underline{r}_A, t) \, \tilde{F}_n^*(\underline{r}_A', t') \qquad 3.33$$

If $V_B > V_A$, the corresponding second order term would contain $(V_B + V_A \tilde{g}_B V_B)$, where $\tilde{g}_B$ is defined in an analogous way as

$$\tilde{g}_B(\underline{r}_A, t; \underline{r}'_A, t) = -i\ \theta(t-t') \sum_j \tilde{G}_j(\underline{r}_A, t)\ \tilde{G}^*_j(\underline{r}'_A, t') \qquad 3.34$$

These second order matrix elements were investigated by Mapleton[56] who demonstrated that the integrals concerned diverged. To see why this is so consider the scattering of a fully stripped ion by a hydrogen ion, so that $V_A = -Z_A/r_A$ and $V_B = -Z_B/r_B$. The second order term is explicitly

$$(-i)^2 \int_{-\infty}^{\infty} dt \int_{-\infty}^{t} dt' \sum_n \langle\tilde{G}_j(t)|V_A(t)|\tilde{F}_n(t)\rangle\langle\phi_n^A(t')|V_B(t')|\phi_i^A(t')\rangle$$

$$\exp\left[i(\epsilon_n^A - \epsilon_i^A)t'\right] \qquad 3.35$$

The sum over n contains a term in which n=i. This term represents an elastic scattering in the initial state followed by a rearrangement amplitude and contains the integral

$$I(t) = \left[\int_{-\infty}^{t} dt' \int d\underline{r}_A\ |\phi_i^A(\underline{r}_A)|^2\ \frac{Z_B}{|\underline{r}_A - \underline{R}|}\right] \qquad 3.36$$

Looking at this integral for large t we have

$$I(t) \underset{t\to\infty}{\sim} \int_{-\infty}^{t} dt'\ \frac{Z_B}{R(t')} \qquad 3.37$$

which is logarithmically divergent. It follows that the perturbation series in this form does not exist. It is evident that the corresponding second order matrix elements for excitation or ionisation, based on plane wave intermediate states also diverge. However, as Macek[57] has shown, it is possible to use plane wave intermediate states by identifying and removing the divergences systematically, and this may be useful because the boundary corrected second order term may be difficult to evaluate. Mapleton's result was not widely known and it was rediscovered by Dewangan and Eichler,[49] who were able to show that the difficulty could be completely avoided by using the correct Coulomb boundary conditions.

At high energies it would seem permissable to replace $\tilde{g}_A$ or $\tilde{g}_B$ by the free-particle Green's function g_o obtained by replacing F_n by a plane wave, $F_n \rightarrow (2\pi)^{-\frac{3}{2}}\exp[i|\underline{K}.\underline{r}_A - K^2t/2)]$. An amplitude based on the matrix element of the operator $[V_A + V_A\tilde{g}_oV_B]$ is usually called the second order OBK amplitude, while that based on the operator $[(V_A+W_{AB}) + (V_A+W_{AB})\tilde{g}_o(V_B+W_{AB})]$ is often called the second-order J-S, or just the second order Born amplitude. The amplitudes containing g_o, rather than $\tilde{g}_A^+$ or $\tilde{g}_B^+$ do not diverge . The second order OBK amplitude, which has been evaluated numerically[58,59] (in the quantum mechanical rather than the impact parameter formulation) for p + H → H + p and p + He → H + He^+, produces poor results, as might be expected from the preceeding discussion. In fact below 3 MeV for p + H and 5 M e V for p + He the calculated cross section is greater than the first order OBK cross section which is itself larger than the experimental data. The second-order Jackson-Schiff cross section is also expected to be poor, except for the p + H system, in which case it coincides with the corresponding boundary corrected approximation. In this special case, the second order cross section has been evaluated at intermediate energies[60] (0.1 to 2.5 MeV). The results are compared in Table 2 with the first J-S cross section for a few energies. The results seem quite reasonable and are not inconsistent with the measured total cross section. This, however, is only available at the lowest energy and includes capture into all excited states. At the higher energies capture is, in fact, mainly into the 1s ground state and capture into levels with principal quantum number n can be estimated, from the J-S approximation, to decrease like n^{-3}.

3.2 The High Energy Form of the Cross Section

The extreme high energy limit of the non-relativistic cross section is very interesting. We shall consider only ground state capture in the forward direction of electrons by protons from atomic hydrogen

$$H^+ + H(1s) \rightarrow H(1s) + H^+ \qquad 3.38$$

and ignore the contribution from the backward direction (the knock-out process arising from the identity of the nuclei). The first order OBK cross section provides a useful basis of comparison. It behaves at large impact velocities v as

$$\sigma^{(1)}_{OBK} \sim \frac{2^{18}}{5v^{12}} (\pi a_o^2) \qquad 3.39$$

The corresponding first order J-S approximation, which coincides with the first order boundary corrected cross section for this system is

Table 2 The cross section for $H^+ + H(1s) \rightarrow H(1s) + H^+$ (units of a_o^2)

E(MeV)	J-S I	II
0.1	0.231	0.360
0.5	3.16×10^{-4}	2.46×10^{-4}
1.0	9.23×10^{-6}	6.58×10^{-6}
2.0	2.12×10^{-7}	1.43×10^{-7}
2.5	6.09×10^{-8}	4.20×10^{-8}

J-S I Jackson-Schiff first order approximation

J-S II Kramer's second order approximation[60]

$$\sigma^{(1)}_{JS} \sim 0.661\ \sigma^{(1)}_{OBK} \qquad 3.40$$

In the second order J-S approximation originally obtained by Drisko[61] the term proportional to v^{-12} is modified, and an additional term in v^{-11} appears

$$\sigma^{(2)}_{JS} \sim \left[0.294 + \frac{5\pi v}{2^{12}}\right]\sigma^{(1)}_{OBK} \qquad 3.41$$

In the third order J-S approximation, there is a further small alteration in the coefficient of the v^{-12} term, while the term in v^{-11} remains unaltered.[61]

$$\sigma^{(3)}_{J\text{-}S} \sim \left[0.319 + \frac{5\pi v}{2^{12}}\right]\sigma^{(1)}_{OBK} \qquad 3.42$$

Shakeshaft and Spruch have shown that for short range potentials the second order term provides the leading term in the high energy limit. However this has not been shown rigorously for Coulomb interactions, although it seems likely that the second order term in v^{-11} does give

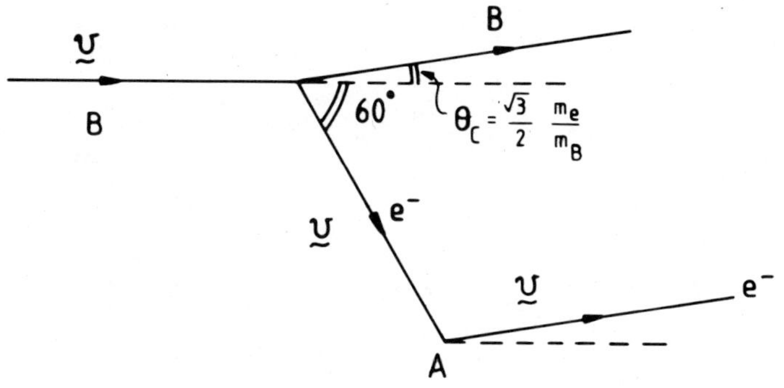

Fig. 8 Thomas double scattering

the correct high energy limit. It should be pointed out that this term does not dominate the cross section until very high energies are reached, > 100 MeV, and at these energies relativistic effects modify the behaviour of the cross section.

The leading term in extremely high energy regions proportional to v^{-11} was originally identified in a purely classical binary collision model by Thomas.[62] Classically the capture results from a double scattering, see Fig. 8. In the first collision the incident proton B strikes the electron which recoils at 60° to the incident direction towards the target proton A, with a speed equal to the speed of B. In the second collision, the electron is scattered without change of speed, again through 60°, so that it emerges in a direction parallel to the proton B. The classical cross section is the product of two Rutherford scattering cross sections for 60° scattering and a 'sticking factor' describing the formation of the final bound state. The fact that classically a double scattering is required to conserve momentum and energy simultaneously, is consistent with the importance of double scattering in the quantum mechanical perturbation series.[63] Although the v^{-11} term in the high energy cross section is dominant only at extremely high energies, the double scattering leads to a peak in the differential cross section at an angle of $(3^{\frac{1}{2}} m_e/2m_P)$ radians at lower energies.[61] This "Thomas" peak has been observed in electron capture by 7 MeV protons from helium.[64] Further peaks can arise in very large angle scattering caused by the inter-ionic potential W_{AB}. These are again due to quasi-classical double scattering in which the first scattering is between the projectile B and the electron, as before, and in the second scattering B is deflected through 60° by the interaction W_{AB}.

3.3 Distorted Wave Approximations

Rather than base perturbation expressions on the functions $F_n^{\pm}, G_j^{\pm}$, which satisfy (1.6a) and (1.6b), we can introduce distorted waves chosen to represent some physical effect, for example to allow for the interaction in the initial and final states. There are clearly infinitely many ways in which to choose distorted waves, but here we shall consider only distorting potentials which are functions of the internuclear separation R. Let us define channel functions $L_n^{\pm}$ and $M_n^{\pm}$ which satisfy

$$\left[-\frac{1}{2}\nabla^2_{r_A} + V_A + \left[W_n^B(R) - \frac{Z_B}{R}\right] - i\frac{\partial}{\partial t}\right]L_n^{\pm}(\underline{r}_A, t) = 0 \qquad 3.13a$$

$$\left[-\frac{1}{2}\nabla^2_{r_A} + V_B + \left[W_n^A(R) - \frac{Z_A}{R}\right] - i\frac{\partial}{\partial t}\right]M_j^{\pm}(\underline{r}_A, t) = 0 \qquad 3.43b$$

The potential $W_n^B(R)$ acts in the direct channels and can vary with the channel index n, while $W_j^A(R)$ acts in the exchange channels and is allowed to vary with the index j. The potentials W_n^B and W_j^A will be taken to be of short range so the Coulomb potentials acting at large R are unmodified. We see that

$$L_n^{\pm}(\underline{r}_A, t) = F_n^{\pm}(\underline{r}_A, t) \exp \mp i \int_{-\infty}^{\pm t} W_n^B(R')\, dt' \tag{3.44a}$$

and

$$M_j^{\pm}(\underline{r}_A, t) = G_j^{\pm}(\underline{r}_A, t) \exp \mp i \int_{-\infty}^{\pm t} W_j^A(R')\, dt' \tag{3.44b}$$

It is straightforward to show that for excitation (or ionisation) the probability amplitude $A_{ni}(\underline{b})$ (see 1.8) can be expressed as

$$A_{ni}(\underline{b}) = -i \int_{-\infty}^{\infty} dt \, \langle L_n^- | V_B - W_n^B + \frac{Z_B}{R} | \psi_i^+ \rangle \tag{3.45}$$

and the corresponding amplitude for charge exchange (see 1.9) is

$$C_{ji}(\underline{b}) = -i \int_{-\infty}^{\infty} dt \, \langle M_j^- | V_A - W_j^A + \frac{Z_A}{R} | \psi_i^+ \rangle \tag{3.46}$$

To develop distorted wave expansions we first obtain the integral equation

$$|\psi_i^+(t)\rangle = |L_i^+(t)\rangle - i \int_{-\infty}^{\infty} dt' \, \theta(t-t') \sum_m |\underline{L}_m^-(t)\rangle\langle L_m^-(t)| V_B - W_m^B + \frac{Z_B}{r} | \psi_i^+ \rangle \tag{3.47}$$

so that the first and second distorted wave amplitudes for charge exchange become

$$C_{ji}^{D1}(\underline{b}) = -i \int_{-\infty}^{\infty} dt \, \langle M_j^- | V_A - W_j^A + \frac{Z_A}{R} | Li^+ \rangle \tag{3.48}$$

$$C^{D2}_{ji}(\underline{b}) = C^{D1}_{i}(\underline{b}) -i \int_{-\infty}^{\infty} dt \int_{-\infty}^{\infty} dt' \langle M^-_j| \left[V_A - W^A_j + \frac{Z_A}{R}\right] g^D_A \left[V_B - W^B_i + \frac{Z_B}{R}\right] |Li^+\rangle$$

3.49

where

$$g^D_A = -i \sum_m \theta(t-t')|L^-_m(t)\rangle\langle L^-_m(t')| \qquad 3.50$$

The corresponding series in terms of the Green's function g^D_B

$$g^D_B = -i \sum_q \theta(t-t')|M^-_q(t)\rangle\langle M^-_q(t')| \qquad 3.51$$

can also be easily obtained.

We can choose the distorting potentials W^B_i and W^A_j to represent as much as possible of the effective interaction between B and $(A+e^-)$ or $(B+e^-)$ + A respectively. The simplest choice is to set $\left[W^B_i - \frac{Z_B}{R}\right]$ equal to the static interaction in the incident channel and $\left[W^A_j - \frac{Z_A}{R}\right]$ equal to the static interaction in the final channel

$$\begin{aligned} W^B_i - \frac{Z_B}{R} &= \langle L^+_i|V_B|L_i{}^+\rangle \\ &= (V_B)_{ii} \end{aligned} \qquad 3.52$$

$$\begin{aligned} W^A_j - \frac{Z_B}{R} &= \langle M^+_j|V_A|M^+_j\rangle \\ &= (V_A)_{jj} \end{aligned} \qquad 3.53$$

In this case, the first order approximation:

$$C^{D1}_{ji}(\underline{b}) = -i \int_{-\infty}^{\infty} dt \langle M^-_j|V_A - (V_A)_{jj}|Li^+\rangle \qquad 3.54$$

is closely related to the first order approximation to the solution of the pair of coupled equations connecting the initial and final channels i and j. This was obtained by Bates[66] in the form

$$C^{DB}_{ji}(\underline{b}) = -i \int_{-\infty}^{\infty} dt \, \frac{\langle M_j^- | V_A - (V_A)_{jj} | L_i^+ \rangle}{1 - |\langle M_j^- | L_i^+ \rangle|^2} \quad 3.55$$

At intermediate and high energies for which the overlap $\langle M_j^- | L_i^+ \rangle \ll 1$, the Bates approximation reduces to the distorted wave approximation (3.54). Another way of viewing (3.54) is to argue that as the initial and final channel functions are not orthogonal, to obtain a well defined transition probability we should first orthogonalise M_j^- to the unperturbed initial channel function L_i^+. The orthogonalised final state function is

$$\bar{M}_j = M_j^- - \langle M_j^- | L_i^+ \rangle L_i^+ \quad 3.56$$

The first order amplitude for the transition caused by the perturbation V_A is

$$C^1_{ji}(\underline{B}) = -i \int_{-\infty}^{\infty} dt \, \langle \bar{M}_j | V_A | L_i^+ \rangle \quad 3.57$$

which is exactly the same as (3.54). The use of the second order Bates approximation has been discussed recently by McGuire[67] and by Rivarola et al.[68] A clear advantage is that the effective perturbations $(V_A - (V_A)_{jj})$ or $(V_B - (V_B)_{ii})$ are short range and difficulties associated with long range Coulomb interactions are avoided.

3.4 The Continuum Distorted Wave Model

A rather different form of distorted wave model was introduced by Cheshire.[69] It aims to use unperturbed functions which allow completely for both the interactions V_A and V_B in the particular case in which there are pure Coulomb potentials

$$V_A = -\frac{Z_A}{r_A}, \qquad V_B = -\frac{Z_B}{r_B} \quad 3.58$$

and treats part of the kinetic energy as the perturbation.

Let us suppose the exact wave function $\psi(\underline{r}_A, t)$ is expressed explicitly as a function of $\underline{r}_A, \underline{r}_B$ and t: $\psi(\underline{r}_A, \underline{r}_B, t)$. We then have the result, using $\underline{r}_B = \underline{r}_A - \underline{R}(t)$:

$$\left[H - i \frac{\partial}{\partial t}\Big|_{r_A}\right]\psi = \left[-\frac{1}{2}\nabla^2_{r_A} + V_A + V_B - i\frac{\partial}{\partial t}\Big|_{r_A}\right]\psi(\underline{r}_A, \underline{r}_B, t)$$

$$\equiv \left[-\frac{1}{2}\nabla^2_{r_A} - \frac{1}{2}\nabla^2_{r_B} - \underline{\nabla}_{\underline{r}_A}\cdot\underline{\nabla}_{\underline{r}_B} + V_A + V_B + i\,\underline{v}\cdot\underline{\nabla}_{\underline{r}_B} - i\frac{\partial}{\partial t}\right]\psi(\underline{r}_A, \underline{r}_B, t) \tag{3.59}$$

where in the last line $\partial/\partial t$ is to be taken with <u>both</u> $\underline{r}_A$ and $\underline{r}_B$ held constant.

Apart from the term $[-\underline{\nabla}_{\underline{r}_A}\cdot\underline{\nabla}_{\underline{r}_B}]$, the Hamiltonian is separable. To make use of this, $[-\underline{\nabla}_{\underline{r}_A}\cdot\underline{\nabla}_{\underline{r}_B}]$ is treated as a perturbation and unperturbed wave functions $\mathscr{L}^{\pm}_n, \mathscr{M}^{\pm}_j$ are defined to be solutions of the equations

$$\left[H_o - i\frac{\partial}{\partial t}\right]\mathscr{L}^{\pm}_n = 0$$

$$\left[\bar{H}_o - i\frac{\partial}{\partial t}\right]\mathscr{M}^{\pm}_j = 0 \tag{3.60}$$

where

$$H_o \equiv -\frac{1}{2}\nabla^2_{r_A} - \frac{1}{2}\nabla^2_{r_B} + V_A + V_B + i\underline{v}\cdot\underline{\nabla}_{\underline{r}_B}$$

and 3.61

$$\bar{H}_o = -\frac{1}{2}\nabla^2_{r_A} - \frac{1}{2}\nabla^2_{r_B} + V_A + V_B$$

The difference between H_o and $\bar{H}_o$ is just due to the fact that we require $\mathscr{M}^{\pm}_j$ to contain the translational factor $\exp(i\underline{v}\cdot\underline{r}_A - \frac{1}{2}v^2 t)$. If the origin is taken at the mid-point of AB rather than at A more symmetrical formulae are obtained. The solutions of (3.60) are of the form of products of two hydrogenic functions, either bound or continuous. The particular solutions $\mathscr{L}^{\pm}_n$, $\mathscr{M}^{\pm}_j$ we require are defined to coincide with the functions $F_n^{\pm}$, $G_j^{\pm}$ (see 1.5) in the appropriate limit. For $t \to -\infty$ we require

$$\mathcal{L}_n^+ \to F_n^+ \; ; \qquad \mathcal{M}_j^+ \to G_j^+ \tag{3.62}$$

and for $t \to +\infty$, we require

$$\mathcal{L}_n^- \to F_n^- \; ; \qquad \mathcal{M}_j^- \to G_j^- \tag{3.63}$$

The functions $\mathcal{L}_n^{\pm}$ and $\mathcal{M}_j^{\pm}$ satisfying these boundary conditions are:

$$\mathcal{L}_n^{\pm} = \phi_n^A(\underline{r}_A) \exp[-i\,\epsilon_n^A\, t] N^{\pm}(\upsilon_B)\, {}_1F_1(\pm i\upsilon_B;\; 1;\; \pm(vr_B + \underline{v}\cdot\underline{r}_B))$$

$$\mathcal{M}_j^{\pm} = \phi_j^B(\underline{r}_B) \exp[-i\,\epsilon_j^B\, t + i(\underline{v}\cdot\underline{r}_A - \tfrac{1}{2}v^2 t)] N^{\pm}(\upsilon_A)\, {}_1F_1(\pm iv_B;\; 1;\; \pm(vr_A + \underline{v}\cdot\underline{r}_A)) \tag{3.64}$$

where $\upsilon_A = Z_A/v$; $\upsilon_B = Z_B/v$ and

$$N^+(\upsilon) = [N^-(\upsilon)]^* = \exp(\pi\upsilon/2)\Gamma(1-i\upsilon) \tag{3.65}$$

From (3.60) and (3.59), the charge exchange probability amplitude can be expressed as

$$C_{ji}(\underline{b}) = \underset{t\to\infty}{\ell t} \langle \mathcal{M}_j^- | \psi_i^+ \rangle$$

$$= i \int_{-\infty}^{\infty} \langle \mathcal{M}_j^- | P^{\dagger} | \psi_i^+ \rangle \, dt \tag{3.66}$$

where

$$P = [\underline{\nabla}_{r_A} \cdot \underline{\nabla}_{r_B}] \tag{3.67a}$$

and the convention is used that P does not operate on the translational factor contained in $\mathcal{M}_j^-$. Explicitly we have

$$P\, M_j^- = \exp[-i\epsilon_j^B t + i(\underline{v}\cdot\underline{r}_A - \tfrac{1}{2}v^2 t)]N^-(\nu_A)\ \times$$

$$\times\ \underline{\nabla}_{r_B}\phi_j^B(\underline{r}_B)\cdot\underline{\nabla}_{r_A}\{{}_1F_1(-i\nu_B;\ 1,\ -(vr_A+\underline{v}\cdot\underline{r}_A))\} \qquad 3.67b$$

As usual $C_{ji}(\underline{b})$ can also be expressed in terms of ψ_j^-, we have

$$C_{ji}(\underline{b}) = i\int_{-\infty}^{\infty} \langle\psi_j^-|P|\mathcal{L}_n^+\rangle\, dt \qquad 3.68$$

The original work of Cheshire was based on a first order approximation to (3.66) of (3.68). Setting $\psi_i^+ \approx \mathcal{L}_i^+$ in (3.66) we find

$$C_{ji}^I(\underline{b}) = i\int_{-\infty}^{\infty} dt\ \langle\mathcal{M}_j^-|P|\mathcal{L}_i^+\rangle \qquad 3.69$$

This approximation called CDWI, has proved to be rather successful in reproducing the total cross sections at high energies, for instance above 100 kev for p + He, above 250 kev/amu for α + He, above 10 kev for p + Li^o and so on. The case of α + He is shown in Fig. 9 as an illustration. Detailed comparisons with experiment for several systems have been given in the reviews by Belkic et al.[5] and by Rivarola.[70] At asymptotically high energies the CDWI approximation behaves like the second order J-S approximation (3.41) for capture into s-states. This is because the approximate wave functions $\mathcal{L}_i$ or $\mathcal{M}_j$ already contain the continuum intermediate states which result in particular in the second order term behaving like v^{-11}.

Theidentity of the asymptotic behaviour of the CDWI and second order J-S approximations does not extend to capture into states with non-zero angular momentum,[71] and the coefficient of the v^{11} differs in the two cases. A related defect of the CDWI approximation is that an unphysical dip occurs at the Thomas angle in the differential capture cross section. These defects can be avoided by going to a second order approximation, which accounts for double scattering (the Thomas mechanism) completely. The wave function ψ_i^+ (or ψ_j^-) can be expanded in terms of the Green's function for the operator $\left[H_o - i\frac{\partial}{\partial t}\right]$ and a perturbation series obtained in the usual way. Two Green's functions can be defined, one based on an expansion in target eigenfunctions $\bar{g}_A$,

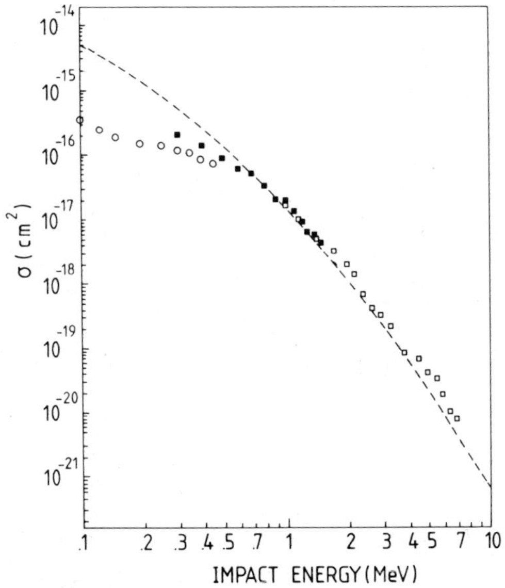

Fig. 9. The total cross section for capture of a single electron from He by alpha particles, calculated in the CDWI approximation. The theoretical and experimental data are taken from Deco et al.[94]

and one on an expansion in projectile eigenfunctions $\bar{g}_B$

$$\begin{aligned} \bar{g}_A &= -i\,\theta(t-t')\sum_n \mathcal{L}^-(\underline{r}_A,t)\mathcal{L}^{-*}(\underline{r}_A',t') \\ \bar{g}_B &= -i\,\theta(t-t')\sum_m \mathcal{M}^+(\underline{r}_A,t)\mathcal{M}^{+*}(\underline{r}_A',t') \end{aligned} \qquad 3.70$$

The second-order expressions for the amplitude are then

$$C^{II}_{ji}(\underline{b}) = G^{(1)}_{i}(\underline{b}) + (i)^2 \int_{-\infty}^{\infty} dt \int_{-\infty}^{\infty} dt' \langle \mathcal{M}^{-}_{j} | P^{+-} g_{A,B} P | \mathcal{L}^{+}_{i} \rangle \qquad 3.71$$

This second order approximation has been derived within a quantum mechanical three-body formalism by Crothers and McCann.[77] Using the target based expansion Crothers[78] has shown that this approximation accounts fully for double scattering and the correct high energy asymptotic limits are attained for all transitions.

In common with most perturbation methods the CDW approximation fails at low energies. As pointed out by Crothers[79] this is in part due to the fact that the functions $\mathcal{L}^{+}_{i}$ or $\mathcal{M}^{-}_{j}$ are not normalised to unity for all t, and in part due to the lack of 'back-couupling' that occurs automatically in a coupled channel model. The approximation is not unitary, that is probability is not conserved, and at sufficiently low energies calculated probabilities greater than unity can be obtained. The lack of unitary and the normalisation problem can be solved by using the funtions $\mathcal{L}_{i}$ and $\mathcal{M}_{j}$ as the basis of a coupled channel expansion. Crothers[80] has discussed a two channel approach in detail and applied it to a number of cases.[81] The wavefunction is represented as

$$\psi^{+}_{i}(\underline{r}_{A}, t) = a_1(\underline{b}, t)\chi_1(\underline{r}_{A}, t) + a_2(\underline{b}, t)\chi_2(\underline{r}_{A}, t) \qquad 3.72$$

with $\chi_1 = \mathcal{L}^{+}_{i}$ and $\chi_2 = \mathcal{M}^{-}_{j}$. Coupled equations of the form (2.13) are obtained with the overlap and potential matrices defined by (2.14) and (2.15) as usual. This model is unitary within the two dimension space of the basis, and should provide reliable results at high energy and results at least as accurate as the two state coupled channel approximation based on the functions F_n, G_j at low energies. In fact for the system $H^+ + H(1s)$ at 25 and 60 kev, the results of the standard two state approximation appear to represent the data rather better than the two state CDW model. However it seems likely that the agreement of the two state standard model with experiment is to some extent accidental, since a realistic model describing excitation , ionisation and charge exchange certainly requires a large basis set.

Before leaving the subject of the CDW model, a related model introduced by Maidagen and Rivarola[81] should be mentioned. In this symmetrical eikonal model (SE), the unperturbed functions are taken to be the large R forms of $\mathcal{L}_n$ and $\mathcal{M}_j$. These are

$$\bar{\mathcal{L}}_n^{+} = \phi_n^A(\underline{r}_A) \exp[-i\epsilon_n^A t] \exp[-i\nu_B \log(\nu r_B + \underline{\nu}\cdot\underline{r}_B)]$$

$$\bar{\mathcal{M}}_j^{-} = \phi_j^B(\underline{r}_B) \exp[-i\epsilon_j^B t + i(\underline{v}\cdot\underline{r}_A - \tfrac{1}{2}v^2 t)] \exp[i\nu_A \log(v r_A + \underline{\nu}\cdot\underline{r}_A)] \qquad 3.73$$

At large R these functions coincide with F_n^+ and G_j^- respectively, and unlike the CDW functions, $\bar{\mathcal{L}}_n$ and $\bar{\mathcal{M}}_j$ are normalised to unity for all t. The first and second order perturbation theory based on these functions has been developed and promising applications have been made to excitation[83] and ionisation[84], as well as charge exchange.[85]

4. CONCLUDING REMARKS

Naturally in this fairly brief review, we have not discussed in any detail the large number of applications of the theoretical models we have introduced. At intermediate energies for systems in which the transitions are due one active electron in the coupled channel model is capable of producing accurate cross sections for excitation, ionisation and electron capture. For the many particle systems in which correlation is important there remains much to be done before an equally confident statement can be made.

At the higher energies, although there has been a long history of false starts, it has become clear that perturbative or distorted wave methods, which are associated with the correct boundary conditions are very promising. Although much remains to be done to demonstrate convergence, even first order CDW, SE and boundary corrected approximations seem able to provide good total cross sections. Again correlation in capture, for example two electron capture, is a challenge which is being taken up.

There are several phenomena which we have not discussed. An important example is the so-called charge exchange into the continuum, which is that part of the ionisation cross section which the motion of the ionised electron is correlated with that of the projectile.[86] Another interesting high energy process is resonant transfer and excitation (RTE) in which a high energy highly charged ion captures an electron from a target atom (or molecule) into an autoionising level which subsequently decays either radiatively or by electron emission

$$B^{Z+} + A \rightarrow [B^{(Z-1)+}]^{**} + [A^{+}]^{*}$$

$$[B^{(Z-1)+}]^{**} \rightarrow [B^{(Z-1)+}]^{*} + h\nu \qquad 3.74$$
$$\searrow [B^{Z+}]^{*} + e^{-}$$

This and related processes have been reviewed recently by Pollack and Hahn.[87]

Finally we should mention that at relativistic velocities radiative electron capture becomes important

$$B + (A+e^{-}) \rightarrow (B+e^{-}) + A + h\nu$$

Ultimately the cross section for this process is just that for the radiative capture of a free electron, the ionic core A acting as a spectator. Measurements have also been made for the non-radiative capture process at relativistic energies. The relativistic second Born approximation has been developed by Humphries and Moisewitsch,[88] following earlier work on the first Born approximation.[89] At high energies the cross section for capture behaves like E^{-1}, in constrast to the E^{-6} or $E^{-5.5}$ behaviour of the non-relativistic theory. Relativistic eikonal models[90] have also been developed and a relativistic version of the CDW method.[91] An interesting review of heavy ion-atom collisions at relativistic energies has been given by Anholt and Gould.[92]

References

1 R. McCarroll, "Atomic and Molecular Collision Theory" Ed. F.A. Gianturco, Plenum, New York, (1982).

2 General accounts of atomic collision theory are given by B.H. Bransden, "Atomic Collision Theory" 2nd Ed., Benjamin, Reading Mass. (1983) and by M.R.C. McDowell and J.P. Coleman "Introduction of the theory of ion-atom collisions", North-Holland, Amsterdam, (1960).

3 B.H. Bransden and R.K. Janev, Adv. Atom. Molec. Phys. 19: 1 (1983).

4 I.M. Cheshire, Proc. Phys. Soc. 84: 89 (1964).

5 Dz. Belkic, R. Gayet and A. Salin, Phys. Repts. 56: 279 (1979).

6 R. McCarroll and A. Salin, J. Phys. B: At. Mol. Phys. 1: 163 (1968).

7 A. Hazi and H.S. Taylor, Phys. Rev. A 1: 1109 (1970).

8 W.H. Bassiches, J.F. Reading and R. Scheerbaum, Phys. Rev. C 11: 316 (1975).

W.H. Bassiches, M. Strayer, J.F. Reading and R. Scheerbaum, Phys. Rev. C 18: 632 (1978).

9 J.F. Reading, A.L. Ford, G.V. Swafford and A. Fitchard, Phys. Rev. A 20: 130 (1979).

10 J.T. Broad, Phys. Rev. A 18: 1012 (1978).

11 B.H. Bransden and C.J. Noble, Phys. Let.. 70A: 404 (1979).

E.O. Fitchard, A.L. Ford and J.C. Reading, Phys. Rev. A 16: 1325 (1977).

12 For a method of evaluation see C.J. Noble, Comput. Phys. Commun. 19: 327 (1980).

13 J.F. Reading, A.L. Ford and R.L. Becker, J. Phys. B: At. Mol. Phys. 14: 1995 (1981).

14 V. Dose and C. Semini, Helv. Phys. Acta 47: 609 (1974).

15 D.F. Gallaher and L. Wilets, Phys. Rev. 169: 139 (1968).

16 T.G. Winter, Phys. Rev. A 25: 697 (1982).

Phys. Rev. A 33: 3842 (1986).

Phys. Rev. A 35: 3799 (1987).

17 D.G.M. Anderson, M.J. Antal and M.B. McElroy, J. Phys. B 7: L118 (1974); J. Phys. B 14: 1707(E) (1985).

M.J. Antal, D.G.M. Anderson and M.B. McElroy, J. Phys. B 8: 1513 (1975).

18 C.D. Lin, T.G. Winter and W. Fritsch, Phys. Rev. A 25: 2395 (1982).

T.G. Winter and C.D. Lin, Phys. Rev. A 29: 567; 3071 (1984), Ibid 30: 3323(E).

19 A.M. Ermolaev and M.R.C. McDowell, J. Phys. B: At. Mol. Phys. 20: in press (1987).

20 J.F. Reading, A.L. Ford and R.L. Becker, J. Phys. B: At. Mol. Phys. 15: 625; 3257 (1982).

21 B.H. Bransden, C.J. Noble and J. Chandler, J. Phys. B: At. Mol. Phys. 16: 4191 (1983).

22 W. Fritsch and C.D. Lin, J. Phys. B: At. Mol. Phys. 15: 1255 (1982).

23 G.C. Angel, E.C. Sewell, K.F. Dunn and H.B. Gilbody, J. Phys. B: At. Mol. Phys. 11: L297 (1978).

24 B. Peart, K. Rinn and K. Dolder, J. Phys. B: At. Mol. Phys. 16: 1461 (1983).

25 K. Rinn, F. Melchert and E. Salzborn, J. Phys. B: At. Mol. Phys. 18: 3783 (1985).

26 W. Fritsch and C.D. Lin, Abstracts XIIIth ICPEAC, p. 502 (1983).

27 M.F. Watts, K.F. Dunn and H.B. Gilbody, unpulished, quoted in ref. 16.

28 B.H. Bransden and J.P. Coleman, J. Phys. B: At. Mol. Phys. 5: 2537 (1972).

29 B.H. Bransden, J.P. Coleman and J. Sullivan, J. Phys. B: At. Mol. Phys. 5: 2061 (1972).

J. Sullivan, J.P. Coleman and B.H. Bransden, J. Phys. B: At. Mol. Phys. 5: 546 (1972).

30 S. Begum, B.H. Bransden and J. Coleman, J. Phys. B: At. Mol. Phys. 6: 837 (1973).

31T. A. Green, H.I. Stanley and Y.C. Chiang, Helv. Phys. Acta 38: 109 (1965).

32 B.H. Bransden, A.M. Ermolaev and R. Shingal, J. Phys. B: At. Mol. Phys. 17: 4515 (1984).

33 R. Shingal, B.H. Bransden and D.R. Flower, J. Phys. B: At. Mol. Phys. 18: 2485 (1985).

34 R. Shingal, B.H. Bransden and D.R. Flower, J. Phys. B: At. Mol. Phys. 20, in press (1987).

35 W. Fritsch and C.D. Lin, Phys. Lett. A in press (1987).

36 K.C. Kulander, K.R. Sandhya Devi and S.E. Koonin, Phys. Rev. A 25: 2968 (1982).

W. Stich, H.J. Lüdde and R.M. Dreizler, J. Phys. B: At. Mol. Phys. 18: 1195 (1985).

37 J.F. Reading, A.L. Ford, J.S. Smith and R.L. Becker, In "Electronic and Atomic Collisions" Eds. J. Eichler, I.V. Hertel and N. Stolterfoht (Elsevier Science Publishers) p. 201 (1984).

38 J.F. Reading and A.L. Ford, J. Phys. B: At. Mol. Phys. (in press) (1987).

Phys. Rev. Lett. 58: 543 (1987).

39 H. Knudsen, L.H. Andersen, P. Hvelplund, G. Astner, H. Cederquist, H. Donald, L. Liljeby and K.G. Rendfelt, J. Phys. B: At. Mol. Phys. 17: 3545 (1984).

41 J.T. Park, Adv. in Atom. and Molec. Phys. 19: 67 (1983).

42 B.H. Bransden, D.P. Dewangan and C.J. Noble, J. Phys. B: At. Mol. Phys. 12: 3563 (1979).

43 B.H. Bransden and D.P. Dewangan, J. Phys. B: At. Mol. Phys. 12: 1377 (1979).

44 We shall follow in part the discussion in D.P. Dewangan and J. Eichler, Comments on At. & Mol. Phys. (in press) (1987).

45 R. Aaron, R.D. Amado and B.W. Lee, Phys. Rev. 121: 319 (1961).

46 J. Macek and R. Shakeshaft, Phys. Rev. A 22, 1441 (1980).
see also
J. Macek and K. Taulbjerg, Phys. Rev. Lett. 46: 170 (1981).

47 J.S. Briggs, J. Phys. B: At. Mol. Phys. 13: L717 (1980).

48 In the Coulomb-Born approximation of Sil and coworkers, the Coulomb boundary conditions in the initial and final channels are obeyed, but the inter-nuclear potential W_{AB} is retained in the lowest-order approximation, see for example

S.C. Mukherjee, N.C. Sil and D. Barn, J. Phys. B: At. Mol. Phys. 12: 1259 (1979).

C.R. Mandal, Shyamal Datta and S.C. Mukherjee, S.C., Phys. Rev. A 24: 3044 (1981)

where other references can be found.

49 D.P. Dewangan and J. Eichler, J. Phys. B: At. Mol. Phys. 18: L65 (1985).

50 J.R. Oppenheimer, Phys. Rev. 32: 361 (1928).

51 H.C. Brinkman and H.A. Kramers, Proc. Acad. Sci. Amsterdam 33: 973 (1930).

52 D.R. Bates and A. Dalgarno, Proc. Phys. Soc. A 65: 919 (1952).

53 J.D. Jackson and H. Schiff, Phys. Rev. 89: 359 (1953).

54 D.P. Dewangan and J. Eichler, J. Phys. B: At. Mol. Phys. 19: 2935 (1986).

Dz. Belkic, R. Gayet, J. Hanssen and A. Salin, J. Phys. B: At. Mol. Phys. 19: 2945 (1986).

C.R. Deco, J. Hanssen and R.D. Rivarola, Phys. Rev A in press (1987).

C.R. Deco, J. Hanssen and R.D. Rivarola, J. Phys. B: At. Mol. Phys. 19: L635 (1986).

Dz. Belkic, Phys. Rev. A in press (1987).

55 Dz. Belkic, S. Saini and H.S. Taylor, Z. Phys. D 3: 59 (1986).

56 R.A. Mapleton, "The Theory of Charge Exchange", Wiley (1972).

57 J. Macek, preprint (1987).

58 P.R. Simony and J.H. McGuire, J. Phys. B: At. Mol. Phys. L737 (1981).

P.R. Simony, J.H. McGuire and J. Eichler, Phys. Rev. A 26: 1337 (1982).

59 J.E. Miraglia, R.D. Piacentire, R.D. Rivarola and A. Salin, J. Phys. B: At. Mol. Phys. 14: 1197 (1986).

60 P.J. Kramer, Phys. Rev. A 6: 2125 (1972).

61 R.M. Drisko, Ph.D. Thesis, Carnegie Institute of Tech. (1955). see also

K. Deltman and G. Liebfried, Phys. Rev. 148: 171 (1966), Z. Phys. 218: 1 (1969).

R. Shakeshaft, J. Phys. B: At. Mol. Phys. 7: 1059 (1974).

62 L.H. Thomas, Proc. Roy. Soc. 114: 561 (1927).

63 R. Shakeshaft and L. Spruch, Rev. mod. Phys. 51: 369 (1974).
J.S. Briggs and L. Dube, J. Phys. B: At. Mol. Phys. 13: 771 (1980).
J.H. McGuire, P.R. Simony, O.L. Weaver and J. Macek, Phys. Rev. A 26: 1109 (1982).

64 E. Horsdal, In "Semi-Classical Descriptions of Atomic and Nuclear Collisions", p. 227, North-Holland (1985).
E. Horsdal-Pedersen, J. Phys. B: At. Mol. Phys. 20: 785 (1987).

65 J.S. Briggs and K. Taulbjerg, J. Phys. B: At. Mol. Phys. 12: 2565 (1979).

66 D.R. Bates, Proc. Roy. Soc. A 247: 294 (1958).

67 J.H. McGuire, J. Phys. B: At. Mol. Phys. 18: L75 (1985).

68 R.D. Rivarola, J.M. Maidagan and J. Hanssen, Nucl. Inst. & Methods, in press (1987).

69 I.M. Cheshire, Proc. Phys. Soc. 84: 89 (1964).

70 R.D. Rivarola, Nucl. Inst. & Methods A 240: 508 (1985).

71 L.J. Dube, J. Phys. B: At. Mol. Phys. 17: 641 (1984).
D.S.F. Crothers, J. Phys. B: At. Mol. Phys. 18: 2874 (1985).

72 Dz. Belkic and R. Gayet, J. Phys. B: At. Mol. Phys. 10: 1923 (1977).

73 S.K. Allison, Phys. Rev. A 27: 2342 (1958).

74 L.I. Pivovar, Tubaev and M.T. Novchov, Sov. Phys. - JETP 15: 1035 (1982).

75 P. Hvelplund, J. Heinemeier, E. Horsdal-Pedersen and F.R. Simpson, J. Phys. B: At. Mol. Phys. 9: 491 (1976).

76 G.R. Deco, J.M. Maidagan and R.D. Rivarola, J. Phys. B: At. Mol. Phys. 17: L707 (1984).

77 D.S.F. Crothers and J.F. McCann, J. Phys. B 17: L177 (1984).

78 D.S.F. Crothers, J. Phys. B: At. Mol. Phys. 18: 2893 (1985).

79 D.S.F. Crothers, J. Phys. B: At. Mol. Phys. 15: 2061 (1982).

80 D.S.F. Crothers and J.F. McCann, J. Phys. B: At. Mol. Phys. 18: 2907 (1985).
D.S.F. Crothers, Physica Scripta T3, 236 (1983).

81 J.M. Maidagan and R.D. Rivarola, J. Phys. B: At. Mol. Phys. 17: 2477 (1984).
G.R. Deco, R.D. Piacentini and R.D. Rivarola, J. Phys. B: At. Mol. Phys. 19: 3727 (1986).

82 R.D. Rivarola, Nucl. Inst. & Methods A240: 508 (1985).

83 G.R. Deco, P.D. Fainstein and R.D. Rivarola, J. Phys. B: At. Mol. Phys. 19: 213 (1986).

84 P.D. Fainstein and R.D. Rivarola, J. Phys. B: At. Mol. Phys. 20, in

85 R.D. Rivarola, G.R. Deco and J.M. Maidagan, Nucl. Inst. Meths. B10/11: 222 (1985).

86 A. Salin, J. Phys. B: At. Mol. Phys. 2: 631, 1225 (1969).
J.H. Macek, Phys. Rev. A 1: 235 (1970).
M.W. Lucas, K.F. Mann and W. Steckelmacher, "Lecture Notes in Physics", vol. 213, Ed. K.O. Groenevell et al., Springer, Berlin (1984).

87 E. Pollack and Y. Hahn, Adv. At. Mol. Phys. 22: 243 (1986).

88 W.J. Humphries and B.L. Moisewitsch, J. Phys. B: At. Mol. Phys. 17: 2655 (1984).
Ibid 18: 2295 (1985).

89 B.L. Moisewitsch and S.G. Stockman, J. Phys. B: At. Mol. Phys. 13: 2975 (1980).

90 R. Anholt and J. Eichler, Phys. Rev. A 31: 3505 (1985).
G.R. Deco and R.D. Rivarola, J. Phys. B: At. Mol. Phys. 20: 317 (1987).

91 G.R. Deco and R.D. Rivarola, J. Phys. B: At. Mol. Phys. 19: 1759 (1986).
Ibid 20: in press (1982).

92 R. Anholt and H. Gould, Adv. At. Mol. Phys. 22: 315 (1986).

93 M.B. Shah, T.V. Goffe and H.B. Gilbody, J. Phys. B: At. Mol. Phys. 12: 3763 (1979).

94 G.R. Deco, J.M. Maidagan and R.D. Rivarola, J. Phys. B 17: L707 (1984).

NON-ADIABATIC MOLECULAR COLLISIONS

V. Sidis

Laboratoire des Collisions Atomiques et Moléculaires
Bâtiment 351, Université Paris-Sud
91405 Orsay Cedex, France

ABSTRACT

Non-reactive atom-molecule (A + BC) collisions in the 1-1000 eV/amu energy range are investigated. The collision is treated in terms of the molecular states of the ABC system and the conditions determining electronically non-adiabatic transitions are considered. The basic quantum mechanical equations that govern the scattering amplitudes for ro-vibronic transitions are reviewed. An outline is given of the theory and practice of adiabatic and diabatic representations. A semi-classical treatment is introduced which lends itself to the application of sudden approximations for rotation and (at high energy) vibration. Recently studied non-adiabatic molecular collisions are discussed.

1. INTRODUCTION

The concept of nonadiabatic behaviour is of crucial interest for the understanding of atomic and molecular collision processes. This concept is forged in two ways :

(i) by reference to the treatment of the time dependent Schrödinger equation,

(ii) by reference to the Born-Oppenheimer approximation of Molecular Physics.

These two references are assumed to be familiar to the reader.

1.1 The notions of adiabatic and sudden behaviours in Quantum Mechanics

Consider an arbitrary quantum mechanical system which is governed by the hamiltonian H_o and which is subjected at time t_i to a time dependent perturbation $V(t)$ that lasts for the interval $T = t_f - t_i$. We are tought by basic textbooks [1, 2] that if T is considerably larger than the typical periods associated with the internal motions of the system ($T >> (E_j - E_k)^{-1}$) then the system behaves *adiabatically* ; in other words the system follows continuously the eigenstate of the time dependent hamiltonian $H = H_o + V(t)$ that correlates with the initially prepared eigenstate of H_o. Otherwise, the system behaves *non-adiabatically* i.e. it has some probability of undergoing transitions among several eigenstates of H and thereby to be formed in a superposition of eigenstates of the final hamiltonian $H_o + V(t_f)$. In the extreme limit of a perturbation that is set up in very short time intervals ($T << |E_j - E_k|^{-1}$) we have a situation of *sudden behaviour* ; in other words the system may barely follow the effect of the perturbation and thus remains in the very eigenstate of H_o (ϕ_I^o) in which it has been prepared initially. The probability of finding the system in an arbitrary eigenstate ψ_F of $H_o + V(T)$ is thereby given by : $|<\psi_F|\phi_I^o>|^2$. Thus non-adiabatic (and in the limit sudden) behaviour is the mediator of transitions among states of quantum mechanical systems that are perturbed temporarily.

1.2 Atomic and molecular collisions viewed as processes of temporary formation and breakup of super molecules

In order to retrieve the concept of adiabatic behaviour in molecular collisions one should realize that when the collision partners approach at small distances a sort of supermolecule is formed. It is thus tempting to treat (as far as possible) slow molecular encounters along the same lines as those followed in standard Molecular Physics. To do so it should be remarked that, owing to the smallness of the electron mass to the nuclear mass ratio ($m_e/m_N < 1/2000$) electrons move much faster than nuclei at *comparable energies of motion*. In other words the typical periods associated with the motion of the electrons are much smaller than those characterizing the motion of the nuclei. Regarding the motion of the nuclei as being the source of slow perturbations it may be expected (according to the preliminary discussion of Sec. 1.1) that, as the reactants approach (or as the products separate), the system follows continuously *one* specific eigenstate of the electronic hamiltonian that is built by clamping the nuclei at the successive positions that they occupy in the course of the collision. The way this is done follows closely the Born-Oppenheimer[3] approximation in Molecular Physics. One first solves a Schrödinger equation for the (fast) electrons whereby the (slow) nuclei are held fixed. The eigenvalues of this equation thus depend parametrically on the relative nuclear coordinates. This introduces the invaluable notion of adiabatic electronic energies (determined for fixed position of the nuclei) which actually constitute the potential energy hypersurfaces from which derive the forces that govern the motion of the nuclei.[1, 4] This is the basis of what may be called the *electronically adiabatic version* of the molecular treatment of slow heavy particle collisions. It is in this framework that most theoretical treatments of vibro-rotational excitation[5] and reactive scattering[6] in molecular collisions at very low energies ($E < 1$ eV/amu) have been carried out so far. Still, if all molecular collisions were electronically adiabatic, processes like : transfer of electronic excitation, chemiluminescence and quenching, charge exchange and chemi-ionization, collision induced predissociation, associative ionization and Penning ionization, etc... could never occur since they involve a change of electronic state somewhere somehow in the course of the collision.[7] The patent existence of such processes at both low and moderate energies ($1 \text{ eV} \leq E \leq 1 \text{ keV}$) is a clear indication that the pure electronically adiabatic point of view breaks down in certain circumstances that we are about to discuss.

1.3 Status of the field

Despite the practical importance of electronically nonadiabatic phenomena for elementary energy transfer and rearrangement processes, *systematic* theoretical calculations are only conceivable to date for atom-atom collision systems. This situation is of course attributable to the formidable computational task that the simplest atom + molecule collision represents. It is therefore not surprising that most of the tremendous activity invested during the past two decades in the understanding and actual treatment of non adiabatic collisional phenomena have benefited to the field of atom-atom collisions. Yet, since a few years several conditions are met for a transfer of "know-how" from the field of atom + atom collisions to that of non-reactive atom + molecule collisions. Some aspects of this endeavour are presented in this series of lectures. Although interesting attempts have been made to solve some reactive collision processes involving non-adiabatic phenomena the topic is still too awkward to lend itself to a plain and unconditional presentation.

2. BASIC FORMALISM

2.1 Coordinate systems

The quantum mechanical hamiltonian that governs the motion of an $A^{(+)}$ + BC collisional system writes, in the laboratory reference frame :

$$H_{Lab} = -\frac{1}{2}\left(\frac{\Delta_A}{m_A} + \frac{\Delta_B}{m_B} + \frac{\Delta_C}{m_C} + \sum_{i=1}^{N} \Delta_i\right) + \mathcal{V} \tag{1}$$

where an index i is associated to each electron in the system. After removal of the center of mass (c.m.) motion and the choice of a set of relative coordinates, e.g. :

$$\vec{r} = \vec{C} - \vec{B}\ , \qquad \vec{R} = \vec{A} - \frac{m_B \vec{B} + m_C \vec{C}}{M}$$

$$\vec{q}_i = \vec{i} - \left(p\,\vec{A} + q\,\frac{m_B \vec{B} + m_C \vec{C}}{M}\right) \ ; \qquad p + q = 1, \quad M = m_B + m_C \tag{2}$$

the total hamiltonian in the center of mass frame writes :

$$H = -\frac{1}{2}\left(\frac{\Delta_{\vec{R}}}{\mu} + \frac{\Delta_{\vec{r}}}{m} + \sum_{i=1}^{N} \Delta_{\vec{q}_i}\right) - \left(\frac{p}{m_A} - \frac{q}{M}\right) \vec{\nabla}_{\vec{R}} \cdot \sum_i \vec{\nabla}_{\vec{q}_i} + \mathcal{V}(\{\vec{q}_i\}\ ;\ R,\ r,\ \gamma)\ ;$$

$$\mu = \frac{m_A M}{m_A + M}\ , \quad m = \frac{m_B m_C}{M} \tag{3}$$

where mass polarisation terms $(\frac{p^2}{m_A} + \frac{q^2}{M}) (\sum_i \vec{\nabla}_{\vec{q}_i})^2$ have been dropped. Clearly, the choice :

$$p = m_A \, (m_A + m_B + m_C)^{-1} \quad , \quad q = (m_B + m_C) \, (m_A + m_B + m_C)^{-1}$$

cancels the cross derivative term in eq. (3), which suggests the choice of the nuclear center of mass G as origin of electronic coordinates, this will be our actual choice throughout (unless stated otherwise). In the foregoing equations the potential function $\mathcal{V}$ depends only on the set of relative coordinates specified in fig. 1.

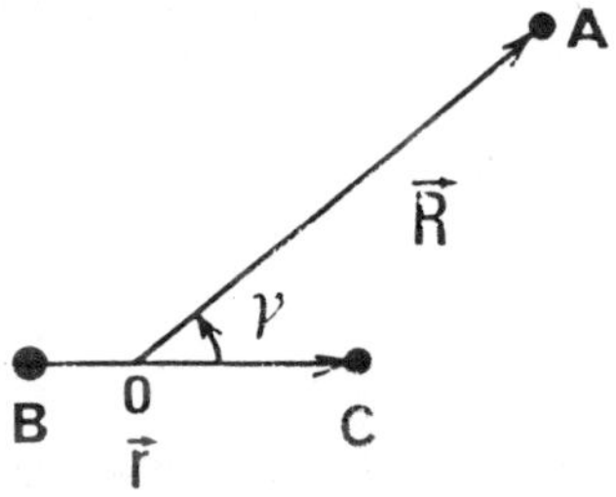

Fig. 1 - Internal coordinates of the atom A and molecule BC in the ABC triatomic plane.

As in atom-atom collisions,[8] it is more convenient to express the electronic coordinates in the reference frame (fig. 2) that rotates with

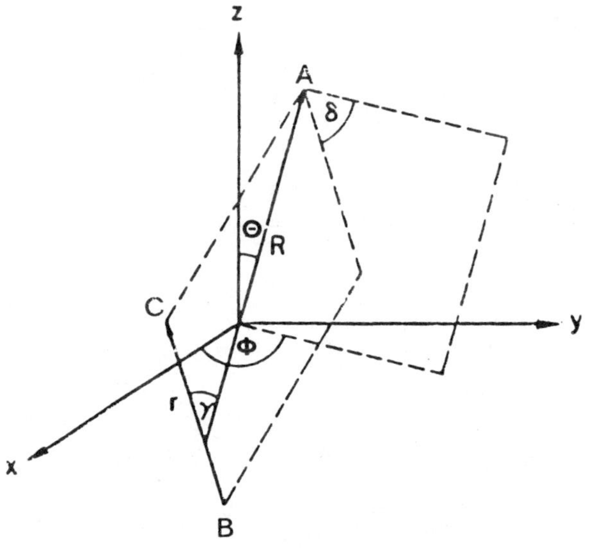

Fig. 2 - Space-fixed and body-fixed systems of coordinates.

the intermolecular vector $\vec{R}(R,\theta,\Phi)$ about the center of mass of the nuclei G (which coincides to a good approximation with the center of mass of all particles). We therefore introduce the reference frame S whose Z axis lies along R and whose X axis is such that the plane X G Z contains the bond vector $\vec{r}$.[9] S constitutes the body fixed reference frame that tumbles with the triatomic plane $(\vec{R}, \vec{r})$ during the collision. The transformation from space fixed coordinates S' : (X', Y', Z') to body fixed coordinates S : (X, Y, Z) is given by :

$$\begin{bmatrix} X_i \\ Y_i \\ Z_i \end{bmatrix} = \begin{pmatrix} \cos\delta & \sin\delta & 0 \\ -\sin\delta & \cos\delta & 0 \\ 0 & 0 & 1 \end{pmatrix} \begin{pmatrix} \cos\theta & 0 & -\sin\theta \\ 0 & 1 & 0 \\ \sin\theta & 0 & \cos\theta \end{pmatrix} \begin{pmatrix} \cos\Phi & \sin\Phi & 0 \\ -\sin\Phi & \cos\Phi & 0 \\ 0 & 0 & 1 \end{pmatrix} \begin{bmatrix} X'_i \\ Y'_i \\ Z'_i \end{bmatrix} \tag{4}$$

Since this transformation has the effect of introducing dependences of the coordinates expressed in the body fixed frame upon θ, Φ an δ, the operators $\partial/\partial\theta|_{S'}$, $\partial/\partial\Phi|_{S'}$ and $\partial/\partial\delta|_{S'}$ arising in the kinetic energy operators associated with $\vec{R}$ and $\vec{r}$ (see eq. (6)) are seen to transform according to the following expressions :

$$\begin{aligned} \frac{\partial'}{\partial\theta} &= \frac{\partial}{\partial\theta} - i\,(\sin\delta\, L_X + \cos\delta\, L_Y) \\ \frac{\partial'}{\partial\Phi} &= \frac{\partial}{\partial\Phi} - i\cos\theta\, L_Z + i\sin\theta\,(\cos\delta\, L_X - \sin\delta\, L_Y) \\ \frac{\partial'}{\partial\delta} &= \frac{\partial}{\partial\delta} - i\, L_Z \end{aligned} \tag{5}$$

where the primed and unprimed operators express that the derivative is effected by keeping the coordinates of the electrons fixed in the reference frame S' or S (resp.). $\vec{L}$ is the electronic angular momentum relative to the center of mass of the nuclei in the body fixed frame S.

Thence, the kinetic energy operator T_N associated with the $\vec{R}$, $\vec{r}$ motions writes :

$$T_N = -\frac{1}{2\mu R^2}\left[\frac{\partial}{\partial R}R^2\frac{\partial}{\partial R} + \frac{1}{\sin\theta}\frac{\partial'}{\partial\theta}\sin\theta\frac{\partial'}{\partial\theta} + \frac{1}{\sin^2\theta}\left(\frac{\partial'^2}{\partial\Phi^2} + \frac{\partial'^2}{\partial\delta^2} - 2\cos\theta\frac{\partial'^2}{\partial\Phi\partial\delta}\right)\right]$$

$$-\frac{1}{2mr^2}\frac{\partial}{\partial r}r^2\frac{\partial}{\partial r} - \left(\frac{1}{2mr^2} + \frac{1}{2\mu R^2}\right)\left(\frac{1}{\sin\gamma}\frac{\partial}{\partial\gamma}\sin\gamma\frac{\partial}{\partial\gamma} + \frac{1}{\sin^2\gamma}\frac{\partial'^2}{\partial\delta^2}\right)$$

$$+\frac{1}{2\mu R^2}\left[2\frac{\partial'^2}{\partial\delta^2} - i\frac{j_+ + j_-}{\sin\theta}\left(\frac{\partial'}{\partial\Phi} - \cos\theta\frac{\partial'}{\partial\delta}\right) + (j_+ - j_-)\frac{\partial'}{\partial\theta}\right] \qquad (6)$$

with :

$$j_\pm = e^{\pm i\delta}\left(\cot\gamma\frac{\partial'}{\partial\delta} \pm \frac{\partial}{\partial\gamma}\right), \qquad \gamma = (\widehat{\vec{r}, \vec{R}})$$

2.2 The close coupling equations

The most general solution of the Schrödinger equation for 3 nuclei and N electron :

$$H\psi(\{\vec{q}_i\}, \vec{r}, \vec{R}) = E\psi(\{\vec{q}_i\}, \vec{r}, \vec{R}) \qquad (7)$$

may be sought in the form of an expansion :

$$\psi(\{\vec{q}_i\}, \vec{r}, \vec{R}) = \sum_n f_n(\vec{R}, \vec{r})\,\phi_n(\{\vec{q}_i\}; \ldots) \qquad (8a)$$

$$f_n(\vec{R}, \vec{r}) = \sum_v F_{nv}(\vec{R})\,G_v(\vec{r}\,; \ldots) \qquad (8b)$$

where G_v and ϕ_n form complete basis sets in the specified variables. We have left points of suspension in the expansions (8) to draw attention to the fact that the G_v and ϕ_n basis sets may contain some additional parametrical dependences. Indeed it is customary in Molecular Physics to use basis sets of the form : $G_v(\vec{r};\vec{R})$ and $\phi_n(\{\vec{q}_i\}\,;\,\vec{r}, \vec{R})$. In the body fixed frame discussed in Sec. 2.1, the mentioned parametrical dependences appear as : $G_v(\vec{r};R)$ and $\phi_n(\{\vec{q}_i\}\,;\,r, \gamma, R)$.

Inserting eq. (8) into eq. (7) and projecting the result onto each ϕ_n provides using eq. (3) the following set of coupled equations :

$$\left(- \frac{1}{2\mu} \Delta_{\vec{R}} - \frac{1}{2m} \Delta_{\vec{r}} + W_{nn} - E \right) f_n =$$

$$- \sum_{k \neq n} \left(W_{nk} - \frac{\vec{D}_{nk} \cdot \vec{\nabla}_R}{\mu} - \frac{\vec{d}_{nk} \cdot \vec{\nabla}_r}{m} \right) f_k \tag{9}$$

$$W_{nk} = V_{nk} + \tau_{nk} , \tag{10a}$$

$$V_{nk} = \langle \phi_n | H_{el} | \phi_k \rangle, \ \tau_{nk} = \langle \phi_n | - \frac{1}{2\mu} \Delta_{\vec{R}} - \frac{1}{2m} \Delta_{\vec{r}} | \phi_k \rangle \tag{10b}$$

$$H_{el} = \sum_i - \frac{1}{2} \Delta_{\vec{q}_i} + \mathcal{V}(\{\vec{q}_i\}, R, r, \gamma) \tag{10c}$$

$$\vec{D}_{nk} = \langle \phi_n | \vec{\nabla}_R | \phi_k \rangle , \quad \vec{d}_{nk} = \langle \phi_n | \vec{\nabla}_r | \phi_k \rangle \tag{10d}$$

For concision we have used in eq.(9) the condensed notations W_{nk}, $\vec{D}_{nk}.\vec{\nabla}_R$ and $\vec{d}_{nk}.\vec{\nabla}_r$ for the matrix elements of all operators appearing in eqs. (5) (6). The operators $\Delta_{\vec{R},\vec{r}}$ and $\vec{\nabla}_{R,r}$ do not act outside the brackets $\langle | | \rangle$. $\vec{D}_{nn}$ and $\vec{d}_{nn}$ clearly vanish when the $|\phi_n\rangle$ wavefunctions are normalized and real since :

$$0 = \vec{\nabla}_{R,r} \langle \phi_n | \phi_n \rangle = 2 \langle \phi_n | \vec{\nabla}_{R,r} | \phi_n \rangle \tag{11}$$

In addition, if the $|\phi_n\rangle$ form an orthogonal set, then :

$$\vec{D}_{nk} = - \vec{D}_{kn} \quad ; \quad \vec{d}_{nk} = - \vec{d}_{kn} \tag{12}$$

The terms τ_{nk} are often much smaller than all other terms and are usually dropped. However, it is known from the treatment of predissociation problems [10] that : (i) in some cases of strong non-adiabatic coupling these terms are not so negligible (see Sec. 3.1), and, (ii) they combine with the second and third terms in the right hand side of eq. (9) to yield (as it should be) hermitian couplings (see eq. (12)).

Equation (9) may still be projected onto the $|G_v\rangle$ functions. However, before doing so, let us remark that since one may choose the orthonormal set $\{G_v\}$ in so many different ways, one is entitled to introduce an additional index n : $\{G_v^n\}$ which indicates that for any preset choice of ϕ_n there could exist some specific choice of $\{G_v\}$ which is most appropriate. Of course in this case we have :

$$\langle G_v^n \mid G_{v'}^m \rangle \neq 0 \qquad \text{for } n \neq m \tag{13}$$
$$= \delta_{vv'} \quad \text{for } n = m$$

The projection onto the G_v with their extended definition entails :

$$\left(T_{\vec{R}} + \langle G_v^n \mid -\frac{1}{2\mu}\Delta_{\vec{R}} - \frac{1}{2m}\Delta_{\vec{r}} + W_{nn} \mid G_v^n \rangle \right) F_{nv} =$$

$$= -\sum_{v'} \left(\langle G_v^n | -\frac{1}{2\mu}\Delta_{\vec{R}} - \frac{1}{2m}\Delta_{\vec{r}} + W_{nn} | G_{v'}^n \rangle - \frac{1}{\mu} \langle G_v^n | \vec{\nabla}_R | G_{v'}^n \rangle \cdot \vec{\nabla}_R \right) F_{nv'}$$

$$- \sum_{\substack{k,v'' \\ k \neq n}} \left(\langle G_v^n | W_{nk} - \frac{\vec{D}_{nk} \cdot \vec{\nabla}_R}{\mu} - \frac{\vec{d}_{nk} \cdot \vec{\nabla}_r}{m} | G_{v''}^k \rangle + \right.$$

$$\left. - \frac{1}{\mu} \langle G_v^n | \vec{\nabla}_R | G_{v''}^k \rangle \cdot \vec{\nabla}_R \right) F_{kv''} \tag{14}$$

Again, as in eqs. (10) the operators in the brackets <||> do not operate on the functions outside.

With these equations and the related definitions and notation one may start wondering on how one may actually tackle the collision problem. First it may be noted that an obvious constraint on the selected basis sets is that they tend to stationary states of the separated partners. The simplest expression for such a constraint is e.g.

$$\phi_n (\{\vec{q}_i\} \; ; \vec{r}, \vec{R} \to \infty) \to \phi_\alpha^A \chi_\beta^{BC} \tag{15}$$

$$G_v^n (\vec{r} \; ; \vec{R} \to \infty) \qquad \to G_{v_\beta}^{n,BC} (\vec{r}) \tag{16}$$

where G_v is seen to be a vibro-rotational wavefunction of the molecule BC in the electronic state β while the atom A is in the state α. Of course we have assumed in eqs. (15), (16) that a Born-Oppenheimer separation of the electronic and vibrorotational motion is a valid approximation for the isolated BC molecule. It may be remarked right away that if the wavefunctions ϕ_n and G_v^n are to be treated in the body fixed reference frame, the electrons (and nuclei B, C) are viewed in a frame that is (and keeps) rotating with the internuclear axis even at the begining (or at the end) of the collision. Hence proper angular momentum algebra has to be performed to couple the internal motion to the intermolecular axis rotation in the begining of the collision and to decouple it at the end of the collision.[6-9, 11-13] Of course "begining" and "end" are irrelevant expressions within a (time-independent) stationary description of the collision, they are just meant here to designate the initial and final asymptotic conditions. The mentioned angular momentum algebra is rather straighforward although it is quite lengthy and requires quite careful notation and definitions. Readers interested in this topic are referred to refs. 6-9, 11-13.

With the constraints (17 a, b), equation (16) is seen to determine the functions $F_{nv}(\vec{R})$ whose behaviours when $R \to \infty$ yield the probability amplitudes associated with the states $\phi_n G_v^n \sim \phi_\alpha^A \chi_\beta^{BC} G_{v_\beta}^{BC}$ when the nuclei are scattered in the direction θ, Φ i.e. : *the scattering amplitudes.*

3. ELECTRONIC REPRESENTATIONS

As a first introductory remark to this section, it should first be recognized that despite the aforementioned analogies between the formalism associated with the description of bound states of molecules and that relevant to the description of slow molecular collisions, the latter displays a few characteristic features that require a specific approach. Indeed, contrary to the case of bound states of triatomic molecules where the *slow* nuclear motion is confined in a *small* $\vec{R}$, $\vec{r}$ domain, in an atom + molecule collision problem the nuclei explore wide ranges of R and r. R varies from the separated A, BC limit ($R \to \infty$) down to the turning point which can be as small as a fraction of bohr radius. In addition, when vibrational excitation or even dissociation of the molecule are induced by the collision, r may vary considerably. Moreover, the nuclei may evolve over a wide velocity scale as one considers low ($E \simeq 1$ eV/amu) moderate ($E \simeq 100$ eV/amu) or high ($E \simeq$ several keV/amu) energy collisions. It is easily conceivable that a unique method of treatment of all these conditions at once may hardly be found and therefore the adiabatic representation that we present first may not constitute a "panacea".

3.1 Adiabatic representation

The best known basis set of ϕ_n functions is that suggested by the Born-Oppenheimer approximation.[3] It consists of solving the eigenvalue equation :

$$H_{el}\,\phi_n^{ad} = E_n\,(R,r,\gamma)\,\phi_n^{ad} \tag{17}$$

for each geometry R, r, γ of the three nuclei (A, B, C). Accordingly the terms V_{nk} in eq. (9) disappear and V_{nn} is replaced by E_n. The use of such a representation will be best justified when the remaining coupling terms in the right hand side of eq. (9) are so small that one or only a few terms in eq. (8a) need to be considered. In this case $E_n(R,r,\gamma)$ represents the actual potential from which may be derived the instantaneous forces that govern the motion of the nuclei A, B, C. Clearly, if the right hand side of eq. (9) were strictly zero eq. (8a) would reduce to a single term thereby excluding electronic excitation (eq. (15)). In this case only vibro-rotational excitation (and possibly reactive scattering) would be left.

In order to have a feeling of how important the $\vec{D}_{nk}$ and $\vec{d}_{nk}$ terms may be, we consider first the following matrix elements $\partial/\partial\alpha$: α = R, r, γ. From the commutation relation

$$\left[\,\partial/\partial\alpha,\ H_{el}\,\right] = (\partial H_{el}/\partial\alpha) \tag{18}$$

it is seen that wavefunctions satisfying eq. (17) exactly obey the Hellman-Feynman theorem [4, 14, 15]

$$<\phi_n^{ad}|\ \frac{\partial H_{el}}{\partial\alpha}\ |\phi_m^{ad}> = \frac{\partial E_m}{\partial\alpha}\ \delta_{nm} + (E_m - E_n)\ <\phi_n^{ad}|\ \frac{\partial}{\partial\alpha}\ |\phi_m^{ad}> \tag{19}$$

Eq. (19) shows that except for specific symmetry reasons making $<\phi_n^{ad}|\ \partial H_{el}/\partial\alpha\ |\phi_m^{ad}>$ vanish identically, the off diagonal matrix elements of $\partial/\partial\alpha$ (and thereby the $\vec{D}_{nm}$ and $\vec{d}_{nm}$ terms) are particularly large for closely lying energy levels. Moreover, owing to the general relation :

$$<\phi_n|\ \frac{\partial^2}{\partial\alpha^2}\ |\phi_m> = \frac{\partial}{\partial\alpha}\ <\phi_n|\ \frac{\partial}{\partial\alpha}\ |\phi_m> + \sum_k\ <\phi_n|\ \frac{\partial}{\partial\alpha}\ |\phi_k>\ <\phi_k|\ \frac{\partial}{\partial\alpha}\ |\phi_m> \tag{20}$$

it is seen that τ_{nk} in eqs. (9), (10a) may also become large when $\partial/\partial\alpha$ is large and varies rapidly. It may also be remarked that the $\vec{D}_{nm}$ and $\vec{d}_{nm}$ matrix elements are multiplied by the $\vec{\nabla}_R/\mu$ and $\vec{\nabla}_r/m$ operators which roughly correspond to the classical velocity ($v_{\vec{R}}$, $v_{\vec{r}}$) of the nuclear motions in $\vec{R}$ and $\vec{r}$. A rough estimate based on first order pertutbation theory may provide some deeper insight into the combined role of $\vec{v}_{\vec{R}}.\vec{D}_{nk}$, $\vec{v}_{\vec{r}}.\vec{d}_{nk}$ (noted Ω_{nk} below) and energy separation :

$$\phi_n^{pert.} \simeq \phi_n + \sum \frac{\Omega_{mk}}{E_n - E_k}\ \phi_k \tag{21}$$

It is seen that the reduction of eq. (8a) to a single term is valid only when :

$$\left| \frac{\Omega_{nk}}{E_n - E_k} \right|^2 << 1 \tag{22}$$

which, by the way, is the criterion of validity of the adiabatic theorem.[1]

Electronically non-adiabatic processes are likely to occur when condition (22) breaks down. As pointed out in the introduction, obvious cases of breakdown occur at very high collision velocities ($v_{\vec{R}}$) when the collision duration is "short". But what is really "short" ? Eq. (22) actually provides the answer : "short with respect to the relevant transition periods" $T_{nk} = |E_n - E_k|^{-1}$. Accordingly, at thermal and low energies the collision duration may be comparable with or shorter than spin-orbit splittings in the isolated atom and/or molecule which may result in inter and intra multiplet transitions at such low energies. In the same energy range the collision duration may be comparable to the energy spacing between two nearly degenerate charge exchange levels which may thereby favour electron transfer, etc... Moreover, since $E_n - E_k$ varies with the conformation of the molecule, the collision system may have an electronic adiabatic behaviour for several geometries of approach whereas it may behave non adiabatically close to near degeneracies of energy levels which are known to occur around so-called conical intersections or avoided crossing seams.[2, 16-22] The qualitative prediction and understanding of non-adiabatic transitions in low energy molecular collisions depends on our ability to locate near degeneracies and (pseudo) crossings of potential energy hypersurfaces.

It is well-known that the adiabatic energy levels of molecular systems obey the "non-crossing rule".[2, 16-22] This rule specifies that the intersection of energy levels of two states of the *same symmetry* is a hypersurface of *at most* ($\mathcal{N}$-2) dimensions *, whereas for states having *different symmetries* the intersection is a hypersurface of ($\mathcal{N}$-1) dimensions. For the considered A B C triatomic systems ($\mathcal{N}$-2) and ($\mathcal{N}$-1) hypersurfaces are curves and surfaces respectively. This reference to the symmetry of the considered states offers the opportunity to review some properties of adiabatic electronic wavefunctions.

---

* Footnote : $\mathcal{N}$ is the number of nuclear degrees of freedom.

3.2 Miscellanous properties of adiabatic electronic wavefunctions

Since H_{el} is a real hermitian operator its eigenfunctions are real (except possibly for phase factors), orthogonal (or may be chosen so) and may be normalized. According to fundamental theorems of quantum mechanics and group theory, the eigenfunctions of H_{el} transform according to the irreducible representations of the group of symmetry operations commuting with H_{el} (i.e. leaving H_{el} invariant). The most general symmetry group of a triatomic system is C_s whose elements are the identity (I) and the reflection σ in the plane containing the three nuclei (triatomic plane). The irreducible representations A' and A" of this group obey :

$$\sigma A' = A' \quad , \quad \sigma A'' = - A'' \tag{23}$$

since : H_{el}, $\partial/\partial R$, $\partial/\partial r$, $\partial/\partial\gamma$ commute with σ (i.e. are invariant in a reflection about the triatomic plane), the matrix elements of these operators are non zero only for states of the same symmetry : A' - A' or A" - A".

We have already pointed out in Sec. 2.1 that the operators $\partial/\partial\theta|_{S'}$ and $\partial/\partial\Phi|_{S'}$, that appear in the components of $\vec{\nabla}_R$, generate in the body fixed reference frame, additional operators involving the electronic angular momentum components L_X, L_Y, L_Z. These operators are parts of the non adiabatic couplings $\vec{D}_{nk}.\vec{\nabla}_R/\mu$ and $\vec{d}_{nk}.\vec{\nabla}_r/m$. Their symmetry properties are such that

$$\sigma L_X = - L_X \sigma \quad , \quad \sigma L_Y = + L_Y \sigma \quad , \quad \sigma L_Z = - L_Z \sigma \tag{24}$$

Hence, L_X and L_Z (which correspond to swinging motions of the triatomic plane around the X and Z axes respectively) couple A' and A" states. On the other hand, L_Y (which corresponds to a sliding motion of the triatomic plane onto itself without deformation) couples states of same symmetry : A' - A' or A" - A".

It is worth noticing that for a general A B C system there always exists a particular conformation of the nuclei, namely : the linear conformation. In this limiting geometry the symmetry group of the molecule is $C_{\infty v}$. As in the atom-atom case we have the selection rules :

(i) $\Delta\Lambda = 0$ (e.g. $\Sigma^{\pm} - \Sigma^{\pm}$, $\Pi_{X,Y} - \Pi_{X,Y}$, etc...) for $\partial/\partial R$, $\partial/\partial r$ and $\partial/\partial\gamma$ or ($\Sigma^{\pm} - \Sigma^{\pm}$, $\Pi_{X,Y} - \Pi_{Y,X}$, etc...) for L_Z

(ii) $\Delta\Lambda = 1$ (e.g. $\Sigma^+ - \Pi_X$, $\Sigma^- - \Pi_Y$, etc...) for $\partial/\partial\gamma$ and L_Y, or,
($\Sigma^+ - \Pi_Y$, $\Sigma^- - \Pi_X$, etc...) for L_X.

where we have used the notation $\Pi_{X,Y}$ to designate respectively the symmetric and antisymmetric combinations of the degenerate Π states, in a reflection about the triatomic plane. It should be noticed that $\partial/\partial\gamma$ breaks the linear geometry and thus pertains to both (i) and (ii).

When BC is a homonuclear diatomic molecule another particular conformation arises namely : the isoceles triangle conformation. In this new limiting geometry the symmetry group of the molecule is C_{2v} whose elements are : the identity (I), two reflections (one in the triatomic plane : σ, the other σ' in the plane perpendicular to σ) and a rotation of $\pi/2$ about the intersection of σ and σ' (Z axis) the corresponding selection rules are then

$A_i \leftrightarrow A_i$,	$B_i \leftrightarrow B_i$ $(i = 1, 2)$	for	$\partial/\partial R$, $\partial/\partial r$ and $\partial/\partial\gamma$
$A_1 \leftrightarrow B_1$,	$B_2 \leftrightarrow A_2$	for L_X	
$A_1 \leftrightarrow B_2$,	$B_1 \leftrightarrow A_2$	for L_Y and $\partial/\partial\gamma$	
$A_1 \leftrightarrow A_2$,	$B_1 \leftrightarrow B_2$	for L_Z	

Again $\partial/\partial\gamma$ breaks the C_{2v} geometry and thereby obeys two selection rules.

When all three nuclei are identical a third special conformation arises, namely : the equilateral triangle conformation. The symmetry group of such a molecule is D_{3h}. The operators $\partial/\partial R$ and $\partial/\partial r$ correspond to deformations of the equilateral triangle towards a C_{2v} geometry and generally produce mixings of $A'_{1,(2)}$ and $E'_{Z,(X)}$ states whereas $\partial/\partial\gamma$ corresponds to deformations towards C_s and cause couplings between A'_1, A'_2 and $E'_{Z,X}$. Similarly A''_2 states couple with E" states. Here again E'_Z and E'_X designate the linear combinations of the degenerate E' states that are symmetric and anti-symmetric with respect to the plane perpendicular the triatomic plane.

As a convention for later discussions we will call the non-adiabatic couplings due to the motions along R, r and γ *internal dynamic deformation couplings*, and when further specification is needed : R-coupling, r-coupling or γ-coupling. The other couplings arising from the rotation of the triatomic system without deformation will be called *rotational or Coriolis couplings*.

This brief reminder of the symmetry properties of the eigenfunctions of H_{el} immediately reveals an invaluable tool for tracking some important regions of non-adiabatic coupling. These regions are expected to be located in a more or less close neighbourhood of the degeneracies and/or allowed curve crossings that occur in the limiting geometries. For instance, R- and γ-coupling is likely to occur near Σ-Π ($C_{\infty v}$), $A_1 - B_2$ or $B_1 - A_2$ (C_{2v}) crossings or E' (D_{3h}) degeneracy. Similarly Coriolis coupling may be expected near a Π ($C_{\infty v}$) or E' (D_{3h}) degeneracy or near a Σ-Π ($C_{\infty v}$) or $A_1(B_2) - B_1$ crossing. In these cases one may speak of *exploiting underlying spatial symmetries* to pinpoint non-adiabatic coupling regions.

So far we have not considered the electron spin. If spin-orbit interactions V_{so} are completely excluded the Wigner "spin-conservation rule" holds and non-adiabatic transitions may only take place between states of same spin-multiplicity. Otherwise, V_{so} may be incorporated to H_{el} when solving the eigenvalue equation (17) or may be treated as terms of type V_{nk} in eqs. (9), (10a) (see Sec. 3.3.2e). In both cases the Coriolis type couplings (eqs. (5), (6), (9), (10d)) should be corrected by replacing the electronic orbital angular momentum $\vec{L}$ by the total electronic angular momentum[23] $\vec{J} = \vec{L} + \vec{S}$.

3.3 Diabatic representations

The central idea of this section is that since non-adiabatic transitions between electronic states arise precisely in cases of breakdown of the Born-Oppenheimer approximation which constitutes the foundation of the adiabatic representation, there is no obligation to stick to electronic wavefunctions generated by eq. (17) especially when such a representation may cause computational and/or conceptual problems. In particular, it should always be kept in mind that evaluation of the internal dynamic deformation couplings in the adiabatic representation requires multiplication of ab-initio calculations in order to determine the wavefunctions at ($R_o + \delta R$, r_o, γ_o), (R_o, $r_o + \delta r$, γ_o) and (R_o, r_o, $\gamma_o + \delta\gamma$) as needed to perform the relevant numerical differentiations. The number of necessary points may increase dramatically when the internal deformation couplings vary rapidly. Moreover, the handling of such rapidly varying quantities may cause serious troubles in actual resolution of the close coupling equations (9), (14). One should thereby stand back from the seeming "easiness" of systematic adiabatic calculations.

The first departures from the customary adiabatic description arose in treatments of atom-atom collision problems during the thirtees. The notion of diabatic states appeared in 1935 when Hellman and Syrkin [24] recognized that one could take advantage of the arbitrariness of an expansion like eq. (8) to build a representation where the $\vec{D}_{nk}$ couplings could vanish. Yet, the concept was already virtually present in 1932 works dealing with the ionic-covalent ($A^+(^1S) + X^-(^1S) \leftrightarrow A(^2S) + X(^2P)$) crossings of alkali-halide (AX) systems [25] which entailed the well-known Landau-Zener linear model of the two-state curve crossing problem. [26, 27] It was Lichten, [28] however, who triggered off the spreading of the concept of diabatic behaviour by revealing that the adiabatic representation limited to a few states could fail in the description of collision problems. His, analysis of the resonant charge exchange process in the He^+ + He collision forced him to introduce *diabatic states that could run freely through an infinite series of crossings* (fig. 3) in contradistinction to adiabatic states that are bound by the non-crossing rule. Another particular merit of this work is that it has demonstrated the importance of molecular orbitals, configuration state functions (CSF) and correlation diagrams in tracking the important regions were electronic transitions may occur and as essential notions related to the concept of diabatic states.

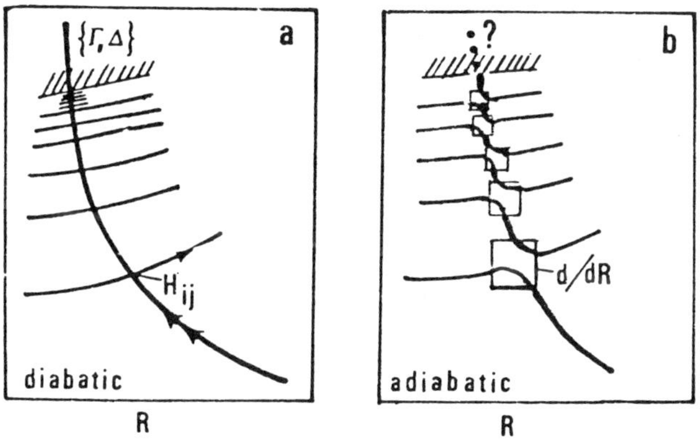

Fig. 3 - Schematic diagram of diabatic (a) and adiabatic (b) potential energy curves. A diabatic state may run freely through an infinite series of crossings and may cut into a continuum. At each diabatic crossing (a) the states are weakly coupled thereby justifying two-state approximations. At each adiabatic avoided crossing (b) the states interact via large and rapidly varying dynamic couplings (e.g. d/dR) thereby necessitating consideration of an infinite number of states.

3.3.1 Formal definitions

No practical calculations involving the foregoing ideas could actually be performed in the absence of a related formalism. The proper theoretical formulation of the problem emerged in the late sixtees in two complementary works by O'Malley [29] and Smith. [30] Both formulations aimed at avoiding as far as possible large and rapid variations of the electronic wavefunctions with the nuclear coordinates which frequently occur in the adiabatic representation near pseudo-crossing seams. Smith [30] proposed for the atom-atom problem that :

$$\langle \phi_m^{dia} | \, \partial/\partial R \, | \phi_n^{dia} \rangle \equiv 0 \tag{24}$$

and derived the generating equation :

$$\frac{d\mathbf{C}}{dR} + \mathbf{t}_R \mathbf{C} = 0 \tag{25a}$$

where $\mathbf{C}(R)$ is the transformation matrix between an arbitrarily selected basis set $\{\psi_\alpha\}$ (e.g. the adiabatic set) and the sought diabatic basis ϕ_n^{dia}

$$\phi_n^{dia} = \sum_\alpha \psi_\alpha \, C_{\alpha n}(R) \tag{25b}$$

and :

$$(\mathbf{t}_R)_{\alpha\beta} = \langle \psi_\alpha | \frac{\partial}{\partial R} | \psi_\beta \rangle \tag{25c}$$

The work of O'Malley [29] contained a different way of achieving the prescription of eq. (24) namely the use of a projection operation technique. Accordingly, if the considered ϕ_m and ϕ_n states are constrained to belong to orthogonal subspaces, the independent variational (adiabatic-like) determination of the states in each subspace, for each R, should closely satisfy the criterion of eq. (24) (since a state in one subspace cannot mix with any state in the other, as would be the case in the presence of different symmetries).

Equations (24), (25) may be generalized to polyatomic systems. [31] For the particular atom-diatomic case Baer [9] showed that the extension of these equations writes :

$$\vec{\overline{\omega}} \, \mathbf{C} + \vec{\mathbf{t}} \, \mathbf{C} = 0 \tag{26a}$$

where :

$$\vec{\overline{\omega}} = (\partial/\partial R, \ \partial/\partial r, \ \partial/\partial \gamma) \tag{26b}$$

$$\vec{\mathbf{t}}_{\alpha\beta} = \langle \psi_\alpha | \, \vec{\overline{\omega}} \, | \psi_\beta \rangle \tag{26c}$$

and $\mathbf{C} = \mathbf{C}(R,r,\gamma)$ has a similar definition to that given in eq. (25b). Necessary and sufficient conditions for a solution to exist is that $\mathbf{C}$ be an analytic function of R, r and γ and that the components of $\vec{\mathbf{t}}$ obey the commutation relations :

$$\frac{\partial \mathbf{t}_x}{\partial y} - \frac{\partial \mathbf{t}_y}{\partial x} = \left[\mathbf{t}_x, \ \mathbf{t}_y \right] \qquad x, y = R, r, \gamma \tag{27}$$

The integral equation corresponding to eq. (26a) reads : [9]

$$\mathbf{C}(R,r,\gamma) = \mathbf{C}(R_o,r_o,\gamma_o) - \int_{R_o}^{R} dR' \, \mathbf{t}_R (R',r,\gamma)\,\mathbf{C}(R',r,\gamma) -$$

$$- \int_{r_o}^{r} dr' \, \mathbf{t}_r (R_o,r',\gamma)\,\mathbf{C}(R_o,r',\gamma) - \int_{\gamma_o}^{\gamma'} d\gamma' \mathbf{t}_\gamma (R_o,r_o,\gamma')\,\mathbf{C}(R_o,r_o,\gamma') \tag{28}$$

In a two-state problem, equation (28) takes a particularly simple form since $\mathbf{C}$ is determined by a single function $\eta(R,r,\gamma)$:

$$\mathbf{C} = \begin{pmatrix} \cos\eta & \sin\eta \\ -\sin\eta & \cos\eta \end{pmatrix} \tag{29}$$

with :

$$\eta(R,r,\gamma) = \eta(R_o,r_o,\gamma_o) - \int_{R_o}^{R} dR' \, t_R(R',r,\gamma) - \int_{r_o}^{r} dr' \, t_r(R_o,r',\gamma) -$$

$$\int_{\gamma_o}^{\gamma} d\gamma' t_\gamma(R_o,r_o,\gamma') \tag{30}$$

These equations mean that to propagate $\mathbf{C}(R_o,r_o,\gamma_o)$ to some point $\mathbf{C}(R,r,\gamma)$ one has to integrate with respect to γ at fixed R_o, r_o, then with respect to r at fixed R_o, γ, and finally with respect to R at fixed r, γ. [9] However complicate these equations might seem, their existence is somewhat at variance with an opinion of Tully [7] that '*the concept of diabatic states will not be as useful in molecular collision problems as it has been for atom-atom processes' owing to the increased dimensionality of the problem.* Moreover, as in the atom-atom case, there are numerous ways of satisfying to a large extent the condition

$$\langle \phi_n^{dia} | \vec{\omega} | \phi_m^{dia} \rangle \cong 0 \tag{31}$$

without resorting to the actual resolution of eqs. (26)-(28). The main ideas to do so stem from the works of Lichten [28] and O'Malley [29] while keeping in mind that eq. (31) tells us qualitatively that diabatic behaviour means *preservation* of the main characteristics of a wavefunction independently of the nuclear geometry. Beyond the technical aspects that have been or will be discussed, it is important to always keep in mind that molecular collision problems at low and moderate energies present a dual aspect : they are slow enough for a molecular (adiabatic) behaviour to show up and they are rapid enough for some characters of the system to be preserved.

3.3.2 Practical procedures for generating diabatic states

a) Preservation of molecular configuration

One obvious characteristic that may be preserved as the A B C triangle is deformed during the collision is the electronic configuration of the system. It was in particular pointed out in ref. 32 that since, $\vec{\omega}$ acts on a Slater determinant as a one electron operator : $\vec{\omega} = \sum_i \vec{\omega}_i$, *its matrix elements vanish* if ϕ_n and ϕ_m differ by at least two spin-orbitals. This property provides the means of building "quasi-diabatic" representations in the sense prescribed by eq. (31). An example of such a treatment is provided by the determination of the diabatic multicrossing pattern (fig. 4) of the $He^+ + H_2$ collisional system.[33] Still, a description based on single configuration state functions (CSF) built from self consistent field (SCF) molecular orbitals (MO) - including improved virtual orbitals (IVO)[33] may turn out to be insufficient to account for some subtle features of the potential energy surfaces (potential wells, barrier heights, proper dissociation, etc..) which often require large scale configuration interaction (CI) calculations. According to the suggestion of O'Malley[29] such CI calculations may be performed in subspaces spanned by the representative quasi-diabatic CSFs.[33, 34-36] The method thus described may be used to treat problems like charge exchange excitation e.g.

$$A^+ + BC \rightarrow A + BC^{+*} \; (\rightarrow A + B + C^+)$$

excitation transfer :

$$A^* + BC \rightarrow A + BC^*$$

Penning and associative ionization :

$$A^* + BC \rightarrow \begin{cases} (ABC)^+ + e^- \\ A + BC^+ + e^- \end{cases}$$

which clearly imply two-electron rearrangement processes.

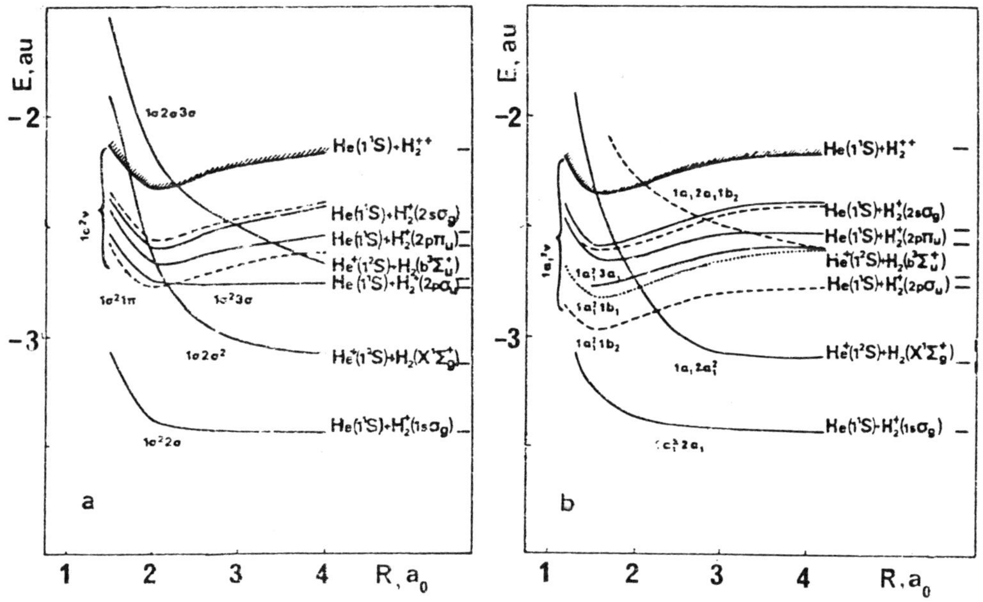

Fig. 4 - Cuts of a few diabatic potential energy hypersurfaces E(R,r,γ) of the $(He-H_2)^+$ molecular ion for a fixed H_2 bond length (r = 1.4 a_o) and varying the $He-H_2$ distance (R) for γ = 0 (a) and γ = π/2 (b). The curves are taken from ref. 33.

b) Preservation of the isolated partner characters

Contrary to the foregoing cases a large class of molecular collision processes involve direct one-electron transitions. The best known examples are

(i) Near-resonant charge exchange

$$A^+ + BC \rightarrow A + BC^+$$

with a small exo- or endo-ergicity between reactants and products, and

(ii) Chemi-ionization (ion-pair formation)

$$A^{(*)} + BC \rightarrow A^+ + BC^-$$

In this wide class of problems CSF built from SCF-MO do not constitute the best starting point. Indeed, since the mentioned processes occur at large R values ($\gtrsim$ 5 a_o), a valence bond (VB) description involving CSF built from orbitals of the isolated collision partners should provide a good first guess of the relevant diabatic states since they are determined independently of R, r and γ. Still, the considered VB-CSF should be properly orthogonalized. This task may of course be achieved by orthogonalization of the orbitals originating from the isolated partners as done in the projected-VB (PVB) method.[37] Some precautions ought however to be taken in doing so since a relation exists between the orthogonalization procedure and the coordinate origin which is to be held fixed during the $\partial/\partial R$ and $\partial/\partial\gamma$ differentiations.[38, 39] Again, if such single CSF may enable one to adequately determine total cross section for the considered processes, they are likely to fail in the detailed description of differential scattering. Indeed raw PVB-CSF lack the proper adiabatic distortions (polarisation and exchange) caused by *energetically distant levels which do not participate directly in the considered non-adiabatic transition*. Again, this defect may be cured by CI in subspaces spanned by PVB-CSF having different characteristics. e.g. : ionic or covalent, reactant-like or product-like...

c) Frozen orbitals and underlying symmetries

From the above two examples it appears that practical diabatization procedures may easily be set up after recognition of specific atomic or molecular characteristics that ought to be preserved under the penalty of introducing large and *rapidly varying internal dynamic deformation couplings*. This is particularly obvious in the vicinity of (conical) intersections or (Jahn-Teller) degeneracies, occuring in limiting geometries (e.g. : $C_{\infty v}$: A-B-C, C_{2v}: $A - \begin{smallmatrix}B\\ \vdots\\ B\end{smallmatrix}$, D_{3h} $A \lessdot \begin{smallmatrix}A\\ \vdots\\ A\end{smallmatrix}$, $D_{\infty h}$: B-A-B), where the γ- or R- or r- coupling may vary quite abruptly (displaying discontinuous shapes or δ-function like behaviours, see Sec. 3.3.2.e).[22] In such cases the characteristics of the states that should be preserved is their symmetry in the limiting geometry (ies). Before proceeding further, it should be stressed that the discussed features near conical intersections (or Jahn-Teller degeneracies) or more general avoided crossings between (many electron) CSF often reflect underlying avoided crossings between the (one-electron) MO by which the CSF differ (figs. 5, 6).[40] It is thus clear that the basic diabatic CSF references (that serve in defining the subspaces in which CI is to be subsequently performed) should be built from *diabatic orbitals* obeying eq. (30) as far as possible.

We have already discussed the case when the diabatic orbitals are essentially the frozen orbitals of the isolated partners. This idea may be extended and actually constitutes the grounds of the so-called frozen orbital (FO) method.[41] In the first step of the FO method, the

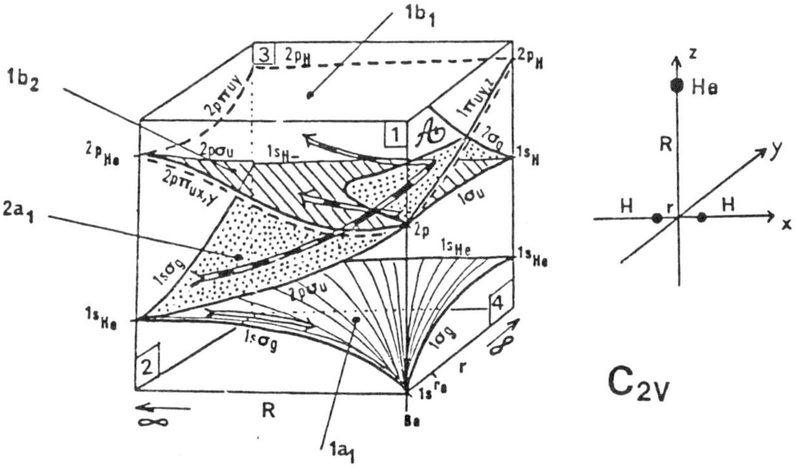

Fig. 5 - Triatomic correlation diagram of the $He^{(+)}$-H_2 system near the limiting C_{2v} conformation. The intersection of the $2a_1$ (dotted) and $1b_2$ (hatched) MO surfaces as well as the regions of contact of the $1b_2$ and $1b_1$ (dashed-transparent) surfaces indicate regions of important non-adiabatic coupling (the figure is borrowed from ref. 40).

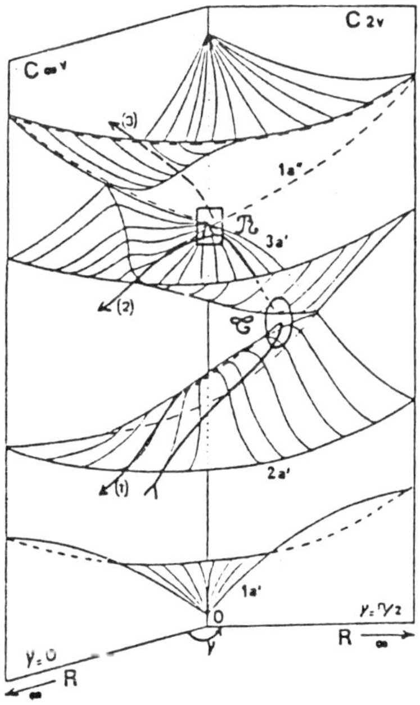

Fig. 6 - Triatomic correlation diagram of the $He^{(+)}$-H_2 system in C_s conformation for $r \simeq r_e$. The arrows represent examples of elastic scattering (1) excitation of the H_2 2pσ MO (2) and excitation of the H_2 2pπ MO (3) via Coriolis coupling (this figure is taken from ref. 40).

expansion coefficients of the diabatic orbitals over a basis set, that is constructed independently of R, r and γ (for example : a basis of contracted gaussian orbitals on each center A, B, C), are determined well outside the critical non-adiabatic zone (e.g. using an SCF calculation) and are kept the same irrespective of the nuclear arrangement. The technicalities then proceed as in the aforementioned PVB method. In cases where one deals with crossings between states of different symmetries in a limiting geometry the freezing procedure may be somewhat relaxed since it has only to be applied to the coordinate(s) that break (or lower) the symmetry of the considered limiting geometry. This point of view is nicely supported by the calculations of Rebentrost and Lester,[42] on the $F(^2P)+H_2$ ($1^2A'$, $2^2A'$ and $1^2A''$) collisional system. These authors showed that the two $^2A'$ diabatic states that remove the badly behaved R- and γ- couplings of the adiabatic basis in the $F+H_2$ problem are merely determined by frozen orientations along Z and X of the in-plane p-hole of $F(^2P)$.

Single CSF's built from more or less frozen orbitals may be rightly judged of insufficient accuracy by to date's Quantum Chemistry and Theoretical Molecular Physics standards. Nevertheless, we reiterate that once well characterized diabatic prototypes have been brought out using the aforementioned procedures they may always be improved by CI in orthogonal subspaces.[33-36]

d) Maximum overlap between adiabatic states and diabatic prototypes

An alternative method has recently been proposed[43] in order to solve some dilemmas which might be encountered when one cannot decide whether a state (or a group of states) should better be used for distortion of one or another diabatic prototype. The method starts by a standard determination of adiabatic states (or orbitals) $\{\phi_\alpha^{ad}\}$ to the desired level of accuracy. One then selects the few, say n, adiabatic states (or orbitals) that are thought to be adequate for the description of the considered non-adiabatic collisional process. Having determined, according to the preceeding discussions, n diabatic protoypes $\{\chi_q^d\}$ representing configurations (or orbitals) with chemical significance one may define n projections :

$$\bar{\chi}_q = \sum_{\alpha}^{n} \phi_\alpha^{ad} \langle \phi_\alpha^{ad} | \chi_q^d \rangle \quad (32)$$

which are non-orthogonal images of the χ_q^d prototypes in the selected n x n adiabatic subspace. Symmetric orthogonalisation of the $\bar{\chi}_q$ projections :

$$\phi_i^{(d)} = \sum_q \overline{\chi}_q \, (\sigma^{-1/2})_{qi} \quad ; \quad \sigma_{pq} = \langle \overline{\chi}_p | \overline{\chi}_q \rangle$$

$$\phi_i^{(d)} = \sum_\alpha \phi_\alpha^{ad} \sum_q \langle \phi_\alpha^{ad} | \chi_q^d \rangle \, (\sigma^{-1/2})_{qi} \tag{33}$$

provides the orthonormal vectors $\phi_i^{(d)}$ that resemble most (i.e. have the largest overlap with) the initial χ_i^d prototypes and defines a unitary transformation which achieves a n x n representation that is equivalent to the selected adiabatic basis. The very appealing aspect of this procedure is that it achieves a nice compromise between ("adiabatic") accuracy, chemical significance and the smoothest behaviour compatible with the selected size (n x n) of the adiabatic subspace. Yet, however attractive this method might seem, it does not definitively settle down the adiabatic versus diabatic debate since it postulates that an arbitrary collision problem may always be treated in terms of a small set of adiabatic states which, as pointed out above was shown to be untrue already twenty years ago.[28]

e) An exotic example : spin-orbit coupling

As mentioned in Sec. 3.2, when spin-orbit coupling :

$$V_{so} \approx \sum_{i=1}^{N} \alpha_i \, \vec{l}_i \cdot \vec{s}_i \tag{34}$$

is considered its effect as an additional source of non-adiabatic coupling may be handled in two different ways.

(i) V_{so} may be incorporated to H_{el} in the eigenvalue equation (17). The corresponding non-adiabatic transitions then arise from internal deformation dynamic couplings (which as already discussed, may cause computational troubles).

(ii) V_{so} may be kept outside H_{el} and is thereby treated as a coupling operator inducing electronic transitions.

Here again the problem of choosing a representation for the treatment of the dynamics (eqs. (9), (14)) arises. Although no specific reference is made to a particular pseudo-crossing problem an adiabatic (i) or a diabatic (ii) approach may be selected. To illustrate the relations between this example and the preceeding discussion consider the case of fine structure transitions within a $^2P_{J,M}$ manifold in an atom-molecule collision :[42, 43]

$$A(^2P^o_{JM}) + B_2(X^1\Sigma^+_g) \rightarrow A(^2P^o_{J'M'}) + B_2(X^1\Sigma^+_g)$$

To treat the ^2P atomic state one may consider as usual a p-electron (e.g. for an excited alkali-atom) or a p-hole (e.g. for a halogen atom). We have already pointed out that for non-adiabatic processes occuring at large R, the orbitals of the isolated partners constitute a good guess of the diabatic orbitals related to the R-motion. Moreover, selection of orbitals $a'p_x$, $a''p_y$ and $a'p_z$ that have fixed directions in the body fixed reference frame automatically provides diabatic orbitals for the γ-motion as well.[42] In linear geometry the $a'p_x$ and $a''p_y$ orbitals constitute the two components of the doubly degenerate π-state which splits apart from the $a'p_z$ (Σ-state) at finite R. In C_{2v} geometry the three components are split apart : $a'p_z$ (A_1) , $a'p_x$ (B_2) and $a''p_y$ (B_1). H_{el} couplings arise between the $a'p_z$ and $a'p_x$ states (away from the two limiting geometries $C_{\infty v}$ and C_{2v}) owing to the approach of the BC molecule at finite R. The trouble with the considered three states is that they are coupled by V_{so} even when $R \rightarrow \infty$. Nevertheless, since a constant (i.e. independent of R, r and γ) unitary transformation of diabatic states preserves their "diabaticity", one may remove the unwanted asymptotic couplings by diagonalizing $H_{el} + V_{so}$ at $R \rightarrow \infty$. Moreover, since the resulting $|J,M\rangle$ states are doubly degenerate, one may combine the two components of each $|J \pm M\rangle$ doublet to construct symmetry adapted basis functions that are symmetric or antisymmetric with respect to the ABC plane (Sec. 3.2). Following ref. 45 we define the symmetry adapted functions as :

$$|J,M\rangle + i\varepsilon\,(-1)^{J-M}\,|J,\pm M\rangle \quad , \ \varepsilon = \pm 1 \quad , \ J = 3/2,\ 1/2 \quad , \ |M| \leq J$$

Introducing : $p_x = (p_- - p_+)/\sqrt{2}$, $ip_y = (p_- + p_+)/\sqrt{2}$ and the spin functions, $\varphi_s = (-\alpha + i\beta)/\sqrt{2}$ and $\varphi_a = (\alpha + i\beta)/\sqrt{2}$ we get :

$$
\begin{bmatrix} 3/2,\ (1)_s \\ 3/2,\ (2)_s \\ 1/2,\ (3)_s \\ 3/2,\ (1)_a \\ 3/2,\ (2)_a \\ 1/2,\ (3)_a \end{bmatrix} = \begin{pmatrix} i/\sqrt{2} & i/\sqrt{2} & 0 & 0 & 0 & 0 \\ i/\sqrt{6} & 1/\sqrt{6} & -2/\sqrt{6} & 0 & 0 & 0 \\ i/\sqrt{3} & 1/\sqrt{3} & 1/\sqrt{3} & 0 & 0 & 0 \\ 0 & 0 & 0 & 1/\sqrt{2} & +i/\sqrt{2} & 0 \\ 0 & 0 & 0 & -i/\sqrt{6} & -1/\sqrt{6} & -2/\sqrt{6} \\ 0 & 0 & 0 & -i/\sqrt{3} & -1/\sqrt{3} & +1/\sqrt{3} \end{pmatrix} \begin{bmatrix} p_x & \varphi_s \\ p_y & \varphi_a \\ p_z & \varphi_s \\ p_x & \varphi_a \\ p_y & \varphi_s \\ p_z & \varphi_a \end{bmatrix} \tag{35}
$$

The above procedure is general and should be used each time a diabatic basis gives rise to asymptotic couplings.

The $H_{el} + V_{so}$ matrix in each of the 3×3 blocks thus brought out writes :

$$
\begin{pmatrix} \dfrac{E_x + E_y}{2} + \dfrac{\alpha}{2} & \dfrac{h+i\ (E_x - E_y)}{\sqrt{12}} & \dfrac{h+i\ (E_x - E_y)}{\sqrt{6}} \\ \dfrac{h-i\ (E_x - E_y)}{\sqrt{12}} & \dfrac{E_x + E_y + 4\ E_z}{\sqrt{6}} + \dfrac{\alpha}{2} & \dfrac{E_x + E_y - 2\ E_z - 3\ ih}{\sqrt{18}} \\ \dfrac{h-i\ (E_x - E_y)}{\sqrt{6}} & \dfrac{(E_x + E_y - 2\ E_z) + 3\ ih}{\sqrt{18}} & \dfrac{E_x + E_y + E_z}{\sqrt{3}} - \alpha \end{pmatrix} \tag{36}
$$

where $E_{x,y,z}$ are the energies of the states labelled $p_{x,y,z}$ and h is the $p_x - p_z$ interaction for $\gamma \neq 0, \pi/2$. This particular example illustrates the extreme simplicity of the result (eq. (36)) as opposed to the extreme difficulty of determining *numerically* the $\partial/\partial R$ and $\partial/\partial\gamma$ couplings in the adiabatic basis, especially if a crossing point R_c exists (e.g. in $C_{\infty v}$ geometry : $E_x\ (R_c, \gamma = 0) = E_y\ (R_c, \gamma = 0) = E_z\ (R_c, \gamma = 0)$ as actually found in the F + H_2 collisional system).[42, 43z]

Actually, if the internal dynamic deformation couplings $\vec{\omega}$ are still needed (now that the $H_{el} + V_{so}$ matrix is set up) it may be derived from the relation between the adiabatic basis and the equivalent diabatic basis (i.e. spanning the same space)

$$
\phi_\alpha^{ad} = \sum_i \phi_i^d < \phi_i^d | \phi_\alpha^{ad} > \tag{37a}
$$

$$< \phi_\alpha^{ad} | \vec{\bar{\omega}} | \phi_\beta^{ad} > = (E_\beta - E_\alpha)^{-1} \sum_{i,j} < \phi_\alpha^{ad} | \phi_i^d > (\vec{\bar{\omega}} V_{ij}) < \phi_j^d | \phi_\beta^{ad} > , \alpha \neq \beta \quad (37b)$$

where V_{ij} is the diabatic $H_{el} + V_{so}$ matrix, e.g. eq. (36). Of course eq. (37b) is general and may be used when spin-orbit interactions are not considered. In the particular two-state case eq. (37b) writes :

$$< \phi_I^{ad} | \vec{\bar{\omega}} | \phi_{II}^{ad} > = \frac{V_{12} \vec{\bar{\omega}} (V_{22} - V_{11}) - (V_{22} - V_{11}) \vec{\bar{\omega}} V_{12}}{(E_{II} - E_I)^2} \quad (38)$$

For the linear model of *a conical intersection* at $R = R_c$, $\gamma = \gamma_c = 0$ one has [20, 39]

$$V_{22} - V_{11} \approx a (R - R_c) \quad (39a)$$

$$V_{12} \approx b \gamma \quad (39b)$$

$$< \phi_I^{ad} | \frac{\partial}{\partial R} | \phi_{II}^{ad} > = \frac{a b \gamma}{a^2 (R - R_c)^2 + 4 b^2 \gamma^2} \quad (39c)$$

$$< \phi_I^{ad} | \frac{\partial}{\partial \gamma} | \phi_{II}^{ad} > = \frac{-a b (R - R_c)}{a^2 (R - R_c)^2 + 4 b^2 \gamma^2} \quad (39d)$$

Analysis of this particular example (which is left to the reader) may help having a feeling of how abruptly such couplings may behave near the apex of the conical intersection ($R = R_c$, $\gamma = 0$).

Hence, with the above example we have illustrated the role of symmetry in the constitution of diabatic prototypes and with eqs. (38)-(39) we have had the opportunuity of verifying how troublesome the so familiar adiabatic basis may be in circumstances which actually occur quite frequently. Finally we have also seen how the triatomic aspect of the problem including spin-orbit interactions introduces couplings between the $|3/2, \pm 3/2>$ components and the couple $|3/2, \mp 1/2>$ and $|1/2, \mp 1/2>$, a feature that does not appear in atom-atom-like problems (where $E_x \equiv E_y$ and $h = 0$).

4. SEMI-CLASSICAL TREATMENT OF MOLECULAR COLLISIONS

Scanning the literature over the past two decades reveals that, except for a very small number of attempts restricted to punctual problems,[13, 46, 47] calculations based on the quantum mechanical formalism depicted in Sec. 2 are virtually inexistent. This situation is essentially due to the fact that when a molecular collision starts to involve a few electronic states a huge number of related vibro-rotational channels are open. For example the number of rotational states is already as large as a few hundreds for an electronically-vibrationally elastic collision at $E \simeq 0.5$ eV/amu (this problem was not considered in ref. 47).

Until quite recently, the most widely spread method for the treatment of non-adiabatic molecular collision processes has been the so-called surface hopping trajectory (SHT) method.[7, 48] The reason why the method is so popular is that it treats without distinction reactive or non-reactive processes. The method has several variants and is well documented. Owing to space constraints its description will be limited to a brief reminder. As in the standard classical trajectory treatment of electronically adiabatic collisions[5] it consists of treating classically the motion of the three nuclei (A, B, C) whereas the electrons are treated quantally and determine the 4-dimensional hypersurfaces $E_n(R,r,\gamma)$ which govern the classical nuclear motion :

$$\mathring{P}_i = - \partial H/\partial Q_i \qquad \mathring{Q}_i = \partial H/\partial P_i \tag{40}$$

H is the classical hamiltonian, Q the relevant coordinates and P the conjugate momenta. In order to handle the motion along the set of potentials $E_n(R, r,\gamma)$ implied in the considered electronically non-adiabatic processes the SHT method requires for each pair of potentials E_n and E_m, the *a priori* knowledge of 3-dimensional hopping seams : $S_{nm}(R,r,\gamma) = 0$. These seams characterize strong non-adiabatic coupling regions (frequently pseudo-crossings)[7] where the trajectory splits into two branches each weighted by a certain probability (P_{nm} and $1 - P_{nm}$). In order to render the calculations tractable the P_{nm} probabilities are assigned functional dependences inspired from elementary models.[26, 27] Sometimes, much cruder methods are used - as e.g. random 0 or 1 values - based on the argument that it is essentially the number of passes through S_{nm} that determines the ultimate value of the transition probability rather than the actual value of P_{nm} at each pass.[7] The method meets with difficulties especially when the hopping seams are ill-defined (which may occur when different types of transition mechanisms come into play simultaneously for a pair of states).

It is the author contend that the most significant future progress in the field of non-reactive molecular collisions at extrathermal energies is likely to arise by following similar tracks as those which have brought the field of atom-atom collisions to maturity. One of these tracks is the semi-classical treatment of relative motion. The justification for resorting to such an approach lies in the fact that two heavy particles (or sets of particles) moving with relative velocities $v > 5.10^{-3}$ a.u. ($\simeq 10^7$ cm/s) have associated de Broglie wavelengths ($\lambdabar = 1/\mu v$) which are much smaller than characteristic atomic and molecular dimensions ($\simeq 1\ a_0$). Thus, to a large extent, the $\vec{R}$-motion may be treated classically while the motions along the other coordinates are kept quantal. This idea dates to the thirtees[49, 50] and various derivations have been proposed since.[51, 56] An outline of the (sufficient) conditions under which the semi-classical equations are derived is given below.

4.1 Classical path and common trajectory methods

Since semi-classical $\vec{R}$-motion is invoked the corresponding JWKB conditions should hold :[2, 56]

(i) the "local" radial wavelength $\lambdabar_l(R)$ (in each channel) is small

$$\lambdabar_l(R) = K_l^{-1}(R) = \lambdabar_l(\infty)\left[1 - \frac{U}{E} - \frac{(l + 1/2)^2}{\mu^2 v^2 R^2}\right]^{-1/2} \ll 1 \qquad (41a)$$

(here U is the relevant potential and l the orbital angular momentum of relative motion appearing when the functions $F_{nv}(\vec{R})$ of Sec. 2.2 are expanded over a basis of spherical harmonics, see e.g. eq. (49)).

(ii) the variations of $\lambdabar_l(R)$ are small :

$$\left|\ \lambdabar_l(R)^{3/2}\ \frac{d^2}{dR^2}\left[\lambdabar_l(R)\right]^{1/2}\ \right| \ll \frac{d\,\lambdabar_l(R)}{dR} \ll 1 \qquad (41b)$$

(iii) $l \gg 1$ so that : $\Theta\, l \gg 1$, and $(\pi - \Theta) l \gg 1$ (41c)

Θ being the scattering angle

The latter condition is often met since $1 \gtrsim \mu v a_o$ ($\mu \gtrsim 2000$ a.u., $v \gtrsim 5.10^{-3}$ a.u.).

Under these conditions it may be shown that provided that :

(iv) the turning points of classical motion R_T (such that : $K_1(R_T) = 0$) are well removed from the regions where (rovibronic) transitions are most likely to occur,

and that :

(v) the wavenumber differences for an arbitrary pair of considered coupled states obey :

$$|K_1(R) - K_1'(R)| \ll \frac{1}{2}(K_1(R) + K_1'(R)) \approx (K_1(R)\, K_1'(R))^{1/2} \approx \overline{K}_1(R) \tag{42}$$

the quantum mechanical problem of Sec. 2 is equivalent to solving the time-dependent-like Schrödinger equation for the internal (ro-vibronic) motion :

$$i \frac{\partial}{\partial t} \psi(\{\vec{q}_i\}, \vec{r}, \vec{R}(t)) = (-\frac{1}{2m} \Delta_{\vec{r}} + H_{el})\, \psi \tag{43}$$

where $\vec{R}(t)$ is determined by the classical motion in an average potential (eq. (42)) :

$$\mu \overset{\circ}{R}(t) = \pm \overline{K}_1(R) \quad ; \quad \overset{\circ}{\theta}(t) = -(1 + 1/2)/\mu R^2 \quad ; \quad t = \pm |t| \tag{44}$$

It turns out in fact, that the detailed knowledge of the trajectory is not of crucial importance as will be shown below in relation with eqs. (50)-(59).

In the classical $\vec{R}$-path method (eqs. (43)-(44)) ψ is expanded over a basis of ro-vibronic wavefunctions :

$$\psi = \sum_{nv} C_{nv}(t)\, \phi_n\, G_v^n \exp\left[-i \int^t U_{nv}(t')\, dt'\right] \tag{45}$$

where $\phi_n\, G_v^n$ are the same as those discussed in Sec. 2.1 and :

$$U_{nv}(t) = \langle G_v^n | -\frac{\Delta_{\vec{r}}}{2m} + \mathcal{H}_{nn} | G_v^n \rangle \tag{46}$$

with : $$\mathcal{H}_{nk} = \langle \phi_n | -\frac{\Delta_{\vec{r}}}{2m} + H_{el} | \phi_k \rangle_{\{\vec{q}_i\}} \tag{47}$$

Projection of eq. (44) onto each rovibronic state of eq. (45) entails the following set of close coupling equations

$$i \frac{d}{dt} C_{nv} = \sum_{v'} \langle G^n_v | \mathcal{H}_{nn} - i \frac{\partial}{\partial t} | G^n_{v'} \rangle \times$$

$$C_{nv'} \exp \left[- i \int^t (U_{nv'} - U_{nv})\, dt' \right] +$$

$$+ \sum_{\substack{k,v'' \\ k \neq n}}' \langle G^n_v | \mathcal{H}_{nk} - i \langle \phi_n | \frac{\partial}{\partial t} | \phi_k \rangle - i \frac{\partial}{\partial t} - \frac{\vec{d}_{nk} . \vec{\nabla}_{\vec{r}}}{m} | G^k_{v''} \rangle \times$$

$$C_{kv''} \exp \left[- i \int^t (U_{kv''} - U_{nv})\, dt' \right] \qquad (48)$$

where $\langle \phi_n | \frac{\partial}{\partial t} | \phi_k \rangle = \overset{\circ}{\vec{R}} . \vec{D}_{nk}$

Clearly the structure of the right hand side of eq. (14) has been preserved. Therefore, the origin of non adiabatic couplings in the body fixed reference frame is the same as discussed in Secs. 2.3. Note that there is one set of equations of this type for each impact parameter $b = (1 + 1/2)\, \lambda\!\!\!{}^{-}(\infty)$.

To extract the scattering matrix within the above framework one needs to know the relation between the C_{nv} probability amplitudes and the scattering amplitudes $F_{nv}(\vec{R})$. To do so, one has to expand the $F_{nv}(\vec{R})$ functions over spherical harmonics. As mentioned in Sec. 2.2 this requires in principle the proper treatment of relative and internal angular momentum couplings.[6-9, 11-13, 55] However, for simplicity we will completely circumvent this step and admit (for conciseness and simplicity) the following approximation :

$$F_{nv}(\vec{R}) \approx \sum_{1} R^{-1} A^1_{nv}(R)\, P_1(\cos\theta) \qquad (49)$$

The required asymptotic behaviour of the radial wavefunctions A^1_{nv} is as usual :[56]

$$A^1_{nv}\,(R \to \infty) \sim k_{nv}^{-1/2} \left| A^{1(-)}_{nv}\, e^{-i(k_{nv} R - \frac{1\pi}{2})} - A^{1(+)}_{nv}\, e^{+i(k_{nv} R - \frac{1\pi}{2})} \right| \qquad (50)$$

$(k_{nv} = K_{nv1}(R \to \infty))$. Equation (50) defines the scattering matrix **S** as :

$$A^{1(+)}_{nv} = \sum_{k,v''} S^1_{nv,kv''} A^{1(-)}_{kv''} \qquad (51)$$

Now the form of $A^1_{nv}(R)$ used to derive eqs. (43)-(48) is :

$$A^1_{nv}(R) = K^{-1/2}_{nvl}(R) \left| a^{1(-)}_{nv}(R)\, e^{-i\,\Phi^1_{nv}(R)} - a^{1(+)}_{nv}(R)\, e^{+i\,\Phi^1_{nv}(R)} \right| \tag{52a}$$

with :

$$\Phi^1_{nv} = \int_{R_T^{nvl}}^{R} K_{nv}(R')\,dR' + \pi/4 . \tag{52b}$$

Identification with eq. (50) entails :

$$A^{1(\pm)}_{nv} = a^{1(\pm)}_{nv}(\infty)\, e^{\pm i\,\eta^1_{nv}} \tag{53}$$

where η^1_{nv} is merely the textbook JWKB phase shift :[56]

$$\eta^1_{nv} = \lim_{R\to\infty} \left| \Phi^1_{nv}(R) - k_{nv} R + \frac{1\pi}{2} \right| \tag{54}$$

Choosing the lower bound of the integral in eq. (45) so that the relation between $a^{1(\pm)}_{nv}$ and $C_{nv}(b,t)$, for each $b = (1 + 1/2)/\overline{k}$ writes :

$$C_{nv}(b,t) = a^{1(\pm)}_{nv}(R) \qquad \text{for} \quad t = \pm\,|t| \tag{55}$$

the condition :

$$C_{nv}\,(b, t \to -\infty) = \delta_{nn_i}\,\delta_{vv_i} \tag{56}$$

entails

$$S^1_{nv,n_iv_i} = C_{nv}\,(b, t \to +\infty)\, \exp\left| i\,(\eta^1_{nv} + \eta^1_{n_iv_i}) \right| \tag{57}$$

It should be stressed that this expression is determined by the (arbitrary) choice : $t = 0$ of the lower bound of the phase integral in eq. (45). This is the most natural choice for a didactic presentation. *However, it is often more convenient to choose the lower bound of the discussed phase integral* to be $t \to -\infty$. It is easily shown then that the phase $\eta^1_{nv} + \eta^1_{n_iv_i}$ in eq. (57) becomes $2\eta^1_{nv}$.

The differential (σ) and total (Q) cross sections are then determined from the expressions :

$$\sigma_{nvn_iv_i}(\theta) = \frac{1}{4\,k^2_{n_iv_i}} \left| \sum_l (2l+1)\,(S_{nv,n_iv_i} - \delta_{nn_i}\,\delta_{vv_i})\,P_l\,(\cos\theta) \right|^2 \tag{58}$$

$$Q_{nvn_iv_i} = \frac{\pi}{k^2_{n_iv_i}} \sum_l (2l+1)\,\left| S_{nv,n_iv_i} - \delta_{nn_i}\,\delta_{vv_i} \right|^2 \tag{59}$$

Equations (50)-(57) show, as announced, that the actual scattering in each channel is not to much constrained by the prescribed common average trajectory since the only explicit uses of eqs. (42), (44) are made in the phase differences appearing in eq. (48) :

$$\pm\,(\Phi^l_{nv} - \Phi^l_{n'v'}) \approx \int_o^t (K_{nvl}(R) - K_{n'v'l}(R))\,\frac{\overline{K}_l(R)}{\mu}\,dt'$$

$$\approx -\int_o^t (U_{nv} - U_{n'v'})\,dt \qquad\qquad t = \pm\,|t| \tag{60}$$

and in the expression of $\langle\phi_n|\,\partial/\partial t\,|\phi_{n'}\rangle$. Hence, the common trajectory appears as a device to extract the probability amplitudes which are essentially determined by the semiclassical action differences (60) particularly in regions where eq. (42) holds. The specific relations between scattering angle and impact parameter, which may be deduced by stationary phase evaluation of eq. (58), clearly depend on the actual reactant and product channels through the JWKB phases (eq. (57)). This discussion justifies to some extent the frequent uses of the so-called impact parameter method where the trajectory is as simple as a straight line :

$$\vec{R} = \vec{b} + \vec{v}\,t \quad ; \quad \mathring{R} = \pm\,v\,\left(1 - \frac{b^2}{R^2}\right)^{1/2} \quad ; \quad \mathring{\theta} = -\,\frac{vb}{R^2}\ , \ t = \pm\,|t| \tag{61}$$

More realistic suggestions for the choice of the trajectory are generalisations of eq. (42) to several states (i.e. arithmetic or geometric averages) or the trajectory governed by the average potential :[54]

$$\overline{U} = \langle\psi|\ -\frac{\Delta_{\vec{r}}}{2m} + H_{el}\ |\psi\rangle \tag{62}$$

with or without consideration of the cross terms arising from eq. (45).

4.2 Sudden approximations for rotation and vibration

So far consideration of eq. (48) instead of eq. (14) has the advantage of providing via the $C_{nv}(b,t)$ amplitudes a view of the evolution of the system as the collision proceeds. Another advantage is that only the needed row (or column) $n_i v_i$ of the scattering matrix is determined thereby achieving some reduction in computation effort. Still, the number of coupled channels to be handled remains as large as in the quantum mechanical treatment. Nevertheless, it should be reminded that if as many as several hundreds to several thousands of channels are open (as is the case for the rotational channels associated with each vibronic state) this is due to fact that the energy separation between two states is tiny compared to the collision energy. We have already mentioned in Sec. 1.1 that this circumstance is characteristic of sudden behaviour. It is easily seen that normal rotation periods of most molecules are such that their orientation $\hat{r}$ in a space fixed reference frame may barely change while the collision partners (A and BC) cover a distance of several a_o at energies $E \gtrsim 1$ eV/amu. Similarly, the vibration periods of molecules are such that the bond length r may barely vary over the duration of a collision at E > 500 eV/amu. Let us first consider the latter case.

4.2.1 - Frozen vibration-rotation at "high" energies

Let us reconsider eqs. (43), (45) where we choose for convenience a strictly diabatic electronic representation ϕ_n^{dia}. Moreover let us choose a basis of vibrorotational wavefunctions quantized in fixed space and obeying :

$$\left(-\frac{1}{2m}\Delta_{\vec{r}} + V_{nn}^T\right) g_v^n(\vec{r}) = U_{nv}^T\, g_v^n(\vec{r}) \tag{63}$$

with $V_{nn}^T = V_{nn}(\vec{r}, \vec{R}(-T))$. It easy to obtain by projecting eq. (45) onto ϕ_n only :

$$\sum_v i\, \mathring{C}_{nv}\, g_v^n\, e^{-i\int_{-T}^{t} U_{nv}^T dt'} = (V_{nn} - V_{nn}^T) \sum_v C_{nv}\, g_v^n\, e^{-i\int_{-T}^{t} U_{nv}^T dt'} +$$

$$\sum_{k\neq n} \left(V_{nk} - i \langle\phi_n^{dia}| \frac{\partial^c}{\partial t} |\phi_k^{dia}\rangle\right) \sum_{v''} C_{kv''}\, g_{v''}^k\, e^{-i\int_{-T}^{t} U_{kv''}^T dt'} \tag{64}$$

The $\partial/\partial t$ term acting on g_v^n does not appear owing to the definition of the vibrorotational wavefunctions. Moreover, all couplings between the considered diabatic states are concentrated in the $V_{nk} = \langle\phi_n^{dia}|\ H_{el}\ |\phi_k^{dia}\rangle$ term with the exception of Coriolis coupling which is left as $\langle\phi_n^{dia}|\ \partial^c/\partial t\ |\phi_k^{dia}\rangle$. It may be noticed, in passing, that the use of the frozen vibrorotational wavefunctions $g_v^n(\vec{r},T)$ extends the discussion of section 3.3 to vibro-rotational representations. Another obvious remark is that, in the $g_v^n(\vec{r},T)$ basis, vibrational and rotational excitation in the manifold n arises from the $\vec{r}$ dependence of the perturbing term : $V_{nn}(\vec{R},\vec{r}) - V_{nn}^T(\vec{r})$.

We define short collision times T with respect to vibration-rotation by requiring that the ro-vibrational kinetic energy $U_{nv}^T - V_{nn}^T$ for an arbitrary g_v^n level is so small that, for $t \in [-T, T]$:

$$\int_{-T}^{t} (U_{nv}^T - V_{nn}^T)\ dt' \ll 1 \qquad \forall\ v \tag{65}$$

With this condition it is seen that, with the change of functions : $a_{nv}(\vec{r},t) = C_{nv}(t) \exp\left[-i \int_{-T}^{t} V_{nn}^T(\vec{r})\ dt'\right]$, eq. (64) becomes

$$i \frac{d}{dt} \sum_v a_{nv}\, g_v^n = V_{nn} \sum_v a_{nv}\, g_v^n + \sum_{k\neq n} (V_{nk} - i \langle\phi_n^{dia}|\ \frac{\partial^c}{\partial t}\ |\phi_n^{dia}\rangle) \sum_{v''} a_{kv''}\, g_{v''}^k \tag{66}$$

It is now possible to lighten the notation by the replacement :

$$\sum_v a_{nv}\, g_v^n = A_n(\vec{r},t)\, G_I(\vec{r}) \qquad , \ \forall\ n \tag{67}$$

which entails, for $t \in [-T, T]$:

$$i \mathring{A}_n = V_{nn} A_n + \sum_{k\neq n} (V_{nk} - \langle\phi_n^{dia}|\ \frac{\partial^c}{\partial t}\ |\phi_k^{dia}\rangle)\ A_k \tag{68}$$

Eq. (68) has the familiar form of the close-coupling equations arising in atom-atom collisions. However, this set is to be solved for a 3-dimensional grid of fixed distances r and orientations θ_r, Φ_r of the diatomic molecule. By projecting eq. (67) onto g_v^n at t = + T we find that the probability amplitude C_{nv} for ro-vibronic excitation writes :

$$C_{nv}(+T) = \langle g_v^n \mid A_n(\vec{r}, +T) \exp\left(i \int_{-T}^{T} V_{nn}^T\ dt'\right) |G_I\rangle \tag{69}$$

In particular if :

$$C_{nv}(-T) = \delta_{nn_i}\ \delta_{vv_i} \tag{70}$$

then, G_I is merely replaced by $g_{v_i}^{n_i}$. These results enable one to discuss various cases.[57, 58]

Firstly, it is seen that the probability of an electronic transition $n_i \to n$ irrespective of the final vibro-rotational state is :

$$\sum_v |C_{nv}(+T)|^2 = \langle G_I| A_n(\vec{r}, +T)|^2 |G_I\rangle \tag{71}$$

as may be obtained by using eq. (69) together with the closure relation in the basis g_v^n. Equation (71) expresses that the summed probability is merely equal to the average of the electronic transition probability over the initial wavepacket G_I.

Secondly, eq. (69) provides conditions for the applicability of the Franck-Condon principle to vibronic transitions. It is thus seen that the *sudden collision condition* eq. (65) has to be complemented by two conditions :

(i) The (near) independence of $A_n(\vec{r};T) \exp(i \int_{-T}^{T} V_{nn}^T \, dt')$ upon bond length

$$\sum_j |C_{nv,j}(+T)|^2 = \langle j_i||A_n(\hat{r};T)|^2|j_i\rangle \langle v_n|\tilde{v}_I\rangle^2 \tag{72}$$

Here we have expanded the notation : j designates collectively the rotation quantum numbers of the BC molecule whereas v_n designates specifically its vibration quantum number in electronic state n. $\tilde{v}_I$ is not necessarily a stationary state of vibration ; it represents the superposition of vibration states, defined in eq. (63), that has been prepared prior to reaching the considered time interval $[-T, +T]$. Hence, the second condition is :

(ii) The identity of $\tilde{v}_I$ and v_n with asymptotic states of vibration i.e.

$$\tilde{v}_I \equiv v_{n_i}^{\infty} \quad \text{and} \quad v_n \equiv v_n^{\infty} \tag{73}$$

Here, the upper index (∞) means that the vibrational states are eigenfunctions of the relevant equation of the type of eq. (63) with $T \to \infty$.

The Lipeles model[59] and its variants are cases of departure of the latter condition. The Lipeles model expresses that the system follows *adiabatically* a single state of vibration determined by the potential $V_{nn}(\vec{r},\vec{R})$ until $\vec{R}(-T)$ where it undergoes a rapid electronic transition (eq. (65)).

Variants of the model [58] express the fact that when *vibrationally non-adiabatic transitions* are likely to occur prior to reaching $\vec{R}(-T)$ the system does not necessarily remain is the same adiabatic state.

Thirdly, it may well occur that the potential energy curve $V^{\infty}_{nn}(r)$ of the electronic state populated by the non-adiabatic transition is repulsive thereby entailing dissociation of the molecule with a center of mass energy ε. In such cases the gross features of the transition probability may be studied using a δ-function approximation of the vibration continuum : [57] $\delta(r - r^*(\varepsilon)) \, |dV_{nn}/dr|^{-1/2}_{r^*(\varepsilon)}$ (where : $r^*(\varepsilon)$ is the classical turning point at energy ε in the relevant potential V_{nn}). Accordingly, the preceeding treatment (eqs. (63)-(69)) yields the corresponding dissociation spectrum $s_n(\varepsilon,\hat{r})$ as a function of the molecular orientation $\hat{r}$:

$$s_n(\varepsilon,\hat{r}) = \left| \frac{A_n(\vec{r}, +T)}{|dV_{nn}/dr|^{1/2}} \, g^n_v(r) \right|^2_{r=r^*(\varepsilon)} \tag{74}$$

where eqs. (70a) and (73) have been assumed.

Fourthly, the reduction of the rovibronic problem to the treatment of electronic transitions for fixed $\vec{r}$ (eq. (68)) enables one to use the familiar two-state models that have been built to explain electronic transition processes in atom-atom collision problems. For example, near-resonant charge exchange problems may be discussed by making reference to the Demkov model [60] where two states have parallel energy levels and are coupled by an exponential (exchange) interaction $\mathcal{E}\exp(-\Lambda R)$. Disregarding molecular orientation and interferences, the corresponding charge transfer probability P_{fv_f,iv_i} is thus expressible as :

$$P^{Demkov}_{fv_f,iv_i} = \frac{1}{2} \left| \langle v_f | \operatorname{sech} \frac{\pi \Delta V(r)}{\Lambda(r)\, v_R} | v_i \rangle \right|^2 \tag{75}$$

where : $\Delta V(r)$ is the assumed R-independent energy difference $|V_{ii}(r) - V_{ff}(r)|$ for fixed r, $\Lambda(r)$ the exponent of the exchange interaction and v_R the radial velocity $\overline{K}_1(R)/\mu$ in the critical non-adiabatic transition zone R_c defined as :

$$\mathcal{E}(r)\, e^{-\Lambda(r) R_c} = \frac{\Delta V(r)}{2} \tag{76}$$

Equation (75) shows that strong departures from the condition of applicability of the Franck-Condon principle specified in eq. (72) will arise at low velocities v_R when ΔV varies significantly with r. This may be seen from the variation with ΔV of the hyperbolic secant in eq. (75) which exhibits a maximum at $\Delta V(r) = 0$ (crossing or degeneracy in r) and a width at half maximum proportional to Λv_R. Another example is provided by the curve crossing problem. For simplicity, let us restrict the discussion to the two internal variables R and r. A diabatic crossing seam in this problem is a relation $R_c(r)$. Since the transition probability is known to be a rapidly rising function of impact parameter b near $b = R_c(r)$,[61] strong non-Franck-Condon transitions are expected to occur with variation of b in the onset range of the transition probability.[62] (see figure in opposite page).

It should be kept in mind that the neglect of phases in the aforementioned examples is an expedient that was only motivated by our will of illustrating some salient features. Normally these phases are important and should be taken into account. A clear illustration of the role of phases is provided by the *electronically elastic* problem (i.e. one value of n in eq. (45)). In such cases the relevant potential ($\mathcal{H}_{nn}$ or V_{nn} in eqs. (47) or (63) resp.) is determined by the adiabatic electronic wavefunction ϕ_n^{ad}. Equation (68) for one term is easily solved :

$$A_n^{elastic}(t) = \exp\left(-i \int_{-T}^{t} V_{nn}(\vec{r}, \vec{R}(t'))\, dt'\right) \tag{77}$$

Disregarding molecular orientation ($\hat{r}$) effects, eqs. (69)-(70) write :

$$C_{nv'}^{elastic}(+T) = \langle v'|\, \exp\left|-i \int_{-T}^{T} (V_{nn}(r,\vec{R}(t)) - V_{nn}^{T}(r))\, dt\right| |v\rangle\,. \tag{78}$$

When v is the vibrational ground state of the electronic state n, an approximate expression of eq. (78) may be found by expanding the phase, noted $\Delta\eta$, near the equilibrium distance r_e of the BC molecule :

$$\Delta\eta(r) = \Delta\eta(r_e) + (r-r_e)\left.\frac{\partial \Delta\eta}{\partial r}\right|_{r_e} + \frac{(r-r_e)^2}{2}\left.\frac{\partial^2 \Delta\eta}{\partial r_e^2}\right|_{r_e} \tag{79}$$

Explicit evaluation of the r-integral in eq. (78) may be achieved by approximating the vibration wavefunctions v = 0, v' by harmonic oscillator wavefunctions :[63]

$$|C_{nv'}^{elastic}(+T)|^2 = \frac{\exp\left(-2c^2/(1+d^2)\right)\, |d|^{v'}}{2^{v'}\, v'!\, (1+d^2)^{(v'+1)/2}} \cdot \left|H_{v'}\; c/(d^2+id)^{1/2}\right|^2 \tag{80}$$

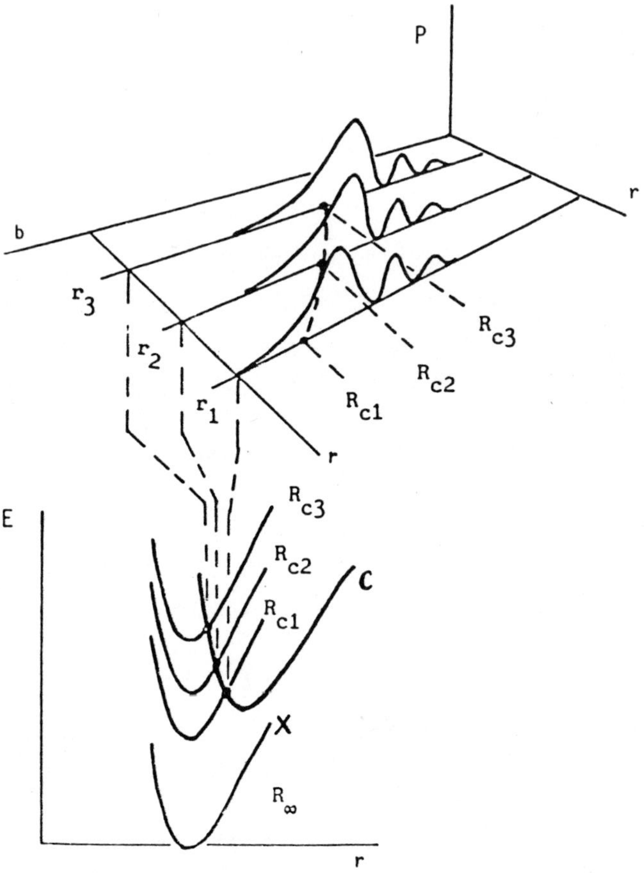

Scheme of the possible r-dependence of a crossing pattern between a repulsive (X) and a flat (C) energy surface. Lower figure : slices of the X and C potential energy surfaces at fixed $R(R_1 > R_2 > R_3)$ as functions of r. Upper figure : expected impact parameter dependence of the electronic transition probability at fixed values of r (r_1, r_2, r_3). The figure illustrates the displacement of the onset range of the probability curves with the decrease of r (R_{c1}, R_{c2}, R_{c3} are the crossings shown in the lower figure). The figure is taken from ref. 62.

where $H_{v'}$ is the Hermite polynomial, $c = (\partial\Delta\eta/\partial r)/\alpha$, $d = (\partial^2\Delta\eta/\partial r)/\alpha^2$, $\alpha = m\,\omega$ and ω is the vibration frequency of the harmonic oscillator. When $d \to 0$ equation (80) reduces to the familiar Poisson distribution : $(2c^2)^{v'} \exp(-2c^2)/v'!$. To date, the role of phases in inelastic vibronic phenomena within the sudden approximation (eq. (69)) has not received the attention it deserves.

4.2.2 - Frozen rotation at "low and moderate" energies

In the 1-100 eV/amu collision energy range the molecule may perform several vibrations during the characteristic time of the collision. On the other hand the BC axis may hardly rotate so that a sudden approximation for the rotation is a reasonable one. From the preceding discussion we know that the related treatment essentially consists of solving a pure vibronic problem for a grid of fixed orientations $\hat{r}$ of the molecule (in fixed space). The corresponding equations are given in eq. (48) with the restrictions that : $\Delta_{\vec{r}}$ is replaced by $r^{-2}\,\partial\,(r^2\,\partial\,G_v^n/\partial r)/\partial r$, $G_v^n = G_v^n(r)$ and $d_{nk}\cdot\vec{\nabla}_r$ is reduced to its radial part.

It should be remarked that since $\hat{r} = (\theta_r, \Phi_r)$ is kept fixed, the angle $\gamma = (\widehat{\vec{r}, \vec{R}})$ varies with $\vec{R}(t)$. For a planar trajectory (Φ = constant) : deriving from a central average potential we have :

$$\cos\gamma = \sin\theta\,\sin\theta_r\,\cos(\Phi - \Phi_r) + \cos\theta\,\cos\theta_r \qquad (81a)$$

$$\frac{d\theta}{dR} = -\,\frac{(1 + 1/2)}{R^2\,K_1(R)} \qquad (81b)$$

Of course this remark holds also for Sec. 4.2.1.

The ro-vibronic excitation cross sections are obtained, as in eqs. (69)-(70) from the matrix elements, between the initial and final rotational states of the $\hat{r}$-fixed amplitudes. Similarly to eq. (71), results summed over all final rotational states are obtained as the average of the $\hat{r}$-fixed results over the initial distribution of $\hat{r}$.

4.3 Model case studies

Since a few years it has become feasible to perform calculations on vibronic networks (eq. (45)) involving a few ($\lesssim$ 5) electronic states each associated with a set of ($\lesssim$ 15) vibrational states [64-71] the majority of calculations carried out up to now have resorted to one or both of the following simplifying assumptions :

(i) neglect of anisotropy (γ-dependence)

$$H_{nk} = H_{nk}(r,R) \tag{83}$$

(ii) neglect of bond-length dependences of the non-adiabatic couplings

$$H_{nk} = h_{nk}(R,\bar{r}) + \delta_{nk} W_n(r) \tag{84}$$

In some cases [71] approximation (i) has been somewhat relaxed by carrying out the calculations at a few *fixed values of* γ and then performing a weighted average of the results. This procedure refers to the "infinite order sudden" (IOS) approximation which is commonly used with some success in the treatment of rotational excitation.[6] However, the physical image conveyed by the IOS approximation is that the $\vec{r}$-axis follows the $\vec{R}$ axis rotation which is conceivable but at lower energies than considered here.

Despite the crudeness of these approximations they have revealed new aspects of molecular collisions some selected examples of which are presented below.

4.3.1 - The classical and quantal views of the bond stretching phenomenon

This phenomenon is invoked in discussions of ion pair formation in alkali-atom (Cs,K)+ molecule (O_2) collisions.[73] The classical SHT description of the phenomenon is as follows. Consider the picture of fig. 7 and assume that eq. (84) for the covalent potential surface H_{cc} is such that $h_{cc} \equiv 0$ whereas the ionic potential surface is such that : $h_{ii} = -1/R$. Moreover let $h_{ic} = \mathcal{E} \exp - \Lambda R$. As the collision partners approach a hop from the covalent A + BC potential energy surface to the ionic surface at $r=r_e$, $R=R_{xe}$ forms the negative ion BC^- on the repulsive wall of the W_{ii} potential (fig. 7a). Since the crossing seam $R_x(r)$ is an increasing function of r (fig. 7b), the stretching of the BC^- molecule subsequent to the hop increases the crossing radius. It is thus seen that only a certain class of trajectories obeying a specific vibration tuning will be able to yield reneutralisation. Indeed, trajectories for which the time lag between two passages accross the seam differs from an integer number of BC^- vibration periods will have crossing radii $R_x(r) > R_{xe}$. Owing to the rapid decrease of the interaction h_{ic}, this case will prevent reneutralization. Varying the impact parameter causes variation of the mentioned time lag and provides a way of tuning through the condition for neutralization thereby causing dips in the differential cross section for ion-pair formation.

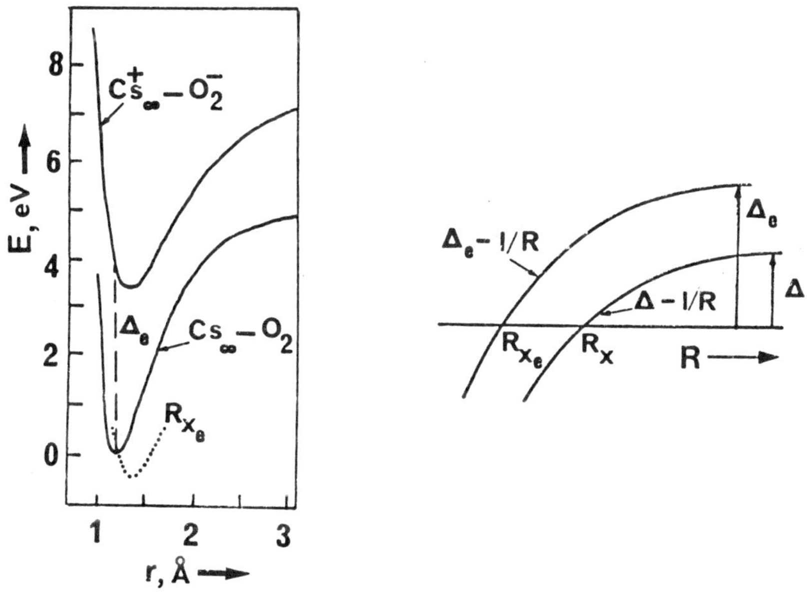

Fig. 7 - (a) Relative energy disposition of the separated neutral reactants and (ion-pair) products in the Cs + O_2 collisional system. Δ_e indicates the vertical energy separation of the curves at the equilibrium distance of O_2. (b) With the decrease of R the energy surface correlating with the ion-pair moves downwards according to a $- 1/R$ law.

One may now wonder how does this phenomenon manifest itself (if at all) in a treatment where the vibration motion is treated quantum mechanically.[67] The relevant pictures are shown schematically in figs. 8a-c. The incident covalent vibronic level $|c\ g_o^c>$ crosses a few ionic levels $|i\ g_v^i>$ and eq. (48) is solved subject to the conditions (83), (84). The basis is assumed to be diabatic in all respects (see Secs. 3 and 4.2.1). The (vibronic) ion-pair formation probabilities shown in figs. 8a-b exhibit a phenomen of quantum beats that are the quantum mechanical equivalent of the aforementioned bond stretching effect. The phenomenon results from an interference between three waves (fig. 8c). The two large phase differences associated with the covalent path and either of the ionic paths when the system evolves between the corresponding crossings, in the incoming or outgoing stages of the collision, cause standard Stueckelberg oscillations.[56, 61] The small phase difference that develops between two adjacent ionic paths (associated with g_v^i and g_{v+1}^i) causes a modulation of the Stueckelberg interference pattern in the

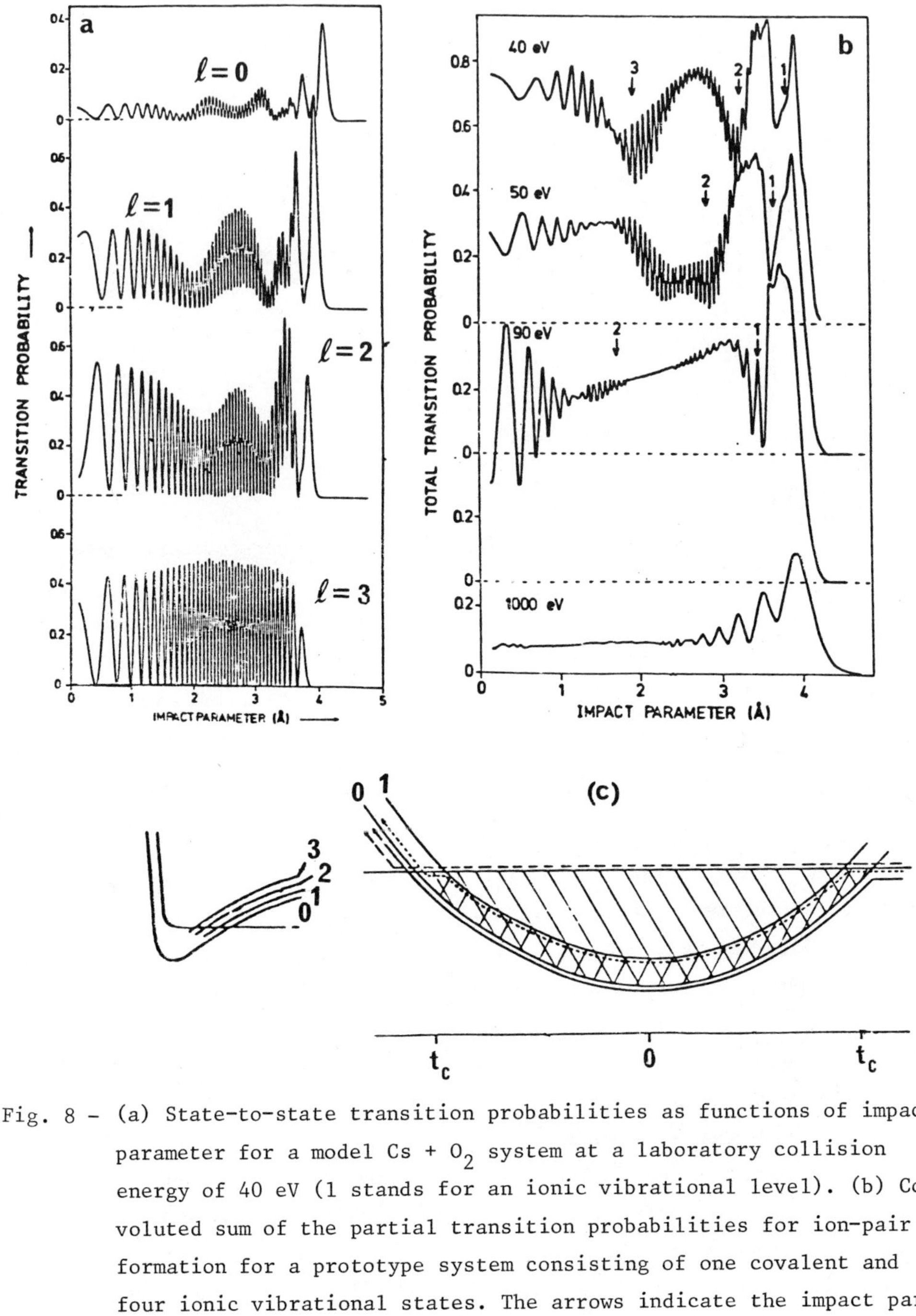

Fig. 8 – (a) State-to-state transition probabilities as functions of impact parameter for a model Cs + O_2 system at a laboratory collision energy of 40 eV (1 stands for an ionic vibrational level). (b) Convoluted sum of the partial transition probabilities for ion-pair formation for a prototype system consisting of one covalent and four ionic vibrational states. The arrows indicate the impact parameters for which the O_2^- ion can perform complete vibration during the collision. (c) Scheme of the three paths relevant to the discussion of the interference phenomena. (Figures are taken from ref. 67).

g_v^i probability. It is easily shown [67] that this small phase difference writes :

$$\omega . t_{coll} = 2 \pi \; t_{coll}/T_{vib} \tag{85}$$

where t_{coll} is the time lag between the entrance and the exit crossings and ω is the energy separation between v and v + 1 vibrational levels in the BC^- molecule. Hence equation (85) entails the same tuning condition (t_{coll}/T_{vib} = N) as that which is arrived at in the classical bond stretching model. The new aspect thus revealed lies in the relationship between the classical vibration and both quantum mechanical phases and vibronic interferences. In particular one sees that if t_{coll} becomes tiny compared to T_{vib} the discussed heating effect disappears which means that the vibration may barely occur during the collision (a point discussed in Sec. 4.2.1).

4.3.2 - Vibronic phenomena in the $(Ar, N_2)^+$ collisional system

Another problem which has recently been the subject of important activity is the charge exchange process :

$$(1) \quad N_2^+(X^2\Sigma_g^+, v) + Ar(^1S_0) \leftrightarrow (2) \quad N_2(X^1\Sigma_g^+, v') + Ar^+(^2P_{3/2,\ 1/2})$$

All theoretical studies were carried out using the vibronic network approach (Sec. 4.2.2). We will focus here on some salient features of this process and refer the reader to refs. 68-70, 45 for details. The simplest approximation of the corresponding H matrix (disregarding spin-orbit coupling as well as polarisation and charge quadrupole interactions) is provided by eq. (84) with

$$h_{11} \simeq 0 \quad , \quad h_{22} \simeq \Delta \quad , \quad h_{12} \simeq \mathcal{E} e^{-\Lambda R} \qquad (\Lambda \simeq 1) \tag{86}$$

as is assumed in the Demkov model [60] (Diabatic potential energy curves and interactions having such characteristics may be obtained using the methods described in section 3.3.2.b,d). [74] Since the relevant BC^+ and BC potentials ($W_1(r)$ and $W_2(r)$ respectively) are nearly identical the matrix elements $\langle g_{v_1}^1 | H_{12} | g_{v'_2}^2 \rangle = h_{12} \langle v_1 | v'_2 \rangle$ are small for $v \neq v'$ (e.g. smaller than 0.5 for $|v - v'| = 1$ and smaller than 0.1 for $|v - v'| > 1$). Hence the transitions to be investigated in the first instance are those having v = v'. In this case the problem is identical to that encountered in an ion-atom charge exchange collision characterized by eq. (86). Estimates of the corresponding total cross section may thenceforth be derived from the Olson scaling law.[75] One finds that the characteristic dimensionless parameter (of the Demkov model) that enters into this law : $\xi = 2\,\Lambda v/\pi\Delta$ (v being the collision velocity) is smaller than 0.5 for center of mass collision energies below 15 eV and hence the total cross section is a few percents of the geometrical cross section πR_c^2 ($R_c \simeq 3$ Å). This result means that the *purely electronic* non adiabatic transitions are so weak at the mentioned energies that an adiabatic 2-state basis diagonalizing the h matrix (eq. (86)) should in principle be preferred for physical significance. The question that arises then is : how does charge exchange occur at such low energies. A glance at fig. 9 immediately provides the answer : charge exchange

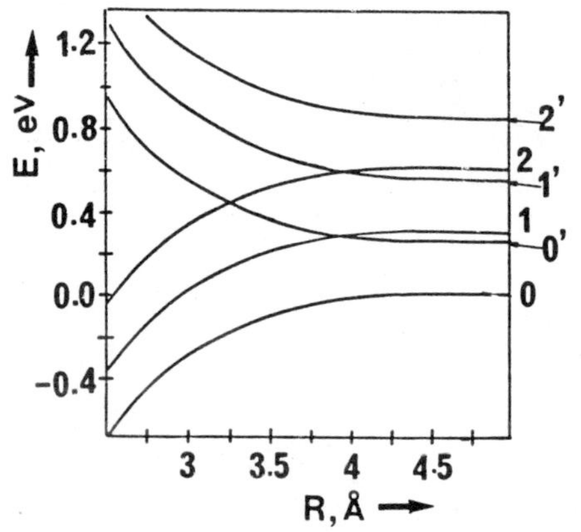

Fig. 9 - Schematic diagram illustrating the behaviour of two sets of vibronic levels (1v, 2v') obtained after diagonalization of the electronic h matrix in a system like $(Ar\text{-}N_2)^+$. v+1 → v' transitions are seen to be possible at curve crossings of vibronic states. (The diagram is inspired by the results of refs. 68, 69 and 45).

transitions will be favoured at vibronic network crossings. The vibronic energy curves shown in fig. 9 are generated as follows.[45] Let c(R) and s(R) be the coefficients of the transformation that diagonalises the h matrix in (eq. (86)) :

$$\phi_1^{ad} = c\,\phi_1^{dia} + s\,\phi_2^{dia} \qquad (87a)$$

$$\phi_2^{ad} = -s\,\phi_1^{dia} + c\,\phi_2^{dia} \qquad (87b)$$

the vibronic states may be defined as :

$$|1, v_1> = c\ \phi_1^{dia}|g^1_{v_1}> + s\ \phi_2^{dia}|g^2_{v_2}> \qquad (88a)$$

$$|2, v_2'> = -s\,\phi_1^{dia}|g^1_{v_1'}> + c\ \phi_2^{dia}|g^2_{v_2'}> \qquad (88b)$$

The corresponding matrix elements of the vibronic hamiltonian (see Sec. 4.2.2 and eqs. (43), (46)) $H_{el + vib}$ are :

$$U_{1v_1} = c^2\ \varepsilon_{1v_1} + s^2\ \varepsilon_{2v_2} + E_-\ (R)\ \underset{-}{\sim}\ \varepsilon_v + E_- \qquad (89a)$$

$$U_{2v_2'} = s^2\ \varepsilon_{1v_1'} + c^2\ \varepsilon_{2v_2'} + E_+\ (R)\ \underset{-}{\sim}\ \varepsilon_{v'} + E_+ \qquad (89b)$$

where :

$$E_{\pm} = \frac{(h_{11} + h_{22}) \pm \left|(h_{11} - h_{22})^2 + 4\ h_{12}^2\right|^{1/2}}{2} \qquad (89c)$$

and ε_{iv_i} (i = 1, 2) are the vibrational energies of the BC^+ and BC molecules referred to the minimum of their respective potentiels $W_1(r_e^+)$ and $W_2(r_e)$. Owing to the close resemblance of W_1 and W_2 one has $\varepsilon_{1v_1} \underset{-}{\sim} \varepsilon_{2v_2} \underset{-}{\sim} \varepsilon_v$. Curve crossings are seen to arise when :

$$\varepsilon_v - \varepsilon_{v'} = E_+ - E_- \qquad v > v' \qquad (90)$$

and the charge exchange transitions are induced by the coupling matrix element (see eq. (48))

$$<1v_1|H_{el + vib} - i\frac{\partial}{\partial t}\ |2\ v_2'> = (c^2 <g^1_{v_1}|g^2_{v_2'}> - s^2 <g^2_{v_2}|g^1_{v_1'}>)\ h_{12} \qquad v \neq v' \qquad (91)$$

The $\partial/\partial t$ contribution actually vanishes owing to the two orthogonality conditions $<\phi_1^{dia}|\phi_2^{dia}> = 0$ and $<g^i_{v_i}|g^i_{v_i'}> = \delta_{vv'}$. Consequently it may be said that the two crossing vibronic sets $|1v_1>$ and $|2v_2'>$ (v > v') are strictly diabatic although built on adiabatic electronic states. Since the Franck-Condon overlaps are very small for $\Delta v > 2$ transitions, the above discussion provides an explanation of the propensity for $v_1 = v_2' + 1$ charge exchange transitions in the N_2^+ + Ar collision at low energies.

A somewhat similar reasoning to that developped above may also be made for spin-orbit transitions

$$(2)\ Ar^+\ (^2P_{3/2}) + N_2\ (v') \leftrightarrow (3)\ Ar^+\ (^2P_{1/2}) + N_2\ (v'')$$

involved in excitation, quenching or population sharing accompanying charge transfer. An interesting aspect of spin-orbit transitions in the discussed system is that they do occur despite the fact that there exists no *direct coupling* between the *diabatic* $^2P_{3/2}$ and $^2P_{1/2}$ states. For instance, though eq. (36) in $C_{\infty v}$ geometry is nearly diagonal ($h = 0$, $E_x = E_y \underset{=}{\sim} E_z$) numerical close-coupling calculations (eq. (48)) yield $^2P_{1/2} - {}^2P_{3/2}$ transitions.[67-69] Actually, a coupling of the two states occurs via their interaction with the charge exchange channel $Ar + N_2^+$; this may be shown in the adiabatic basis of eq. (87a). For instance in $C_{\infty v}$ geometry, the transformation given in eq. (35) may also be applied if p_z is not the diabatic separated atom state of Sec. 3.3.2e but the adiabatic state of eq. (87a) (i.e. : $-$ s $3\sigma g$ (N_2) + c 3p (Ar)). In this case E_z in eq. (36) is replaced by the E_+ eigenvalue of eq. (89c). Thence forth the previous $(\sqrt{2}\,(E_{x,y} - E_z)/3)$ coupling between the two spin-orbit states is replaced by $\sqrt{2}\,(E_{x,y} - E_+)/3$ while their energy separation writes : $|(E_{x,y} - E_+)/3 - 3\alpha/2|$. Note that because $\alpha < 0$ for Ar^+ (more than half filled p shell) and $E_+ > E_{x,y}$ a crossing may be anticipated between the two spin orbit states correlating with |3/2, 1/2> and |1/2, 1/2>. These two states are diabatic since they are built as a *constant linear* combinations of the Σ state, arising from eq. (87b), and the $p_{x,y}$ Π state. The characteristics of the interaction between these two states makes that it comes under the Nikitin exponential model[76] rather than the Landau-Zener linear model.[26, 27] Contrary to the charge exchange process characterized by eq. (86), the existence of a crossing between the considered (purely electronic) spin-orbit states favours Ar^+ (3/2) $\leftrightarrow$ Ar^+ (1/2) transitions which conserve the vibration quantum number of the N_2 molecule.

The charge exchange and spin-orbit transitions have been disconnected in the above presentation for didactic purposes only. In actual determinations of detailed state-to-state cross sections[68, 69] these processes are treated at once and both vibrational excitation and deexcitation of the molecular partner in the electronically elastic channel are considered. In addition, the most recent calculations[69] have considered non-adiabatic effects induced by the $N_2^+(A^2\Pi)$ + Ar channel. Finally, it is worthwhile to remark that the discussed features of the $N_2^+ + Ar \leftrightarrow N_2 + Ar^+$ system constitute examples of situations where the familiar SHT approach will get into strong difficulties (whatever adiabatic or diabatic potential energy surfaces are used).

4.3.3 - Dissociative near resonant charge exchange

Another interesting case that has lent itself to modelling is the direct dissociative near-resonant charge exchange (DD-NRCE) process, a prototype of which is the reaction :[77]

$$(1)\ H_2^{+}(X,v) + Mg \rightarrow (2)\ H_2^{*}(b^3\Sigma_u^+, \varepsilon) + M_g^{+}$$

where, in the products, the notations $H_2^{*}(b^3\Sigma_u, \varepsilon)$ means H + H fragments dissociating with energy ε in the repulsive $b^3\Sigma_u^+$ potential of H_2. The interesting aspect of this problem is that the set of vibronic states associated with the products *form a continuum*. At high collision energies (several keV) this feature does not cause major difficulties as discussed in sec. 4.2.1 in relation with eq. (74). At low energies (1-100 eV) a difficulty arises, namely : the set of close coupling equations (48) for the products is infinite. However, the similarity of structure between this set of equations and those appearing in ionization problems in atom-atom collisions [78, 79] suggests the use of similar methods of treatment. The most popular method is based on the *local complex potential* approach in which the incident discrete state $|1, g_v^1>$ acquires a complex energy $U_{1v_1} - i\,\Gamma_{1v}/2$ owing to coupling with the continuum $|2, g^2(\varepsilon)>$ in which it is embedded ; Γ_{1v} is called the width of the state $|1, g_v^1>$:

$$\Gamma_{1v} = 2\,\pi\,\rho(\varepsilon)\ |<g_v^1|V_{12}|g^2(\varepsilon)>|^2 \qquad (92)$$

where $\rho(\varepsilon)$ is the density of states of the continuum (and depends on its normalization). The time evolution of the discrete state according to this method is then such that :

$$C_{1v}(t) = \exp\left(-\frac{1}{2}\int_{-\infty}^{t} \Gamma_{1v}\, dt'\right) \qquad (93)$$

and the probability amplitudes associated with the continuum states obey :

$$i\dot{C}_2(\varepsilon,t) = \int_{-\infty}^{t} dt'\ <g^2(\varepsilon)|\ V_{21}\ |g_v^1>\ C_{1v}(t')\ e^{-i\int_{-\infty}^{t'}(\varepsilon - U_{1v})\,d\tau} \qquad (94)$$

($\rho(\varepsilon)\ |C_2(\varepsilon, +\infty)|^2$ constitutes the spectrum of the dissociation products formed in the DD-NRCE process). Hence the infinite set of coupled equations is reduced to eqs. (93) and (94) ; the latter has to be solved for as many ε values as needed to calculate the dissociation spectrum.

Another interest of dissociative processes is that they constitute nice probes of the effect of molecular orientation on non-adiabatic transitions. Indeed, since the molecule axis may barely rotate during both the collision and a direct dissociation (as opposed to predissociation), fragments ejected at a given angle θ_r, Φ_r (with respect to the incident direction) correspond to the class of molecules having this spatial orientation prior to the collision. These processes thereby constitute a strong incitement to study the role of anisotropy on non-adiabatic processes.

When the bound continuum interaction in eqs. (92), (94) has an exponential behaviour :[77] $\mathcal{A}_{1v,2}(\varepsilon)\ R^m \exp(-\Lambda R)$ (m positive integer), it may be shown that, in *the weak coupling* case : $C_{1v}(t) \simeq 1$ (representative of large impact parameters), and, for a straight line trajectory (eq. (61)), eq. (94), with $\varepsilon - U_{1v} = \Delta$ = constant, yields :[77]

$$C_2\ (\varepsilon,\ t \to \infty) = i\mathcal{A}_{1v,2}(\varepsilon)\ (-\quad)^{-m}\ \frac{\partial^m \beta}{\partial \Lambda^m} \tag{95a}$$

$$\beta = \frac{2u\ K_1(u)}{\Lambda v\ (1 + \frac{\Delta^2}{\Lambda^2\ v^2})} \tag{95b}$$

$$u = \Lambda\ b\ (1 + \frac{\Delta^2}{\Lambda^2\ v^2})^{1/2} \tag{95c}$$

(b is the impact parameter and v the collision velocity)). These equations take a particularly simple form when the Bessel function $K_1(u)$ is replaced by its asymptotic expression for $u \gg 1$.[77] The above results are not restricted to dissociative charge exchange but may also apply to arbitrary charge exchange problems in the weak coupling limit with the aforementioned exponential interaction and constant energy defect Δ. Particularly interesting is the fact that β^2 as a function of Δ is symmetric about the maximum at $\Delta = 0$ and has a full width at half maximum $w = 2 \ln 2\ (\frac{\Lambda}{b})^{1/2}\ v$. This means that as v increases and/or b decreases a broader range of Δ values is available to the charge exchange transition and, for DD-NRCE, the spectrum of dissociation fragments broadens.

Considering, next an interaction of the form :

$$\langle g_v^1 |\ V_{12}\ | g^2(\varepsilon)\rangle = \mathcal{A}_{1v,2}(\varepsilon)\ R^m \exp(-\ \Lambda\ R) \cos^p \gamma \tag{96}$$

in the above weak coupling case, with $p \leq m$, and :

$$\cos\gamma = \frac{b \sin\theta_r \cos\Phi_r + vt\cos\theta_r}{R} \tag{97}$$

(i.e. eq. (81a) for a straight line trajectory) one finds :

$$C_2(\varepsilon, \theta_r, \Phi_r, t \to \infty) = i\mathscr{S}_{1v,2}(\varepsilon)\left[\sum_q^p \binom{p}{q}\left(-\frac{b\sin\theta_r\cos\Phi_r}{\Lambda}\right)^{p-q} \times \right.$$
$$\left.(-\,i\,v\cos\theta_r)^q \frac{\partial^{m-p}}{\partial\Lambda^{m-p}}\frac{\partial^q}{\partial\Delta^q}\beta\right](-)^{m-p} \tag{98}$$

It is amazing to see that the spectrum at $\theta_r = 0$, $\forall\,\Phi_r$ is related to the pth derivative of the spectrum at $\theta = \pi/2$ ($\Phi_r = 0$). This feature entails the property that odd values of p give rise to a dip at $\Delta = 0$ in the spectrum at $\theta_r = 0$ owing to the fact that β is a function of Δ^2 (fig. 10). The vanishing of the integral determining $C_2(\varepsilon, \theta_r = 0, \forall\Phi, t \to \infty)$ (eq. (94)) for odd values of p arises from the antisymmetry of the integrand, for $C_{1v}(t) \simeq 1$ and $\Delta = 0$, when t changes sign ($t \in]-\infty, +\infty[$). When the weak coupling approximation breaks down ($C_{1v}(t) \neq 1$) exact cancellation between the $t \leq 0$ and $t \geq 0$ contributions to the integral in eq. (94) does not occur anymore and the aforementioned dip in the spectrum at $\theta_r = 0$ for $\Delta = 0$ and p odd disappears (see fig. 10).

When Δ varies with R (instead of remaining constant, as assumed above : $\Delta = \Delta_{1v}(\varepsilon, R)$. The integral appearing in eq. (94) may be evaluated using the stationary phase approximation.[56, 61, 78, 79] The points of stationary phase R_x are defined by $\Delta_{1v}(\varepsilon, R_x) = \varepsilon$ thereby establishing a relation between R_x and ε. There are two such points : one for the entrance path $t \leq 0$ and one for the exit path $t \geq 0$. Hence, the system may undergo a transition to the same state of the continuum $g^2(\varepsilon)$ either in the entrance part or in the exit part of the collision thereby giving rise to an interference pattern in the dissociation spectrum (fig. 11). When the two points of stationary phase coalesce, i.e. : when, for a given ε, $R_x(\varepsilon) \simeq b$, the standard procedure[56, 61, 78, 79] yields an expression of $C_2(\varepsilon, t \to +\infty)$ which stems from the properties of the Airy function : $Ai(z) = \int_{-\infty}^{+\infty} \exp(izt + t^3/3)\,dt$. For interactions of the form given in eq. (96) (and with eq. (97)) one finds :[77]

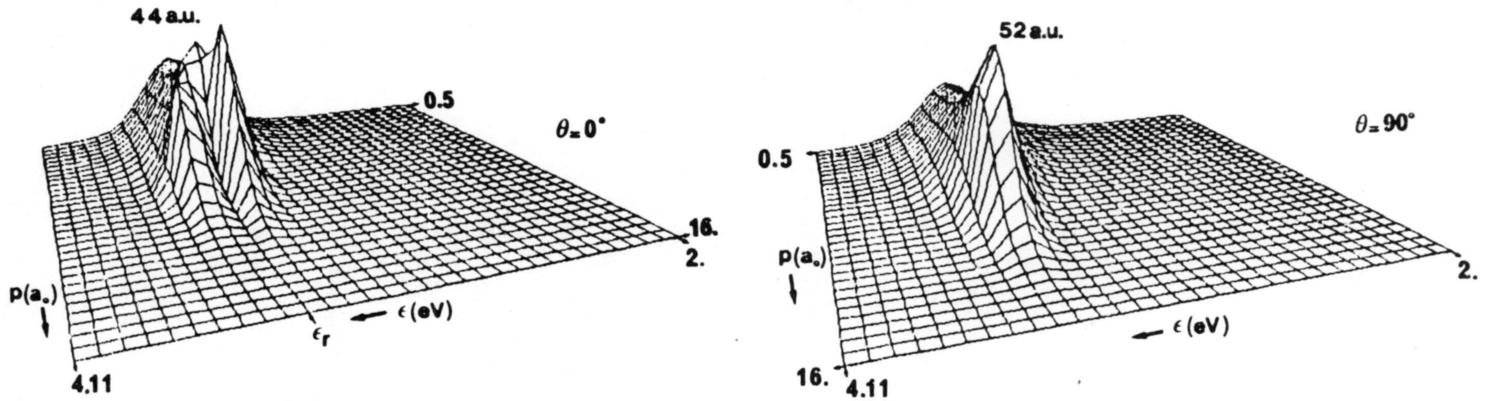

Fig. 10 - Dependence of the DD-NRCE spectrum upon orientation θ of the diatom BC with respect to the incident direction. The anisotropy factor $f(\gamma) = \cos\gamma$, the resonance energy defect Δ is constant and the considered prototype system is H_2^+ + Mg at E_{cm} = 10 eV. Note that p in this figure stands for the impact parameter (the figure is taken from ref. 77).

$$C_2(\varepsilon, t \to +\infty) = 2\pi \mathscr{S}_{1v,2}(\varepsilon) \exp\left(-\frac{1}{2}\int_{-\infty}^{0} \Gamma_{1v_1} dt\right) \xi^{-1/3} \times$$

$$\sum_q^p \binom{p}{q} (b \sin\theta_r \cos\Phi_r)^{p-q} (-iv\cos\theta_r)^q \frac{\partial^{m-p}}{\partial\Lambda^{m-p}} \frac{\partial^q}{\partial D^q} B \qquad (99a)$$

with :

$$B = \exp(-\Lambda b) \, \mathrm{Ai}\left(\frac{D}{\xi^{1/3}}\right) \qquad (99b)$$

and :

$$D = \Delta_{1v}(\varepsilon, b) \quad ; \quad \xi = \frac{v^2}{R} \frac{\partial \Delta_{1v}}{\partial R}(\varepsilon, R)\Big|_{R=b} \qquad (99c)$$

Again, one sees that the spectrum at $\theta_r = 0$ ($\forall\ \Phi_r$) is related to the pth derivative of the Airy function that characterizes the spectrum at $\theta_r = \pi/2$ ($\Phi_r = 0$) (see fig. 11).

Hence the above examples provide illustrations of some aspects of the role of anisotropy in molecular collisions. Moreover, if (as mentioned in Sec. 4.2.1 in relation with eq. (74)) one uses the δ-function approximation of the continuum wavefunction, one may get a hint at the role of the r-coordinate since in this case :[77]

$$\langle g_v^1 \mid V_{12} \mid g^2(\varepsilon)\rangle \simeq \left| V_{12} \right| dV_{22}/dr \left|^{-1/2} g_v^1(r) \right|_{r=r^*(\varepsilon)} \qquad (100)$$

which emphasizes the role of the nodal structure of the incident state $g_v^1(r)$.

5. CONCLUDING REMARKS

The complete quantum mechanical treatment of non-reactive non-adiabatic molecular collisions in the 1-1000eV/amu energy range is out of reach of present computational capabilities owing to the tremendously large number of ro-vibronic channels involved. At lower energies, where quantum mechanical calculations are unavoidable a γ-fixed treatment along the same lines as described in the IOS approximation [6] of rotational excitation could be used. In the 1-1000 eV/amu energy range where non-adiabatic transitions are not the exception but the rule, a semi-classical approach involving a common trajectory for the relative motion and a sudden approximation for the molecular rotation prove to be of great use-

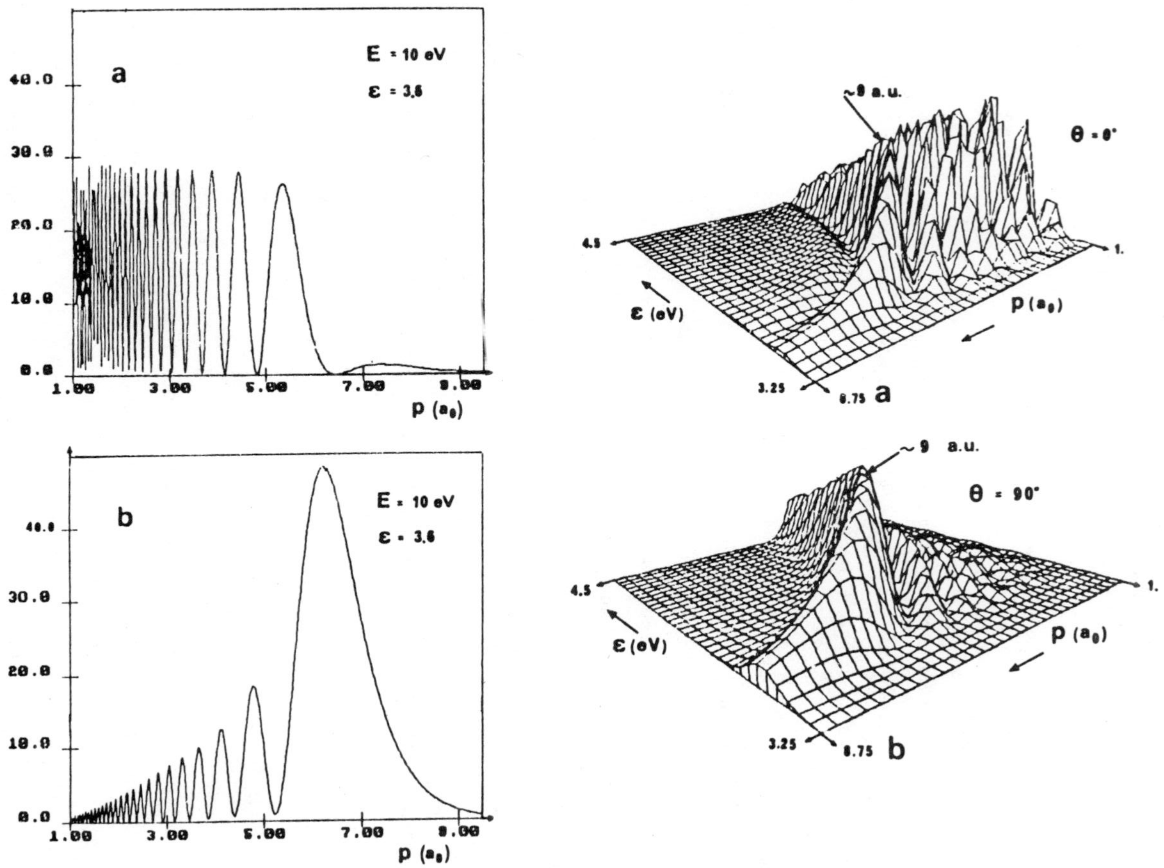

Fig. 11 - Right : same as fig. 10 except that $\Delta = \Delta_\infty + 4\ \exp(-0.5\ R)/3R$ and that the coupling strength in weaker by a factor 10. Left : Slice of the corresponding perspective view (on the right) at $\varepsilon = 3.6$ eV (the figure is taken from ref. 77).

fulness from the point of view of computational convenience and as regards possibilities of modelling.

As for what has appeared in the literature as the adiabatic versus diabatic debate, it should first be mentioned that most numerical close-coupling calculations to date have been carried out in diabatic or diabatic-like bases primarily because of computational convenience. In this context it should be reminded that it is not necessary in principle to solve eq. (17) in order to achieve a good solution of eq. (7) (except asymptotically, when the collision velocity tends to zero). Other zero-th order electronic wavefunction may well be more suitable especially when the non-adiabatic (internal dynamic deformation) couplings hinder the use of the Born-Oppenheimer approximation. The pragmatic conclusion that emerges from the practice of the so-called molecular treatment of heavy particle (atoms and molecules) collisions since the early 70's has already been summarized in sec. 1.1 of the introduction : treat adiabatically (the farthest possible) all attributes of the (ABC) supermolecule that vary systematically over shorter time scales then the characteristic collision time, treat suddenly all those which vary over longer time scales and use as convenient a description as dictated by computational constraints and chemico-physical significance for motions and interactions varying over comparable time scales.

REFERENCES

1. A. Messiah, *Quantum Mechanics*, Vol. 2 (North-Holland, Amsterdam and Interscience, New York, 1963), Chap. XVII.
2. L.D. Landau and E.M. Lifschitz, *Quantum Mechanics*, (Pergamon Press, London 1959).
3. M. Born and J.R. Oppenheimer, Ann. Phys. 84, 457 (1927).
4. R.P. Feynman, Phys. Rev. 56, 340 (1939).
5. D.L. Bunker, Methods Comput. Phys. 10, 287 (1971).
6. F.A. Gianturco, in *Atomic and Molecular Collision Theory* (Ed. F.A. Gianturco, NATO ASI, Plenum, New York and London, 1982).
7. J.C. Tully, in *Dynamics of Molecular Collisions* (Ed. W.H. Miller, Plenum, New York, 1976).
8. W.R. Thorson, J. Chem. Phys. 34, 1744 (1961).
9. M. Baer, Chem. Phys. 15, 49 (1976).
10. M.C. Van Hemert, M. Dohman and S.D. Peyerimhoff, Chem. Phys. 110, 55 (1986)
11. A.M. Arthurs and A. Dalgarno, Proc. R. Soc. A 256, 540 (1960).

12. F.H. Mies, Phys. Rev. A7, 942 (1973).

13. F. Rebentrost and W.A. Lester Jr., J. Chem. Phys. 67, 3367 (1977).

14. W. Hobey and A.D. Mc Lachlan, J. Chem. Phys. 33, 1577 (1960).

15. L.I. Kleinman and M. Wolfsberg, J. Chem. Phys. 59, 2043 (1973).

16. J. Von Neumann and E.P. Wigner, Physik Z. 30, 467 (1929).

17. E. Teller, J. Chem. Phys. 41, 109 (1936).

18. H.C. Longuet-Higgins, Adv. Spectrosc. 2, 429 (1961).

19. G. Herzberg and H.C. Longuet-Higgins, Discuss. Faraday Soc. 35, 77 (1963).

20. T. Carrington, Acc. Chem. Res. 7, 20 (1974).

21. E.E. Nikitin, in *Chemische Elementarprozesse* (Ed. Hartmann, Springer, Berlin, 1968).

22. M. Desouter-Lecomte, C. Galloy and J.C. Lorquet, J. Chem. Phys. 71, 3661 (1979).

23. R.L. Kronig, *Band Spectra and Molecular Structure* (Cambridge University Press, New York, 1930).

24. H. Hellmann and J.K. Syrkin, Acta Physica Chemica, USSR 2, 433 (1935).

25. London, Z. Physik 74, 143 (1932).
O.K. Rice, Phys. Rev. 38, 1943 (1931).

26. L.D. Landau, Physik Z. Sowjet Union 2, 46 (1932).

27. C. Zener, Proc. Roy. Soc. (London) A 137, 696 (1032).

28. W. Lichten, Phys. Rev. 131, 229 (1963).

29. T.F. O'Malley, J. Chem. Phys. 51, 322 (1969).
T.F. O'Malley, Adv. Atom. Molec. Phys. 7, 223 (1971).

30. F.T. Smith, Phys. Rev. 179, 111 (1969).

31. X. Chapuisat, A. Nauts and D. Dehareng-Dao, Chem. Phys. Lett. 95, 139 (1983).

32. V. Sidis and H. Lefebvre-Brion, J. Phys. B 4, 1040 (1971).

33. C. Kubach, C. Courbin-Gaussorgues et V. Sidis, Chem. Phys. Letters 119, 523 (1985).

34. V. Sidis, M. Barat and D. Dhuicq, J. Phys. B 8, 474 (1975).

35. C. Gaussorgues, V. Sidis and M. Barat, *Proc. 10th Int. Conf. on the Phys. of Electronic and Atomic Collisions* (Paris : CEA, 1977), Abstracts p. 706.

36. V. Lopez, A. Macias, R.D. Piacentini, A. Riera and M. Yanez, J. Phys. B 11, 2889 (1978).

37. C. Kubach and V. Sidis, Phys. Rev. A 14, 152 (1976).

38. R.W. Numrich and D.G. Truhlar, J. Phys. Chem. 82, 168 (1978).

39. V. Sidis, C. Kubach and D. Fussen, Phys. Rev. A 27 (1983).

40. V. Sidis and D. Dowek, in *Electronic and Atomic Collisions* (Eds. J. Eichler, I.V. Hertel and N. Stolterfort, North-Holland, Amsterdam - Oxford - New York - Tokyo, 1984), p. 403.

41. J.P. Gauyacq, J. Phys. B 11, 85 (1978).
J.P. Gauyacq, J. Phys. B 11, L217 (1978).
C. Courbin-Gaussorgues, V. Sidis and J. Vaaben, J. Phys. B 16, 2817 (1983).

42. F. Rebentrost and W. A. Lester Jr., J. Chem. Phys. 64, 3879 (1976).

43. R. Cimiraglia, J.P. Malrieu, M. Persico and F. Spiegelmann, J. Phys. B. 18, 3073-3084.

44. L.D. Thomas, W.A. Lester Jr. and F. Rebentrost, J. Chem. Phys. 69, 5489 (1978).

45. E.E. Nikitin, M. Ya Orchinnikova and D.V. Shalashilin, Chem. Phys. 111, 313 (1987).

46. S.I. Chu and A. Dalgarno, J. Chem. Phys. 62, 4009 (1975).

47. C.H. Becker, J. Chem. Phys. 76, 5928 (1982).

48. J.C. Tully and R.K. Preston, J. Chem. Phys. 55, 562 (1971).

49. N.F. Mott, Proc. Cambridge Phil. Soc. 27, 523 (1931).

50. H.S.W. Massey and R.A. Smith, Proc. Roy. Soc. (London) A 143, 142 (1933).

51. R.J. Cross, J. Chem. Phys. 51, 5163 (1969).

52. D.R. Bates and D.S.F. Crothers, Proc. Roy. Soc. (London) A 315, 465 (1970).

53. J.B. Delos, W.R. Thorson and S. Knudson, Phys. Rev. A 6, 709 (1972).

54. K.J. Mc Cann and M.R. Flannery, J. Chem. Phys. 69, 5275 (1978).

55. C. Gaussorgues, C. Lesech, F. Masnou-Seeuws, R. Mc Carroll and A. Riera, J. Phys. B 8, 239 (1975).

56. M.S. Child, *Molecular Collision Theory* (Academic Press, 1974).

57. V. Sidis and D.P. de Bruijin, Chem. Phys. 85, 201 (1984).

58. D. Dhuicq, J.C. Brenot and V. Sidis, J. Phys. B 18, 1395 (1985).
E.A. Gislason and G. Parlant, Comm. At. Mol. Phys.
A.E. de Pristo and S.B. Sears, J. Chem. Phys. 77, 298 (1982).

59. M. Lipeles, J. Chem. Phys. 5, 1252 (1969).

60. Yu N. Demkov, Soviet Phys. JETP 18, 138 (1964).

61. E.E. Nikitin and S. Ya Umanskii, *Theory of slow atomic collision* (Berlin Springer, 1984).

62. D. Dhuicq and V. Sidis, to be published.

63. F.S. Collins and R.J. Cross, J. Chem. Phys. 65, 644 (1976).

64. D.R. Bates and R.H.G. Reid, Proc. Roy. Soc. (London) A 310, 1 (1969).

65. A.F. Hedrick, T.F. Moran, K.J. Mc Cann and M.R. Flannery, J. Chem. Phys. 66, 24 (1977).

66. M.R. Spalburg and U.C. Klomp, Computer Phys. Comm. 28, 207 (1982).

67. M.R. Spalburg, V. Sidis and J. Los, Chem. Phys. Letters 96, 14 (1983).
J. Los and M.R. Spalburg, in *Electronic and Atomic Collisions* (Eds. J. Eichler, I.V. Hertel and N. Stolterfoht, North-Holland, Amsterdam - Oxford - New York - Tokyo, 1984), p. 393.

68. M.R. Spalburg, J. Los and E.A. Gislason, Chem. Phys. 94, 327 (1985).

69. M.R. Spalburg and E.A. Gislason, Chem. Phys. 94, 339 (1985).

70. G. Parlant and E.A. Gislason, Chem. Phys. 101, 227 (1985).

71. T.F. Moran, M.R. Flannery and D.L. Albritton, J. Chem. Phys. 62, 2689 (1975).
T.F. Moran, M.R. Flannery and P.C. Cosby, J. Chem. Phys. 61, 1261 (1974).
T.F. Moran, K.J. Mc Cann and M.R. Flannery, J. Chem. Phys. 63, 3857 (1975).

72. T.F. Moran, K.J. Mc Cann, M. Cobb, R.F. Borkman and M.R. Flannery, J. Chem. Phys. 74, 2325 (1981).
A.E. de Pristo, J. Chem. Phys. 78, 1237 (1983)
A.E. de Pristo, J. Chem. Phys. 79, 1741 (1983).

73. A.W. Kleyn, V.N. Khromov and J. Los, Chem. Phys. 52, 65 (1980).
A.W. Kleyn, J. Los and E.A. Gislason, Physics Repts 90, 1 (1982).

G. Parlant, M. Schröder and S. Goursaud, Chem. Phys. Letters 80, 526 (1981).

74. P. Archirel and B. Levy, Chem. Phys. 106, 51 (1986).

75. R.E. Olson, Phys. Rev. A 6, 1822 (1972).
T.R. Dinterman and J.B. Delos, Phys. Rev. A 15, 463 (1977).

76. E.E. Nikitin, Adv. Quant. Chem. 5, 135 (1970).

77. V. Sidis and C. Courbin-Gaussorgues, Chem. Phys.

78. W.H. Miller, J. Chem. Phys. 52, 3563 (1970).

79. V. Sidis, J. Phys. B 6, 1183 (1973).

RECENT PROGRESS IN HIGH RESOLUTION PROTON-MOLECULE CHARGE TRANSFER SCATTERING STUDIES

M. Noll and J.P. Toennies

Max-Planck Institut für Strömungsforschung

D-3400 Göttingen, Federal Republic of Germany

ABSTRACT

Vibrational state resolved scattering experiments on the charge transfer collisions of H^+ with H_2, CO_2, O_2, and H_2O target molecules are reported at collision energies of about 30 eV. The results are discussed in terms of simple models, which are based on the simultaneous availability of the scattered H^+ and H total angular distributions as well as their time-of-flight spectra in dependence on the scattering angle, $0° \leqq \theta \leqq 15°$. They provide a detailed insight into the vibrational interaction dynamics with respect to the relative importance of energy resonance and the role of the Franck-Condon factors in determining the charge transfer transition probabilities to the final vibrational states. In addition, the dependence of these probabilities on the particular vibrational excitation of the neutral molecule by the time-dependent proton electric field can be studied in great detail.

I. INTRODUCTION

Investigation of the dynamics of proton-molecule vibrational interactions at low collision energies, 10 eV $\lesssim$ E $\lesssim$ 30 eV, has recently received a considerable impetus from the measurements of vibrationally resolved charge transfer spectra for a number of different systems and over a wide range of scattering angles. Besides the interesting information on the charge transfer process itself these spectra elucidate the earlier investigated ro-vibrational inelastic scattering processes from a different viewpoint which has not been considered in detail before. The results contribute to the field of non-adiabatic charge transfer vibronic coupling phenomena,[1] where very little information based on scattering experiments with vibrational state-to-state resolution is available.[2-4]

In an electronically non-adiabatic collision, where two potential energy surfaces are involved, the coupling region will be passed twice; first along the incoming branch of the trajectory and second along the outgoing one. This generally leads to four different collision pathways depending on whether the system stays on the diabatic or adiabatic energy surface on the in- and outgoing way. One of the attractive aspects of the present proton scattering experiments is that the collision energies fall in a range where the target molecules are practically stationary during the time for traversing a single, well-localized charge transfer coupling region ($t \approx 10^{-15}$s), while the molecular vibrations ($t_{vib} \approx 10^{-14}$s) take place during the time which elapses between the crossing of one coupling region to the next ($t_{coll} \approx 10^{-14}$s). This implies near 'vertical' electron transitions between the two participating energy surfaces and, consequently, a dominant role of the Franck-Condon factors in determining the relative importance of the final vibrational states. Deviations from the Franck-Condon factors are expected, however, if e.g. polarization interaction or other forces exerted on the vibrational coordinate distort the molecule prior to the actual charge transfer transition.[5,6]

A systematic investigation of the charge transfer process for a number of simple systems at comparable collision energies and scattering angles, as it is presently undertaken in this laboratory, seems also valuable in further clarifying the competitive role of energy resonance conditions and the Franck-Condon factors. The importance of both has long been pointed out in the literature.[7] However, a deeper understanding of their interrelation largely suffered from the lack of adequate experimental work with sufficient energy and angular resolution to allow for vibrational state-to-state differential cross section measurements. Recently, we have succeeded in determining such cross sections for the charge transfer processes in H^+-H_2, CO_2, O_2, and H_2O collisions at a typical energy of 30 eV. In this work the essential experimental results will be briefly summarized and discussed with respect to the Franck-Condon factors, the energy resonance condition, and the vibrational interaction dynamics on the single potential energy surface in the close approach region.

II. CROSSED BEAM APPARATUS

The experiments have been carried out on the crossed beam time-of-flight apparatus,[4,8] which is schematically shown in Fig. 1. Briefly, the compact rotatable ion source consists of a hydrogen gas discharge cell for ion generation. The ion beam, formed by extraction through a small hole in the anode, subsequently passes through a 60° magnetic and 127° electrostatic sectorfield for mass- and energy selection, respectively. Depending on the mode of operation (angular or time-of-flight distribution measurements) either a continuous or pulsed beam is employed to collide with the target atomic or molecular species which is generated in a low pressure nozzle beam expansion. The scattered particles are detected by an open electron

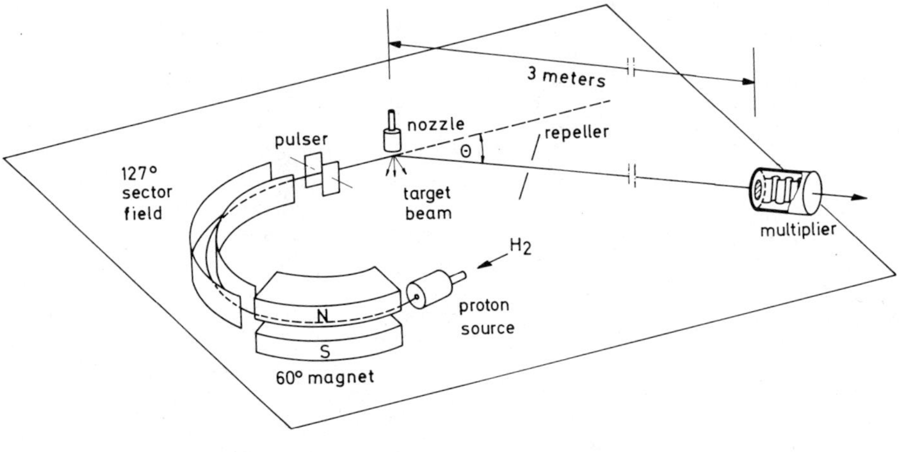

Fig. 1. Perspective diagram of the crossed beam apparatus

multiplier after a drift path of approximately 3m. The multiplier also serves as a simultaneous detector for the fast neutral H atoms originating in charge transfer collisions of the reactant protons. The atoms can be distinguished from the protons by application of an electric field between scattering region and the multiplier.

Although this method has already been exploited in a variety of similar experiments,[9] its novel application at the relatively low kinetic energies of the present work, $E_{H^+} \approx$ 20-30 eV, results in a particularly high energy resolution for time-of-flight measurements of the neutral products. The high angular and energy resolution of the apparatus can be seen from Fig. 2 (left and right part, respectively) showing some typical results for elastic and charge transfer scattering of H^+ by Kr atoms.[10]

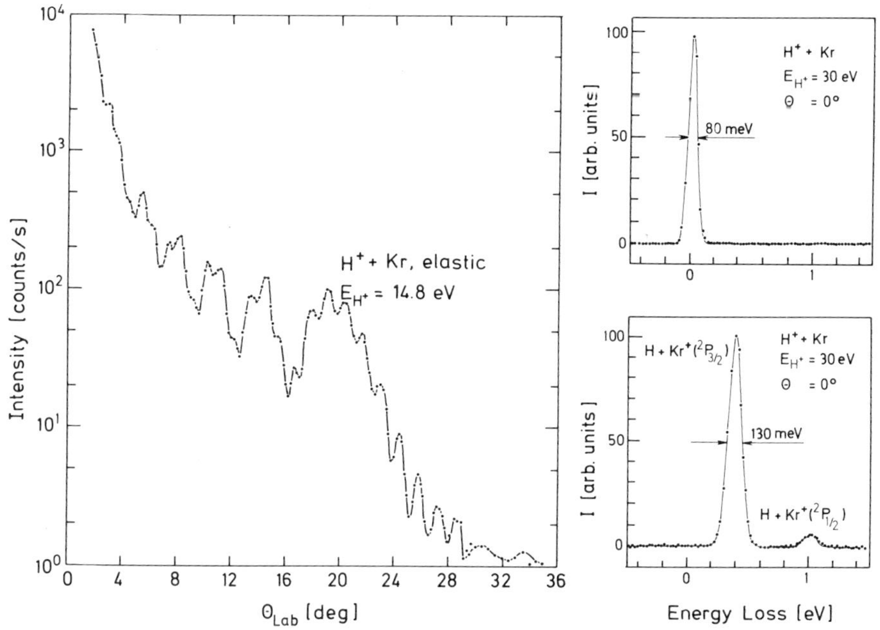

Fig. 2. Illustration of the experimental angular resolution by means of the highly structured relative elastic differential cross section for H^++Kr collisions (left).[10] The spectra on the right indicate the typical energy resolution in the H^+ and H time of flight measurements (top and bottom, respectively).

III. RESULTS: ANGULAR AND ENERGY LOSS DISTRIBUTIONS OF H^+(H) RELEASED IN PROTON-MOLECULE INELASTIC (H^+) AND CHARGE TRANSFER (H) COLLISIONS

III.1 The H^++H_2 System

The presentation of our experimental results starts with the simple H^++H_2 system, for which charge transfer is highly endothermic, ΔE_{CT}=1.83 eV, and therefore rather unlikely ($\sigma_{CT} \approx 0.4$ Å^2 at E_{cm} = 10 eV[11]). As a two electron system, its collision dynamics has already attracted great interest compared to other bimolecular heavy particle systems and, in particular, extensive effort has been undertaken to investigate the rotational and vibrational inelastic scattering of H^+ by H_2.[12-15] With the well-known potential surfaces experimental results on these processes proved to be extremely valuable for rigorous tests of various model calculations.[14,16,17] In comparison with the non-reactive inelastic scattering, however, information on the charge transfer process[18-20] is rather scarce and no state resolved results from scattering experiments were so far available.

In our measurements vibrationally resolved translational energy distributions of the product H atoms at E_{cm}=20 eV and center-of-mass angles $0° \leqq \vartheta \leqq 18°$ have been obtained in addition to the total charge transfer angular distribution. Typical experimental results from which the basic conclusions can be derived are presented in Fig. 3. The left part (Fig. 3a) displays the two total angular distributions for the inelastic scattering process (H^+ detection, upper curve) and the charge transfer scattering process (H detection, lower curve). At ϑ=12° the underlying relative translational energy distributions of the H^+ and H products are shown on the right (Fig. 3b, top and bottom spectrum, respectively).

The results for H^+ are in good agreement with earlier experiments at about the same collision conditions.[14] First, it is noted that the angular distribution shows a pronounced rainbow maximum at $\theta_{Lab} \approx 6\text{-}7°$ (ϑ=12-14°). Second, the H^+ energy loss distribution exhibits strong vibrational excitation of the neutral H_2 molecule (see top arrows). Transitions up to $v_f(H_2)$=5 can still be observed in an extended intensity scale. The non-Gaussian peak forms in the H^+ spectrum (see the deviations from the solid line Gaussian-fits) as well as their relatively large halfwidth ($\Delta E_{fwhm} \approx$ 220 meV) are attributed to the poorly defined kinematic conditions, which result from the low mass of the H_2 target and the use of an unskimmed H_2 beam.

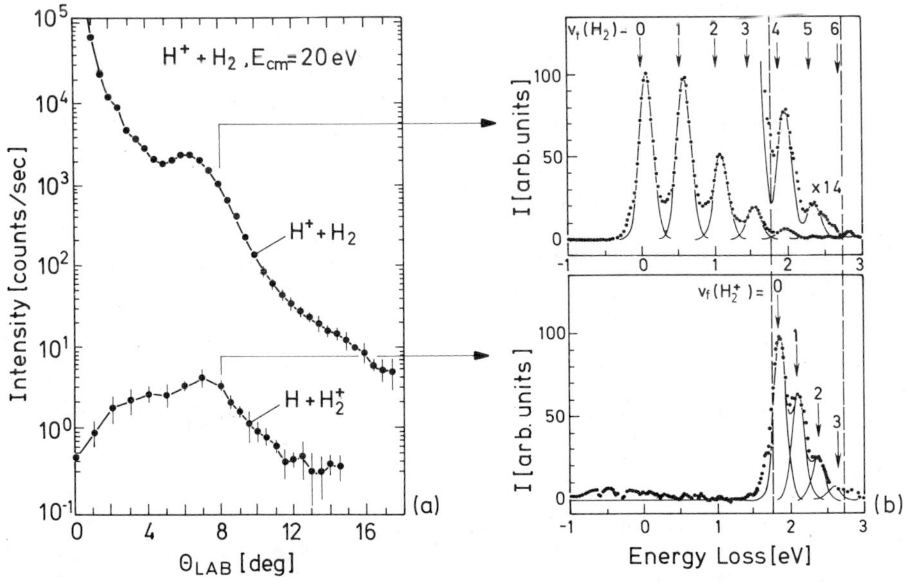

Fig. 3. Total angular distributions for (in-)elastic (H^++H_2) and charge transfer (H+H_2^+) scattering of H^+ by H_2 at E_{cm}=20 eV (a). At θ_{Lab}=8° the relative translational energy distributions of H^+ and H are shown in the upper and lower part of (b), respectively. The arrows indicate the energy levels of the H_2(H_2^+) product vibrational states.

The novel results for the $H^+ + H_2$ charge transfer process, shown in the lower parts of Fig. 3a and 3b, indicate the following: (i) The total H atom angular distribution is peaked at the rainbow angle of the corresponding H^+ curve ($\theta_{Lab} \approx 7°$) and both curves yield a similar fall-off at large angles. (ii) A significant increase of the total H signal has to be compared to a steep decrease for H^+ in the low scattering angle region. (iii) The H signal at a given angle is considerably smaller than that of H^+ even if the lower H detection probability, $P(H/H^+) \approx 0.03$ [18], is taken into account. (iv) The H product translational energy distribution exhibits resolved structures which correlate with the v_f=0,1,2, and 3 vibrational states of H_2^+ (the vertical arrows label the respective energy levels). (v) The charge transfer vibrational transition probabilities, $P(v_f)$, reveal a monotonic decrease with increasing quantum number. An additional important observation from the entire series of measurements at different angles is that the $P(v_f)$ are practically constant for the charge transfer process (variations of less than 10-20%), while they show drastic variations in the case of the neutral H_2 vibrational excitation.

All these observations are in reasonable agreement with a simple dynamical model based on the strong H_2 vibrational excitation on the lower $H^+ + H_2$ potential energy surface[16]. Since practically no coupling between the two charge transfer states is possible in the incoming channel due to the large energy difference and, in addition, the $H^+ + H_2$ surface has a deep well in contrast to the repulsive $H + H_2^+$ surface, the system remains on the lower surface until the collision is nearly completed. At that time, however, the H_2 molecule is already excited and the energy barrier to the lowest charge transfer state, $v_f(H_2^+)=0$, is therefore significantly reduced for the higher H_2 vibrational states. In the case of $v_f(H_2)=4,5$ near energy resonance with v_f=0,1,2,3 of H_2^+ is achieved (see the region between the dashed vertical lines in Fig. 3b) and a largely increased electron transition probability between these states becomes conceivable. The small charge transfer differential cross section at low scattering angles, Fig. 3a, is thus attributable to the glancing collisions in which the population of the higher H_2 vibrational states is less likely. The maximum at $\theta_{Lab} \approx 7°$, on the other hand, occurs due to rainbow scattering on the common $H^+ + H_2$ potential energy well region, which is clearly reflected by the similar shape compared to the angular distribution of the scattered protons. Since at a given angle only few collisions lead to $v_f(H_2) = 4,5$ it is also reasonable that the measured total H atom intensities are generally much smaller than the H^+ intensities. Excitation of a quite narrow band of H_2^+ product vibrational states, v_f=0-3, compared to the much broader Franck-Condon distribution[21] agrees with the negligible vibrational excitation probabilities for $v_f(H_2) \geqq 6$. A more detailed presentation of the experimental data and a comparison with the results from classical trajectory surface hopping calculations is given in a forthcoming paper[22].

III. 2 The $H^+ + CO_2$ System

Characteristic results from the (in)elastic and charge transfer collisions of H^+ with CO_2 are presented in Fig. 4 at a proton energy of 30 eV. As in Fig. 3 the respective total angular distributions (Fig. 4a) are plotted together with the relative translational energy distributions of H^+ and H at $\theta_{Lab} = 9°$ (Fig. 4b). From both types of measurements the following results are clearly revealed: (i) The two angular distributions are almost perfectly parallel to each other at $\theta_{Lab} \gtrsim 2°$ with a marked rainbow structure in the region $6° \lesssim \theta_{Lab} \lesssim 15°$. (ii) They even have about the same absolute intensities if the different detection probabilities, $P(H/H^+) \approx 0.03$, are taken into account. (iii) The energy loss spectra of both the scattered protons and the product H atoms show a markedly selective excitation of the asymmetric stretch mode (00n) of CO_2 and CO_2^+, respectively. (iv) The transition probabilities, P(000→00n), are nearly the same for both product channels at a given scattering angle. In summary one can say that practically no differences between the (in-)elastic and charge transfer state-to-state differential cross sections are observed in the present H^++CO_2 scattering experiments.

The origin of this behavior is attributable to the combination of several factors: (a) The identical rainbow structures in Fig. 4a indicate that most of the trajectories leading to charge transfer also sample the lower H^+ + CO_2 potential energy surface before the electron transition takes place, and (b) the H+CO_2^+ surface, which the system experiences after the electron transition took place, has a negligible effect on the total scattering angle. (c) Selective vibrational excitation of the (00n) mode of CO_2 via interaction with the proton is induced by a combination of electrostatic and valence bond forces.[23,24] (d) A projection of the CO_2(00n) vibrational excitation pattern in a one to one ratio onto CO_2^+ in the actual charge transfer process can be explained by the Franck-Condon selection rule $\Delta n = 0$, which is valid for the asymmetric stretch modes of all linear symmetric molecules in vertical electronic transitions[25]. This is even valid if the fundamental frequencies are quite different like in the present case of CO_2 and CO_2^+. A confirmation is provided by the photoelectron spectrum of CO_2 which shows no measurable intensities for the CO_2^+ (00n) transitions[21].

The outcome of the present charge transfer collisions has thus been identified as a vertical (Franck-Condon) electron transition with extremely large modifications in the vibrational state populations depending on the interaction dynamics between the proton electric field and the vibrations of the neutral CO_2 molecule. A more detailed data analysis as well as a comparison with the results for the exothermic charge transfer process in the related H^++N_2O system will appear soon.[26]

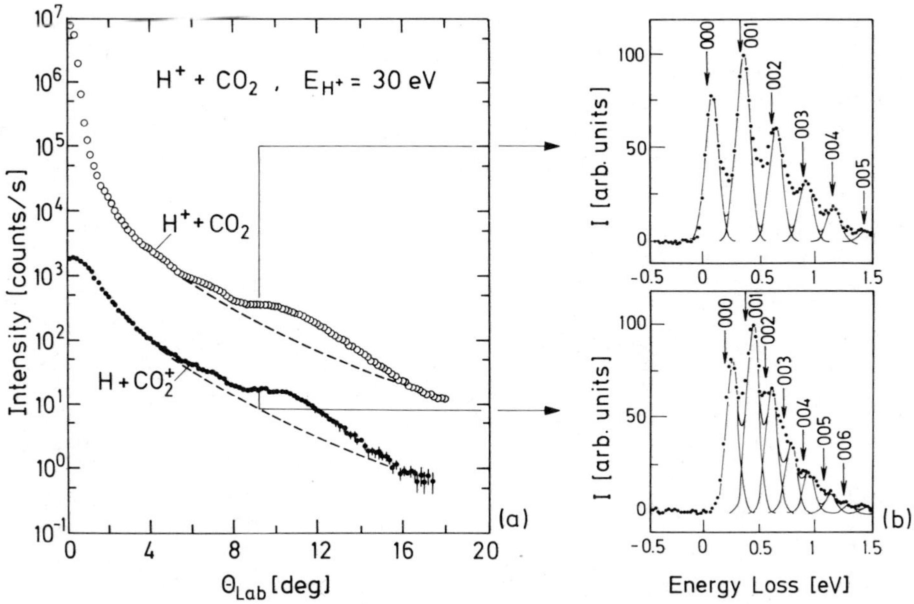

Fig. 4. Same as in Fig. 3, but for CO_2 instead of H_2 target molecules. Note the surprisingly identical results for the (in-)elastic (H^++CO_2) compared to the charge transfer (H+CO_2^+) scattering process.

III. 3 The $H^+ + O_2$ System

After having discussed the results for a highly endothermic charge transfer system ($H^+ + H_2$, ΔE_{CT} = 1.83 eV) and a near resonant system ($H^+ + CO_2$, ΔE_{CT} = 0.2 eV) we now turn our attention to an exothermic system ($H^+ + O_2$, ΔE_{CT} = -1.53 eV). Results are plotted in Fig. 5 at a proton energy of 23.7 eV. The data presentation of the total H^+ and H angular distributions (a) and their relative translational energy distributions at $\theta_{Lab} = 4°$ (b) is analogous to Figs. 3 and 4. The discernible experimental observations are[4]: (i) Parallel angular distributions with a weak indication of a rainbow maximum at $\theta_{Lab} \approx 10°$. (ii) Comparable H^+ and H intensities if $P(H/H^+) \approx 0.01$ is accounted for. (iii) Small vibrational excitation of the neutral O_2 molecule compared to the charge transfer O_2^+ product ion. (iv) Excellent agreement between the measured charge transfer vibrational transition probabilities and the respective Franck-Condon factors[21] for a vertical transition from $O_2(v_i=0)$ to the vibrational states of O_2^+ at the special scattering angle of $\theta_{Lab} = 4°$ (see Fig. 5a). (v) At smaller angles drastic deviations from the Franck-Condon factors are observed in favor of an enhancement of the more resonant states, $v_f(O_2^+) \geqq 3$[4]. (vi) In addition, increasingly large deviations from the Franck Condon factors are found at $\theta_{Lab} > 4°$, again favoring the higher O_2^+ product states.

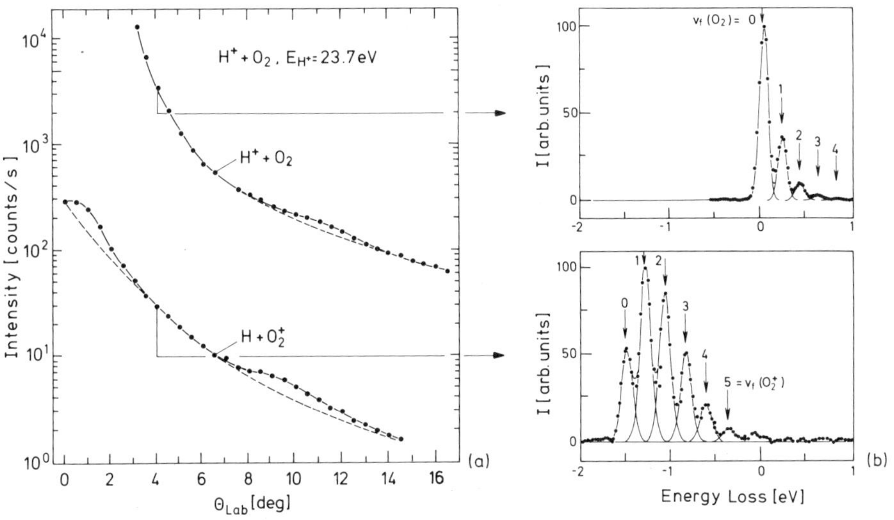

Fig. 5. Total angular (a) and translational energy distribution measurements (b) for inelastic and charge transfer collisions between H^+ and O_2 at E_{H^+} = 23.7 eV (θ_{Lab} = 4° in (b)). Note that the peak heights in the lower right charge transfer spectrum agree within the experimental accuracy with the Franck-Condon factors for the respective O_2^+ vibrational states.

The latter effect has been interpreted in terms of the Lipeles model[5] employing a polarization interaction between H^+ and O_2 which leads to a small stretching of the O_2 bond distance prior to the charge transfer process. This modifies the actual Franck-Condon factors and good agreement with the measured change of the relative charge transfer probabilities, $P(v_f)$, has been obtained.[4] The observed strong enhancement of the more resonant states at the smallest scattering angles, $\theta \to 0°$, is associated with the small amount of impulsive momentum and energy transfer at the corresponding large impact parameters. Consequently, the distribution of non-adiabatic charge transfer vibrational transition probabilities has to be centered at ΔE_{cm} = 0eV (resonant charge transfer) with an energy minimum width inversely proportional to the collision time as suggested by the uncertainty principle. Weighting the Franck-Condon factors with this distribution qualitatively explains the observed prevalence of the more resonant charge transfer states when approaching $\theta=0°$.

III.4 The $H^+ + H_2O$ System

Most recently, experiments have been carried out to study the inelastic and charge transfer vibrational excitation processes within the more complicated $H^+(D^+) + H_2O$ system[27,28]. Our choice of this particular system has been stimulated by the great chemical and biological importance of the three species H_2O, H_2O^+, and H_3O^+ (H_2DO^+), which are all involved in the collision process.

A typical charge transfer product D atom translational energy distribution at $\theta = 1°$ and E_{cm} = 30.1 eV is shown in the lower part of Fig. 6 in comparison with a high resolution photoelectron spectrum of H_2O[29] at the top. In both cases (charge transfer and photoionization) contributions from the two lowest electronic states of H_2O^+, $\tilde{X}^2B_1$ and $\tilde{A}^2A_1$, can be distinguished (the $\tilde{B}^2B_2$ state is also observed in photoionization at IP ≧ 17.2 eV, but is completely missing in the charge transfer collisions). Despite the lower energy resolution in the present scattering experiment, which is partly attributed to rotational broadening, the observed structures indicate that the same type of vibronic transitions are excited as in photoionization. Obviously, however, the corresponding vibronic transition probabilities are fairly different. In comparison with the photoionization probabilities (Franck-Condon factors) the charge transfer probabilities clearly show an enhancement of the quasi-resonant states in the vicinity of ΔE_{cm} = 0 eV. The minimum at ΔE_{cm} = 0 eV is consistent with the negligible Franck-Condon factor of the respective state.

The physical reason for this resonance enhancement is probably the same as has been discussed for $H^+ + O_2$ in the preceding subsection: for the charge transfer process the Franck-Condon factors have to be weighted by a distribution of non-adiabatic transition probabilities, which, at small scattering angles (large impact parameters) has to be roughly centered at ΔE_{cm} = 0 eV because of the small momentum transfer. Excitation of the same vibrational modes in photo-ionization and in small angle charge transfer scattering can be also understood in terms of the large impact parameters in the latter type of collisions. This implies that, like in photoionization, the particular vibrational states within an electronic band of H_2O^+ are primarily determined by the symmetry and bonding properties of the orbital from which the electron is removed ($1b_1$ nonbonding orbital and $3a_1$ bonding orbital in the case of the $\tilde{X}^2B_1$ and $\tilde{A}^2A_1$ states, respectively).

IV. SUMMARY AND CONCLUSIONS

A short summary of our recent experimental work on charge transfer collisions between H^+ and H_2, CO_2, O_2, and H_2O has been presented and the results were briefly discussed. For all systems state resolved information on the vibrational transition probabilities was obtained from the product H atom translational energy distributions. In addition, the total H atom angular distributions as well as the translational energy and angular distributions of the scattered protons proved to be particularly valuable for deriving simple dynamical interpretations of the observed charge transfer vibrational excitation patterns.

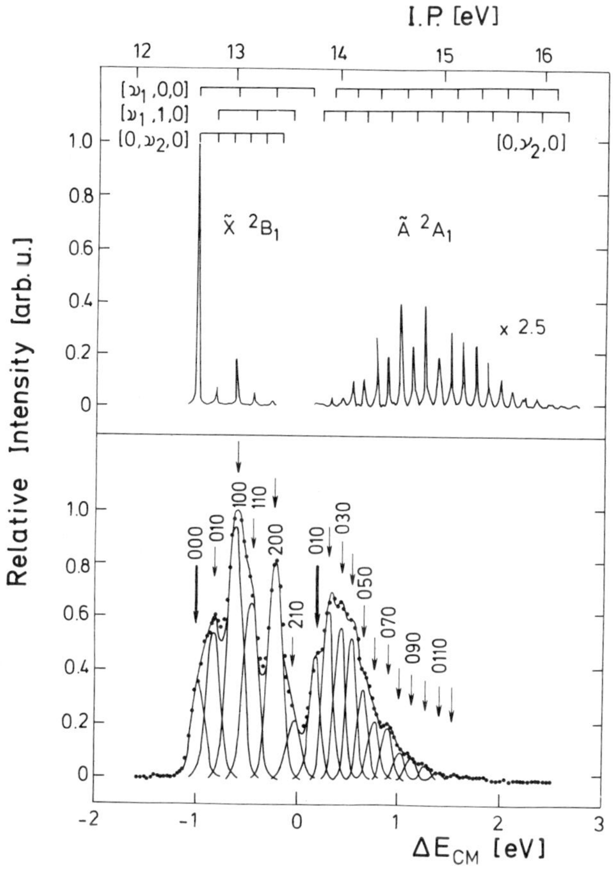

Fig. 6. Comparison between the low energy photoelectron spectrum of H_2O (top)[29] and the product D atom translational energy distribution produced in charge transfer collisions between D^+ and H_2O at $\theta = 1°$ and $E_{cm} = 30.1$ eV (bottom). In both cases contributions from two electronic states of H_2O^+, the ground state $\tilde{X}^2B_1$ and the first excited state $\tilde{A}^2A_1$, can be distinguished. Although the same H_2O^+ vibrational modes seem to be excited within a given electronic state the vibronic transition probability distributions show drastic differences between the two spectra.

Three main mechanisms could be shown to be responsible for the final observations: (i) the Franck-Condon mechanism, which assumes a 'vertical' electron transition between the proton and the neutral molecule; (ii) the energy resonance mechanism, which favors the vibrational states close to the recombination energy of the proton; and (iii) the interaction dynamics outside the charge transfer coupling region, which leads to a distortion of the target molecule. The relative importance of these mechanisms, however, is quite different for the different systems and also depends on the scattering angles. For $H^+ + H_2$ the vibrational excitation of H_2 by the strong interaction with the proton is necessary to surmount the rather large charge transfer energy barrier. Once excited high enough, $v_f(H_2) \geqq 4$, charge transfer takes place to the quasi-resonant vibrational states of H_2^+, $v_f = 0-3$, although these states have small Franck-Condon factors. For $H^+ + CO_2$ the situation is quite different because of the much smaller asymptotic energy mismatch between the initial ground state and the final charge transfer state. Nevertheless, the strong vibrational interaction between H^+ and CO_2, leading to a selective excitation of the asymmetric stretch mode (00n), is of particular importance for the charge transfer process. The results indicate that the neutral CO_2 vibrational excitation pattern as a whole is imprinted on the product CO_2^+ ion independent of the quantum number n. This suggests the validity of a quite strong $\Delta n = 0$ selection rule when the electron is transferred from CO_2 to H^+. Such a strong selection rule is indeed known to apply in the case of vertical electronic transitions within linear symmetric molecules like CO_2.[25]

As opposed to H_2, $CO_2 + H^+$ the vibrational excitation of O_2 by H^+ is obviously of very little importance for the final charge transfer relative vibrational transition probabilities within this sytem. On the contrary, the strong distortion of the target molecule in a temporary charge transfer collision seems to be the only mechanism being able to explain the unusually large vibrational energy transfer to O_2 compared to similar collisions of H^+ with N_2, CO, and NO.[30] The outcome of the actual $H^+ + O_2$ charge transfer collisions is close to the Franck-Condon prediction, but, depending on the scattering angle, significant deviations are still observed. At the smallest angles, $\theta < 2°$, the enhancement of the more resonant states compared to the most probable Franck-Condon states is clearly visible and attributed to the low momentum transfer at the respective large impact parameters. Deviations at the larger angles have been found to be in reasonable agreement with the assumption of a stretched O_2 molecule due to polarization interaction with the proton electric field prior to the electron transition.

Rather complicated charge transfer spectra, involving contributions from two electronic states and several vibrational modes, have been measured in the case of $H^+(D^+) + H_2O$ collisions. These results are in reasonable agreement with the symmetry and bonding properties of the molecular orbitals from which the electron is removed as well as with the resonance criterion, which favors charge transfer product vibrational states close to the recombination energy of $H^+(D^+)$ if small scattering angles are considered.

In conclusion, the present data of inelastic and charge transfer scattering of protons by simple di- and triatomic molecules provide extremely detailed information on the respective vibrational state-to-state relative differential cross sections.

ACKNOWLEDGEMENT

Valuable collaboration with our colleagues B. Friedrich and G. Niedner is gratefully acknowledged.

REFERENCES

1. e.g.: E.E. Nikitin and S.Ya. Umanskii, "Slow Atomic Collisions", Springer, Berlin (1984); A.W. Kleyn, J. Los, and E.A. Gislason, Phys. Rep. 90, 1 (1982)
2. B. Friedrich, W. Trafton, A. Rockwood, S. Howard, and J.H. Futrell, J. Chem. Phys. 80, 2537 (1984)
3. K. Birkinshaw, A. Shukla, S. Howard, and J.H. Futrell, Chem. Phys. 113, 149 (1987)
4. M. Noll and J.P. Toennies, J. Chem. Phys. 85, 3313 (1986)
5. M. Lipeles, J. Chem. Phys. 51, 1252 (1969)
6. E.A. Gislason and E.M. Goldfield, Phys. Rev. A25, 2002 (1982)
7. e.g.: J. Glosik, B. Friedrich, and Z. Herman, Chem. Phys. 60, 369 (1981)
8. U. Gierz, M. Noll, and J.P. Toennies, J. Chem. Phys. 83, 2259 (1985)
9. e.g.: D. Dowek, D. Dhuicq, J. Pommier, V.N. Tuan, V. Sidis, and M. Barat, Phys. Rev. A24, 2445 (1981)
10. M. Baer, R. Düren, B. Friedrich, G. Niedner, M. Noll, and J.P. Toennies, Phys. Rev. A, in press
11. G. Ochs and E. Teloy, J. Chem. Phys. 61, 4930 (1974)
12. H. Udseth, C.F. Giese, and W.R. Gentry, Phys. Rev. A8, 2483 (1973)
13. V. Hermann, H. Schmidt, and F. Linder, J. Phys. B: At. Mol. Phys. 11, 493 (1978)
14. R. Schinke, H. Krüger, V. Hermann, H. Schmidt, and F. Linder, J. Chem. Phys. 67, 1187 (1977)
15. K. Rudolph and J.P. Toennies, J. Chem. Phys. 65, 4483 (1976)
16. C.F. Giese and W.R. Gentry, Phys. Rev. A10, 2156 (1974)
17. R. Schinke, Chem. Phys. 24, 379 (1977)
18. R.K. Preston and J.C. Tully, J. Chem. Phys. 54, 4297 (1971)
19. J.C. Tully and R.K. Preston, J. Chem. Phys. 55, 562 (1971)
20. J.R. Krenos, R.K. Preston, R. Wolfgang, and J.C. Tully, J. Chem. Phys. 60, 1634 (1974)
21. D.W. Turner, C. Baker, A.D. Baker, and C.R.Bundle, "Molecular Photoelectron Spectroscopy", Wiley-Interscience, London (1970)
22. G. Niedner, M. Noll, J.P. Toennies, and Ch. Schlier, J. Chem. Phys., in press
23. J. Krutein and F. Linder, J. Phys. B: At. Mol. Phys. 10, 1363 (1977)
24. G. Bischof, V. Hermann, J. Krutein, and F. Linder, J. Phys. B: At. Mol. Phys. 15, 249 (1982)
25. G. Herzberg, "Molecular Spectra and Molecular Structure", Vol. III, v. Nostrand, Princeton (1967)
26. G. Niedner, M. Noll, and J.P. Toennies, J. Chem. Phys., in press
27. B. Friedrich, G. Niedner, M. Noll, and J.P. Toennies, J. Chem. Phys., in press
28. B. Friedrich, G. Niedner, M. Noll, and J.P. Toennies, to be published
29. J.E. Reutt, L.S. Wang, Y.T. Lee, and D.A. Shirley, J. Chem. Phys. 85, 6928 (1986)
30. F.A. Gianturco, U. Gierz, and J.P. Toennies, J. Phys. B: At. Mol. Phys. 14, 667 (1981)

MOLECULAR BEAM STUDIES OF IONIZATION PROCESSES IN COLLISIONS OF EXCITED RARE GAS ATOMS

Brunetto Brunetti and Franco Vecchiocattivi

Dipartimento di Chimica dell'Università

06100 Perugia, Italy

INTRODUCTION

Rare gas atoms in their lowest excited levels exhibit interesting properties mainly due to their high electronic energy and their long lifetime. Actually their energy goes from 19.82 eV for the He atom in the 2^3S state and 8.32 eV of the Xe atom in the $^3P_2$ state, while the lifetime ranges from $4.2x10^3$ s for the former atom to 0.078 s for the latter. These characteristics make possible several processes during the collisions with atomic and molecular targets: electronic energy transfer, ionization reactions or chemical reactions leading to the formation of an excimer molecule are some examples. Because their importance in several applications such as plasmas, electrical discharges or laser systems, these processes are extensively studied both in bulk systems [1] and under single collision conditions [2-4]. The long lifetime of these species makes possible the study of collisional processes of excited rare gas atoms by the molecular beam technique.

In this paper we report some recent experimental results about ionization processes occurring in the collisions between metastable rare gases and some atomic and molecular targets. The experiments have been carried out in a crossed beam apparatus which allows to measure the cross sections, for each ionization channel, as a function of the collision energy in the thermal range.

THE CROSSED BEAM APPARATUS

The crossed beam experimental set up, schematically shown in Fig.1, has been described in detail elsewhere [5-8]. Only a general description is given here.

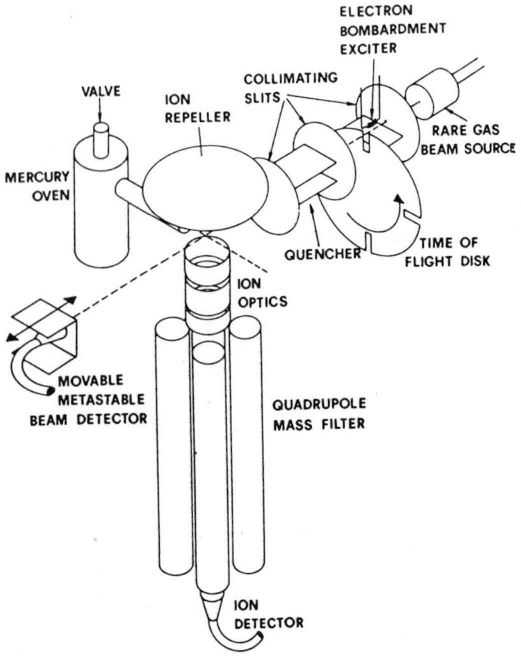

Fig.1 A schematic view of the experimental set up.

The metastable rare gas atom beam is produced by electron bombardment of an effusive rare gas beam. The ions and the high Rydberg state rare gas atoms which also are produced in the electron impact excitation region are deflected off the beam by an electric field of about 1.5 KV/cm. The target beam is produced by a microcapillary array beam source and crosses the metastable atom beam at 90°. The temperature of both beam sources can be adjusted from room temperature up to about 900 K. The metastable atom beam is monitored by a channel electron multiplier located in a box which can be moved into the beam crossing center. The product ions are extracted from the scattering volume by an electric field, focused by an electrostatic lens system, mass analyzed by a quadrupole filter and finally detected by another channel electron multiplier.

The velocity selection is accomplished by time of flight technique: the metastable beam is pulsed by a rotating slotted disk and the pulses for the metastable atoms or for the product ions are counted as a function of the delay time with respect to the slit opening. The time of flight spectra are obtained sending the pulses to a computer controlled CAMAC unit data acquisition system based on a high-speed multichannel scaler and a PDP 11/23 unit. The time of flight spectra for the product ions are then corrected for the ion flight time and the trasmission throughout the mass filter.

For each collision velocity, i.e. for each delay time, the relative cross sections are obtained from the metastable atom intensity, the relevant product ion count rate, and the relative collision velocity.

IONIZATION OF ATOMS

Two ionization processes are possible in the collisions between a metastable rare gas atom, R^*, and an atomic target, A:

$$R^* + A \rightarrow R + A^+ + e^-$$
$$\rightarrow RA^+ + e^-$$

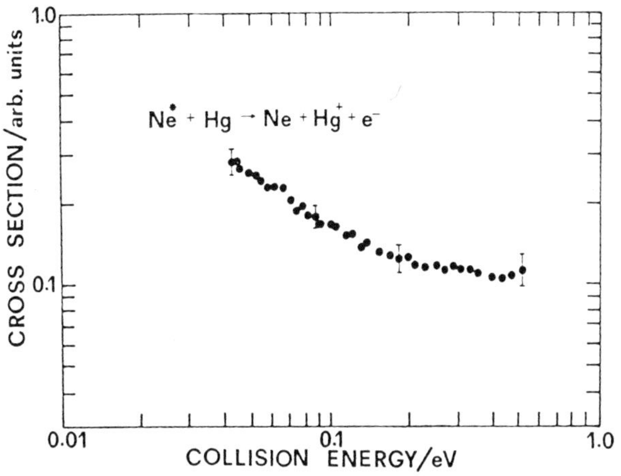

Fig.2 Collision energy dependence of the Ne^*-Hg Penning ionization cross section.

The first process is usually called "Penning Ionization" (PI), the second one "Associative Ionization" (AI). The cross sections for both processes have been studied as a function of the collision energy in the thermal energy range for Ne^*-Ar, Kr and Xe systems and for Ne^*, Ar^* and Kr^*-Hg. In all cases the PI process has been found to be dominant, although the cross sections for the AI channel is of the same order of magnitude as the PI one in most cases.

For all rare gas atomic target cases [6,7] the AI to total ionization branching ratio has been found to decrease as a function of the collision energy. In these systems the final diatomic ion $NeAr^+$, $NeKr^+$ and $NeXe^+$ can be formed in three possible states [9]. An analysis

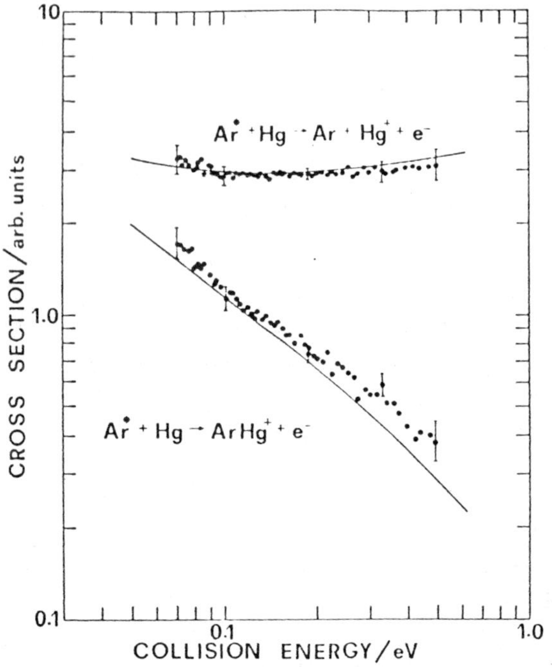

Fig.3 Collision energy dependence of the Ar^*–Hg Penning and associative ionization cross sections.

of the experimental cross sections in terms of an optical potential model has been accomplished [6,7] by using potential energy curves for the ionic states estimated from photoionization spectroscopy and theoretical calculation [9]. This analysis has shown that in these systems the AI and PI cross sections are mainly determined by the potential energy curve of the ground state of the product molecular ion in the high collision energy limit.

Interesting effects are also shown during the ionization of mercury by metastable rare gas atoms. In Fig.2 the PI cross sections for Ne^*–Hg collisions, as measured in our laboratory, are reported as a function of the collision energy. In this case the AI has been found to have only an average yield of less than ~3% of the total ionization. In Fig.3 the AI and PI cross sections for Ar^*–Hg collisions are reported; the AI yield is much larger than for Ne^* and the AI to total ionization branching ratio shows the usual energy dependence already observed in the ionization of rare gas atoms [6,7].

For the ionization of Hg by metastable krypton atoms a peculiar trend has been observed. The AI/total ionization cross section ratio for Kr^*–Hg collision is plotted in Fig.4, two extrema are evident in the

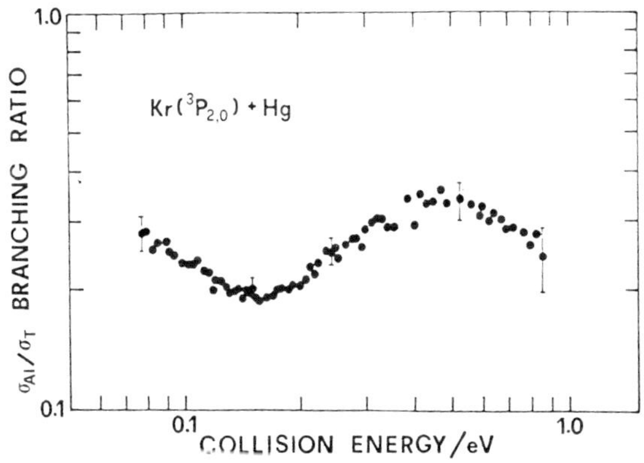

Fig.4 Energy dependence of the associative to total ionization branching ratio for the Kr*-Hg collisions.

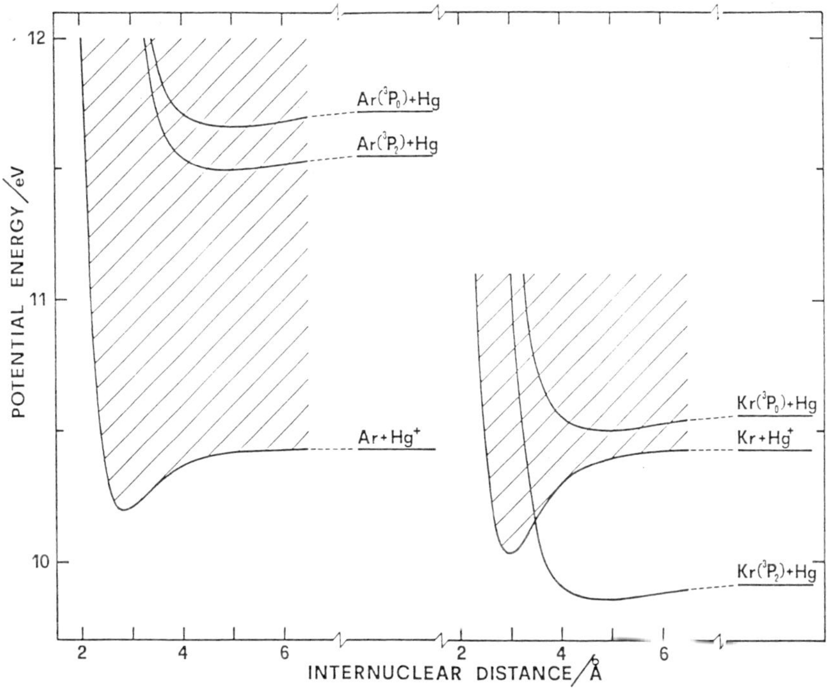

Fig.5 Schematic plot of the potential energy curves involved in Ar*-Hg and Kr*-Hg ionization processes.

collision energy dependence. These two extrema are correlated to the threshold energies for the ionization of Hg due to the $^3P_2$ component of Kr^*. This becomes evident from the potential energy curves involved in these processes, plotted in Fig.5.

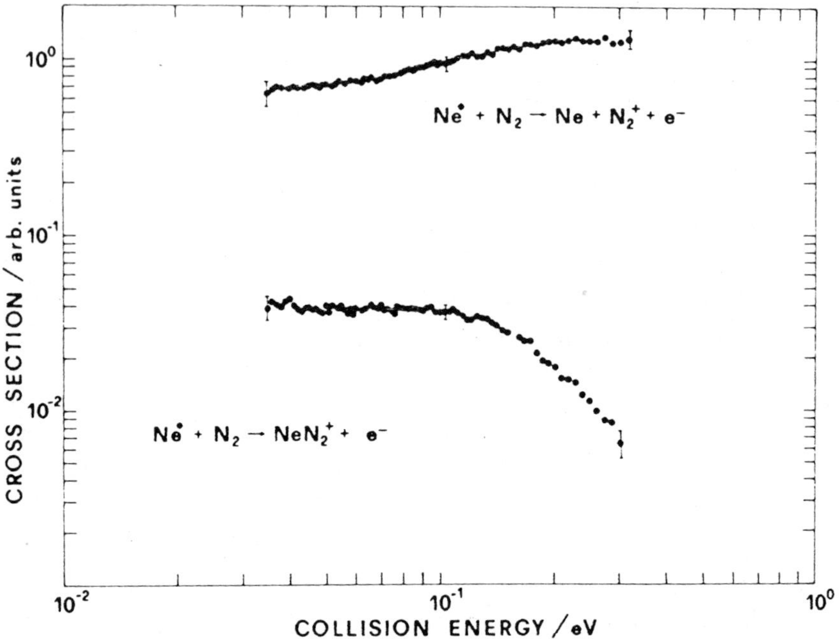

Fig.6 Energy dependence of the Penning and associative ionization cross sections for Ne^*-N_2 collisions.

The curves for the interaction between an Hg atom and $Ar^*(^3P_0)$, $Ar^*(^3P_2)$ and $Kr^*(^3P_0)$ atoms are always embedded in the ionization continuum, while that one representing the interaction with $Kr^*(^3P_2)$ penetrates the ionization continuum only at shorter distances and above a threshold energy. Therefore the $Kr^*(^3P_2)$ atoms can ionize mercury only for energies larger than such a threshold, and, at low energies, produce only $KrHg^+$ ions. When the collision energy becomes larger than the dissociation level of $KrHg^+$, also Hg^+ ions can be produced.

IONIZATION OF MOLECULES

Results for ionization of some diatomic molecules are presented in this section. The ionization of N_2, O_2, CO and NO by collision with metastable neon atoms has been studied. Rearrangement ionization reactions are energetically impossible and therefore only AI and PI processes occur. In Fig.6 and Fig.7 the collision energy dependence of

AI and PI cross sections respectively for the Ne^*-N_2 and Ne^*-CO systems are reported.

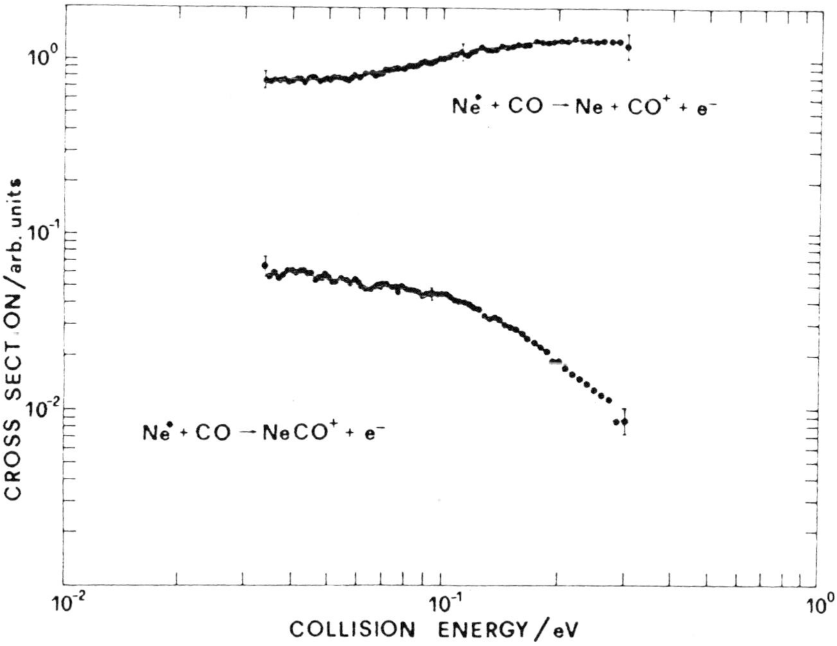

Fig.7 Energy dependence of the Penning and associative ionization cross sections for Ne^*-CO collisions.

The energy dependence of these cross sections is very similar for both systems and moreover is also similar to the AI and PI cross section energy dependence of the Ne^*-Ar system [6]. This is not surprising when one considers that Ar, N_2, and CO have very similar polarizability and therefore one could expect similar interactions between Ne^* and these partners. However also O_2 and NO have a polarizability very close to the one of Ar, N_2, and CO, but in these cases the ionization shows different characteristics. For Ne^*-O_2, NO systems the average AI to PI cross section ratio has been found to be approximately one order of magnitude smaller than that one of Ne^*-N_2, CO systems. The intensities of the NeO_2^+ and $NeNO^+$ ions were so small that only the measurement of the PI cross section energy dependence has been possible in our experiment. These cross sections are plotted in Fig.8 and Fig.9. They appear to decrease slightly with collision energy, while the PI cross sections for N_2 and CO appear to increase slightly (see Fig.6 and Fig.7).

The different behavior between the ionization characteristics of

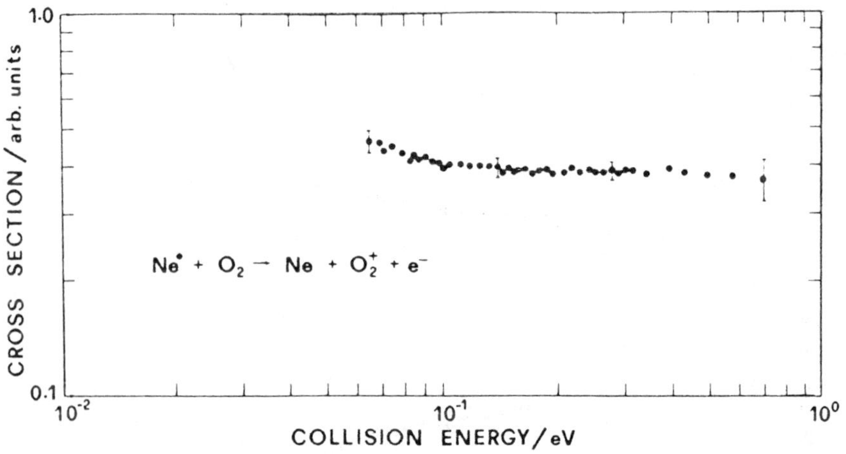

Fig.8 Energy dependence of the Penning ionization cross sections for Ne^*-O_2 collisions.

Ne^*-N_2, CO and those of Ne^*-O_2, NO, can be explained assuming that, at thermal energies, an ionic intermediate, Ne^+-O_2^- or Ne^+-NO^-, can be formed by a "harpooning" mechanism [10]. For Ne^*-N_2, CO this appears much less probable because of their negative electron affinity.

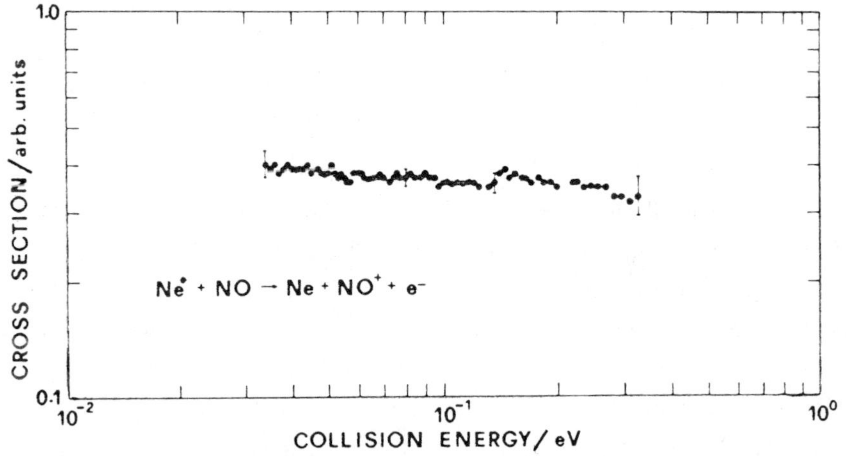

Fig.9 Energy dependence of the Penning ionization cross sections for Ne^*-NO collisions.

The formation of such intermediate ionic complexes is well documented in the literature [11-15] for collisions between metastable rare gas atoms and molecules having a positive electron affinity. At hyperthermal collision energies the product of the direct dissociation of the intermediate complex [16-20] have been observed (see Ref.20 for the Ne^*-O_2 system). Such a mechanism can explain the different behaviour in the ionization by Ne^* of O_2 and NO, with respect to N_2 and CO. In fact the different interaction between the collision partners leads to a different energy dependence of the ionization cross sections. Moreover the presence of an ionic interaction leads to a large increase of the kinetic energy of the two collision partners, which does not occur for N_2 and CO. This makes more difficult the formation of bound NeO_2^+ or $NeNO^+$ ions, explaining also the lower AI/PI branching ratio in these systems.

REFERENCES

1 - J.H.Kolts and D.W.Setser, in "Reactive Intermediates in the Gas Phase", ed. D.W.Setser, Academic, New York 1979, p.152.

2 - A.Niehaus, Adv.Chem.Phys. 65 (1981) 399.

3 - H.Haberland, Y.T.Lee, and P.E.Siska, Adv.Chem.Phys. 65 (1981) 487.

4 - P.E.Siska, Comm.At.Mol.Phys. 15 (1984) 155.

5 - A.Aguilar-Navarro, B.Brunetti, M.Cardinalini, F.Vecchiocattivi, and G.G.Volpi, Gazz.Chim.Ital. 113 (1983) 711.

6 - A.Aguilar-Navarro, B.Brunetti, S.Rosi, F.Vecchiocattivi, and G.G.Volpi, J.Chem.Phys. 82 (1985) 773.

7 - B.Brunetti, F.Vecchiocattivi, and G.G.Volpi, J.Chem.Phys. 84 (1986) 536.

8 - L.Appolloni, B.Brunetti, J.Hermanussen, F.Vecchiocattivi, and G.G.Volpi, Chem.Phys.Lett. 129 (1986) 287.

9 - B.Brunetti, F.Vecchiocattivi, A.Aguilar-Navarro, and A.Solè, Chem.Phys.Lett. 126 (1986) 245.

10 - see for instance: R.D.Levine and R.B.Bernstein, "Molecular Rections Dynamics", Oxford University, New York 1974.

11 - W.Goy, V.Kohls, and H.Morgner, J.Electr.Spectr.Rel.Phen. 23 (1981) 383.

12 - W.Goy, H.Morgner, and A.J.Yencha, J.Electr.Spectr.Rel.Phen. 24 (1981) 77.

13 - O.Leisin, H.Morgner, and W.Müller, Z.Phys. A 304 (1981) 23.

13 - W.Kischlat and H.Morgner, Z.Phys. A 312 (1983) 305.

14 - P.E.Siska, Comm.At.Mol.Phys. 15 (1984) 155.

15 - O.Leisin, H.Morgner, W.A.Müller, H.Seiberle, and J.Steigmaier, Mol.Phys. 54 (1985) 1101.

16 - K.T.Gillen, T.D.Gaily, and D.C.Lorents, Chem.Phys.Lett. 57 (1978) 192.

17 - T.M.Miller and K.T.Gillen, Phys.Rev.Lett. 44 (1980) 776.

18 - K.T.Gillen and T.M.Miller, Phys.Rev.Lett. 45 (1980) 624.

19 - H.Schall, Th.Beckert, J.M.Alvariño, F.Vecchiocattivi, and V.Kempter, Nuovo Cim. B 63 (1981) 378.

20 - J.M.Alvariño, C.Hepp, M.Kreiensen, B.Staudenmayer, F.Vecchiocattivi, and V.Kempter, J.Chem.Phys. 80 (1984) 765.

PART III

ADIABATIC PROCESSES IN MOLECULAR SCATTERING

INELASTIC MOLECULAR COLLISIONS AT THERMAL ENERGIES

Franco A. Gianturco

Department of Chemistry, University of Rome

Città Universitaria, 00185 Rome, Italy

ABSTRACT

Low-energy collision processes that involve the excitation and/or de-excitation of rotational and/or vibrational internal states of at least one of the colliding partners are discussed and general patterns of distributions are analysed. The relation of state-to-state differential cross sections to the anisotropy of the interaction potentials and their classification in terms of rotational rainbows are presented. Predissociating processes originating from weakly bound states are also analysed in terms of half-collisions for atom-diatomics cases and their behaviour is related to the anisotropy of the full potential energy surface.

1. INTRODUCTION

The general area of collisional processes which cause changes in the distribution of the various forms of internal energy available in a simple molecule when it interacts with either a structureless partner (an ion or a neutral atom) or another molecule is of fundamental importance to many, apparently unrelated, fields of study. Thus, even by limiting the amount of energy available in the thermal bath to fractions of an electron volt, the rotational energy transfer (RET) and vibrational energy transfer (VET) processes already provide a tremendously large body of data which is becoming experimentally available today for many atom-diatomics systems. The detailed understanding of their dynamics is important for transport properties like diffusion, viscosity and heat conductivity, determines the final state distribution in chemical reactions and excites radiative energy levels in laser media or in interstellar clouds. Moreover, it is widely known that the collisional processes governing the lineshapes of molecular spectra are essentially the same as those which prevail in the rotational relaxation of molecular gas mixtures.

From the theoretical viewpoint, the rotational momentum transfer during a single collision is conceptually the simpler of the above two mechanisms and it is also one of the most efficient processes in molecular dynamics. As a consequence, its detailed investigation has interested many researchers, both experimentalists and theoreticians, over a long period of time and considerable success has been achieved in recent years as far as its detailed understanding is concerned for several atom-molecule and ion-molecule systems. Many more questions, on the other hand, are still open for the (VET) process and our theoretical treatment of the detailed dynamics for even the simplest cases is still far from being fully satisfactory. In the present chapter we will therefore concentrate on the collisional processes that lead to rotational excitations and/or de excitation of the molecular target and will examine in detail some of the more recent developments that have brought theoretical treatments to a very close understanding of a broad range of experimental findings. The subjects discussed here can be viewed as a follow-up to the general theoretical models discussed a few years back in the previous NATO Advance Study Institute on the same subject[1].

On the theoretical side, it is well known that two problems have to be solved to realistically tackle (RET) processes. First, one needs to know the relevant potential energy surface (PES) or surfaces over a wide range of nuclear geometries and then scattering cross sections need to be obtained from the equations of nuclear motion.

Although a broad range of recipes is currently available for calculating cross sections via quantum, semiclassical and classical methods[2], and various prescriptions that reduce the full dimensionality of the ensuing equations have been put forward and tested[3] on several systems, the ab initio computation of reliable potential energy surfaces is still a big problem. Except for a few simple examples, therefore, the usual procedure has been to rely on experimental data for getting detailed results from which to extract the underlying nonspherical interaction potentials[4]. In the following Section we will therefore show how this has been made possible by accurate experimental data and how much it has helped us to progress from general, semiempirical procedures which can yield broad rules that govern rotational energy partitioning among molecular levels to detailed knowledge that reproduces very well state-to-state angular distributions for specific systems at specific energies.

Furthermore, in Section 3 an internal energy transfer process, rotational predissociation (RP), which has received considerable attention in the last few years is described from a theoretical point of view and some general features that are typical of Van der Waals (VdW) complexes are reported and analysed.

2. ENERGY PARTITIONING IN ATOM-DIATOMICS ROTATIONAL EXCITATIONS

If one were to suppose that a complete set of state-to-state rate constants had been obtained in a number of chemically similar systems, then to study the systematic variations of these constants with the changes of the systems under study requires the analysis of 10^5-10^6 rate constants per system, even if we were to restrict our attention to thermally averaged, level-to-level rate constants for inelastic processes in an atom-diatom system at a single temperature, and only to those levels which might be populated at room temperature. If, moreover, we wanted to know state-to-state differential cross sections for about a dozen different velocities for the same system, the number of data to handle would easily reach 10^{10}. It is therefore clear that to relate the relative probabilities of ending up in each of all the possible final rotational states to microscopic features of the system like the nature of the interaction potential and its influence on each dynamical event becomes a very difficult task. In fact, because of the inherent averaging processes involved it has proved virtually impossible to derive direct information on the PES from rate constant measurements. The information content of the above data is therefore best described by the so called fitting laws, which essentially state that the rates decrease with increasing energy transfer ΔE[5].

On the other hand, a more satisfactory comparison between theoretical results and experiments needs detailed, state-to-state differential cross sections so that the ensuing angular distributions and energy losses can be directly related to the interaction potential as its is known from elastic scattering data[6]. Such comparisons allow us to be immediately confronted, and try to answer, with the following questions: what is the underlying physical mechanism and its most important variable (i.e. statistics, energy gap, dynamical coupling, etc)? Is the observed distribution sensitive to the PES or to parts of it? Is it possible to scale some features of the distribution with respect to parameters of the system like energy, mass, scattering angle, etc? We will show below how some of the above questions have been recently answered and have helped to justify as much as possible the various general distributions or fitting laws that have also appeared in the literature in recent years.

2.1. The quantum dynamics

In order to better discuss the various apparoximations that are most often employed both in the detailed analysis of state-to-state differential cross sections and in the scaling and fitting laws which generate scores of rate constants, let us briefly review the quantum formulation od the space-fixed (SF) coupled equations that describe the N-channel problem[2,3]. One seeks solutions of the following second-order differential equations in one dimension:

$$\underline{\underline{W}}^J \underline{\underline{F}}^J (R) = 0 \tag{1}$$

where:

$$\underline{\underline{W}}^J = \underline{\underline{L}} - V^J (R) \tag{2}$$

$$[L]_{ij} = \{ \frac{d^2}{dR^2} - \ell_i (\ell_i + 1) R^{-2} + k_i^2 \} \delta_{ij} \tag{3a}$$

$$[V^J]_{ij} = \frac{2\mu}{h^2} \sum_\lambda V_\lambda (R) \langle \mathcal{Y}^J_{j_i \ell_i} | P_\lambda | \mathcal{Y}^J_{j_j \ell_j} \rangle \tag{3b}$$

$$[F^J]_{ij} = F^J_{ij} (R) \tag{3c}$$

An element of the continuum matrix function $\underline{\underline{F}}$ is labelled by a channel index i, a particular linearly independent solution j and it reflects the influence of the coupling via the potential matrix V^J which distorts during their propagation from the origin the uncoupled, free-wave solutions arising from the matrix operator $\hat{L}$. Each V_λ coefficient comes from the standard multipolar expansion of the full PES:

$$V(\hat{R},\hat{r}) = \sum_\lambda V_\lambda (R) P_\lambda (\hat{R}\cdot\hat{r}) \tag{4}$$

where the diatomic target is treated as a rigid rotor with fixed internuclear distance $|\underset{\sim}{r}|$. It is clear from eq. (3b) that the distorsion of the free-particle motion and the angular momentum transfer during the interaction depend on both the range of radial action of each V_λ, its 'strength' when compared with the last two terms of eq. (3a) and the number of P_λ's which need to be included in eq. (4) and which couple the angular coefficients $\mathcal{J}^J_{j\ell}(R,r)$ [3]. More on this point will be discussed in each of the subsections below.

It is well known that the dimensionality problem when solving the above equations (1) comes from the rapid proliferation of quantum channels associated with the $(2j + 1)$ degeneracy of the energy levels of the rotating target. The well-known Infinite Order Sudden Approximation (IOSA)[2,3] therefore introduces two drastic approximation for the last two diagonal elements of the $\underline{\underline{L}}$ matrix elements of (3a). Firstly, it assumes an effective orbital angular momentum eigenvalue and 'freezes' the centrifugal potential as being given by:

$$\frac{\ell^2}{2\mu R^2} \simeq L(L + 1)/2\mu R^2 \tag{5a}$$

which is called the centrifugal sudden (CS) approximation in that the relative kinetic energy is sufficiently large with respect to the various effective potentials that the precise value of the centrifugal term is not important. If the interaction is predominantly repulsive, then the rate of change of the turning points as ℓ changes is not large and therefore condition (5a) is expected to be valid. It is also called the j_z - conserving approximation in the sense that only coupling between different rotational states (and not between substates) is allowed to appear in the dynamical treatment (see refs 1-3 for a more detailed discussion and for all the original references of the derivation). Secondly, if one recalls that the channel energy is given by each diagonal wavevector in eq. (3a):

$$k_i \equiv k_{jj'} = [2\mu\,(E - \varepsilon_{j'} - \varepsilon_j)]^{1/2} \tag{5b}$$

where E is the reference energy and the ε's are the rotor level energies, the IOS scheme assumes the channel wavevector to be constant for all coupled channels (Energy sudden, ES, approximation):

$$k^2_{jj'} \simeq k^2_{\bar{j}\bar{j}} \quad \text{for all } j' \tag{5c}$$

The final differential equations are now decoupled and one solves separate equations for each good quantum number L, thereby obtaining new radial wavefunctions:

$$F^J_{ij}(R) \Longrightarrow F^{L\bar{j}}(R,\hat{R}\cdot\hat{r}) \tag{5d}$$

that depend parametrically on the relative orientations of $\underline{R}$ and $\underline{r}$.

The corresponding S-matrix elements for each rotational channel are now obtained by coupling the angular functions of eq. (3b) via the new, approximate, angle-dependent S-matrix:

$$S^{J}_{j\ell j'\ell'}(E) \sim i^{\ell+\ell'-2L} \left\langle \mathcal{Y}^{J}_{j'\ell'} \,\middle|\, S^{\bar{j}L}(\gamma) \,\middle|\, \mathcal{Y}^{J}_{j\ell} \right\rangle \tag{6}$$

The above integration is usually accomplished by standard Gaussian quadrature over γ. The main result of this semplification is that state-to-state (degeneracy averaged) integral and differential cross sections for transitions between two arbitrary levels, j_i and j_f, are given by computing only one row (or one column) of the whole matrix of cross sections and by using simple geometric weighting factors[2,3]:

$$\sigma^{IOSA}_{j_i \to j_f}(E) = \frac{k_o^2}{k_{j_i}^2} \sum_{j''} C^2 (j_i j'' j_f;\, 000)\, \sigma^{IOSA}_{o \to j''}(E) \tag{7a}$$

$$\frac{d\sigma^{IOSA}}{d\Omega}(j_i \to j_f \mid \theta) = \sum_{j''} C^2 (j_i\, j''\, j_f;\, 000) \frac{d\sigma^{IOSA}}{d\Omega}(o \to j'' \mid \theta) \tag{7b}$$

where the energy factor has been dropped in (7b) to be in keeping with the ES assumption (5c).

The C(...|...) are standard Clebsh-Gordan coefficients.

It is interesting to note at this point that the above scaling relationship can also be obtained by introducing only the ES approximation and without the further limitation of the j_z-conserving or CS approximation[7,8]. The physical basis of this less restrictive condition is simply given by the sudden limit of the collision reduced duration, τ, a dimensionless parameter that gives the number of radians of rotation of the diatom during collision:

$$\tau_j = \omega_j \times T_d = \omega_j \times \frac{l_c}{v} \tag{8}$$

where ω_j is the angular molecular velocity when its angular momentum is j, v is the relative velocity of the partners and l_c is a characteristic length over which the interaction affects collision. As $\tau_j \to o$, the diatom does not rotate during collision and the rotational energy ε_j must satisfy the following condition:

$$\varepsilon_j << \frac{I}{2\mu \ell_c^2} \left(\frac{1}{2} \mu v^2 \right) \tag{9}$$

In the case of a heavy molecule and a light perturber.

This condition is likely to be satisfied and does not require the additional CS simplification[5].

The effect of the ES approximation on the CC equations (1) can be briefly seen if one rewrites more explicitely the coupling matrix elements (3b):

$$[V^J]_{ij} = \frac{2\mu}{\hbar^2} \int Y^*_{\ell' m_{\ell'}}(R)\, Y^*_{j' m_{j'}}(\hat{r})\, V(R,\hat{R},\hat{r})\, Y_{\ell m_\ell}(\hat{R}) \times Y_{j m_j}(\hat{r})\, d\hat{R} d\hat{r} \tag{10}$$

If one now replaces the j-state wavevector, for all j values, with a constant value k_j^2, the above matrix elements, and the unknown radial functions of eq. (3c), can be rewritten in a form which does not contain any more the couplings of $Y_{j'}$ and Y_j but depends explicitely on the SF orientation of the molecule, $\hat{r}$ [9].

$$[V^J]^{ES}_{ij} \simeq \int d\hat{R}\; Y^*_{\ell' m_{\ell'}}(\hat{R})\; V(R,\hat{R},\hat{r})\; Y_{\ell m_\ell}(R) \tag{11}$$

Since the potential only depends on the <u>relative</u> orientation $\hat{r} \cdot \hat{R}$, one can now define a set of equations in which the R-dependence of the unknown continuum functions is separated from their $\hat{r}$ dependence by bringing the molecular orientation to a fixed position with respect to the original SF orientation of the impinging atom, $\hat{R}$. This orientation is defined by m_ℓ and gives the name of ℓ_z-conserving approximation to the ES simplification[8,9]:

$$[V^J]^{ES}_{ij} = V^{\bar{m}}_{\ell\ell'}(R) = 2\pi \int_{-1}^{1} Y^*_{\ell' \bar{m}_\ell}(\gamma,o)\; V(R,\gamma)\; Y_{\ell \bar{m}_\ell}(\gamma,o)\; d(\cos\gamma) \tag{12}$$

The significant difference in using this approximation as opposed to the full IOS, resides in the wide range of values of $\bar{m}_\ell$ for which coupled equations need to be solved, since many $\{\ell,\ell'\}$ values are used in the partial wave expansion of heavy particles. Thus, while the CS approximation needs to solve at the most $(j_{max} + 1)$ equations, the ℓ_z--conserving scheme goes further and requires the coupling of $(2j_{max} + 1)$ different ℓ values via the potential matrix elements of eq.(12)[10]. One can therefore say that the ES approximation is somewhat complementary to the CS reduction scheme since it is expected to work for heavier systems and for long range anisotropic PES, which are incorrectly treated by the physical simplifications implied by the j_z -conserving scheme.

In conclusion, starting from the full CC equations (1), we have briefly reviewed the hierachy of approximations which have been most generally tested in the last few years to obtain both individual, partial differential cross sections and state-to-state rate constants for (RET) processes. In the following subsection we will then describe some of the scaling and fitting laws which have been suggested by employing in part the same dynamical approximations discussed here.

2.2. General energy distribution laws

Most of the molecular systems in which (RET) processes need to be studied present, as mentioned before, too large an array of state-to-state rate constants for the full ab initio approach to be feasible for them. Even if simplifications as the ES or IOS schemes are introduced, the scaling relationships still require the calculation of the basis cross sections out of the reference state j=o, a task which is further complicated by our lack of knowledge of the relevant PES's, or single surface, needed to guide the dynamics.

It is important, at this point, to distinguish between the use of scaling laws that incorporate the dynamical simplifications discussed before but still require the direct computation of a subset, however large, of cross sections and the use of pure fitting laws which aim at expressing the entire matrix in terms of a set of parameters which do not come from a specific rate constant but are adjusted to an entire ensemble of either experimental or theoretical data (if they exist). In both cases one obviously tries to find the physical origin on which such relationships are based, although dynamical scaling laws are the only ones for which this has been done at all convincingly. We will here concentrate mainly on the latter laws and refer the reader to earlier reviews for the work based on statistical considerations[11,12].

If one takes advantage of the simplifications occurring when the reduced duration goes to zero, then the most direct scaling law to obtain the full matrix of rate constants starting from a set of basis rate constants $K_{o\to\ell}$ and making use of eq. (7a) is:

$$K_{j_i\to j_f} = (2j_f + 1)\sum_\ell \begin{bmatrix} j_i & j_f & \ell \\ o & o & o \end{bmatrix}^2 K_{o\to\ell} \tag{13}$$

where the large parentheses represent now 3-j symbols[10].

The above equation, as discussed earlier, can be simply understood in classical terms when one assumes that the diatom does not rotate during the collision (i.e. τ_{j_i}<<1), as shown by specific calculations of classical trajectories[13].

When one applies the above relationship to thermally averaged rate constants at a given temperature T, the above expression becomes:

$$K^{IOS}_{j_i \to j_f} = (2j_f + 1) \exp \left(\frac{\varepsilon_{j_i} - \varepsilon_{j_>}}{K_B T} \right) \sum_{\ell} \begin{bmatrix} j_i & j_f & \ell \\ o & o & o \end{bmatrix}^2 (2\ell + 1) K_{\ell \to o} \tag{14}$$

where $K_B T$ is the average relative translational thermal energy and $j_>$ is the greater of j_i and j_f.

The detailed balance is guaranteed by the two factors in front of the summation over ℓ. The above law was found in good agreement with observation for systems where the target atom is much lighter than the diatom and the ES approximation is expected to hold[14].

The above restriction is obviously rather limiting and it is important to generalize the scaling relationship (14) to systems close to thermal equilibrium for which the sudden condition on the rotational motion is not valid any longer.

By invoking a semiclassical time-dependent approach, one can write the first-order transition moment for going from state |i> to state |f> as given by:

$$M_{i \to f} \; \alpha \int_{-\infty}^{\infty} dt \; \exp(-i \; \frac{\Delta E}{\hbar} t) \; V_{i \to f} \; (t) \tag{15}$$

where $\Delta E = \varepsilon_f - \varepsilon_i$ and $V_{i \to f}$ is that part of the intermolecular potential which couples,during relative motion, state |i> with state |f> . It is worth noting at this point that the largest potential contributions are centered around the classical turning point and will extend around it depending on the ℓ_c value of eq. (8). Thus, as the collision time increases the exponential in (15) will oscillate rapidly relative to any variation of $V_{i \to f}$ (t) and the transition moment will be small. This situation is usually called 'adiabatic' and is related to the well-known Massey adiabatic parameter η:

$$\eta \equiv \frac{\Delta E}{\hbar} \times \frac{\ell_c}{v} = |\Delta j| \times \tau_j \tag{16}$$

and if $|\Delta j|$ is large the collision can be adiabatic ($\eta > 1$) even if dinamically sudden ($\tau_j \ll 1$). Thus, for coupling potentials that are either strongly anisotropic (large λ indeces in eq. (4)) or that act over a large range ℓ_c, the sudden condition will not correctly describe the process at hand.

One recent way of introducing adiabatic corrections has been to expand the exponential term in eq. (15) by assuming[15] $V_{i \to f}$ (t) to be simmetric about the time of closest approach and therefore obtaining the first non-zero correction to be second order in $\omega = \Delta E/h$:

$$M_{i\to f} \,\alpha \int_{-\infty}^{\infty} dt \;\; V_{i\to f}(t) + \int_{-\infty}^{\infty} dt(-\omega^2 t^2) \;\; V_{i\to f}(t) + \ldots$$

$$= \int_{-\infty}^{\infty} dt \;\; V_{i\to f}(t) \;\; (1 - \omega^2 T^2 + \ldots) \tag{17}$$

Since the above expression has an unphysical sign change when $\omega T^2 > 1$, it can be replaced by the following expression[15]:

$$(1 - \omega^2 T^2) \to 1/(1 + \omega^2 \ell^2 /_v 2) \tag{18a}$$

where ℓ now depends on the impact parameter b which appears in $V_{i\to f}$ [$\underline{R}(t)$, r(t)] when the integration is carred out along each trajectory contributing to the collision. If one now assumes that adiabatic corrections mostly come from the distant collisions dominated (for neutral systems) by R^{-6} terms in the coupling potential, then the range parameter ℓ can be replaced by a characteristic length, b_c, assumed to be independent of j_i and j_f and the classical velocity will also contain the exponent of the attractive tail:

$$(1 + \omega^2 T^2) \to \frac{1}{\left(1 + \omega^2\right) \times b_c^2/_{6v^2}} \tag{18b}$$

Since this factor can be taken out of the integral (17), then the square of the corrected transition moment will give us the corresponding Energy corrected sudden (ECS) rate constant[15].

Once the correction factor is applied both to the basis rate constants $K_{o\to\ell}$ and to the final rate constant $K_{j_i \to j_f}$ of eq. (14), one modifies the IOS scaling low into the ECS scaling law:

$$K^{ECS}_{j_i \to j_f} = (2j_f + 1) \exp\left(\frac{\varepsilon_{j_i} - \varepsilon_{j_>}}{K_B T}\right) \sum_\ell \begin{bmatrix} j_i & j_f & \ell \\ o & o & o \end{bmatrix}^2 (2\ell + 1) \;\; A(j_i j_f, \ell)^2 \times$$

$$\times K_{\ell \to o} \tag{19}$$

where the adiabatic correction factor is given by:

$$A(j_i j_f, \ell) = \frac{1 + \tau_\ell^2/6}{1 + \tau_{j_>}^2/6} \qquad (20)$$

Calculations on many systems have confirmed the utility of the ECS scaling law[5] and a recent example for the Ar-CO system[16] is shown in Fig. 1, where the correct rates were computed via quasiclassical trajectories and the IOS and ECS curves are shown for comparison. Several values of b_c are reported and were estimated by a weighted average impact parameter leading to inelastic processes from j=o. The accord with the accurate rates is quite good.

The earliest example of a fitting law is the well-known exponential gap law (EGL) which was suggested for the interpretation of chemiluminescence experiments[17] and simply proposed that the efficiency of the (RET) process for a particular channel decreases exponentially as the amount of energy exchanged increases in comparison to the average kinetic energy:

$$K^{EGL}_{j_i \to j_f} = C(E)\; K^{(o)}_{j_i j_f}\; \exp\left(-\theta\, |\Delta E|/E\right) \qquad (21)$$

where $E = \frac{1}{2}\mu\, v^2 + B(j_i + 1)j_i$. $C(E)$ and θ are energy dependent empirical parameters obtained from a best fit of the data. Moreover:

$$K^{(o)}_{j_i j_f} = (2j_f + 1)\; \frac{\langle (E_t - \Delta E)^{1/2} \rangle}{\langle (E_t)^{1/2} \rangle} = (2j_f + 1)\; R\,(\Delta E) \qquad (22)$$

is the prior statistical rate constant with E_t being the relative translational energy distribution. It has been suggested from surprisal analysis[18,19] but it still remains unclear, at the microscopic dynamical level, why should inelastic cross sections be so insensitive to $V(\underline{R}, \underline{r})$ as to satisfy such a simple fitting law. On the other hand, the agreement with measured and computed data is only true in a broad sense and often fails as the energy gap increases[20].

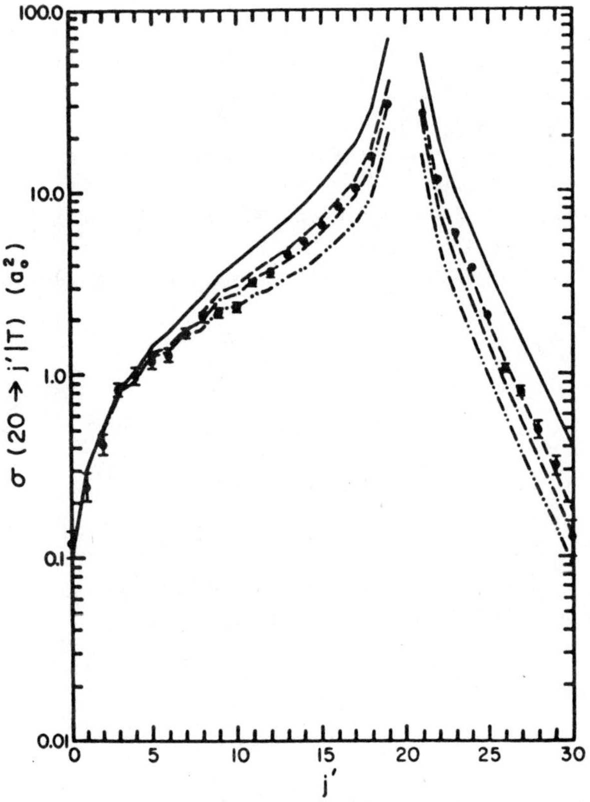

Fig. 1. Thermally averaged cross sections for the rotational excitation of CO (j = 20) by collision with Ar at 500 K. The solid curve is the IOS result while the ECS approximation is shown for b_c = 2.5 Å (-..-), b_c = 1.9 Å (-.-) and b_c = 1.5 Å (--). (From ref. 16).

A modification of the EGL expression was later provided with some statistical justification that assumed the average T-matrix element to depend on ΔE as a simple power law[21]. It was called the statistical power gap law (SPG) and was written as:

$$K^{SPG}_{j_i \to j_f} = a \left| \frac{\Delta E}{B_{rot}} \right|^{-\alpha} \times N_\lambda \times R(\Delta E) \qquad (23)$$

where B_{rot} is the rotational constant and a, α and λ are variable parameters; N_λ is a spin degeneracy factor from the statistical weight of the levels involved[21]. In order to extend the above expression, a later modification included it with the IOS factorisation and generated a new fitting expression from the scaling law that was now called the Infinite order sudden power law[22] (IOS-P). It provides a prescription of exponential form for the basis rate constants of eq. (19):

$$K_{\ell \to 0} = A\,[\ell\,(\ell + 1)]^{-\Gamma} \qquad (24a)$$

which has now an energy gap dependence similar to the SPG law. This fitting procedure worked reasonably well for light projectiles, He and H_2, in collision with Na_2^* but failed for the heavier projectiles[14,22]. Typically the values of the parameter Γ fall in the range 1.1 - 1.3.

Here one finds that the fitting law receives some justification from the scaling properties of the ES approximation and therefore one could further extend its usefulness by applying the energy correction discussed above, thus introducing the (ECS-P) fitting law that contains an adiabatic factor like that of eq. (20) but generates basis rate constants from eq. (23). This last parametrisation has been remarkably successful in fitting a wide variety of experimental data[5], although no clear justification exists for the two-parameter exponential fit of eq. (23).

It was further found in the I_2 + He system[23] that the (RET) rates fell off rapidly in the high Δj channels and therefore could not be fitted by the previous (ECS-P) procedure. If one now assumes that a possible breakdown of the above fitting law occurs because of the existence of short-range repulsive cores in any realistic PES, then it is reasonable to expect a limit to the closest approach that would provide in turn an upper limit to the maximum angular momentum which can be transferred during a given trajectory[5]. Thus, one can represent the effect of the large $|\Delta j|$ region by adding an exponential factor to the previous power law, the (ECS-EP) fitting law:

$$K_{\ell\to o} = A[\ell(\ell+1)]^{-\Gamma} \exp \frac{-\ell(\ell+1)}{\ell^*(\ell^*+1)} \tag{24b}$$

where ℓ^* is interpreted as the maximum tranferable angular momentum, Δj_{max}, which can be classically exchanged during a given collision. This interpretation has been confirmed by experiments[24], where it was concluded that the new parameter ℓ^* is consistent with the assumption of angular momentum constraints. Thus, the earlier justification of fitting laws as coming from energy conservation requirements have gradually evolved into the explicit inclusion of angular momentum conservation, as in eq. (24), and into a more explicit connection with possible features of the potential energy surfaces.

An example of the quality of fits that can be obtained with (ECS-EP) procedure is shown in Figure 2, where the results for the I_2 - He[23] are shown for two different initial rotational states. The value of ℓ^* was 24 which, under the experimental conditions, corresponds to an impact parameter b = 3 Å.

A recent proposal of a fitting law that draws its justification from dynamical approximation is the AON model[25,26]. The physical conditions assumed in the model are that (i): the anisotropic interaction is strongly dominated by a single V_λ coefficient in the expansion (4) and (ii): that the short-range part of $V_\lambda(R)$ is responsible for the large $|\Delta j|$ transfers during RTE processes. A further assumption, as in previous fitting laws, is the validity of the ES condition for rate constants factorisation. The model is derived from a semiclassical infinite order expansion of the T-matrix, in which the actual transition comes from an infinite sum of virtual excitations, as originally proposed in the theory of collisional line broadening[27]. If one writes down the basis rate constant from the semiclassical expression of the relaxation, tensorial operator[27]:

$$K_{\ell\to o} = 2\pi N \int_o^\infty bdb \int_o^\infty u\, f(u)du \times \langle oo|S(u,b)|\ell o\rangle \langle oo|S(u,b)|\ell o\rangle^* \tag{25}$$

where N is the relative number density of the perturbing particles, u the relative velocity, f(u) a velocity distribution and S(u,b) the classical path S-matrix for a velocity u and impact parameter b.

The further assumption of sudden regime removes the integration over u, to be replaced by an average velocity $\bar{u}$:

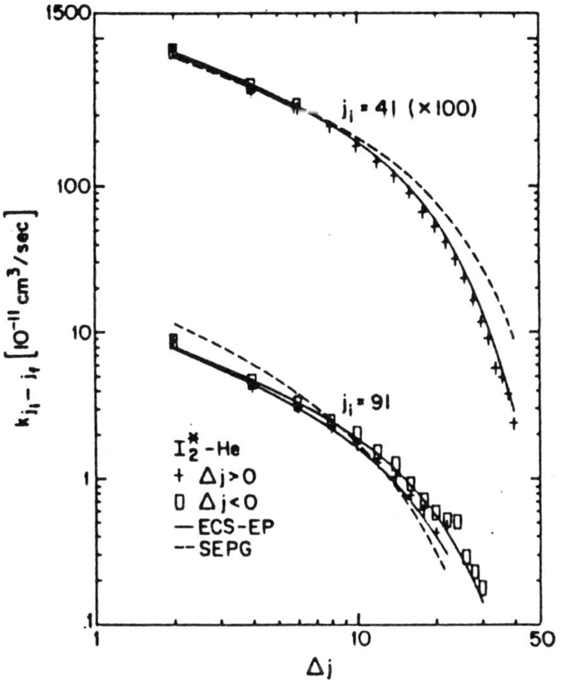

Fig. 2. Power law fits to the He - I_2 data of ref. (23) using both ECS-EP (solid line) and SEPG (dashed line). From ref (23) .

$$K_{\ell \to o} = 2\pi N \bar{u} \int_o^{\infty} bdb <oo|S_{\bar{u}}(b)|\ell o> <oo|S_{\bar{u}}(b)|\ell o>^* \tag{26}$$

Finally, the dominance of a single V_λ term produces a simplification in the above integral and the introduction of a factor, K_λ, which corresponds to the classical path integral over the dominant anisotropy:

$$K_\lambda(\bar{u},b) = (2\alpha_\lambda/\hbar) \int_o^{\infty} V_\lambda [R(t)]\, dt \tag{27}$$

and which appears directly in the final expression for the basis rate constant[28-30]:

$$K_{\ell \to o} = 2\pi N \bar{u} \int_o^{\infty} bdb\ J_n^2 (K_\lambda) \tag{28}$$

where $\alpha = 1$ if $\lambda = 1$ and 3/4 if $\lambda = 2$ and n, the order of the Bessel function J_n, is given by ℓ/λ.

If the anisotropic coefficient is assumed to be given by an exponentially repulsive expression, then:

$$K_\lambda(\bar{u},b) = a \exp(-b/Ro) \tag{29}$$

with $a = 2^{3/2} \alpha_\lambda V_\lambda(R_c) Ro/h\,\bar{u}$, where R_o is the range of the repulsive exponential and R_c is distance of closest approach at $E = \frac{1}{2}\mu \bar{u}^2$. The quantity a also represents the maximum value of angular momentum transferred in a collision[31], simular then to ℓ^* parameter of eq. (24). From the approximate expressions for Bessel functions and from restricting integration to $K_\lambda >> n$ which implies dominance of close impact collisions to the RET process[25], the basis rate constant of eq. (28) is given by:

$$K_{\ell \to o} = \frac{2N\bar{u}}{a} \int_o^{Ro\,\ell n\,(a/n)} b \exp(b/Ro)\, db \tag{30}$$

which finally yelds the AON fitting law:

$$K^{AON}_{\ell \to o} = C\{(a/n)\ln(a/n) - a/n + 1\} \tag{31}$$

where parameters a and n have been defined before and $C = (2N\bar{u}/a)Ro^2$.

The sets of basis constants are therefore given in terms of physical conditions $(\bar{u},N,T)$ and two parameters which depend on the strength of the interaction at the classical turning point, $V_\lambda(R_c)$ and on the range of the exponential repulsion, Ro/h.

If a and C are taken as free parameters, the above quantities could be evaluated by fitting experimental data with eq. (31):

$$Ro = (a\,c/_{2N\bar{u}})^{1/2} \quad ; \quad V_\lambda(R_c) = (h\,\bar{u}/_{2\alpha_\lambda})\,(N\bar{u}a/_c)^{1/2} \tag{32}$$

In conclusion, the AON law is very similar to the SPL or (ECS-P) fitting law, the main difference being the finite, maximum value of angular momentum transfer built in as a constraint for the integration in eq. (30) and not as an explicit third parameter ℓ^* as in eq. (24) for the (ECS-EP) law.

A further attempt at obtaining a dynamically based fitting law was presented recently[32] by using a semiclassical version of the effective potential method. By assuming an anisotropic repulsive interaction:

$$V(R,\gamma) = A \exp(-R/r_o)\,[1 + B\,P_2\,(\cos\gamma)] \tag{33}$$

The author obtained a fitting law of the form:

$$\sigma(j,j') = (2j'+1)\,(2j+1)\,a\,(1/_{|\Delta j|} - b) \tag{34}$$

The parameter a is related to the distance of closest approach for head-on collisions, r_o:

$$a = (^{8K_BT}/_{\pi\mu})^{1/2}\,\pi\,r_o^2 \tag{35}$$

and plots of statistically weighted cross sections versus $|\Delta j|^{-1}$ gave reasonable agreement for the I_2^* + He data[32].

Since the two parameters a and C of the AON fitting law depend on the relative velocity, one can write:

$$a(E_\ell) = \alpha\,E_\ell^{-1/2} \tag{36}$$

where: $E_\ell = 1/2\,\mu\,u_\ell^2$ is the kinetic energy in the ℓ channel, given by: $\frac{1}{2}\,\mu\,\bar{u} - B\ell(\ell+1)$. The eq. (31) can then be rewritten by including explicit dependence on E_ℓ:

$$K^{AON}_{\ell \to o}(E_\ell) = (2 N \bar{u} R^2_{o/n}) \left\{ \ell n \left[\frac{a(E_\ell)}{n} \right] -1 + \frac{n}{a(E_\ell)} \right\} \tag{37}$$

which must be applied always to exothermic transitions[33]. A comparison of AON fits and AON(E) fits is shown in Fig. 3 for the CSH-H_2 system and the energy-dependent expression shows very good agreement with IOS-P results[33].

2.3 Features of state-resolved differential cross sections

As was discussed before, the scaling laws and fitting laws for RET processes under collisional conditions do not allow us to obtain microscopic information an the interaction forces and an their orientational dependence, even for the simplest atom-homonuclear molecule case.

The problem of accurately knowing the full PES has therefore been tackled in recent years by essentially following a combination of two different approaches. The first of them determines the intermolecular potential by ab initio methods, if possible by including correlation forces, and then state-selected integral cross sections as well as temperature-dependent rates are calculated:

$$V(R,\gamma) \quad Q(j \to j' \mid E) \to K(j \to j' \mid T) \tag{38}$$

The bottleneck of this flow diagram, however, is the calculation of accurate potential energy surfaces, especially for the weak Van der Waals interactions of the species which have been experimentally studied[34].

The second approach starts from precise measurements of state-resolved differential cross sections (DCS) $\frac{d\sigma}{d\Omega}(j \to j' \mid \theta, E)$ and determines the potential surface either by inversion techniques or by comparison of measured and calculated data based on model potentials[6].

The problem of this approach is the preparation of the input data which are as close as possible to various regions of the PES, i.e. state-to-state DCS at several energies and for a broad range of angles and final j' states. The progress in the experiments, however, has been rather remarkable in the last five years or so and therefore we now possess accurate intermolecular potentials for a selected number of systems, namely the H_2- rare gas[35,36], the H_2 dimers and their isotope variations[37-39], the Na_2 - rare gas systems[40,41] and the O_2-He[4] and N_2-He[42] systems.

The combination of the above two approaches therefore tries to compare accurate experimental data with computed quantities where the employed PES has been obtained with as much theoretical input as possible and with a wider analysis of other experiments that are sensitive to different regions of the interaction and therefore allow one to reliably test as much as possible the PES employed to compute state resolved DCS[43].

The crucial point is now to see how specifically such experimental data are directly related to either dynamical properties or to features of the non-spherical interactions.

Generally speaking, small angle scattering is caused by attractive and large angle scattering be repulsive forces. In elastic scattering angular distributions, the classical rainbow angle separates these two regions very clearly[1]. Systems with small reduced masses and shallow potential wells are dominated by diffraction oscillations.

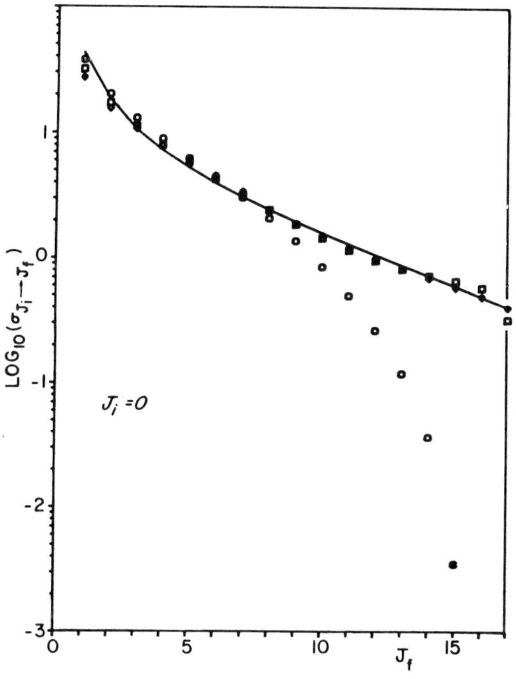

Fig. 3. Semi-log plots of the σ ($j_i \rightarrow j_f$) for CsH - H_2 as function of j_f for j_i = o. Solid line: IOS-P; ooo AON; □□□ AON (E) (From ref. 33).

The angular positions of these oscillations give a direct measure of the diameter of the repulsive wall, Ro, to better than 1%. Their amplitudes are sensitive to the form of the potential between the minimum and the distance σ where it passes through zero. The large angle scattering data are determined by the steepness of the repulsive wall and can determine this property to within 20%, better if data at several collision energies are available[34].

The anisotropy of the PES can be broadly classified into the weak coupling case and the strong coupling case. In the former, inelastic cross sections are smaller than the elastic ones and usually only one (or two) coefficientes are sufficient to represent eq. (4) and the excitation cross sections follow eq. (15) with only one V_λ as $V_{i\to f}$ and wavefunctions only coming from V_o scattering. Thus the inelastic cross sections are roughly proportional to $|V_\lambda|^2$ around the classical turning point.

When the anisotropy is stronger, and if one can still assume that IOS conditions hold, then a new effect appears in the energy loss cross sections, which can now be even larger than the elastic cross section. This is called rotational rainbow[34] and manifest itself in the appearance of a maximum in the differential cross section as a function of the final angular momentum j' at a fixed deflection angle θ[44].

One way of discussing the physics underlying such an effect is via the IOS approximation, which has the advantage of a great and fairly direct power of interpretation, especially in the semiclassical limit when the action is large compared with $\hbar$[45,46].

If one assumes that the diatom is not rotating (j_i = o) the familiar expression for the IOS scattering amplitude associated to a given rotational transition is given by[2,3]:

$$f(o\to j_2|\theta) = \frac{i}{2}(-)^{j_2}(k_o k_{j_2})^{-1/2}\sum_\ell (2\ell+1)\,P_\ell(\cos\theta)\times T_\ell(o,j_2) \tag{39}$$

where the T-matrix element is obtained by quadrature over the internal angle γ of the fixed-angle elastic scattering phase shifts for that particular partial wave ℓ:

$$T_\ell(o,j_2) = \frac{1}{2}(2j_2+1)^{1/2}\int_o^\pi d\gamma\sin\gamma[1-\exp(2\,i\,\eta_\ell(\gamma))]\;P_{j_2}(\cos\gamma) \tag{40}$$

and the corresponding state-to-state DCS is obviously given by:

$$\frac{d\sigma}{d\Omega}(o\to j_2|\theta) = \frac{1}{4k_o^2}\left|\sum_\ell (2\ell+1)\,P_\ell(\cos\theta)\,T_\ell(o,j_2)\right|^2 \tag{41}$$

where the postcollision choice ℓ = ℓ' has been made in the derivation[2] and $k_j = [2\mu(E - B(j+1)j)]^{1/2}$.

The semiclassical limit of the above expressions, valid for large values of ℓ and j_2, is derived in the same way as done in the atom-atom scattering by a spherical potential[2]: the sum over ℓ is therefore replaced by an integral, the Legendre polynomials are replaced by their large index limit and the resulting two dimensional integrals are evaluated by the method of stationary phases. In the case of atom-diatom inelastic scattering, the primitive semiclassical differential cross section for an homonuclear molecule and a PES with a single turning point (i.e. for a repulsive, or mostly repulsive, $V(R,\gamma)$ at fixed γ) is thus given by[47]:

$$\frac{d\sigma}{d\Omega}(o \to j_2|\theta) = 2\,(k_o^2 \sin\theta)^{-1}\,\{F_1^2 + F_2^2 + 2\,F_1\,F_2 \sin(\phi_1 - \phi_2)\} \tag{42}$$

for a classically allowed transition, while the classically forbidden transitions yield and exponential decaying inelastic DCS[44]:

$$\frac{d\sigma}{d\Omega}(o \to j_2|\theta) = ?\,(k_o^2 \sin\theta)^{-1}\,|F|^2 \exp(-2\,im\,\phi) \tag{43}$$

where:

$$F_\nu = [(\ell_\nu + 1/2)\sin\gamma_\nu \,/\, |D(\ell_\nu,\gamma_\nu)|]^{1/2}\} \tag{44}$$

and

$$\phi_\nu = 2\,\eta(\ell_\nu,\gamma_\nu) - \theta\,(\ell_\nu + 1/2) - \gamma_\nu\,(j_2 + 1/2) \tag{45}$$

The angular momentum ℓ_ν and angle γ_ν are the real/complex solutions of the stationary phase equations for a coupled two-dimensional case:

$$\chi(\ell,\gamma) \equiv 2\,\partial\eta(\ell,\gamma)/\partial\ell = \theta \tag{46a}$$

$$J(\ell,\gamma) \equiv 2\,\partial\eta(\ell,\gamma)/\partial\gamma = j_2 + 1/2 \tag{46b}$$

They relate the scattering angle and the final rotational angular momentum with the classical deflection function χ and the classical excitation function J[44], respectively. The D determinant in eq. (44) is the Jacobian determinant of the sistem and is given in terms of the above χ and J functions:

$$D(\ell,\gamma) = (\partial\chi/\partial\ell)_\gamma \times (\partial J/\partial\gamma)_\ell - (\partial\chi/\partial\gamma)_\ell \times (\partial J/\partial\ell)_\gamma \tag{47}$$

When eqs (46) have real solutions, then the $(o \to j_2)$ transition is allowed at the angle Θ, while it is classically forbidden if the solutions are complex. The interference terms in eq. (42) originate from the quantal superposition principle and are omitted in the classical expression.

The couplet (ℓ, γ) in the Jacobian (47) can be regarded as a set of coniugate coordinates of Θ, the center-of-mass scattering angle and j, the final molecular rotational angular momentum. Rainbow singularities in both the classical and primitive semiclassical cross sections therefore occur whenever the Jacobian becomes zero for a particular set of (ℓ_R, γ_R) values which then correspond to a set of (Θ_R, j_R) values.

In general, it is not possible to classify rainbows into types that depend on which one of the derivatives in (47) is zero since the rigorous rainbow point in the (Θ, j) physical space happens because the full determinant vanishes. However, for most realistic cases the deflection function χ is a slow function of γ and the excitation function is a slow function of ℓ, especially for repulsive potentials and at scattering angles well above $|\chi_{min}|$. Thus, the rainbow effects can be due either to extrema in the deflection function at fixed orientation as the angular momentum varies $(\partial\chi / \partial\ell = o)$ or to extrema of the excitation function as the internal orientation varies for a given impact parameter $(\partial J/\partial\gamma = o)$. For systems with one turning point the first rainbow condition is not present as it originates from the familiar interference condition for mixed-spherical potential scattering[2], while the second condition is called a <u>rotational rainbow</u> and occurs only for inelastic DCS from, obviously, anisotropic PES. For heteronuclear AB molecules the two branches of $\dot{J}(\gamma)$, which appear as maxima in the quantal treatment, are no longer equivalent and therefore they cause two different rainbow patterns for the approach on either atom A or atom B[44].

The rainbow condition on the Jacobian therefore defines, through the stationary phase equations (46), the rainbow curve $j_R(\Theta)$ or its inverse $\Theta_R(j_2)$. A typical example, taken from ref. (44), is shown in Fig. 4. $j_R(\Theta)$ is a monotonically rising function of Θ and separates the classical from the non-classical region. The classical IOS cross section is singular along this curve and vanishes in the classically forbidden region, while the quantal Airy-uniform IOS cross section has a maximum either at fixed final angular momentum or at a fixed scattering angle. This maximum is shifted into the classically forbidden region. The quantal curve of $j_{max}(\Theta)$ is also shown in Figure 4 below.

One also sees from the Figure that cross sections can either be represented as distributions over final rotational states at a fixed angle or as angular distributions for a fixed final rotational state. In the first instance, the cross sections show oscillations in the classical region, have a broad maximum at the rainbow transition $j_{max}(\Theta)$ and tunnel, exponentially decreasing, into the classically forbidden region of large $|\Delta j|$. The value of j_{max} obviously increases with Θ. In the case of angular distributions at fixed j_2, the cross section increases rapidly with increasing Θ, reaches a maximum at Θ_R as it leaves the non-classical region and then rapidly declines in the backward scattering region that is classically allowed and where additional oscillations, the supernumerary rainbows, usually appear. Also the value of Θ_{max} is an increasing function of j_2.

For impulsive conditions with repulsive potentials, it was found to be a good approximation[48] to write down the state-to-state DCS as the product of a j-independent DCS, which monotonically decreases with scattering angle, and an angle dependent excitation probability:

$$\frac{d\sigma}{d\Omega}(o \to j_2 \mid \Theta) = \frac{d}{d}(\Theta) \times \mid T^{\ell}_{\Theta}(o \to j_2) \mid^2 \qquad (48)$$

which is a convenient way to classify in terms of rainbow behaviour the weak coupling and strong coupling cases mentioned before.

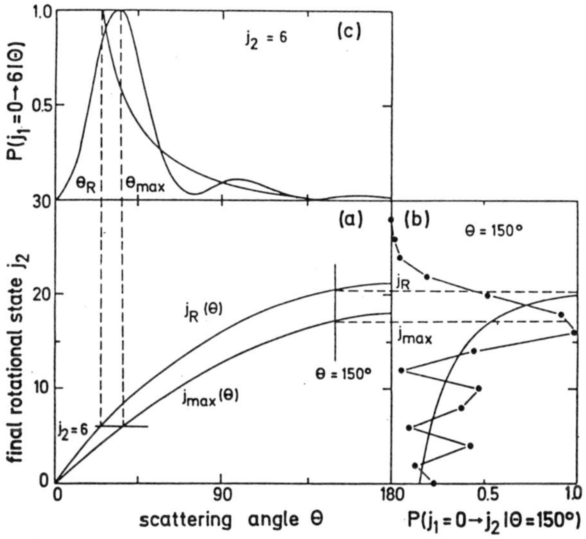

Fig. 4. Classical and quantal rainbow curves, $j_R(\Theta)$ and $f_{max}(\Theta)$, for the He-Na_2 example (from ref. (44)). The smooth curves in the quadrants (b) and (c) represent the classical cross sections at fixed angle and at fixed final state j_2, respectively.

In the weak coupling case, in fact, the inelastic cross sections are much smaller that the elastic ones and therefore the rainbow maxima, which are increasing functions of Θ, have to rise over a decreasing background that forces the rotational rainbow to appear mostly in the backward direction.

In the case of strong coupling states with large j_2 values are classically allowed and rainbows appear at small angles being controlled by the maximum allowed rotational momentum transfer rather than by the maximum energy transfer. This situation resembles qualitatively the angular momentum constraints discussed by the (ECS-EP) and (AON) fitting laws of the previous subsection. In this instance all rotational rainbows are usually clearly seen and supernumerary rainbows are also present[41].

If one approximates the PES as a hard-core where:

$$V(R,\gamma) = o \quad \text{for} \quad R > \bar{R}(\gamma)$$
$$V(R,\gamma) = \infty \quad \text{for} \quad R > \bar{R}(\gamma) \tag{49}$$

and chooses for the contour function $R(\gamma)$ the hard ellipsoid:

$$\bar{R}(\gamma) = B(1 - \varepsilon^2 \cos^2\gamma)^{-1/2} \tag{50}$$

where:

$$\varepsilon = (A^2 - B^2)^{1/2}/A \tag{51}$$

is the numerical excentricity. In the limit of small anisotropy ($\varepsilon^2 \ll 1$) eq. (50) becomes:

$$\bar{R}(\gamma) = B\left(1 + \frac{\varepsilon^2}{2} \cos^2 \gamma\right) \tag{52}$$

One can now show that the j_R dependence on θ can be given classically by a simple analytic expression[47,49]

$$j_R(\theta) = 2K(A - B) \sin \theta/2 \tag{53}$$

with $k = (2\mu E)^{1/2}$ and A,B the ellipsoid parameters which therefore measure somehow the anisotropy of the chosen potential. Thus, the above relation becomes extremely useful for relating observed rainbows, at fixed scattering angles, to kinematic factors like $\sqrt{E}$ and $\sqrt{\mu}$ and to structural factors of the potential through the linear dependence of j_R on the (A - B) feature.

The analysis of a large amount of data for the Na_2 - rare gas RET collisions[40,41,50,51] revealed the important use of rotational rainbows for elucidating, at a microscopic level, the interplay of dynamical factors and features of the anisotropic potential. In particular:

(i) The j_{max} value only depends on the transferred angular momentum Δj and not on the transferred energy ΔE. As discussed before, it can be rationalized via the scaling relationships and implies that pure RET processes do not depend on the initial rotational state;

(ii) Eq. (53) gives a very realistic description of the data and was further tested on N_2 O_2, HF and LiH interacting with He[52] and Ar[53]. The linear dependence on $\sin\theta$ often allows one to separate kinematic factors and anisotropic factors[52]. It should be noted that the rotational rainbow maximum depends only very slightly on the slope of the potential in the repulsive region sampled by the model and therefore possible energy dependence of such effects are hard to isolate for realistic potentials, in spite of earlier rationalization for fitting laws[26].

Figure 5 shows the very good agreement that has been achieved[41] between measured quantities and computed rainbow maxima for the Na_2 + Ne system.

It is interesting to point ont that, after the number of cases that have been examined in recent years, (for a recent review experiments and models see ref. 34) rotational rainbows turn ont to be an ubiquitons phenomenon in RET processes. Since they are so pronounced and since are so directly connected with features of the interaction, their observation in energy loss spectra, together with the study of elastic, or total differential cross sections, allows one a rather stringent test of molecular interactions that can be probed by resolving the rotational rainbow structure, even if no state-to-state resolution is achieved.

Finally it is worth mentioning that, the classical expression for energy transfer at the rainbow maximum can be obtained analytically for the model potential of eq. (50)[54]:

$$\left(\frac{\Delta E}{E}\right)_R = \frac{2Q}{1+Q^2}\left[1 + Q\sin^2\theta - \cos\theta\,(1 - Q^2\sin^2\theta)^{1/2}\right] \qquad (54)$$

where the key factor is:

$$Q = B_{rot}\,\frac{2\mu}{\hbar^2}\,(A - B)^2 \qquad (55)$$

since the maximum energy transfer at θ = 180° is given by :

$$(\Delta E/_E)_{max} = 4Q = \frac{4\mu\,(A-B)^2}{I} = \frac{4\mu\,(A-B)^2}{(2\mu B_{rot}/\hbar^2)} \qquad (56)$$

where I is the moment of inertia of the molecule and the above quantity is something like the ratio of a dynamical moment of inertia of the scattering system to the moment of inertia of the target[34]. Such a parameter could therefore be used to qualitatively rationalize the observed energy loss spectra from angular distribution measurements[34].

3. ROVIBRATIONAL PREDISSOCIATION DYNAMICS

Another set of processes which can give us additional information on the detailed dynamics of the internal energy transfer mechanisms and which

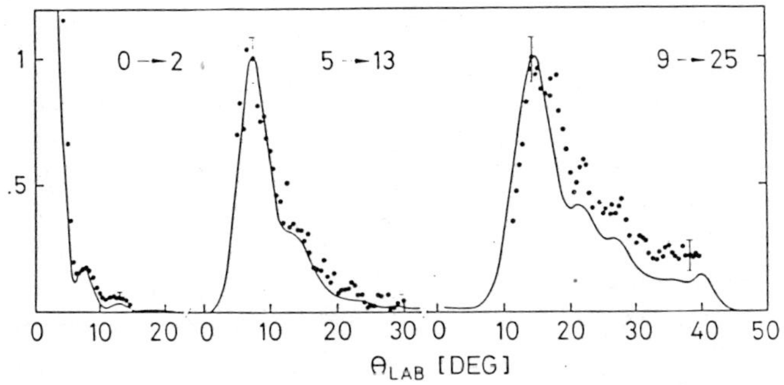

Fig. 5. Measured and computed state-to-state differential cross sections (arb. units) in the lab system for different rotational transitions of Na_2 + Ne at E = 190 mev (ref. 41).

can help us to shed more light on the relative interplay of potential couplings and kinematic factors are the 'inelastic' predissociative processes of weakly bound Van der Waals (V d W)molecules that have received a great deal of attention, both from experimentalists and theoreticions, in the last ten years or so. Because of their strong connections with collisional RET and VET mechanisms, and because of the great similarities between the theoretical tools employed in the previous Sections and those developed for rotational predissociation (RP) of exactly the same systems, we will discuss now in detail what has been found to happen for the simplest case of the atom-diatomics systems.

3.1 Exact quantum treatment

Within the Born-Oppenheimer (B-O) description of the potential energy surface, a weakly bound V d W complex can predissociate through one of two mechanisms. The simplest of them is the tunnelling type rotational predissociation (TRP) of a quasibound level much lies above the dissociation limit but below the tap of a potential barrier on the effective spherical part of the PES. The second mechanism, which is the one of interest here, is associated with the conversion of internal rotational and/or vibrational energy of the diatomic molecule forming the complex with the structureless atom into relative translational energy of the separating fragments. The interest lies on the observation that the (VP), (RP) or (V,RP) processes are induced by the same anisotropy and/or internal stretching coordinate dependence of the PES which causes RET and/or VET by collision, hence they could provide a good way of determining those parts of the PES which drive the collisionally inelastic processes discussed in the previous Section.

In the time-independent scattering theory, a metastable state can be regarded as a resonant feature in the energy dependence of one or more continuum wavefunctions, which are associated with the existence of closed scattering channels.

A full scattering experiment probes the asymptotic behaviour of the continuum wavefunctions which is summarized by the corresponding S-matrix. Hence, knowledge of all relevant S-matrices together with the correct boundary conditions is sufficient to determine all observables of the experiment. These experiments differ from predissociation studies in that the inelastic processes must occur at energie above the appropriate internal motion excitation threshold and are observed in the types of experiments that we have discussed earlier. Predissociation, on the other hand, usually occurs at energies <u>below</u> this threshold and is mainly observed spectroscopically [55]. If the radiation field can be treated as a weak perturbation, then such experiments probe essentially the same continuum functions that are involved in scattering calculations where the resonant behaviour of the S-matrix can therefore provide the same type of information [56].

The main set of processes examined here can be described as:

$$\mathrm{BC}\ (v,j) \text{---} \mathrm{M} \rightarrow \mathrm{BC}\ (v',j') + \mathrm{M} \tag{57}$$

where $v' < v$ and/or $j' < j$, with Δv usually = 1 and Δ_j = 1,2 or 3 depending on the structural properties of BC and the number of bound states supported by the initial triatomic molecule [57].

After separating out the motion of the centre of mass, the full Hamiltonian can be written as:

$$\hat{H}\ (\underline{R}, \underline{r}) = -\ (h^2/2\mu)\ R^{-1}\ (\partial^2/\partial R^2)\ R + \frac{\hat{\ell}^2}{2\mu R^2} + \hat{V}\ (R,\xi,\gamma) + \hat{H}_d(\underline{r}) \tag{58}$$

where $\underline{R}$ is the axis of the complex, $\underline{r}$ that of the diatom, $\xi = (r-r_{eq})/r_{eq}$ is the stretching coordinate, γ the $\hat{R}\cdot\hat{r}$ internal angle and $\hat{H}_d(\underline{r})$ the vibration-rotation Hamiltonian of the isolated diatom.

The exact eigenfunctions of this Hamiltonian can be expanded as:

$$\Psi_{\alpha}^{JM}(\underline{R}, \underline{r}) = r^{-1} R^{-1} \sum_{v'} \sum_{a'} \chi_{v'a'}^{J\alpha}(R)\, \phi_{v'j'}(r)\, \Phi_{a'}^{J}(\hat{R}, \hat{r}) \tag{59}$$

where J and M are the quantum numbers for the total angular momentum and for its space-fixed projection and α is an index which labels a particular eigenstate of the full system. In general, α is defined by the total energy and by the parity $p = (j + \ell + J)$. For weak-anisotropy systems, however, it may be associated with a set of zero-coupling limit quantum numbers: v, k, n, ℓ and J. The functions $\phi_{vJ}(r)$ are the wavefunctions for the stretching motion of the diatom, the χ^{α} are the radial channel eigenfunctions and the $\Phi_a^J(\hat{R}, \hat{r})$ are a complete set of angular basis functions with quantum numbers collectively identified as a. In the SF representation[3], $a = \{j, \ell\}$ and Φ_a^J is the total angular momentum ergenfunction $\mathcal{Y}_{j\ell}^{JM}(\hat{R}, \hat{r})$ which is defined as the appropriate linear combination of the products of spherical harmonics. Alternatively, in the BF coordinate system[3] $a = \{j, \Omega\}$, the operator $\hat{\ell}^2$ is replaced by $(\hat{J} + \hat{j})^2$ and the angular basis functions are parity-adapted linear combinations of the products $Y_{j\lambda}(\hat{r})\mathcal{D}_{\lambda M}^{J}(R)$, where the $\mathcal{D}$'s are the usual symmetric-top wavefunctions and $h\Omega = |\lambda|h$ is the magnitude of the projection of the total angular momentum onto the axis of the complex. In either case, by applying operator (58) to expansion (59), premultiplying by $(r^{-1}\phi\,\Phi)^*$ and integrating over r and the angular variables one gets the usual set of coupled equations:

$$\begin{aligned}\{-\frac{h^2}{2\mu}\frac{d^2}{dR^2} + U(v,a;\, v\, a;\, J/R) - E_{\alpha}^{J}\}\, \chi_{v\,a}^{J\alpha}(R) \\ = \sum_{v',a'} U(v,a;\, v',a';\, J|R)\, \chi_{v'a'}^{J\alpha}(R)\end{aligned} \tag{60}$$

where E_{α}^{J} is the total energy of the system and the U's are the matrix elements of the angular and diatom-stretching wavefunctions with the full operator $\hat{U}$ defined as:

$$\hat{U}(\underline{R}, \underline{r}) = \hat{H}_d(\underline{r}) + \frac{\hat{\ell}^2}{2\mu R^2} + \hat{V}(R, \xi, \gamma) \tag{61}$$

In either angular basis the matrix elements of $\hat{H}_d(\underline{r})$ are obviously diagonal and equal to the rovibrational energies of the free diatom, $E_d(v,j)$. Similarly, in the SF representation the matrix elements of

$\ell^2/2\mu R^2$ are diagonal in v', j' and ℓ' and simply equal to $\ell'(\ell'+1)^2/2\mu R^2$ (see eq. (3a)).

3.2 The resonant behaviour

In the above close coupling (CC) formulation, predissociating states are treated as scattering resonances and are located by solving the eq.s (60) on a grid of energy values. Near a resonance the phases of the S-matrix elements change rapidly with energy[58]. The formal result for the S-matrix in the neighborhood of a resonance denoted by m is given by[59]:

$$S(E) - S_d(E) - \frac{i\, \underline{g}_m(E)\, \underline{g}_m^+(\underline{E})}{E - E_m(E) + \frac{1}{2} i \Gamma_m(E)} \tag{62}$$

$S_d(E)$ is a complex, unitary symmetric (n x n) matrix representing scattering without metastable states, $\underline{g}_m(E)$ is a complex column vector of order n, and $E_m(E)$ and $\Gamma_m(E)$ are respectively the energy and the FWHM width of the m resonance:

$$\Gamma_m(E) = \sum_i \Gamma_{mi}(E) = \sum_i |g_{mi}(E)|^2 \tag{63}$$

where Γ_{mi} is called the partial width for channel i, and the index i runs over open channels. Each element of the S-matrix therefore contains the product of partial width for the incoming and the outgoing channel and has the obvious physical interpretation of forming a metastable state m from the incoming wave of channel i' during the initial "half collision", followed by its break-up into channel i during the final half-collision.

In the isolated narrow resonance (INR) approximation it is assumed that $E_m(E)$, $\Gamma_m(E)$ ang $\underline{g}_m$ are constant parameters which characterise the resonance m. This is valid provided these quantities do not vary significantly over the width of the resonance, a situation which is expected to arise when the resonance in question both does not overlap significantly with nerghbouring ones and is narrow compared to the energy dependence of the direct scattering.

For only one open channel, the asymptotic phase shift increases by π over the width of the resonance and follows the well-known Breit-Wigner formula[60] plus a linear background phase:

$$\eta(E) = b + CE + \arctan \frac{\Gamma}{2(E_r + E)} \tag{64}$$

A generalization to many open channels is done by introducing the ergenphase sum $\sigma(E)$. Since both $\underline{\underline{S}}(E)$ and $\underline{\underline{S}}_d(E)$ are unitary, symmetric matrices one can write:

$$\underline{\underline{S}}(E) = \underline{\underline{B}}(E)\,\underline{\underline{\Lambda}}^2(E)\,\underline{\underline{B}}^+(E)$$
$$\underline{\underline{S}}_d(E) = \underline{\underline{B}}_d(E)\,\underline{\underline{\Lambda}}_d^2(E)\,\underline{\underline{B}}_d^+(E) \tag{65}$$

here the B's are real orthogonal matrices and the Λ's are diagonal matrices of ergenphases λ and λ^d:

$$[\Lambda]_{ij}(E) = \delta_{ij}\exp[i\,\lambda_i(E)] \tag{66a}$$

$$[\Lambda_d]_{ij}(E) = \delta_{ij}\exp[i\,\lambda_i^d(E)] \tag{66b}$$

It can then be shown[56,58] that one can define the eigenphase sums as:

$$\sigma(E) = \sum_{i=1}^{n}\lambda_i(E) \tag{67a}$$

$$\sigma(E) = \sum_{i=1}^{n}\lambda_i^d(E) \tag{67b}$$

which then lead to a generalized Breit-Wigner formula:

$$\sigma(E) = \sigma_d(E) + \arctan\frac{\Gamma_m(E)}{2[E_m(E) - E]} \tag{68}$$

In practice, one calculates the real symmetric K-matrix and the $\sigma(E)$ are obtained computationally by summing inverse tangents of the eigenvalues of $\underline{\underline{K}}(E)$.

3.3 Partitioning of the Hamiltonian

As mentioned above, the exact treatment of predissociation requires that the energy and width of the metastable state must be deduced from the energy dependence of the total wavefunction. In most practical cases, however, it becomes necessary to simplify the CC solutions by partitioning the full Hamiltonian of the system into two parts. The first of these, H_o, provides a zeroth-order description of the dynamics of the system, and its descrete eigenvalues provide estimates of the energies of all bound and metastable states of the system. The remainder of the Hamiltonian, denoted $\hat{H}' \equiv \hat{H} - \hat{H}_o$, couples the states involved in the predissociative process, i.e. the discrete ergenfunctions of $\hat{H}o$ associated with the metastable levels to the isoenergetic continuum wavefunctions which describe the open channels leading to molecular break-up. It is this part of the mechanism which reminds us of the RET collisions since it is often the anisotropic part of the potential that is used to produce the additional coupling in RP calculations, as is the case in collisional inelasticity. If one further stays within the INR approximation discussed before, the F W H M of a predissociating level is then given by the familiar golden rule expression[60,61]:

$$\Gamma = \sum_i \Gamma_i = 2\pi \sum_i \left| \int \Psi_b \hat{H}' \Psi_i \, d\tau \right|^2 \qquad (69)$$

where Ψ_b is the unit normalized bound-state eigenfunction of $\hat{H}_o$ associated with the metastable state and Ψ_i is a continuum ergenfunction of the same unperturbed Hamiltonian associated with the open channel i and normalized to a delta function of energy. Since im most cases more than one open channel is available for the dissociation, the total width becomes a sum of partial widths Γ_i and the branching ratios are determined by the ratios of the contributing partial widths.

The simplest approximate methods are those in which $\hat{H}_o$ includes only the diagonal matrix elements of $\hat{U}$ of eq. (61), so that $\hat{V}(R, \xi, \gamma)$ is replaced by a set of channel potentials $V(v\ a;\ v\ a;\ J|R)$ which depend only on R. In either the SF or BF representations, this approximation involves the neglect of the coupling terms on the r.h.s. of eq. (60), and hence causes the expansion (59) to collapse to a single term. When this is the only approximation made, the result is the <u>distorsion approximation</u> introduced earlier in the literature[62,63]. The resulting coupling operator $\hat{H}'$ therefore comprises the off-diagonal elements of the interaction of eq. (61) that couple the various channels in eq. (60).

As might be expected, results obtained with the partitioning approximation are quite sensitive to che choiche of angular basis, since they are the wavefunctions which finally appear in the couplings of eq. (69), in the following subsection we will briefly review some of the separable forms of total wavefunctions that have been suggested for treating the RP problem.

3.4 Separable wavefunction approximations

As mentioned before, when a partitioned form of the Hamiltonian is selected, the dominant source of error in the computed widths for the metastable states is the neglect of the coupling between closed channels that comes from the off-diagonal matrix elements of the operator $\hat{U}$. The relative quality of the results therefore depends on how much of the correct physics is already being caught in the chosen partitioning and in the selected basis sets which write the total wafefunction in a separable form.

One important element when discussing RP processes is to consider the two relative motions: one along the V d W bond as a large amplitude stretching motion and the other as an internal rotation of the diatom with respect to the anisotropy angle $\gamma = \arccos(\hat{R}\cdot\hat{r})$. The relative time scales of the two motions and the extent to which they can be considered as separate because of weak anisotropic coupling from the PES in the region of the well, obviously dictate the possible separation strategies.

The simplest approach is to consider the most drastic reduction discussed above as distorsion approximation, whereby one disgregards completely the effect of couplings between V d W stretching and γ - libration of the diatom. The corresponding wavefunction could then be written as a simple product of two functions which depend only on either the stretching variable (R) or the libration variable (γ). At this stage, it is convenient to call it, the diabatic rotational expansion (DRE) for the bound states of the complex[64]:

$$\Psi^{JMP}_{\alpha,v_\alpha}(\underset{\sim}{R},r) = \chi^{JP}_{\alpha v_\alpha}(R)\ \Phi^{JMP}_{\alpha}(\gamma) \tag{70}$$

where J and M are the SF quantum numbers, P is the parity index and the indeces α, v_α refer to the quantum numbers associated with the librational and stretching motions respectively.

The function Φ belongs to a complete orthonormal set of angular basis functions defined in either of the two reference frames. The function χ, because of the presence of only one set of channel indeces, is the solution of the following decoupled equation for the bound states in which an effective potential, defined by averaging over the angular function, is now employed:

$$\{ -\frac{1}{2\mu}\frac{d^2}{dR^2} + V^{\alpha\alpha\ J}_{eff\ \ \ P}(R)\}\ \chi^{J}_{\alpha,v_\alpha\ P}(R) = E_{\alpha,v_\alpha}\ \chi^{J}_{\alpha,v_\alpha\ P}(R) \tag{71}$$

One obvious improvement in the diabatic approach could be attained by employing <u>all</u> the separate solutions of eq. (71) to generate the full wavefunction of the V d W cluster:

$$\Psi^{JMP}_{n}(\underset{\sim}{R},\hat{r}) = \sum_{\alpha,v_\alpha} C^{n}_{\alpha,v_\alpha}\ \Phi^{JPM}_{\alpha}(\gamma) \tag{72}$$

which can then be called a DR expansion plus configuration interaction (DRCIE). The above summation now extends to all the bound states supported by each effective potential of eq. (71). The functions on the r.h.s. of eq. (72) can then be used to construct a matrix representation of the full Hamiltonian, which in turn yields, after diagonalisation, the new bound eigenvalues E_n and eigenstates $|n\rangle$. The improvement produced by such an approach obviously depends on the structural properties of V_{eff} and on the number of $|v_\alpha\rangle$ states supported by it. In the examples below, we will see that the CI approach markedly improves, for Ar-X_2 systems, the estimates yielded by the simpler DRE treatment.

In most cases, however, the drastic separation implied by eq. (70) does not provide a realistic description of internal motion within the VdW clusters. It therefore appears to be more reasonable to think of the total wavefunction for the discrete and metastable states of the complex as given by the product of a function which depends on only one variable, be it either the stretching or the bending variable, times another function which depends only weakly on that variable and more markedly on the other.

If one begins by choosing an angular adiabatic expansion (AAE) one therefore writes:

$$\Psi_{s,\ell}(\underset{\sim}{R}, \hat{r}) = \phi_s(R;\gamma)\, F_{s,\ell}(\gamma) \tag{73}$$

where the discrete eigenfunctions now depend on the quantum numbers s and l, associated with the V d W bond stretching and librational motions respectively. By assuming that the ϕ_s functions only exhibit a weak dependence on γ, then they became 'transparent' to the effects of the l^2 and j^2 operators and solutions of an equation of motion which depends parametrically on the angular variable:

$$\{-\frac{1}{2\mu}\frac{\partial^2}{\partial_R^2} + V(R,\gamma)\}\ \phi_s(R;\gamma) = W_s(\gamma)\ \phi_s(R;\gamma) \tag{74}$$

where the $W_s(\gamma)$ are now adiabatic potential curves supporting bound and metastable angular states described by the $F_{s\ell}$ functions of expansion (73).

Also in the present approximation one can improve on eq. (73) and expand the total wavefunction via a complete basis of angular adiabatic solutions:

$$\Psi_k(\underline{R},\hat{r}) = \sum_{s,\ell} a^k_{s,\ell}\ \phi_s(R;\gamma)\ F_{s,\ell}(\gamma) \tag{75}$$

The ensuing, usual Schrödinger equation can then be solved by writing the full Hamiltonian in that basis and by diagonalising the corresponding matrix. When one calculates all the contributions to the CI matrix elements (e.g. see ref.s 64, 65) and performs the diagonalisation, the final representation of the wavefunction of eq. (75) therefore corresponds to an Angular Adiabatic plus Configuration Interaction Expansion (AACIE) for the discrete levels and metastable levels of the triatomic complex.

Figure 6 reports the adiabatic potential curves for the Ar-O_2 system for the J = 0 and J = 1 values. The quantum number n labels the stretching states associated with the V d W bond, i.e. the $|s>$ states of eq. (74) that has been solved numerically. One sees that several potential curves exist, i.e. several stretching states exhist within each complex. The index m in the Figure labels the librational (γ-bending) states of eq. (74) described by the index ℓ in the expansion of eq. (73) where the F_s functions appear. One also sees that different rotational states of the complex (i.e. different J values) shift the energy levels of the bond librational modes within a given adiabatic potential, thus suggesting that a CI approach within the angular adiabatic separation could be a fruitful improvement for obtaining resonance positions and FWHM values[66].

In the case of the DR expansion, eq. (71) requires the calculation of the effective diabatic potential, and this can be done in either the SF or the BF representation. In the former the $V_{eff}^{\alpha\alpha,JP}(R)$ is given by $V_{\ell\ell'}^{jj'}(J,P|R)$ where the j = j' and $\ell = \ell'$ case provides the diagonal portion appearing in $\hat{H}_o$ while the off-diagonal parts describe the couplings. In the latter frame of reference, the potential becomes $V_{\Omega\Omega}^{jj'}(J,P|R)$. In this case the off-diagonal coupling functions are solely due to the $(\hat{\underline{J}} - \hat{\underline{j}})^2$ term in the total Hamiltonian[67]. Different representations however correspond to different views of the process, as the BFDR sees the complex as a rigid symmetric top while the SFDR representation sees the diatom as freely rotating within the complex. The quality of this approximation therefore depends on the specific characteristic of the PES of the full system[67,68].

When employing the AA expansion of eq. (73), one essentially assumes that the opposite physics is true for the complex: V d W stretching occurs more rapidly than angular librations which therefore provide the angle-dependent effective potentials of eq. (74). The necessary continuum functions are here obtained by writing them within the IOS approximation described earlier[67]:

$$\Psi_{j\Omega\varepsilon\bar{\ell}}^{cont}(\underline{R}, \hat{r}) = \phi_{\varepsilon\bar{\ell}}(R;\gamma) < \hat{R}\,\hat{r} \mid J\ M\ j\ \Omega > \qquad (76)$$

where the ϕ's are solution of the quation:

$$\{-\frac{h^2}{2\mu}\frac{\partial^2}{\partial R^2} + V(R,\gamma) + \frac{\bar{\ell}(\bar{\ell}+1)}{2\mu R^2}\}\,\phi_{\varepsilon\ell}(R;\gamma) = \varepsilon\,\phi_{\varepsilon\bar{\ell}}(R;\gamma) \qquad (77)$$

corresponding to a relative kinetic energy of the fragments given by the $\mathcal{E}$ value.

The RP half-widths for a given $|s,\ell>$ final state are therefore given via eq. (69), where the discrete-continuum couplings, V^{dc}, are given by:

$$V^{dc}_{s\ell\to j\Omega} = < \Psi_{\varepsilon j\Omega} (\underline{R}, \hat{r}) | \hat{H}' | \Psi_{s\ell} (\underline{R}, \hat{r}) > \qquad (78)$$

and their specific expression has been discussed in detail before[67,69].

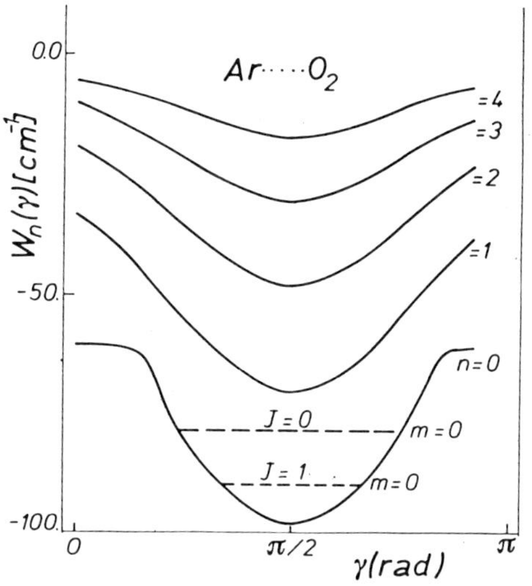

Fig. 6. Rotational adiabatic potential curves for the O_2-Ar system (from ref. 66). See text for the meaning of symbols.

A recent application of this approach to vibrational predissociation of HeI_2[70] has produced computed VP linewidths in rather good accord with experiments, as shown in the results of Fig. 7. The labels WD and ND correspond to two different couplings for the vibrational motion[70] and, in spite of the rather crude potential employed in the calculations, both indicate the same dependence on v' as that shown by experiments.

A further possibility for generating a separable wavefunction, in the adiabatic sense, is to construct an effective potential for the libration variable that changes as the stretching coordinate varies. This would correspond to chosing the latter as the slow variable and therefore is often called the Radial Adiabatic Expansion (RAE)[65] or the Best Local

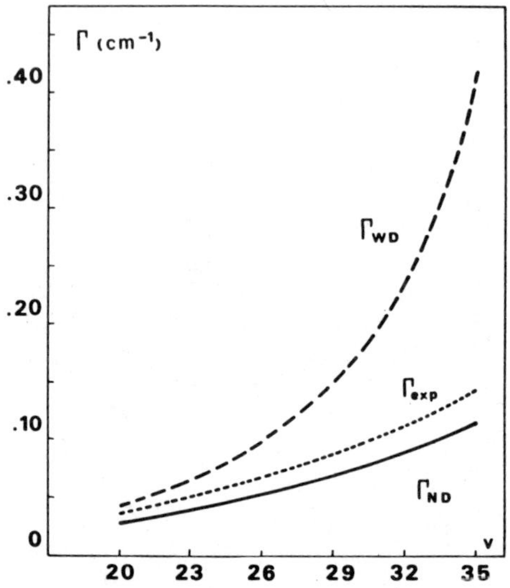

Fig. 7. Computed and measured vibrational predissociation linewidths for the I_E (B) system as function of the vibrational quantum number of I_2 (from ref. 70).

(BLE) expansion[71]. In this approach the angular functions of eq. (59) are taken to be parametrically dependent on R:

$$\phi_a^J(\hat{R}, \hat{r}) \implies \phi_a^J(\hat{R}, \hat{r}; R) \tag{79}$$

and the coupling among various channels arises from matrix elements involving such angular functions and the radial kinetic energy operator

$(-\hbar^2/2\mu)\,\partial^2/\partial R^2$. The ensuing coupling functions have the general form :

$$f^{jj'}_{\ell\ell'}(J|R) + g^{jj'}_{\ell\ell}(J|R)\,\frac{d}{dR} \qquad (80)$$

and therefore their evaluation requires numerical integration over the slow variable the accuracy of which strongly depends on the behaviour of the potential anisotropy as function of R, vis à vis the oscillatory behaviour of the adiabatic functions (79).

4. REFERENCES

1. F. A. Gianturco, Ed., "Atomic and Molecular Collision Theory", Plenum Publ. Co., New York (1980).
2. R. B. Bernstein, Ed. "Atom-molecule collision Theory", Plenum Publ. Co., New York (1979).
3. e.g. see F. A. Gianturco, "The transfer of molecular energy by collisions", Springer Verlag, Berlin (1979).
4. M. Faubel, K. H. Kohl, J. P. Toennies and F. A. Gianturco, J. Chem. Phys, 78:5629 (1983).
5. e.g. see: T. A. Brunner and D. Pritchard, "Fitting laws for rotionally inelastic collisions", in "Dynamics of the excited state", K. P. Lawley Ed., Wiley, New York (1982).
6. F. A. Gianturco and A. Palma, J. Phys B, 18:L 519 (1985).
7. V. Khare, "On the l_z-conserving energy sudden approximation for atom-diatom scattering, J. Chem. Phys., 68:5631 (1978).
8. G. Bergeron, C. Leforestier and J. M. Launay, "On the fixed-nuclei approximation as applied to rotational excitation of molecules by atoms", Chem. Phys. Lett., 59:129 (1978).
9. D. Secrest, "Theory of angular momentum decoupling approximations for rotational transitions in scattering", J. Chem. Phys., 61:710 (1975).
10. S. I. Chu and A. Dalgarno, "The rotational excitation of CO by H impact", Proc. R. Soc. London, A342:191 (1975).
11. R. D. Levine and J. L. Kinsey, in "Atom-molecule Collision Theory", R. B. Bernstein Ed., Plenum Publ. Co., New York (1979).
12. M. Quack and J. Troe, in "Theoretical Chemistry: Advances and perspectives", Vol. 6B, Academic Press, New York (1981).
13. S. S. Bhattacharyya and A. S. Dickinson, J. Phys B 12:L521 (1979).
14. T. E. Brunner, T.A. Scott and D. E. Pritchard, J. Chem. Phys., 76:5641 (1982).
15. E. De Pristo, S. D. Augustin, R. Ramaswamy and H. Rabitz, J. Chem. Phys, 71:850 (1979).

16. S. Green, D. L. Cochrane and D. G. Truhlar, "Accuracy of ECS scaling procedure for rotational excitation of CO by collision with Ar", J. Chem. Phys, 84:3865 (1986).
17. J. C. Polanyi and K. P. Woodall, J. Chem. Phys, 56:1563 (1972).
18. R. D. Levine and R. B. Bernstein, in "Dynamics of molecular collisions", W. H. Miller Ed., Plenum, New York (1975).
19. I. A. Procaccia and R. D. Levine, J. Chem. Phys, 64:808 (1976).
20. F. A. Gianturco, A. Palma and M. Venanzi, Mol. Phys, 56:399 (1985).
21. T. A. Brunner, R. D. Driver, N. Smith and D. E. Pritchard, Phys Rev. Lett., 41:856 (1978).
22. M. Wainger, L. Alagil, T. A. Brunner, A. W. Karp, N. Smith and D. E. Pritchard, J. Chem. Phys, 71:1977 (1979).
23. S. L. Dexheimer, M. Durand, T. A. Brunner and D. E. Pritchard, J. Chem. Phys, 76:4996 (1982).
24. J. Derouard and N. Dadeghi, Chem. Phys, 88:171 (1984).
25. B. J. Whitaker and Ph. Brechignac, Chem. Phys Lett., 95:407 (1983).
26. Ph. Brechignac and B. J. Whitaker, Chem. Phys, 88:425 (1984).
27. E. W. Smith, M. Giraud and J. Cooper, J. Chem. Phys, 65:1256 (1976).
28. A. S. Dickinson and D. Richards, J Phys B, 11:1085 (1978).
29. C. Nyeland and G. D. Billing, Chem. Phys, 60:359 (1981).
30. Y. Alhassid and R. D. Levine, Phys. Rev., A18:89 (1978).
31. Ph. Brechignac, in "Spectral Line Shapes", Vol. 3, F. Rastas Ed., Walter de Gruyter Co, Berlin (1985).
32. C. Nyeland and G. D. Billing, Chem. Phys, 40:103 (1979).
33. Ph. Brechignac, "Fitting form for rotational transitions of large inelasticity", J. Chem. Phys, 84:2101 (1986).
34. U. Buck, "State selective rotational energy transfer in molecular collisions", Comments At. Mol. Phys, 17:143 (1986).
35. U. Buck, F. Huisken, J. Schleusener and J. Schaefer, J. Chem. Phys, 73:1512 (1980).
36. J. Andres, U. Buck, F. Huisken, J. Schleusener and F. Torello, J. Chem. Phys, 73:5620 (1980).
37. U. Buck, F. Huisken, J. Schleusener and J. Schaefer, J. Chem. Phys, 74:535 (1981).
38. U. Buck, F. Huisken, G. Maneke and J. Schaefer, J. Chem. Phys, 78:4430 (1983).
39. U. Buck, F. Huisken, A. Kohlhase, D. Otten and J. Schaefer, J. Chem. Phys, 78:4439 (1983).
40. U. Hefter, P. L. Jones, A. Mattheus, J. Witt, K. Bergmann and R. Schinke, Phys. Rev. Lett., 46:915 (1981).
41. P. L. Jones, U. Hefter, A. Mattehus, J. Witt, K. Bergmann, W. Mueller, W. Meyer and R. Schinke, Phys. Rev., A26:1283 (1982).
42. C. Candori, F. A. Gianturco, F. Pirani, F. Vecchiocattivi, M. Venanzi, A. S. Dickinson and Y. Lee, Chem. Phys,109:417 (1986).
43. e. g. see: F. Battaglia, F. A. Gianturco, P. Casavecchia, F. Pirani and F. Vecchiocattivi, in "Van der Waals molecules", Far. Disc. 73:257 (1982).
44. R. Schinke and J.M. Bowman, in "Molecular Collision dynamics", chapt. 4, J. M. Bowman Ed., Springer, Berlin (1983).
45. L. D. Thomas, J. Chem. Phys, 67:5224 (1977).
46. R. Schinke, Chem. Phys, 34:65 (1978).

47. M. J. Korsch and R. Schinke, "Rotational rainbow: an IOS study of rotational excitation of hard-shell molecules", J. Chem. Phys, 75: 3850 (1981).
48. R. Schinke, W. Müller and W. Meyer, J. Chem. Phys, 76:895 (1972).
49. S. Bosanac, Phys. Rev., A22:2617 (1980).
50. P. L. Jones, E. Gottwald, U. Hefter and K. Bergmann, J. Chem. Phys, 78:3838 (1983).
51. W. P. Moskowitz, B. Stewart, R. M. Bilotta, J. L. Kinsey and D. E. Pritchard, J. Chem. Phys, 80:5496 (1984).
52. F. A. Gianturco and A. Palma, "Computed rotational rainbows from realistic potential energy surfaces", J. Chem. Phys, 83:1049 (1985).
53. F. A. Gianturco and A. Palma, in "Intramolecular Dynamics" pg. 63, J. Jortner and B. Pullman Ed.s, Reidel, Dordrecht (1982).
54. D. Beck, U. Ross and W. Schepper, Z. Phys A, 293:107 (1979).
55. e.g. see: D. M. Levy, Adv. Chem. Phys, 47:323 (1981).
56. C. J. Ashton, M. S. Child and J. M. Hutson, "Rotational predissociation of the Ar HcL V d W complex: close-coupled scattering calculations", J; Chem. Phys, 78:4025 (1983).
57. T. E. Gough, R. E. Miller and G. Scales, J; Chem. Phys: 69:1588 1978.
58. e.g. see: A. U. Hazi, Phys. Rev., A19:920 (1979).
59. R.D. Levine, "Quantum Mechanics of Molecular rate processes", Clarendon, Oxford (1968), Sect. 3.2.2.
60. U. Fano, Phys. Rev., 124:1866 (1961).
61. H. Feshbach, Ann. Phys, 5:357 (1958); 19:287 (1962).
62. R. D. Levine, J. Chem. Phys, 49:51 (1968).
63. J. T. Muckerman and R. B. Bernstein, J. Chem. Phys, 52:606 (1970).
64. J. A. Beswick and A. Requena, J. Chem. Phys, 78:4374 (1980).
65. F. A. Gianturco, A. Palma, G. Delgado-Barrio, O. Rancero and P. Villarreal, Int. Rev. Phys. Chem. 7 : 1 (1988).
66. F. A. Gianturco, G. Delgado-Barrio, O. Roncero and P. Villarreal, "RD dynamics in weakly bound molecular systems", Int. J. Quantum Chem. 21 : 389 (1987).
67. P. Villarreal, G. Delgado-Barrio, O. Roncero, F. A. Gianturco and A. Palma, "Rotational predissociation of strongly anisotropic V d W complexes", Phys. Rev. A, 36 : 617 (1987).
68. R. J. Leroy, G. C. Carey and J. M. Hutson, "Predessociation of V d W molecules", Far. Disc., 73:339 (1982).
69. E. Segev and M. Shapiro, "Energy levels and photo-predissociation of the He-I_2 V d W complex in the IOS approximation", J; Chem. Phys, 78:4969 (1985).
70. G. Delgado-Barrio, P. Mareca, P. Villarreal, A. M. Cortina and S. Miret-Artès, "A close-coupling IOSA study of VP of the HeI_2 V d W molecule", J. Chem. Phys, 84:4268 (1986).
71. S. L. Holmgren, M. Waldman and W. Klemperer, J. Chem. Phys, 67:4414 (1977).

CHAOS AND COLLISIONS:

INTRODUCTORY CONCEPTS

William P. Reinhardt

Department of Chemistry
University of Pennsylvania
Philadelphia, Pennsylvania 19104-6323 USA

ABSTRACT

Recent developments in the theory of non-linear Hamiltonian dynamics have changed the way in which classical phase space should be viewed. This change in viewpoint has been absorbed by workers in semi-classical quantization of the energy levels of systems of several coupled degrees of freedom, but is only recently having an impact on our thinking about atomic and molecular dynamics. What is done in these lectures is to introduce the elementary concepts needed to understand the nature and structure of chaotic phase space, and the role of time independent phase space structures on the dynamical processes controlled by them. Many illustrations involving the dynamics of classical two degree of freedom systems are introduced, and hints at generalization to higher numbers of active degrees of freedom, and to the inclusion of quantum effects are provided.

1. INTRODUCTION

During the past decade a revolution in the natural science has been in progress. By a revolution, I mean not only that new methodologies have been introduced or that new tools are available, but that, in addition, the classes of problems amenable to mathematical description has been enlarged to include descriptions of parts of the static and dynamic properties of nature which were not thought to be "part of the scientific enterprise" before. That is, a new mind set has appeared. For example, suddenly "fractals" are seen everywhere, and are found to be useful for at least cataloging similarities and differences of systems which a decade ago were not even recognized as being part of the domain of discourse. Both the physical phenomena as well as the mathematical tools to describe complexity in terms of "fractals", or complex dynamics in terms of "chaos" have been around for a long time. The phenomena presumably since t=0, and the mathematical tools for a century, or at least almost that long. Housedorff and Weierstrauss knew of the difficulties in describing continuous but non-differentiable functions, and Poincare´, and Einstein understood the full complexity of the dynamics of classical Hamiltonian systems. In more modern times Mandelbrot has forced the fractal upon us, and Lorenz, Smale, KAM, and Ford, among others, have brought to our attention the role of complexity in dynamical systems. Texts, and now even popularizations have now appeared! Chaos is a topic of conversation over coffee or cocktails.

What does this have to do with chemistry? At the macroscopic level, oscillatory chemical reactions can display chaotic behavior, making predictability and control potentially difficult or impossible; chemistry can take place in spaces of unusual dimensionality, requiring the rethinking of phenomenological kinetics itself; surfaces and spongy or porous solids (i.e. zeolites) can be characterized by fractal dimensionalities, as can the light scattering properties of soot.

At the microscopic level of the dynamics of electrons in atoms, or of atomic motion within well defined isolated molecules, once one tries to go beyond separable or "single particle" descriptions, the generic "classical" behavior involves instability and chaotic motions. Thus classical descriptions cannot avoid these "new" issues. While, for low lying energy eigenfunctions, superposing wave mechanical properties on the classical pictures does tend to suppress any signature of chaos, higher states present an opportunity to observe "quantum chaos" (or at least signatures of chaotic classical dynamics in the correspondence principle limit), and at the very least complicate the interpretation of the use of classical dynamics in applications where computational quantum mechanics is inadequate due to the total inadequacy of modern computers to tackle even a four atom polyatomic at an excitation energy of, say, 20,000cm^{-1}. The fact is that the use of classical dynamics provides the interpretive basis for our visualization of the mechanisms for chemical reactions, and thus for much of chemistry. It thus should follow that fundamental advances in our understanding of classical dynamics, and the structure of classical phase space will affect the way we think about certain classes of problems in atomic and molecular collision physics.

It is the purpose of the present lectures to introduce the basic ideas relevant to the role of chaos in atomic and molecular collisions. These will be often illustrated by examples involving two degrees of freedom. For example, the bound motion of a coupled two oscillator system, or a collinear chemical reaction A +BC $\rightarrow$ AB + C. A special property of these two freedom systems (known to Poincare´ in 1912) gives them a special advantage in terms of visualization. A major purpose of the present lectures is to present a "modern" visualization of the structure of classical phase space. However, it should be noted at the onset, that this is only partly the reason for focussing on two freedom systems: the understanding of many degree of freedom systems is not only harder to visualize, but is also far less well developed. Perhaps this is not surprising as the important "modern" aspects of non-linear dynamics to be introduced here have only been in the mathematics and physics literature for three or four years! In the following expository material, only parts of Sections 4.2 and 4.4 have any originality with the author; the main purpose of the lectures and these notes is to bring to students of collision theory a qualitative understanding of concepts which form the classical underpinnings of their science. If access is thereby provided to a rich and rapidly evolving new literature, the purpose will have been achieved.References to such are in Section 6. The view taken here is that generic features of the microscopic dynamics are important for the understanding of individual systems. A more abstract and global approach has been taken by others.

2. REGULAR AND IRREGULAR MOTION

Borrowing the title of an excellent review by M. V. Berry, the concept of chaos, or "irregular motion" is easily qualitatively introduced by contrasting it with "regular" or "integrable" dynamics. In Sections 2 and 3 we will restrict the discussion to conservative, or Hamiltonian, dynamical systems where the positions $\mathbf{q} = (q_1, q_2, q_3, \ldots q_N)$, and momenta $\mathbf{p} = (p_1, p_2, \ldots p_N)$, evolve acording to the canonical (Hamilton) equations

$$dp_i/dt = -\partial H(\mathbf{p},\mathbf{q})/\partial q_i$$

$$dq_i/dt = \partial H(\mathbf{p},\mathbf{q})/\partial p_i$$

where $H(\mathbf{p},\mathbf{q})$ is the Hamiltonian function, which is assumed not to depend explicitly on time. It is a commonplace that for such a dynamical system energy is conserved,

$$E = H(\mathbf{p},\mathbf{q}),$$

and that (Liouville's Theorem) volumes in the 2N dimensional $(\mathbf{p},\mathbf{q})$ phase space are conserved. What about other conserved quantities, or integrals of the motion? Geometric symmetries often give rise to conservation laws, the most familiar example being the relationship between rotational invariance and conservation of angular momentum. The result of such a constant of motion is to reduce the dimensionality of the phase space accessible to a given trajectory, by one. Thus for a one dimensional dynamical system energy conservation reduces the phase space dimensionality of trajectory to one. In a two degree of freedom system if, say, both energy and angular momentum are conserved, an individual trajectory explores a space of dimensionality two; if *only* energy is conserved the trajectory explores the three dimensional "energy shell" in the four dimensional space (q_1, q_2, p_1, p_2). What do these surfaces or subspaces look like, and are there constants of motion in addition to energy and those due to obvious symmetries?

2.1 Regular Motion and Invariant Tori

The existence of conserved quantities seemingly unrelated to obvious geometric symmetries is easily illustrated. Consider the two degree of freedom system investigated by Henon and Heiles in their now infamous studies of the dynamics of elliptical galaxies. The Henon-Heiles Hamiltonian is a two isotropic harmonic oscillator perturbed by a cubic polynomial:

$$H(\mathbf{p},\mathbf{q}) = H^o(\mathbf{p},\mathbf{q}) + q_1{}^2 q_2 - q_2{}^3/3$$

$$H^o(\mathbf{p},\mathbf{q}) = (p_1{}^2 + q_1{}^2)/2 + (p_2{}^2 + q_2{}^2)/2.$$

As this system is not cylindrically symmetric in the co-ordinate plane, H does not have vanishing Poisson bracket with the projection of the angular momentum component perpendicular to the plane. Does this imply that the energy is the only conserved quantity, and that a typical trajectory will occupy a three dimensional energy shell in the four dimensional phase space? Figure 1 shows a typical "quasi periodic" trajectory for the Henon-Heiles system.

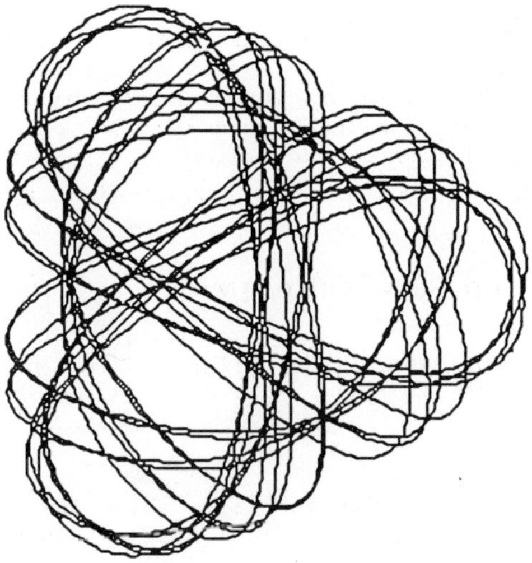

Figure 1. A quasiperiodic orbit for the Henon-Heiles system plotted in the q_1, q_2 co-ordinate space. Most trajectories of this type will evolve indefinitely without closing on themselves. If the trajectory does eventually close on itself, a periodic orbit results. The distinction between periodic and quasiperiodic orbits will be important in the discussions of Section 3.

What volume does this quasiperiodic trajectory evolve in? How do we tell? For two degree of freedom systems a method due to Poincare´, developed for treating a restricted model of the Earth-Sun-Jupiter planitary system, is most useful for visualizing the phase space dynamics. Poincare´ observed that the motion of an individual trajectory conserves energy, and thus only three of the four phase space variables (q_1,q_2,p_1,p_2) are independent. Choosing to eliminate p_1 a trajectory may be visualized as evolving in the 3-space (q_1,q_2,p_2), as shown in Figure 2. Further, a projection onto the (q_2,p_2) plane may be then defined by taking those points in the (q_2,p_2) plane when q_1 passes through a prespecified value. Taking the "sectioning" value of $q_1$to be $q_1=0$, the traditional Poincare´ surface of section for the Henon-Heiles problem is also illustrated in Figure 2.

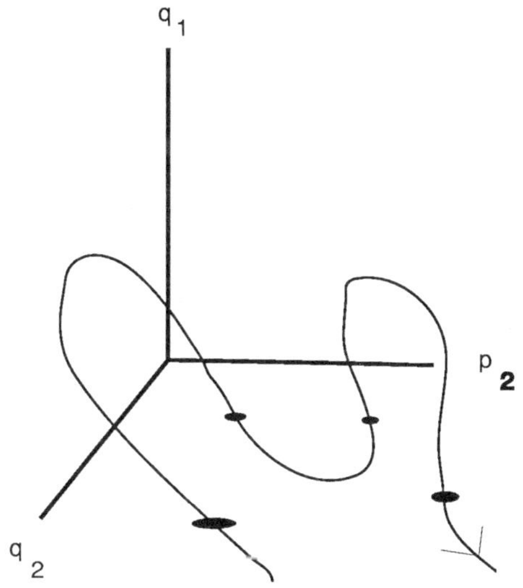

Figure 2. Definition of the Poincare′ surface of section for the Henon-Heiles Hamiltonian system. The classical trajectory moves on the three dimensional energy conserving shell within the four dimensional phase space.Thus any three of the phase space variables may be taken to be independent. We have (arbitrarily, but conventionally) chosen q_1, q_2, p_2. Further to give a display which can be easily shown as an illustration on a two dimensional surface, the trajectory is sampled when it passes through the $q_1 = 0$ sectioning plane. This is indicated by the (enlarged) dots. The collection of these intersections allows easy determination of the phase space behavior of the trajectory.

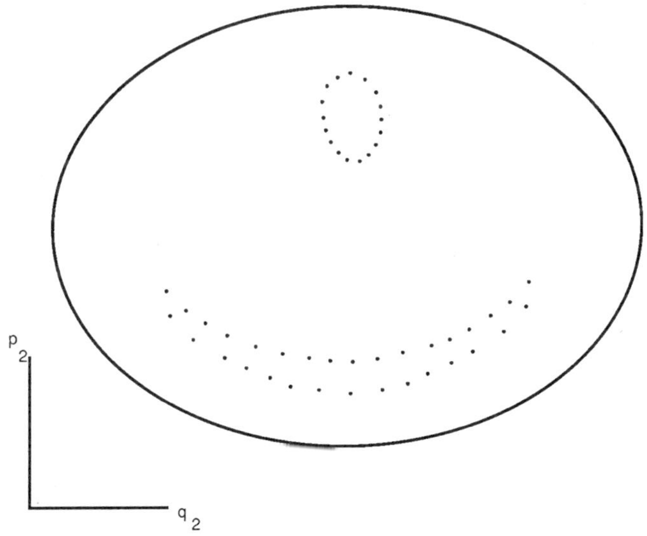

Figure 3. A Poincare′ surface of section of a quasi-periodic orbit of the type shown in Figure 1. The orbit appears to lie on a two dimensional surface in the phase space, and the $q_1 = 0$ sectioning plane cuts this surface in two closed curves. This is consistent with motion on a 2-torus in the 4 dimensional phase space.

The surface of section of the quasiperiodic trajectory of Figure 1 is shown in Figure 3. It is immediately evident that the trajectory is not sampling the whole allowed energy shell. That is, the trajectory is not ergodic in the energy allowed region of phase space. Existence of a "second" constant of motion is thus suggested. In addition to the observation that the trajectory is not ergodic, it is also evident that the sectioning plane "cuts" the phase space trajectory in two closed curves. This is consistent with the fact that, for two degree of freedom systems with *two* constants of the motion, the two dimensional surface which confines trajectories has the topology of a torus. This is illustrated in Figure 4.

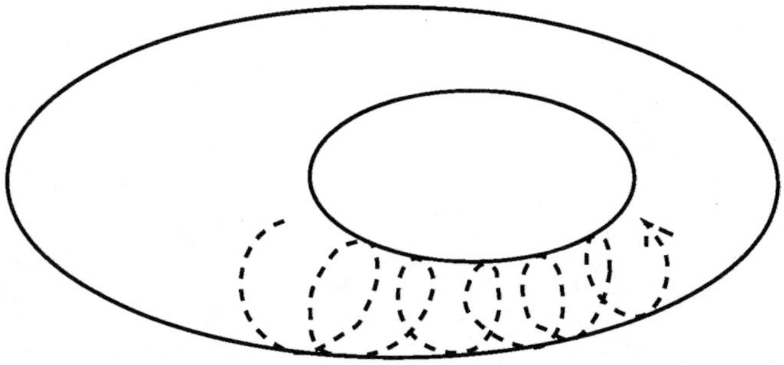

Figure 4. Sketch of a two dimensional torus, with a segment of an evolving trajectory shown as a dashed line. If the orbit in not periodic on the torus, it will eventually fill the surface densely. The motion is then ergodic on the torus, but not ergodic on the energy shell. If the orbit eventually closes on itself the orbit is periodic, and does not, by itself, define a torus. This will emerge, in Section 3, to be an important distinction.

The torus of Figure 4 is a time independent object, and thus is called *an invariant torus*. Trajectories evolve on the torus, and a typical slice through the torus will intersect a sectioning plane in two closed curves, consistent with the surface of section shown in Figure 3. If all of the motion of a Hamiltonian system takes place on such invariant tori, the motion is said to be *integrable,* and a canonical transformation to what are refered to as angle-action variables is possible. In the angle action representation the Hamiltonian itself is written an an explicit function of new generalized momenta (the actions) which are constants of the motion, as their conjugate variables (the angle variables) are absent. Thus

$$H(\mathbf{p},\mathbf{q}) \rightarrow H^{new}(\mathbf{I})$$

where $\mathbf{I} = (I_1, I_2, \ldots I_N)$. The new Hamiltonian is only a function of the N actions, I_i, rather than the 2N original phase space variables p_i, q_i. The missing or *ignorable* variables conjugate to the I_i are traditionally called the angle variables, θ_i. If such a canonical transformation exists it follows at once that each of the actions, I_i, is a constant of the motion as

$$dI_i/dt = -\partial H(I)/\partial\theta_i = 0$$

following from the fact that the angle variable, θ_i, is simply absent form H^{new}. The time dependence of the angle variables is given by

$$d\theta_i/dt = \partial H(I)/\partial I_i$$

which is also independent of the ignorable angle, and we then have

$$\theta_i = \omega(I)\,t + \delta$$

The geometric significance of these new variables is intimately related to the fact that motion for an N degree of freedom system is on an N-torus. The N actions give the dimensions of the various shanks of the torus, and the N angles specify a point on the surface. This is illustrated for two degrees of freedom in Figure 5.

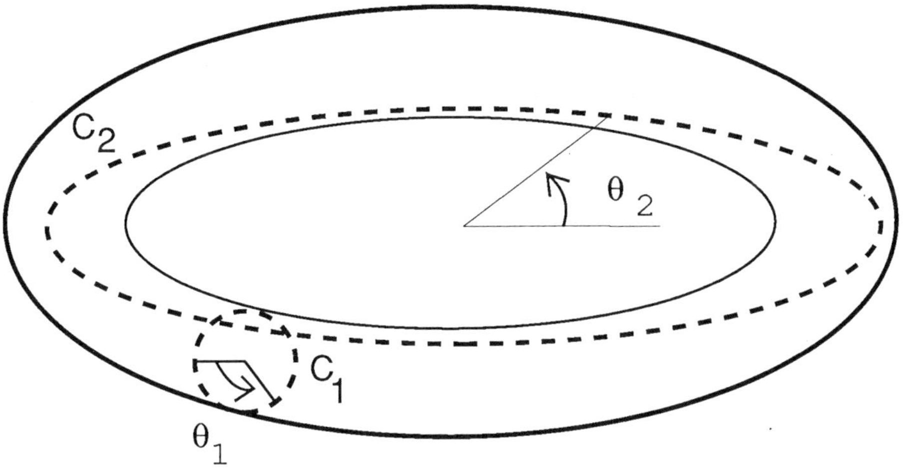

Figure 5. Actions and angles on a 2-torus which confines a trajectory for integrable motion for a two degree of freedom Hamiltonian system. The constant actions are given as $\int p\cdot dq$ along the two indicated paths, which are the two topologically independent 1-cycles on the torus. As pdq is a 1-form, continuous distortions of the integration paths leave the integrals invariant. The angles, θ_1 and θ_2 give a simple co-ordinate system on the torus. These angles evolve linearly in time as a trajectory winds its way around the surface. The generalization to N degrees of freedom is straightforward formally, although visualization is more difficult.

In summary, integrable systems are characterized by the fact that N constants of the motion confine the dynamics of an individual trajectory to an N torus in the 2N dimensional phase space. Setting the values of the actions $I_1, I_2, \ldots$ selects a particular torus; choice of a particular set of the angles $\theta_1, \theta_2, \ldots$ is equivalent to a specific choice of initial conditions on the torus specified by the choice of actions. The role of invariant tori in semi-classical quantization of the energy levels of non-separable bound systems is a well established, but beyond the scope of the present lectures.

2.2 Non-Integrability and Chaos

In Figures 6 and 7 are a trajectory and Poincare surface of section of a second, higher energy, trajectory of the Henon-Heiles system also illustrated in Figures 1 and 3.

Figure 7. An irregular, or chaotic, trajectory for the Henon-Heiles system. Comparison with the quasiperiodic orbit of Figure 1 suggests a qualitative difference, which is yet more forcefully shown in the Poincaré Section, shown below, in Figure 8.

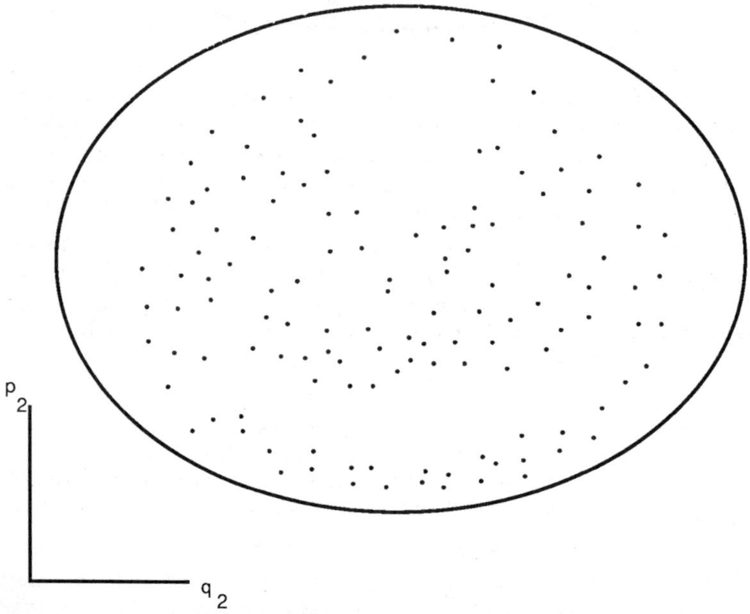

Figure 8. A sketch of the surface of section corresponding to the trajectory of Figure 7. In this case the trajectory is not confined to a torus, and is possibly ergodic in at least part of the energy shell.

The striking difference between the regular dynamics of Figures 1 and 3, and the chaotic or irregular dynamics of Figures 7 and 8 is our first example of deterministic chaos. Such motion actually fills a volume in the phase space if left to run long enough, and shows extreme sensitivity to initial conditions, and to the accuracy of the numerical method of trajectory integration. The Henon-Heiles dynamics illustrated here is generic in the sense that the dynamics of a typical coupled non-linear Hamiltonian system is neither integrable nor fully ergodic: chaotic motion and invariant tori exist in the same system, even existing at the same energy. At a fixed energy changing initial condition can lead from regular to irregular regions of the phase space. The fraction of the phase space which is irregular is usually a function of the energy. There are, of course systems which have no chaos: separable systems are always fully integrable, families of coupled non-linear Hamiltonian systems which are nevertheless integrable are also known, but such systems are exceedingly rare among the class of all such Hamiltonian systems.

Equally rare are systems which have no trajectories evolving on invariant tori! Such truly ergodic systems have only two classes of bounded classical motion: 1) periodic orbits, which through the fact that they repeat indefinitely are not chaotic (although they may be arbitrarily complex with arbitrarily long period and thus hard to distinguish from chaotic motion in a finite computation); and, 2) chaotic orbits which fill all of the available space "between" the (dense) periodic orbits, yet never repeat. As the periodic orbits are of measure zero in such a system, one can choose to ignore them, and note that the remaining part of phase space is occupied by trajectories which are ergodic, and thus time averages approach phase space averages at long enough times, and if this time scale for at least approximate ergodicity is short compared to the dissociation time of a collision complex, one can expect statistical theories to be valid. It is also the case that such fully chaotic systems (which are sometimes said to display "hard chaos") can be attacked semi-classically, using path integral methods.

It goes without saying, that in the case of generic Hamiltonian dynamics, being a mixture of regular and chaotic motion, that no global transformation to angle action variables exists, as these constants of the motion are absent for chaotic trajectories, even though neighboring regions of phase space may be filled with tori, suggesting local integrability. This further complicates the problem of semi-classical energy quantization, as neither the EBK or Phase Space Path Integral methods may be used globally. It also complicates the development of statistical theories of collision dynamics, as we now discuss.

2.3 The Structure of Phase Space and RRKM Theory

Suppose a collisional mechanism between chemical species A and B results in a collision complex, $(AB)^*$:

$$A + B \rightarrow (AB)^* \rightarrow A + B$$

or

$$(AB)^* \rightarrow C + D.$$

Such complexes and their decay have long been treated by making assumptions about the structure of the underlying phase space. An overly simplistic view is illustrated in Figures 9A,B. Implicit in the Figure 9A is that all of the (micro-canonical) phase space of the complex is accessible to any trajectory entering or leaving. An additional assumption, suggested by the initial conditions of 9B, is that the time scale for randomization of the complex is fast compared to decomposition to give either A + B or C + D. If this is the case the specific mechanism of formation will be forgotten, and the kinetics described by statistical theories which suggest that the rate of decomposition will only be a function of the excess energy for decomposition into a given channel.

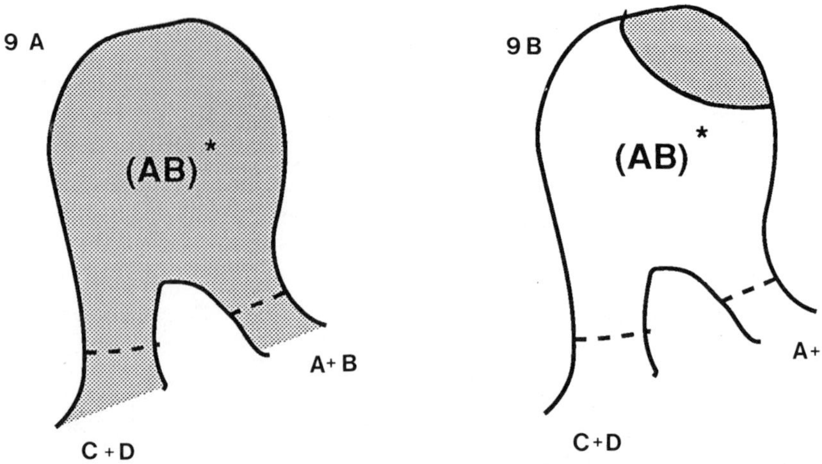

Figure 9. In part A the micro-canonical phase space for the complex is shown as being ergodic, that is all parts of the space may be visited by a trajectory entering or leaving either reaction channel. In B a laser or chemical activation prepares the system in the "corner" of phase space indicated by the shaded area. If the mixing time in the (assumed) ergodic phase space of the complex is fast compared to decomposition into the A,B or C,D channels, state preparation is forgotten and the rates are only a function of energy. {Or of temperature, in a canonical, rather than micro-canonical picture.} Assumptions such as these lead to a statistical theory such as that of RRK or RRKM.

Is the picture of Figure 9 consistent with the microscopic discussions of regular and chaotic dynamics discussed in the previous two sub-sections? Perhaps surprisingly, the answer for two degree of freedom is that such behavior would be exceedingly rare. A more typical (or, again "generic") situation is illustrated in Figure 10.

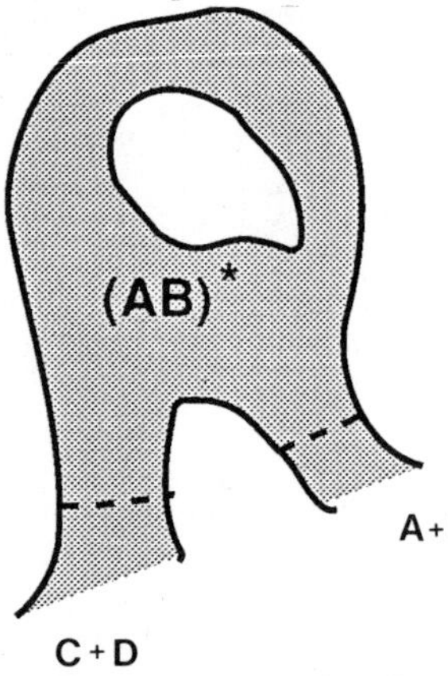

Figure 10. Generic phase space structure: at fixed energy an ergodic region (here connected to both arrangement channels) and a regular region co-exist. The dynamics is thus non-statistical.

2.4 Examples of Physical Systems

As already indicated in Figure 10 the generic dynamical situation for two degree of freedom systems, namely that phase space will be neither entirely regular nor fully ergodic, implies that not all parts of phase space are connected by the allowed dynamics. What does this have to do with actual dynamical observation? We consider two different situations where theoretical calculations have identified specific non-statistical behavior in restricted models of the dynamics of triatomic molecules. In the first example we consider models for the dissociation dynamics of HCC (in 3 dimensions) and then ABA local mode triatom two dimensional systems. In this case it is found, in both these two and three degree of freedom systems, that the trapped regions of regular motion, as shown in figure 10, yield non-statistical behavior in both classical and quantum discussions of dissociative behavior.

Second, we consider a two and three degree of freedom models of the intramolecular dynamics of energy flow in models of the OCS system. In this case it is empirically observed that even for a single trajectory in a chaotic region of phase space, that there are several relevant time-scales This sets the stage for introduction of the idea that even within a single chaotic region of phase space that leaky barriers can yield a multiplicity of different time scales *even in a region of phase space which is ergodic in the long time limit.* This is illustrated in Figure 11.

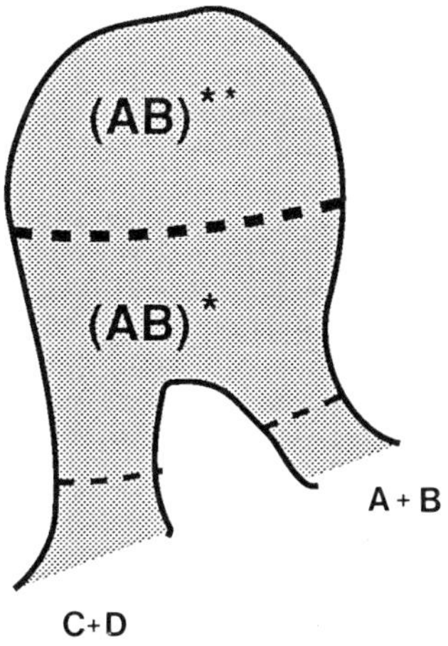

Figure 11. Illustration of the possibility of "intrinsic" non-RRKM behavior. What is indicated by the "barrier" between the two forms of the AB complex (labelled * and **, respectively) is a slow communication time between two parts of the same ergodic region of phase space. Thus even within a chaotic region we will need to consider the possibility of several important time scales.

Hase has referred to this type of multi time scale behavior as "intrinsic" non-RRKM behavior. A possible experimental observation of such behavior in a complex chemical system has been given by Rabinovitch et. al. and is introduced at the conclusion of the Section.

Trapped Trajectories and Vibrational Predissociation

Can volumes of phase space containing regular classical dynamics, not connected directly to regions of phase space allowing dissociation, affect dissociation dynamics? In studies of the dissociation of a three degree of freedom model of the HCC fragment (two stretches and a bend, with no angular momentum) Wolf and Hase discovered large trapped volumes of quasi-periodic motion, not connected to any dissociation channel, as illustrated in Figure

10. This of course implies that the classical micro-canonical dissociation dynamics is incomplete, in direct contrast to the RRKM idea that all micro-canonical phase space is dissociative once such dissociation is energetically allowed. What are the quantum implications? Quantum studies of unimolecular dissociation of three freedom systems at energies well above dissociation are not currently possible for realistic covalently bonded systems: there are simply too many quantum states to allow use of, say, complex co-ordinate methods to determine the distribution of lifetimes. In order to look at such classical trapping effects in models of two freedom systems, Hedges and Reinhardt, following the impetus of the Wolf-Hase work, considered a family of local-mode triatoms including CH_2 and H_2O. Large volumes of non-dissociative phase space well above the classical dissociating limit were found. These volumes correspond to a region of regular dynamics of large enough extent to contain a large number of semi-classical quantum states. Complex co-ordinate and simple, perturbative quantum estimates were made, and suggested the existence of very long lived "doubly excited" vibrational states corresponding precisely to the classically trapped volumes of phase space. It is important to note that in these cases there was no potential barrier forbidding dissociation, rather, dissociation was energetically allowed but dynamically forbidden. This is our first example of the role of time independent phase space structure in determination of molecular dynamics.

At the same time as the existence of long lived ABA vibrational states was being investigated, Kennedy and Carrington observed very long lived and very highly excited states of H_3^+ at energies up to several eV above the energetic threshold for dissociation to give $H^+ + H_2$. Pollack and Taylor, in a joint effort, showed that trapped classical motions are indeed associated with the strongly non-statistical behavior of this very interesting system.

Vibrational Energy Flow in OCS: Evidence for Long Time Trapping

Within a volume of chaotic phase space neighboring trajectories diverge exponentially. It is this exponential divergence which gives rise to the sensitivity to initial conditions characteristic of chaos. Is a chaotic volume of phase space characterized by a single rate of trajectory divergence? More qualitatively we can ask, when a trajectory *is* in a chaotic region, is diffusion in the full chaotic region slow or rapid, and more to the point can the diffusion towards ergodic filling of a chaotic regime be described by a single time scale, independent of choice of initial conditions? That is, *is a chaotic region of phase space homogeneous?*

By the early 1980's there was much empirical evidence that the answer to the above rhetorical questions was no! A connected chaotic phase space is not necessarily simply a homogeneous region governed by a single relaxation time. Early empirical observations of this in the context of chemical dynamics were made by Jaffe, Shirts and Reinhardt, who observed trapping "near" tori, and calculated an intra-molecular relaxation rate between what they called "vague tori"; DeLeon and Berne found similar "long time correlations" in a simple model of isomerization. Perhaps the most striking early results in the molecular physics area were those of Davis and Wagner, who observed trapping of chaotic orbits (vague tori) near stable tori in a two degree of freedom model of the internal dynamics of the OCS molecule, constrained to co-linearity. Further, these workers observed long time trapping in a three degree of freedom model of the same system considered as two local anharmonic local oscillators (the OC and SC bond modes) and the OCS bend.
An example of the type of trapping and multi-timescale behavior observed is illustrated in Figure 12.

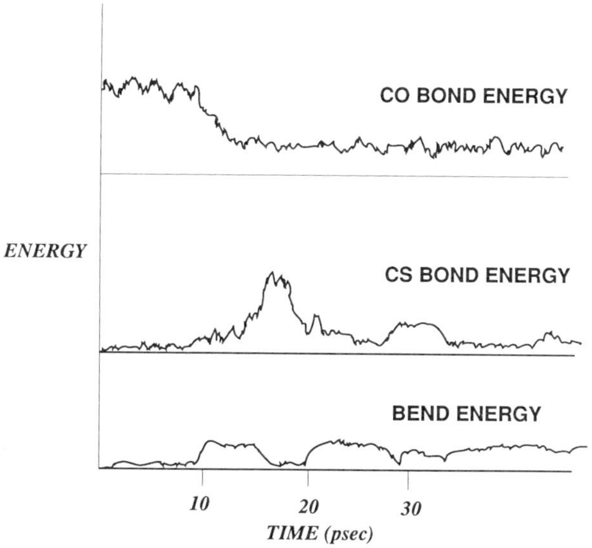

Figure 12. A sketch indicating the type of trapping observed by Davis and Wagner, as discussed in the text. Consider, for example the initial dynamics: for about 8 picoseconds energy is trapped in the CO bond, followed by a "sudden" energy sharing involving all three modes. Time scales for the internal stretches and the bend are one to two orders of magnitude faster than this. Davis and Wagner observed relaxation of ensemble averages of swarms of trajectories on a time scale of up to 45 picoseconds. The question at hand is "how can the high frequency bond modes give rise to dynamics with such long time scales?". What properties of the phase space are responsible for these long time scales?

Energy Trapping in Decene

Trenwith, Rabinovitch, Ostwald and Flowers have carried out chemical activation experiments, and have given an interpretation consistent with a structured phase space of the type seen by Davis and Wagner, and illustrated in Figure 11. What these works actually deduced was a relaxation path illustrated as:

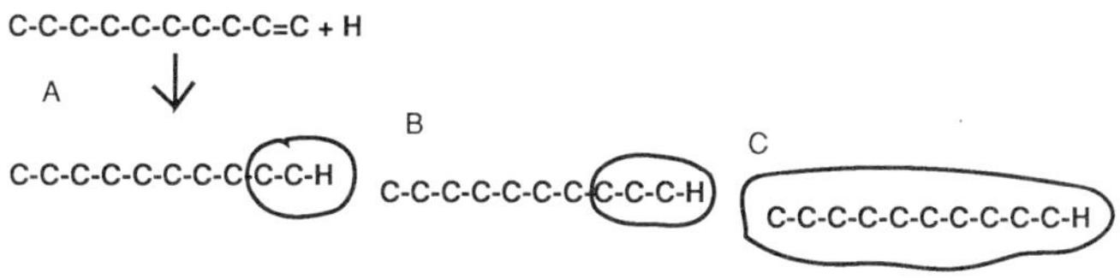

In panel "A" an H is added to the terminal carbon in the decene, presumably giving an adduct with a two carbon moiety "hot", as indicated by the circled region, following the addition. There is indeed a goodly amount of available energy: 42 kcal/mole is released during the H addtion. The subsequent kinetics led Rabinovitch et. al. to conclude that some "prompt", three carbon, dissociation product came from a molecule with localized excitation of the type shown in panel "B", with a slower product coming from the fully vibrationally relaxed "RRKM" excited intermediate "C". The observation of kinetics implying transient existence of the partially relaxed adduct "B", was surprising to these workers, as they had only expected to see products with the RRKM rate expected from the fully relaxed species "C". The trapping "observed" here was on the scale of 1 picosecond, but this is long compared to possible CH or CC relaxation times, which can occur on the 80 to 100 femtosecond times scale. In interpretation of these results the authors refer to the trapping in species "B" as indicative of the existence of a "sequential" relaxation mechanism. We note that such an observation could be consistent with the type of phase space structures leading to a range of relaxation time scales as noted in the previous sub-Sections, and to be further discussed below.

3. DEVELOPMENT OF THEORY

The empirical and qualitative observations of Section 2 set the stage for a more thorough discussion of the theory behind the concept of internal structure within a chaotic volume of phase space. We begin by introducing the concept of a point map as a useful tool in analyzing phase space structures. Using this tool a brief discussion of the origin of resonances and chaos itself is given with emphasis on some of the number theoretic properties of periodic orbits in the phase space. It was the understanding of such properties which led to the famous KAM theory of the existence of tori in non-integrable systems. We will see that number theory determines the stability of dividing surfaces in phase space, and that when such surfaces begin to disintegrate that leaky barriers to diffusion in the phase space are formed. Direct calculation of the flux across such barriers then follows using a succession of periodic orbits to approximate the minimum flux surfaces. What we then have is a phase space transition state theory, which allows calculations of transport properties from an understanding of time independent structure. Examples are given, and the conclusions then summarized.

3.1 Point Maps

Much of the analysis and subsequent theoretical advances in our understanding of several degree of freedom Hamiltonian systems has come, not from the study of such systems themselves, but from analysis and study of a far simpler set of mathematical systems, the non-linear areas preserving point maps. The idea of a point map is to generate a surface of section directly, without the trouble of going through the full solution of Hamilton's equations. Examination of Figure 2, shows at once why this can be profitable: in generating a surface of section by following a trajectory in the 4 dimensional phase space, almost all of the integration time is spent in regions of the phase space "off" of the sectioning plane. Can we find a mathematical process which simply predicts the Nth (p_2,q_2) point given the (N-1)st? Given the Hamiltonian, $H(\mathbf{p},\mathbf{q})$ it is certainly possible to do that for a two degree of freedom system: given the energy, the fact that $q_1 = 0$, and the (N-1)st value of (p_2,q_2) in the sectoining plane defines fully the initial conditions needed to predict the next point in the sectioning plane. However without actually running a full trajectory, it is not in general possible to actually make such a prediction. However, we can introduce point maps directly. For example two such point maps which we will discuss in what follows are the Henon-Siegel map:

$$p_{n+1} = p_n\cos(\alpha) - (q_n - p_n^2)\sin(\alpha)$$

$$q_{n+1} = p_n\sin(\alpha) + (q_n - p_n^2)\cos(\alpha)$$

and the Chirikov, or *standard* map:

$$I_{n+1} = I_n + k\sin(\theta_n)$$

$$\theta_{n+1} = \theta_n + I_{n+1}$$

The first of these is easily interpreted in terms of the phase space surface of section used in Section 2: the point (p_n, q_n) is mapped directly into (p_{n+1}, q_{n+1}). It is easily shown that the Henon-Siegel map preserves areas (the Hessian is unity) and thus that it mimics the dynamics of a Hamiltonian system. If the p^2 terms are absent on the right hand side, the Henon-Siegel map reduces to a circle map: points are mapped around circles of radius $r = (p^2 + q^2)^{1/2}$ advancing by the angle α at each iteration of the map. The result of taking several different values of "r" followed by iteration of the map for the value $\cos(\alpha) = 0.23$ are shown in Figure 13.

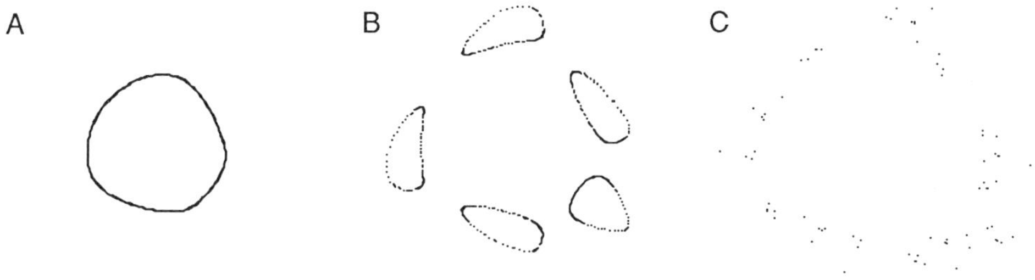

Figure 13.Interactions of the Henon-Siegel point map for three different initial conditions in the (p,q) plane. In panel A, a slightly distorted form of the original circle map results; in B a single initial condition generates five closed curves (a five fold "island chain") each surrounding a periodic orbit, or fixed point, of order five, that is a point which is mapped onto itself after five iterations of the map; panel C shows a chaotic orbit, which for this value of the parameters eventually dissociates, or escapes the region near the (0,0) fixed point. We see at once that the generic types of behavior seen in the Hamiltonian dynamics of Section 2, through examination of surfaces of section for two degree of freedom Hamiltonian systems, are mimicked in this simple non-linear but area preserving point map.

It is evident from the three trajectories of Figure 13 that the Henon-Siegel map can show the point map equivalents of invariant tori, which here are simple invariant curves in the plane; non-linear resonances; and, chaos. For the parameter values shown here, and in Figure 14, below, which shows a composite of the phase space for the Henon-Siegel map, the chaotic trajectories actually dissociate, that is, they move to infinity (never to return) after spending a finite time near the stable part of the map near the (0,0) fixed point. [The point p=0,q=0 is mapped into itself and is thus a fixed point.] We will use this property of the Henon-Siegel map to use it as a model for a two degree of freedom collision complex which dissociates with several intrinsic timescales. See Section 4.

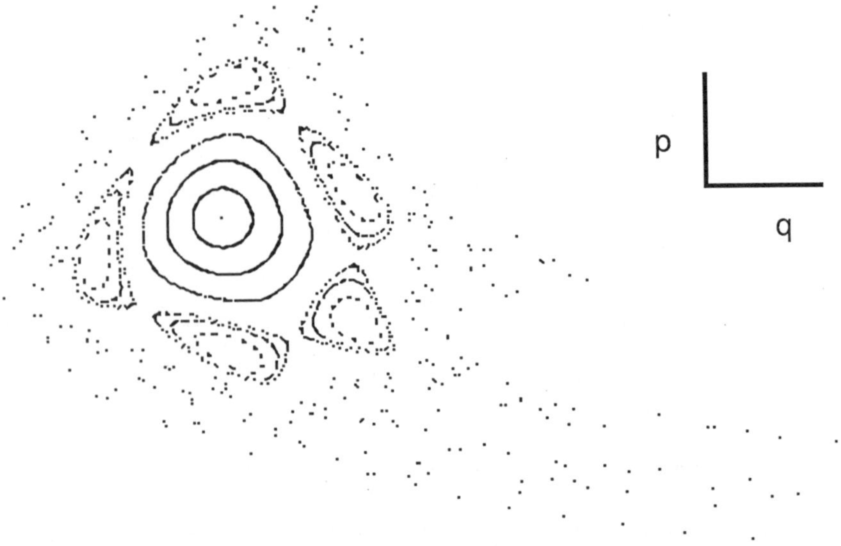

Figure 14. Composite of the iteration of several different initial conditions for the Henon-Siegel map with cos(α)=0.23. The central closed curves in the (p,q) plane are each generated by a single initial condition and correspond to the surface of section which would result from motion on an invariant torus in a two degree of freedom Hamiltonian system. These central curves may be thought of as perturbed versions of the circles of the linearized map obtained by neglect of the quadratic terms. Outside these central "tori" is a *five fold island chain*. As projected here in the plane it might appear that there are five separate tori: in fact all five of the islands are part of a single "resonance" surrounding a periodic orbit of order 5, as further discussed in the text. The (0,0) fixed point is seen as a lone point in the center of the inner invariant circles.

The standard map, defined above, might seem somewhat more abstract at first, as it is written in a sort of unrolled angle-action notation. The k=0, uncoupled, map is

$$I = I_{n+1} = I_n$$

which indicates that the "action" I is conserved but the "angle",θ, evolves as

$$\theta_{n+1} = \theta_n + I.$$

The angle thus advances at a rate proportional to the action. In what follows we will follow evolution of the angle mod(2π), or, thought of in a different way, as evolution around a cylinder. If I is a rational multiple of 2π, say I = (2πn)/m for integer n and m, the orbit will then close on itself, and refer to a fixed point of order n/m, or equivalently a periodic orbit with *winding number* n/m. In Figures 15 through 19 the results of iteration of the standard map for a range of initial values of I are shown, for the values k=0.0, 0.1 ,0.5, 0.8 and 0.95. In each case the initial angle was chosen to be pi, and 500 iterations of the map were carried out for each initial action. These form the basis for the subsequent discussion.

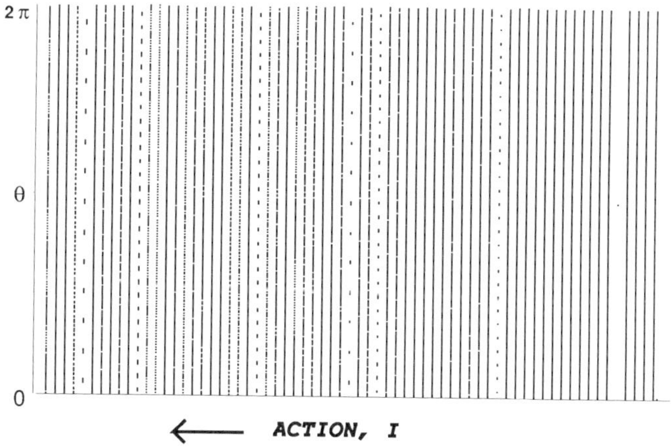

Figure 15. Composite of invariant curves for the standard map at k =0. In each case, as in the following sequnce, 500 iterations were performed for 70 values of initial action, I.

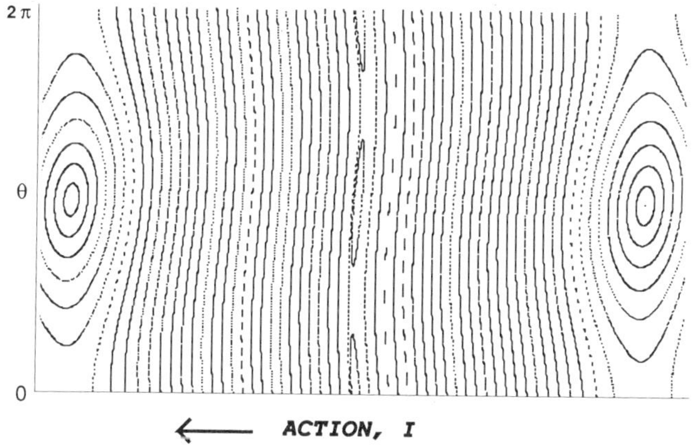

Figure 16. Composite of invarinat curves for the standard map at k = 0.1. Prominant are the non-linear resonances associated with the 1:1 and 1:2 stable fixed points.

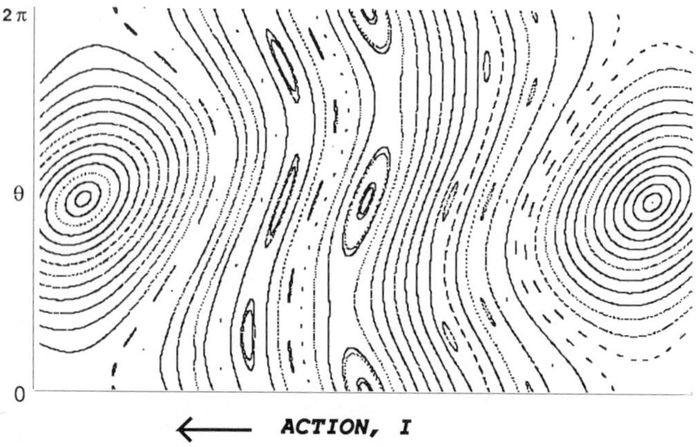

Figure 17.Composite map for k = 0.05 for the standard map. All orbits at this value of k appear to lie on perturbed versions of the original tori, or on island chains associated with low order rational winding numbers.

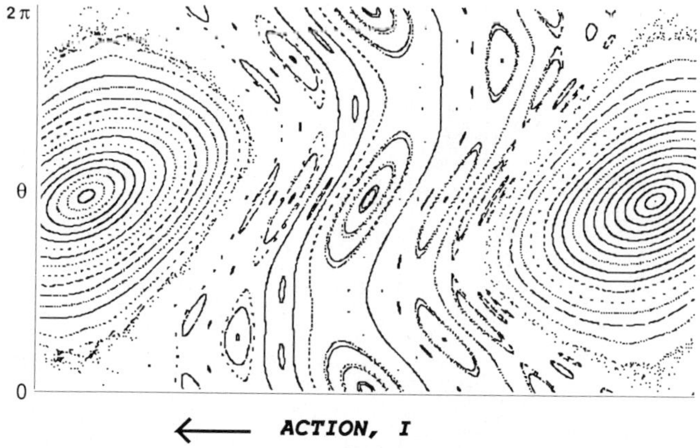

Figure 18. Composite map for k = 0.8 for the standard map. The 1:1 resonances dominate, and the 1:2 are next most prominant. In this picture it appears as if only three for four of the rotational tori, that is those which are continuous curves running from 0 to 2π, still exist.

3.2 The Poincare'-Birkhoff Theorem and the Destruction of Tori

Inspection of the phase spaces of the Henon-Siegel and standard maps, as displayed in the previous sub-section, shows that these deceptively simple recursive systems display considerable complexity. We now give a qualitative description of the origins of this complexity, as a prelude to introducing the important number theoretic concepts underlying the description and origins of chaos.

We begin with the standard map for k=0:

$$I_{n+1} = I_n = I_{initial}$$

$$\theta_{n+1} = \theta_n + I_{initial}$$

This simple mapping has I = constant as an invariant curve for any I. By invariant curve is ment that a point starting on the curve will not move off the curve under any number of iterations of the map. A numerical study of

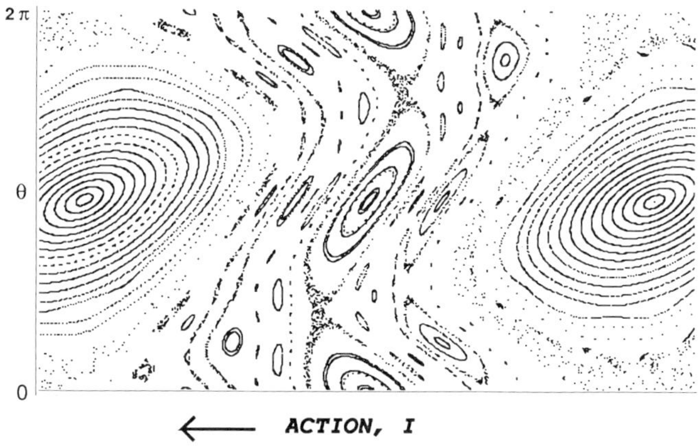

Figure 19. Composite mapping for k = 0.95 for the standard map. Only remnants of rotaional tori appear to remain, although the last actually persists up to a slightly larger value of k.

this seemingly obvious point is shown if Figure 15, where for a range of different initial values of I, 500 iterations of the map were carried out numerically, in each case taking an initial value of pi, for the angle. Inspection of the figure immediately shows that rather than just a series of vertical lines (which would be the invariant curves, or the invariant tori, for the system) the numerical study yields *two distinct types of behavior*. Namely, the 500 iterations of the mapping either seem to fill a line, giving the expected invariant curve to computer screen resolution, or the 500 iterations fall or a series of bands, or even seemingly isolated points. This easily observed behavior leads us to immediately distinguish between invariant curves which have *rational* and those with *irrational* winding numbers. This will be our first, but by no means the last, distinction of this number theoretic type!

Remembering the definition of the winding number (Section 3.1) as being $I/(2\pi)$, we note that any orbit on an invariant curve with a *rational* winding number of the form n/m (n,m integers) will lead to a periodic orbit, and the iterations of the map will yield m isolated points, one of which will correspond to that initially chosen. What about the rest of the invariant curve (which contains the m points, but uncountable many others,too)? The remainder of the invariant curve will be found by continuous variation of the initial condition for the angle yielding, each time, only m points. Thus an invariant curve with action a rational multiple of (2π) is not fully explored by any one trajectory. Conversely, if the winding number is *irrational,* any orbit will eventually explore the whole invariant curve, given a long enough time (i.e. if we iterate the map enough times). Looking back, again, at Figure 15, we note that the 500 iterations used "fill" the various curves to quite differing extents. That is, some of the invariant curves have winding numbers which allow a fairly uniform filling of the curve, while other curves seem to take many more iterations to fill in the gaps: In what follows we will relate this simple observation to the fact that some irrational are easily approximated by rationals, while others are in a sense to be made explicit, very far from rational. It is these latter winding numbers which lead to the more rapid filling of an invariant curve. Conversely if a winding number is very close to rational, an orbit will seem to be periodic for an arbitrarily long time.

What now happens to the invariant curves when the "perturbation" k is varied from its value of 0? A very powerful, and ancient, theorem due to Poincaré and Birkhoff (P&B hereafter) is at hand to help us. A key element of the P&B theorem is the distinction between the curves with the rational winding numbers of order m, i.e. the curves supporting orbits with fixed points of order m, and those with irrational winding number, and thus no fixed points at all. Stated informally the P&B theorem essentially says that the invariant curves with rational winding number are unstable with respect to small perturbations, the form of the instability is such that of the infinite number of mth order fixed points on the k=0 invariant curves becomes finite. The most usual case being that there will be only m stable fixed points (i.e. a single stable periodic orbit with winding number n/m) and m unstable fixed points (an unstable periodic orbit with winding number n/m). According to P&B these stable and unstable fixed points alternate along the original invariant curve. A good physical grasp of the difference between stable and unstable fixed points is given by considering the physical pendulum, shown in Figure 20: small oscillations about the fixed point of part A are stable and lead to small amplitude harmonic motion, a

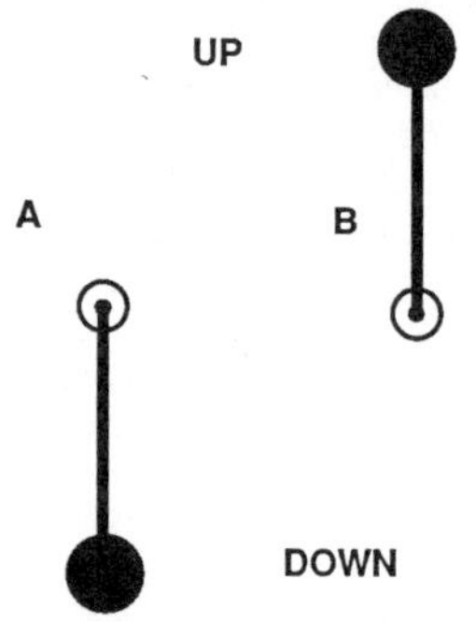

Figure 20. Stable and unstable fixed points (equilibrium points) for a physical pendulum in a gravitational field. A is a stable, and B an unstable equilibrium point, assuming zero initial kinetic energy in each case.

small perturbation from the unstable fixed point of part B of the Figure, leads to large amplitude motion!

What does this imply as k increases from zero in the standard map? A sequence of invariant curves in the neighborhood of a 1:2 (i.e. a very low

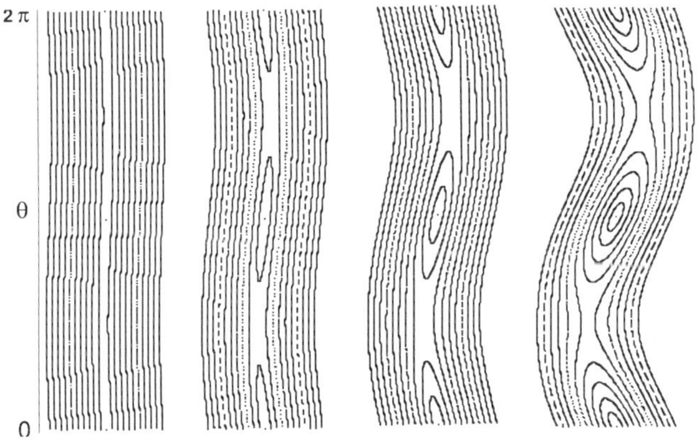

Figure 21. Formation of a 1:2 resonance zone on the standard map. From left to right k = 0.01,0.025.0.05 and 0.1, in a narrow region around I=π.As k increases an ever stronger resonance is created with a 1:2 periodic orbit at the center of the "islands". As θ increases from 0 to 2π from top to bottom, the ends of the invariant curves are to be identified, and in particular, note that the 1:2 orbit only creates 2 resonance islands.

order rational ratio) winding number orbit is shown in Figure 21 for k = 0.01, k=0.025, k=0.05; and k = 0.1. In each case the initial values of I were taken to be centered at (1/2)*(2π) with a range of ±0.1.

What is seen in the Figure is a periodic orbit of order 2 has a greater and greater influence on the dynamics as k increases. At the left of Figure 21, we see the vertical invariant curves associated in a one to one manner with the I=const curves of the k=0 map. However, the fixed point displays itself clearly (remember that $\theta=0$ and $\theta = 2\pi$ are to be identified). As k increases (moving to the right in the Figure) the invariant curves no longer run continuously from 0 to 2π, rather a new type of "resonance" invariant curve, with the remaining stable (elliptic) fixed point, at the center appears. The motion in the resonance zone may be considered as small oscillations around the stable periodic orbit. This is in accord with P&B. What about the unstable periodic orbit also predicted by P&B? This will be a hyperbolic fixed point at the point where the resonance islands touch. This is the analog of the osculations of the separatrix in the motion of the physical pendulum for Figure 20, as illustrated below, in Figure 22. What is seen is that for energies below that needed for free rotation, the phase portrait consists of stable oscillations about the lower equilibrium point. As the amplitude increases, eventually an energy is reached where the motion will bring the pendulum to a stop at exactly the inverted position: this is the unstable fixed point. Such a motion has infinite period, and, in the phase plane separates, and thus is a separatrix, motion bounded in θ, from that which is unbounded. The type of motion in a two degree of freedom system, and in the map of Figure 21, is precisely analogous.

The analogy between the phase portrait of Figure 22 and the resonant and non-resonant invariant curves of Figure 21, is made even more complete if we we to consider the effect of reducing the gravitational field on the physical pendulum: the islands of stability would then narrow and eventually vanish, just as happens for decreasing k in the standard map.

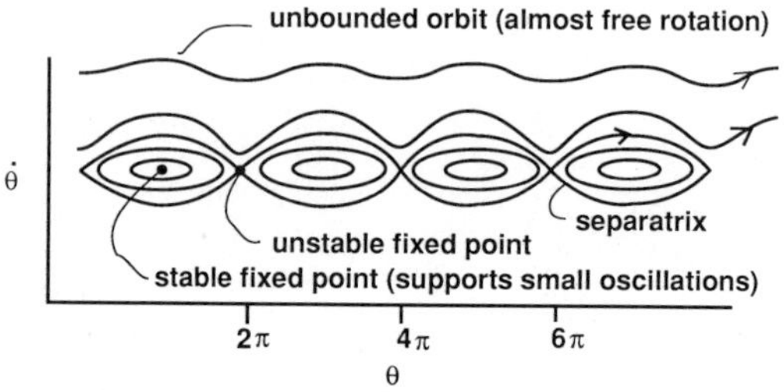

Figure 22. Bounded and unbounded motion of the physical pendulum in a gravitational field. (See Fig 20 for the physical situation.) At high energy the pendulum rotates freely, and the angle is unbounded, increasing to the left or right depending on the sense of rotation. A low energy motion is trapped in one of the "wells" defined by the value of theta, and small oscillations occur about these stable (elliptic) fixed points; at an intermediate energy the separatrix divides the two types of motion. Branches of the separatrix meet at the (hyperbolic) unstable fixed points, one of which is indicated. The origins of the sobriquets "elliptic" and "hyperbolic" for the two type of fixed point are evident from the figure. Note the similarity between this figure and the panels of Figure 21, which are rotated by 90°.

3.3 KAM and the Onset of Chaos

As seen in Section 3.2, invariant curves with rational winding number become unstable if perturbed, and the result is a series of stable and unstable fixed points, or "resonances". An immediate objection to this picture would follow by noting that as the rational numbers are dense on the real line, rationals are everywhere, *so why don't we see islands and resonances on all scales?* Well, why don't we? Why aren't all of the invariant curves destroyed at once, leading to a jumble of islands? Examination of Figures 16 through 19, it appears that many of the non-resonant invariant curves (i.e. those running continuously from 0 to (2π) are destroyed, but that some always remain. Again, we can ask why?

Without going into detail, for which see Section 6, while it is correct that, for a generic non-linear system, island chains form about all orbits with rational winding number, most of the island have essentially zero width: for an orbit with winding number n/m, the width of the island chain will be of order k^m, where k is a measure of the perturbation strength. For small k and high order winding numbers the effect is thus vanishingly small for a given orbit. But, as there are infinitely many rationals between any two irrationals, does the cumulative effect of infinitely many infinitesimal instabilities destroy all of the non-resonant invariant curves. Answering this question is clearly an exercise in a combination of number theory and measure theory, and is non certainly non-trivial. The result is the justly celebrated KAM theorem. KAM stands for Kolmogorov, Arnold and Moser, and their theorem essentially says:

> **KAM theorem: Invariant curves (invariant tori) are not uniformly destroyed by a small perturbation, in fact they survive in finite measure. One should not, therefore, be surprised by observation of invariant curves at finite values of the couplings.**

This theorem is often said to be the first and perhaps foremost result of modern non-linear dynamics of conservative systems. It was, for example, lack of knowledge of the content of this theorem which caused von Neumann, Fermi, Pasta, and Ulam...a pretty talented group... to be shocked by the fact that a non-linear Hamiltonian system of coupled anharmonic oscillators did not behave ergodically, when first studied computationally in the early 1950s. Examination of Figures 16 to 18 clearly displays the content of the P&B and KAM theorems: Many of the unperturbed invariant curves do break up, many types of island chains are seen, but some of the original invariant curves always seem to remain. What about Chaos? The simplest interpretation of chaotic motion is that it occurs in regions of the phase space where an orbit is under the influence of two or more dominant resonances.Each of the resonances tries to drive the orbit at its own frequency,m and the orbit responds erratically. Chirikov has established this idea in the from of a model criterion for the onset of chaos as being do the the overlap of two non-linear resonances, choosing the two to be those of lowest order.

In summary, we see the great significance of the distinction between orbits with rational, as opposed to irrational, winding numbers, in terms of the breakup of invariant curves under the action of a perturbation. It will emerge that the significance of rational numbers in chaotic dynamics goes even farther than this, as we now see.

3.4 Number Theory and Dynamical Processes

The relationship between the existence of time independent, invariant, curves on the Henon-Siegel and standard maps, or of the invariant tori of Hamiltonian systems, is intimately connected with the number theoretic properties of the winding numbers of trajectories on such curves. We have seen that tori with low order rational winding numbers are unstable, and break up according to the P&B theory. We will now indicate that the remaining tori allowed by the KAM theorem are destroyed in a specific order, characterized by the ease of rational approximation to their irrational winding numbers. This sounds quite abstract, but the parts of number theory needed to understand how the KAM invariant curves are destroyed are not difficult, and further, these same parts of the theory will allow us to characterized the properties of the "bottlenecks" which are a prime cause of the need to introduce more than one time scale in discussing the approach to ergodicity, namely the length of time need to insure that time averages may be replaced by averages over the connected chaotic volumes of phase space.

The relevant parts of number theory needed to discuss the stability of KAM surfaces with irrational winding number was used in the 19th century by

Liouville in his investigations of transcendental numbers, that is those irrational numbers which are not the solution of algebraic equations. However, various parts of the result were known to the ancient Egyptians in their approximating pi by 22/7; by the Greeks who introduced the Golden Mean of ideal proportion; and interestingly enough, given that the present School assembled in Pisa, by Leonardo di Pisa, also known as Fibonacci, whose famous numbers, bearing this latter name, give optimal rational approximants to the Golden Mean.

We begin by asking how to systematically approximate real numbers, be they rational or not. Any decimal number may be systematically rewritten uniquely as a continued fraction. Suppose we have a number, σ, which we want to approximate by a rapidly convergent succession of rationals: any such number has a unique expansion of the form

$$\sigma = a_0 + \cfrac{1}{a_1 + \cfrac{1}{a_2 + \cfrac{1}{a_3 + \cfrac{1}{a_4 \cdots}}}}$$

where a_o is a positive or negative integer, and all of the a_i, for $i>0$, are natural numbers, i.e. $a_i = 1,2,3,4,\ldots$.If σ itself is a rational the process truncates after a finite number of "levels" in the continued fraction. If σ is not rational the expansion is infinite. What can then be shown is that

1) the expansion is optimal in the sense that it is quadratically convergent, which is the best generally possible; and,

2) successive truncations give successive upper and lower bounds to the number being approximated.

By quadratic convergence is ment, that if the fraction is truncated by keeping on the a_i for i up to n:

$$\sigma_n = a_0 + \cfrac{1}{a_1 + \cfrac{1}{a_2 + \cfrac{1}{a_3 + \cfrac{1}{a_4 \cdots + \cfrac{1}{a_n}}}}}$$

followed by evaluating the fraction as a ratio of two relatively prime integers, r_n and s_n then

$$\sigma_n = r_n/s_n$$

and the error is bounded by

$$\left| \sigma - \sigma_n \right| = \left| \sigma - \frac{r_n}{s_n} \right| \leq \frac{C}{s_n^{2}}$$

for some positive constant *C*, giving an expansion converging quadratically with increasing order of the denominator integer of the rational, s_n.

Can one do better than such quadratic convergence? This is where things get interesting. Liouville showed that for algebraic numbers, that is irrationals which are the solution of finite order algebraic equations, that the answer is no, quadratic is the best we can hope for in general. He then showed that there existed an infinite class of real numbers for which the bound could be improved, namely that

$$\left| \sigma - \sigma_n \right| = \left| \sigma - \frac{r_n}{s_n} \right| \leq \frac{C}{s_n^{2+\varepsilon}}$$

for $\varepsilon > 0$. As these numbers were not solutions of algebraic equations, for which a quadratic convergence is the best we can expect, he thus deduced the existence of the infinite class of *transcendental* numbers! The odd situation is then that transcendental numbers (i.e. those like "π" and "e" [the base of natural logs] which are not solutions of algebraic equations) are *easier* to approximate than algebraic irrationals. That is, to find a rational approximation good to a prescribed accuracy, smaller integers are needed to give the desired rational expression...thus the surprising goodness of 22/7 as a low order approximation to π. The converse, namely that algebraic numbers are *hard* to approximate is easily illustrated: the *slowest* possible convergence of the continued fraction is given if $a_i=1$ for all $i=1,2,3,4.....$

$$\omega^{Au} = \cfrac{1}{1+\cfrac{1}{1+\cfrac{1}{1+\cfrac{1}{1+\cdots}}}}$$

The number ω^{Au} is evidently algebraic: the self-similarity of the fraction gives

$$\omega^{Au} = \frac{1}{1 + \omega^{Au}}$$

which has the simple (algebraic) solution $(\sqrt{5}-1)/2$, which is the Golden Mean of Greek architecture, defined in terms of the rectangle with sides S (short) and L (long) such that

$$S/L = L/(S+L)$$

and which was thought to give an ideal sense of proportion. For the historically and mathematically inclined it is interesting to note that the rational approximants to the Golden mean are ratios of the successive Fibonacci numbers, which thus play an important role in non-linear dynamics. The slow convergence of the continued fraction expansion for ω^{Au} makes it the most difficult to approximate by rationals, and in this specific sense it is the *most* irrational number, which gives it a very special role in the dynamics of non-linear systems. The Golden Mean is the simplest example of a whole class of hard to approximate irrationals where the $a_i = 1$ for i past a finite value, rather than simply for all i, these from the class of "noble" irrationals.

The important role of the Golden Mean follows from the ideas behind the proof of the KAM theorem: invariant curves or tori with rational winding numbers are unstable via the P&B theorem, KAM indicates that *some* tori with irrational winding numbers survive, and in particular the tori with irrational winding numbers which are, in the sense of rational approximation discussed above, the *farthest* from rationals are the more stable and robust as a perturbation is turned up. In a landmark work Greene, in 1979, made this explicit in an analysis of the k dependence of the standard map. Examination of the series of Figures 15 through 19 appears to indicate that for k up to 0.8, that some of the original, k=0, invariant curves running continuously from 0 to 2π still persist. These surviving curves will be referred to a the rotational tori. However, at k=0.95 (Figure 19) it appears that no such rotational tori persist. At least none appears at the resolution shown. Is there a value of k beyond which no rotational tori exist? Greene made this explicit, he first showed that it is the rotational torus with the Golden Mean winding number which is the last surviving such torus as k increases. This confirms the qualitative idea that winding numbers "far" from low order rational give the most robust structures. Further, Greene showed that for $k > k_{crit}=0.971635...$rotational tori no longer exist.

Two questions now arise:

1) what does all of this have to do with our original motivation, namely the desire to understand the approach to ergodicity in a chaotic phase space; and,

2) what *happens* to the last torus as k passes beyond k_{crit}.

We will now see that these are very closely related issues, and that number theory actually provides the basis for understanding the existence of "bottlenecks" to rapid global transport.

3.5 Barriers & Bottlenecks to Diffusion in Phase Space

In a two degree of freedom Hamiltonian system; the invariant tori are two dimensional surfaces embedded in a three dimensional energy shell. It is a property of closed surfaces of dimension N-1, embedded in a space of dimension N, that they possess and *interior* and an *exterior*. Thus a trajectory starting in the interior of a 2-torus would have to pass through the torus to get to the exterior. If the torus is indeed and invariant surface all points on the surface evolve into points on the surface, and such a crossing is thus impossible. The tori thus form rigorous barriers to diffusion in the phase space. A chaotic region inside a torus is dynamically separated from a chaotic region outside the torus.

The analog of this trapping for the standard map is that the rotational tori, being the invariant curves running continuously from 0 to 2π, rigorously prevent diffusion of individual trajectories from regions of low values of I to high values. As k increases past k_{crit} the rotational tori vanish, and global transport becomes possible. [Note that trajectories starting on the various island tori, on in the small chaotic regions within such tori do not show such global diffusion.] How does the rate of such diffusion depend on $(k-k_{crit})$? Does the global diffusion, once it sets in, show any signature that the Golden Mean rotational torus used to block global diffusion entirely? The sequence of Figures 23 to 27 gives a feeling for this.

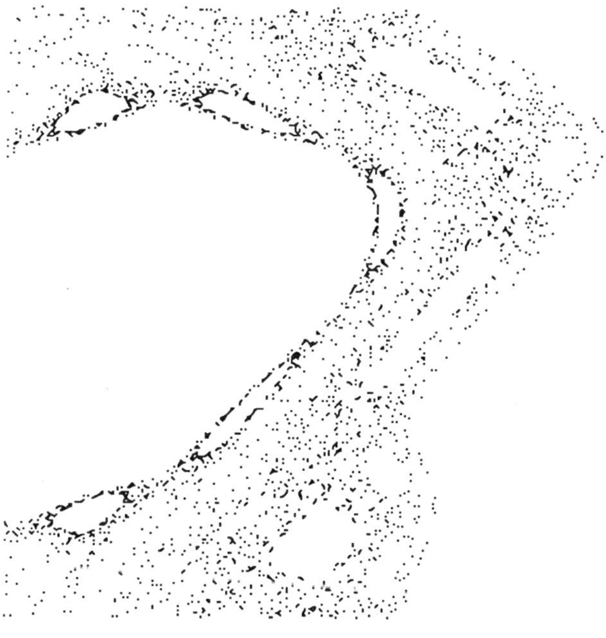

Figure 23. Diffusion in the standard map at k= 1.19. The first 5000 iterations. Co-ordinate axes are as in Figures 15 through 19. Here we see that iteration of a single point (initial condition) rapidly fills a large fraction (about 1/3) of the availalbe phase space, but that it has not yet fully explored the full connected region of chaos dynamically allowed to it. This immediatly suggests existence of more than one time scale for this diffusive process. This is further illustrated in the following sequence of Figures.

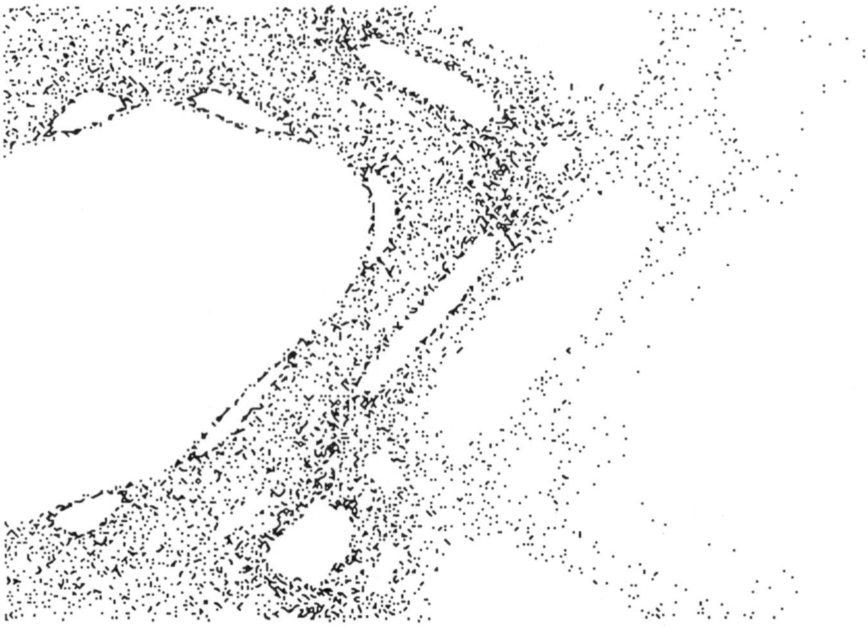

Figure 24. Diffusion in the standard map at k =1.19. The first 10,000 iterations.

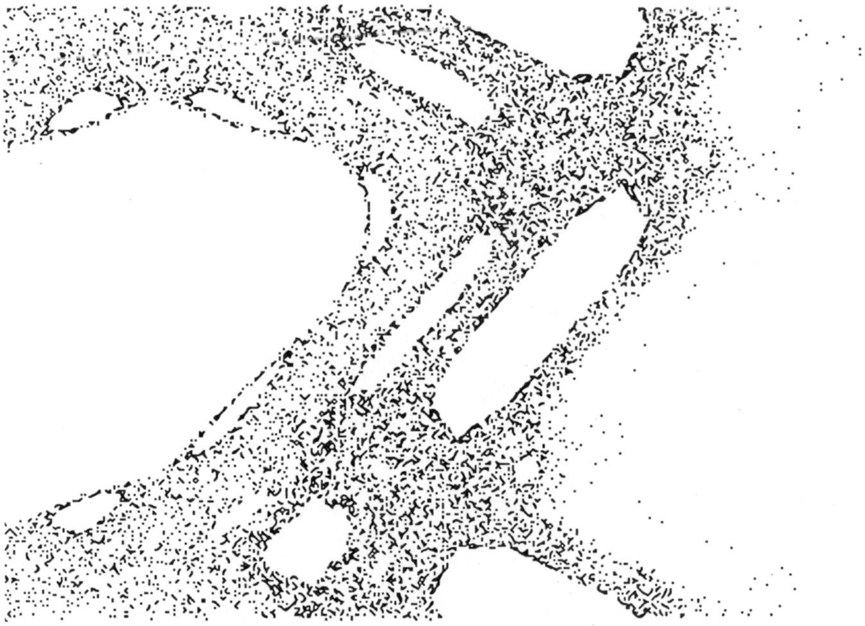

Figure 25. Diffusion in the standard map for k = 1.19. The first 15,000 iterations.

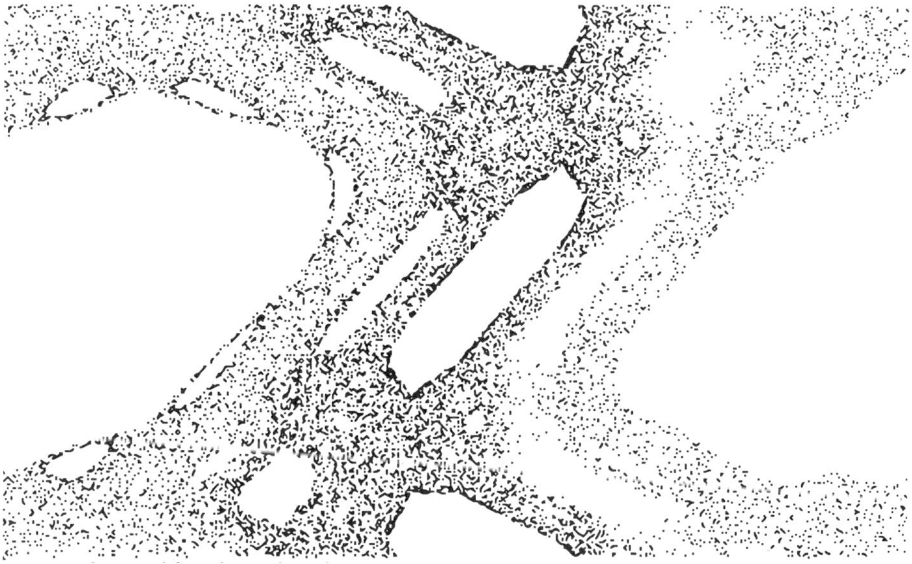

Figure 26. Diffusion in the standard map for k = 1.19. The first 20,000 iterations.

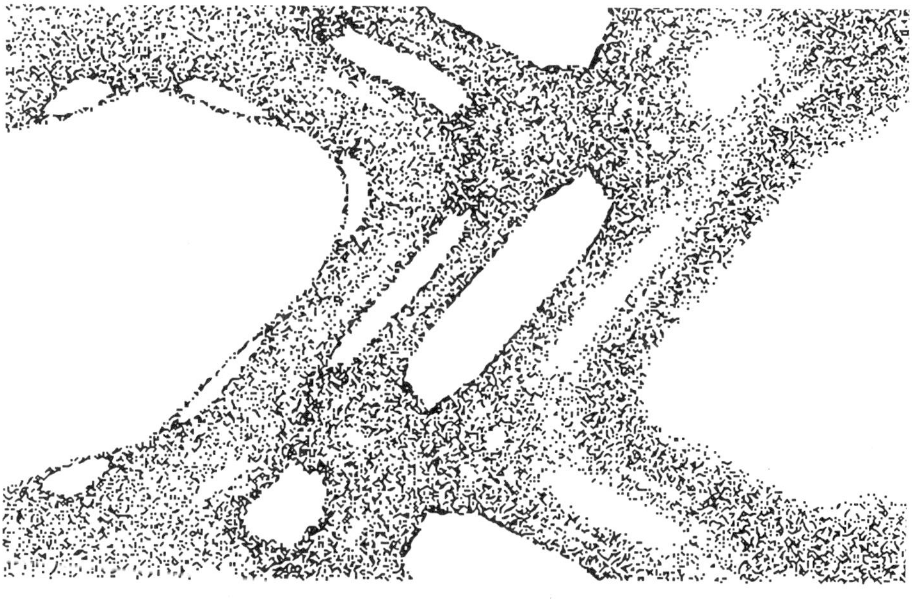

Figure 27. Iteration of a single trajectory for the standard map at k=1.19. 30,000 iterations are shown. The trajectory almost uniformly fills the available region of chaotic phase space at this value of k. The white regions are occupied by "island" or "resonant" tori, which can locally trap a trajectory, but which do not prevent global diffusion within the large chaotic phase space, which appears to be ergodic on this longer time scale. Inspection of Figures 23 through 26 does indicate, however, that considerable time must pass before this micro-canonical equilibrium limit is reached.

Inspection of the sequence of Figures 23 through 27 indicates that chaotic diffusion is indeed possible within a large region of the phase space, but that a longer time scale is needed than for approximating ergodic behavior in smaller volumes of the space. What then is the time independent phase space structure responsible for the "bottleneck" slowing the progress of global diffusion? Here is where the developments of the early 1980s become important, and we face the question of what happens to an invariant torus when it ceases to exist as a function increasing k.

To this point we have discussed the structure of phase space in terms of continuous invariant surfaces (the tori) and fixed points. These are both time independent quantities, and as such are properties of the phase space of the dynamical system, rather than of individual trajectories, which must time evolve on, or around (in the case of chaos), these fixed objects. A third class of invariant surface will now be introduced, recognition of this new type of invariant is associated with names Mather, Aubry, and Percival,and provides a detailed description of how tori break up, and how their remnants form precisely those bottlenecks which set the time scale for global diffusion. What these workers observed was that the invariant curves with noble winding numbers are not only the most robust, with respect to survival under perturbation, but when they break up, such break up is not at all instantaneous, but follows a well defined pattern: that is small gaps open up in the curve, the size of these gaps depending on a power of $(k-k_{crit})$ where k_{crit} is the critical perturbation for the invariant cure at hand.

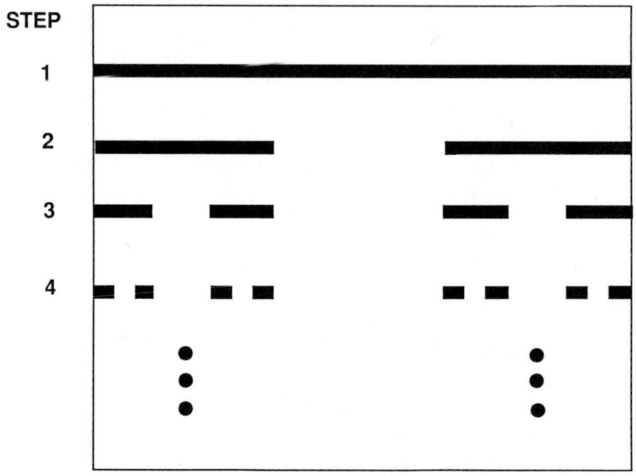

Figure 28. The traditional construction of a non-denumerable set of measure zero, due to Cantor. In Step 1, we consider a unit interval. In Step 2, the middle 1/3 is removed, leaving two segments, each of length 1/3. In step 3, the middle 1/3 of each of the segments in the previous step is removed, and so on. What is left in the limit of an infinite number of steps is an infinite set of points but which is of measure zero, as the lengths removed add to unity. Note that there are gaps on all length scales from 1/3 down to zero. This is suggestive of our discussion above, but for the fact that the remaining set is of zero measure.

However, the remnant (the complement of the gaps) is still an *invariant set*. It follows at once that an infinite number of gaps exist: as this new structure is an invariant set of the map (or more generally of the Hamiltonian system) a gap must be mapped into another gap and, as the winding number is "highly" irrational (i.e. very far from low order rationals), such gaps will appear at an infinite number of places along the curve at once. They will also appear on all length scales. This latter follows from the fact that the sum of the lengths of the gaps and invariant set add to the length of the smooth curve which supports both. That is, the invariant set and gaps both may have finite length or measure. As there will be an infinity of gaps, their lengths must decrease more rapidly than the increase of their number as a function of decreasing length. Visualization is helpful at this point. Figures 28 and 29 show two constructions of *Cantor Sets* one of which has measure zero, and one of which has finite measure. Figure 30 shows the Cantor set nature of the formation of gaps in what was formerly an invariant curve of a dynamical system. Percival has introduced the term *cantorus* for the new type of invariant set, as it is both a Cantor set, and is certainly torus-like. Mathematicians refer to the cantori as *Mather sets*.

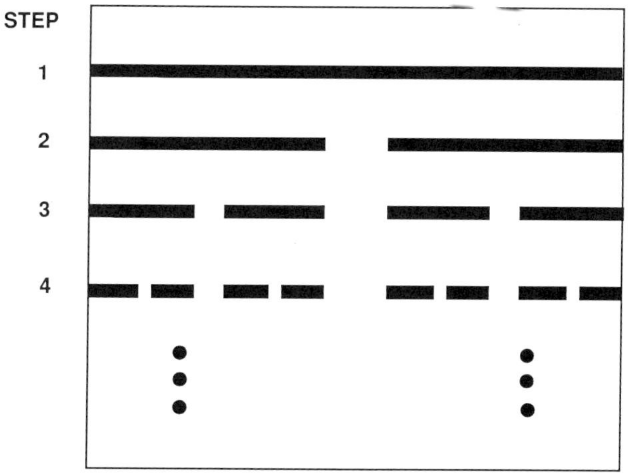

Figure 29. Construction of a Cantor set of finite measure. In this case the sum of the lengths of the gaps is, even after an infinite number of steps, still less than unity. The result is an infinite number of gaps, which exist on all length scales, and a complementary Cantor set which also has segments on all length scales. This is an analogous situation to that occurring as small gaps appear in invariant tori.

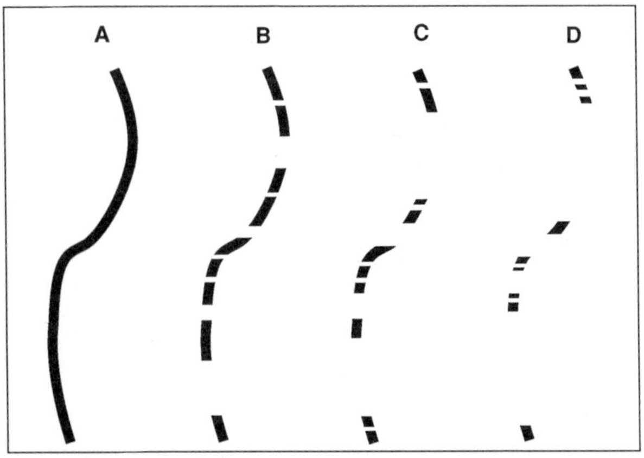

Figure 30. An invariant curve A, grows gaps, B, (shown here on several length scales) which themselves grow, C and D, as k increases beyond k_{crit} which will be different for each curve. These finite measure "cantori" are invariant sets, and are the remnants of the tori. Flux through the gaps allows diffusion as k in increased beyond k_{crit}.

3.6 Examples

The picture emerging from the previous sub-section is that invariant tori provide dividing surfaces, preventing, global chaos and diffusion, but that when couplings increase these tori can become leaky, allowing flux to cross what were formerly impenetrable barriers. In a given region of phase space, following traditional ideas of transition state theory (from Wigner and Keck), a rate through the cantori, acting as a bottleneck, may be found by finding the cantorus with minimal flux. The flux across an individual cantorus may be, in turn, found variationally. The details of the flux calculation, and derivation of the scaling laws for flux through a single cantorus , i.e.

$$\text{flux} \propto (k-k_{crit})^{\varepsilon}$$

for $k > k_{crit}$ with ε being a critical "flux" exponent, are beyond the scope of the present introduction. Readers interested in such detail are referred to the work of MacKay, Meiss & Percival and Bensimon & Kadanoff. We only note here that approximate locations of the cantori with noble winding numbers are very easily and systematically found, as successive truncation of the continued fraction representations of the noble frequencies themselves gives the best possible rational approximants, which in turn are the winding numbers of period orbits which give successively better approximations to the cantori. In particular MacKay, Meiss and Percival give an an analysis of the role of the Golden Mean rotational cantorus in setting the rate for the diffusion shown in the sequence of Figures 23-27.

The first applications of the use of cantori as bottlenecks in phase space as setting an intra-molecular rate were carried out by Davis, in an analysis of the long time correlations found in a two degree of freedom model of the OCS dynamics discussed in Section 2.4. The results of his analysis are graphically summarized in Figure 31.

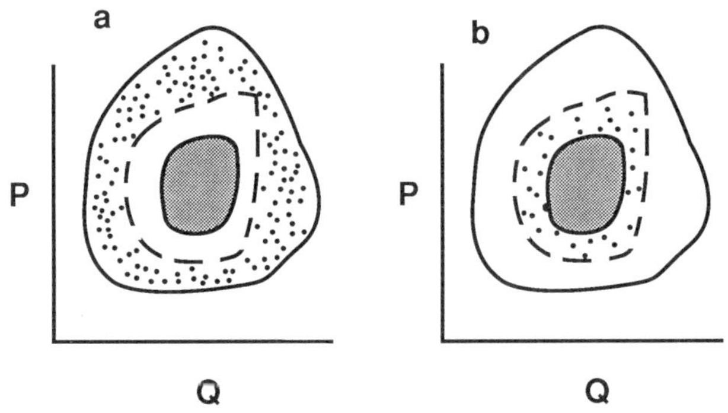

Figure 31. Sketch of the type of results seen by Davis for collinear OCS. A cantorus (shown as the dashed line) was located, and found to be the phase space bottleneck for flow between the two regions of phase space shown in this surface of section. In "a", a trajectory was confined for 24.6 psec in the outer region, before finally penetrating the cantorus, where trapping in the inner region followed, as in "b". A direct calculation of the flux across the bottleneck gave good agreement with the ensemble averaged rate of crossing obtained from a traditional Monte Carlo study of the intra-molecular rate. Note that the shaded central region is inaccessible to the trajectory shown, as its outer boundary is a full torus.

In this case the bottleneck was found to be closely related to an invariant curve with the Golden Mean winding number, substantiating the idea that real Hamiltonian systems do belong to the same family of generic systems as the standard map. This type of analysis was then extended to consideration of the formation of long lived collision complexes by Davis and Gray in a two degree of freedom model of the dynamics of the I_2-He van der Waals complex. Again the various time scales of the dynamics were well explained by the time independent phase space structures discussed here. As a simple illustration of the role of a cantorus in slowing dissociation from a chaotic region of phase space, Figure 32 shows the result of starting two trajectories for the Henon-Siegel map, and iterating each for 10,000 steps. The first defined an invariant curve, while the second took almost 9000 iterations to dissociate, and its outer border clearly defines the cantorus.

3.7 Implications for "Thinking about Phase Space"

During the mid 1980s our understanding of the morphology of Hamiltonian phase space for two degree of freedom systems has been greatly expanded. To summarize these developments, as outlined in Sections 3.1 through 3.6, we have now discussed *three* type of invariant structures:

1) Periodic orbits (fixed points) which may be stable or unstable;

2) Invariant curves (e.g. invariant tori) of KAM fame; and,

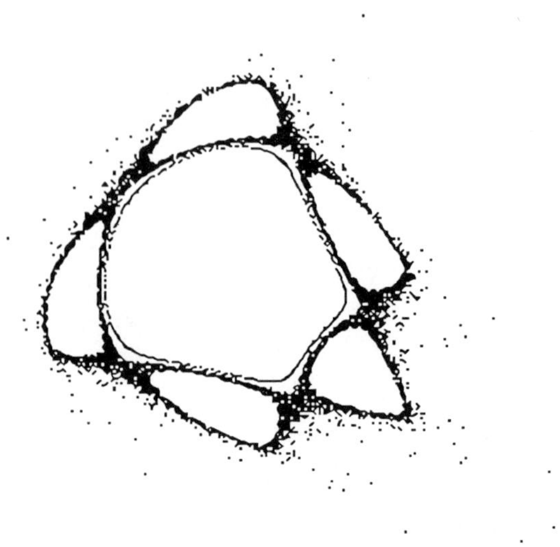

Figure 32. 10000 iterations of the Henon-Siegel map for two initial conditions, one on an invariant curve, the other in a chaotic region, bounded on the inside by the invariant curve, and on the outside by a not-too-leaky cantorus which confined the trajectory for many thousands of iterations. Once "outside" the canotorus, whose location is suggested by the outer edge of the dense chaotic region, most trajectories (including this one) dissociate in 10 to 50 further iterations. A different look at this part of the phase space is taken in Section 4.2

3) Invariant Cantor sets, or cantori

The first of these formed the basis of discussion of Section 2, where two types of non-periodic motion were recognized: regular, quasi-periodic, motion on tori, and chaos. With the addition of the cantori to our repertoire, we now have a tool for understanding structures within chaotic regions of phase space which govern classical pictures of intramolecular dynamics. We also clearly see that a chaotic volume of phase space is not at all a homogeneous region, but that many important time scales may exist, and that some of them may be very long indeed. These long time correlations are seen when phase space bottlenecks formed by partially destroyed tori trap trajectories. The time scales for passage through bottlenecks may be determined by phase space transition state theory, and will depend on the details of non-linear interactions, rather than what might appear to be the fundamental natural frequencies of the zero order system.

4 THE FRACTAL STRUCTURE OF HAMILTONIAN PHASE SPACE

In this final Section we introduce an alternate way of investigating the time independent structure of Hamiltonian phase space. It parallels work carried out for over a decade in the area of dissipative chaos. A dissipative system does not conserve energy, and does not conserve phase space volume. This can be used to advantage in understanding the time independent structure of the phase space, as such systems will relax to fixed points or fixed surfaces or curves called *attractors.* It is not all difficult to find such attractors: almost any initial condition will give a trajectory which will eventually (actually usually quite quickly!) relax to the attractor, which will then display itself. Such maps not surprisingly cannot be inverted: if many initial conditions lead to the same final state, finding the system "on" an attractor implies loss of all information as to the initial state. In this Section we introduce some methods from the theory of algebraic non-conservative point maps. In particular, in Section 4.1, we discuss attractors and basins of attraction. These basins of attraction often have apparently complicated boundaries, which on closer examination appear to be self-similar fractals. Attractors themselves may be complex objects. The vocabulary of such objects is introduced here.

We then apply the methods of Section 4.1 to visualize the global and microscopic structure of dissociative phase space for a Hamiltonian system. The trick used here, in 4.2, is to use the "point at infinity" as an attractor to build up a picture of the phase space, which will be seen to be remarkably complex. Observation of similar complexity in the phase space for atom-diatom collisions is then discussed in Section 4.3. The structures observed directly in Sections 4.2 and 4.3 are then discussed in terms of the homoclinic oscillations originating at the unstable fixed points of low order unstable periodic orbits. This introduction of homoclinic structure completes our discussion of the morphology of chaotic phase space, and we briefly indicate how either diffusion through cantori *or* motion controlled by a combination of initial conditions and a single homoclinc oscillation can give rise to exponential decay of a collision complex.

4.1 Simple and Complex Attractors of Maps

Attractors, Attraction Basins and Fractal Boundaries

The idea of an attractor is easily introduced: The physical pendulum of Figures 20 and 22 has a series of point attractors if a normal dissipative term is added to the dynamics. Such "friction" leads to eventual sowing down and stopping of the pendulum at one of the elliptic fixed points of Figure 22 These points are then referred to as "attractors". [If we think about this problem "mod(2π)" there is only one fixed point and one attractor.] Thus a dynamical system or a non-area preserving map may have a range of initial conditions all of which eventually lead to one or more isolated points. A simple example, which has a very interesting generalization, comes from considering the dissipative map

$$z_{n+1} = z_n^2$$

in the complex plane: $z = x + iy$. Iteration of this very simple dynamical system is easily analyzed, and this simple example is useful for the introduction of terminology. Writing the initial point z_0, as

$$z_0 = re^{i\theta}$$

we have

$$z_n = r^n e^{in\theta}$$

If $r<1$, z_n spirals into the origin and thus $z=0+i0$ is a point attractor; in the same way, if $r>1$, iteration of the map diverges to infinity and the "point at infinity" is also an attractor. The unit circle is neutral in this sense and forms the boundary between the regions controlled by the two distinct attractors. The appropriate terminology is to call the interior of the unit circle the "basin" of the attractor at (0,0), and the exterior the basin of attraction for the point at infinity. The boundary, is then quite naturally called simply the *basin boundary,* or alternatively is referred to as the *Julia Set* of the quadratic point map. The situation is illustrated in Figure 33. Let it be said at once, that it might seem a bit pretentious to introduce all of this terminology in such a simple situation, but be patient for a few moments.

The rationale behind the introduction of so much verbiage will become clear by taking what appears to be an innocuous generalization of the simple z^2 map. Let us consider the general class of maps of the form:

$$z_{n+1} = z_n^2 + c$$

where z is, again a complex number, and c a complex constant. What are the attractors, basins, and Julia Sets? For the specific choice of

$$c = 0.31 + i0.041$$

empirically there are still two simple points of attraction, one near the origin, one at infinity, but the geometry of the attraction basins if

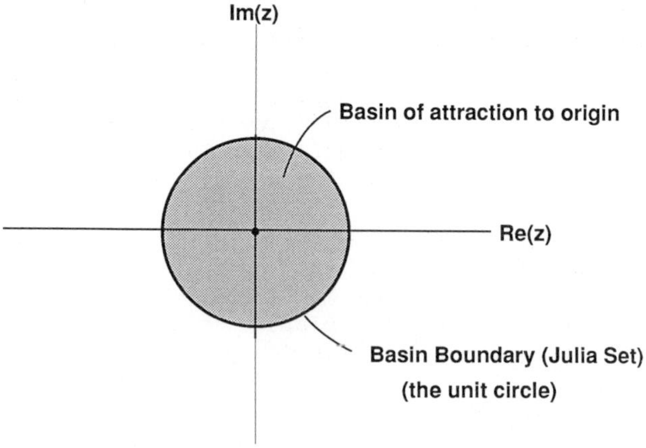

Figure 33. Attractor, and attraction basin(s) for the simple quadratic z^2 map. Any point started strictly within the unit circle (in the cross hatched area) will spiral in towards the origin, $z = 0$ in the complex plane, which is thus an attractor; its attraction basin is thus the interior of the unit disk. The exterior of the unit disk is the attraction basin of the point at infinity. The unit circle is the boundary between attraction basins and is referred to as the Julia Set, or,more simply, as the basin boundary.

another matter. Shown in Figures 34A,B,C are a view of the attraction basin (and its edge the Julia Set) for this map, and two higher magnifications. What was done to construct these figures was simply to iterate the map with initial conditions taken to correspond to each pixel on a computer screen and to color "black" those initial conditions (pixels) which did *not* lead to divergence of z_n in the iterative process. The black area is thus the complement of the attraction basin for the point at infinity, and thus consists of the Julia Set and its interior.

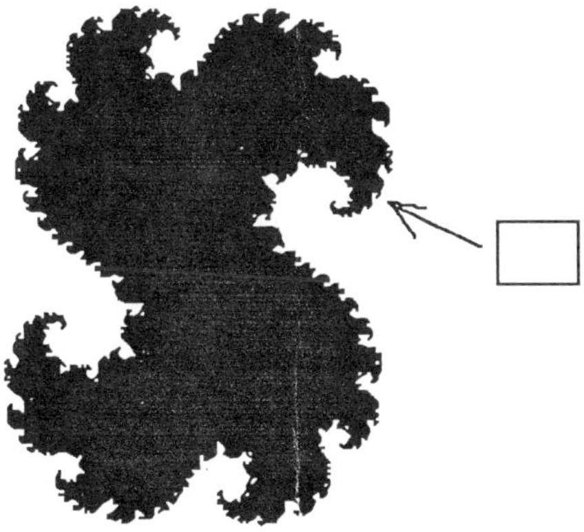

Figure 34A. The basin of attraction for the $z^2 + c$ map discussed in the text along with its boundary, the Julia Set. The Julia Set is evidently a very "rough" object, which is confirmed in the following two figures,B and C. In B, an area the size of the indicated "box" is shown, enclosing the tip of the spiral arm indicated by the arrow.

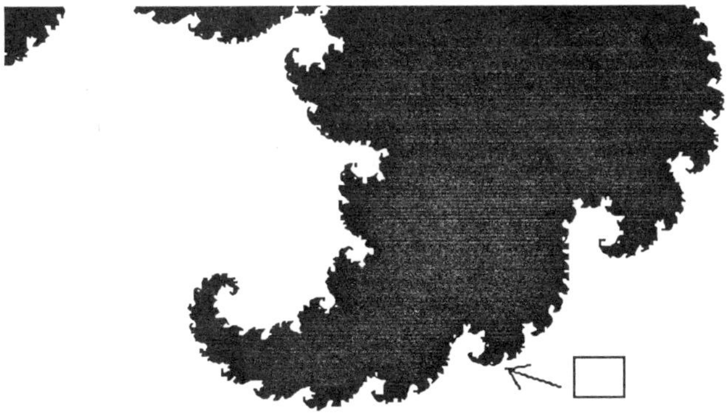

Figure 34B, a magnification of a part of 35A, showing additional and selfsimilar structure. 35C shows a further magnification, see box

Figure 34C. A further magnification of the bounded region of attraction of the map $z^2 + c$, showing a closer view of a region the size of the box in the previous Figure, in the location shown. Another level of self-similar structure is revealed. The Julia Set (basin boundary) is evidently a fractal curve.

It is now evident that the basins of attraction corresponding to simple point attractors are not necessarily bounded by simple curves. In fact, the Julia sets of most such maps are fractals, with the fractal dimensionality depending very sensitively on the chosen value of the complex constant *c*. However, not only are attraction basins capable of great complexity, the attractors themselves need not be points!

Strange Attractors

Very simple systems show attractors of startling complexity: for example the non area preserving Henon [as opposed to Henon-Siegel] map:

$$x_{n+1} = y_n + 1 - ax_n^2$$

$$y_{n+1} = bx_n$$

has an attractor which has a Cantor set type of structure. This is illustrated in Figure 35, for the values a = 1.4, b = 0.3. In this case the attractor is a curve in the x,y plane, but it is a curve of infinite length, and one which displays folds on all length scales. A line which cuts across the attractor will intersect it a an infinite number of points, with gaps of all different lengths scales, namely a Cantor set. This surprising and complex nature of the attractor in the Henon map is summarized by the name "strange attractor" for this type of invariant (or time independent) object.

As Hamiltonian systems do not "relax", they seemingly do not possess attractors of any type much less strange attractors. This has, in fact made the numerical study of Hamiltonian systems more difficult than that of systems which can more easily make use of their own intrinsic properties to display themselves! Nevertheless we will see in Section 4.2 that analogous structures do exist, and that a combination of the techniques introduced in this brief survey of non-area preserving maps will be of use in exposing them.

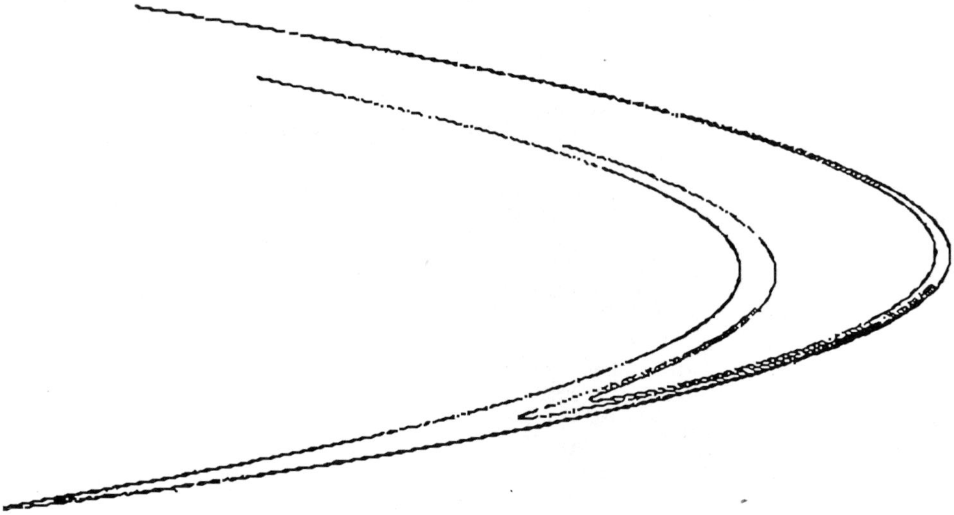

Figure 35A. The Henon attractor in the x,y plane. All initial conditions in the vicinity of this attractor lead to "dissipative" motion ending up on this complex object, with an infinite levels of folded structure. A magnification is indicated in part B.

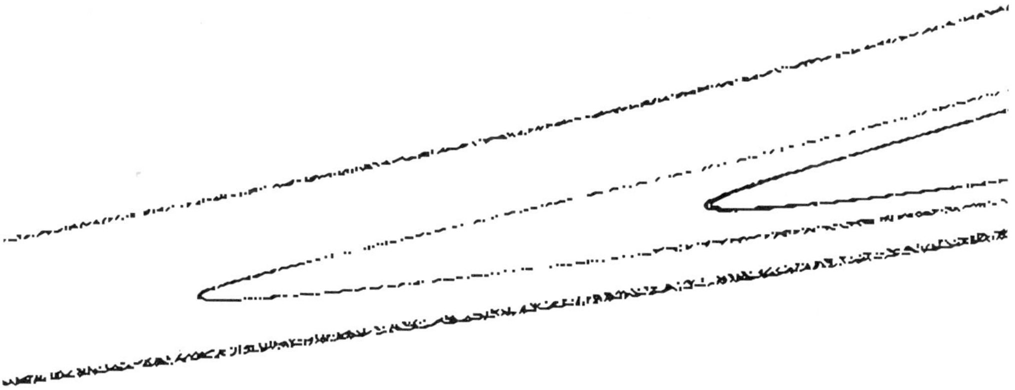

Figure 35B. Magnification of a portion of the Henon attractor, showing that what appear to be curves in part A, are actually bands, with a higher level of unresolved fold structure.

4.2 Julia Sets in Hamiltonian Dissociation

In Hamiltonian systems, or in their counterpart the area preserving maps, attractors do not exist in the sense of the point attractors of the $z^2 + c$ giving attraction basins which exhaust (i.e. *tile*) the plane, with Julia sets as boundaries between the various basins. Nor do such systems display complex attractors, like that of the Henon map, of Figure 35, simply because these systems do not dissipate energy, or its equivalent. Namely they never "settle down" and thus never display their time independent structure in an an automatic and simple manner. However, various tricks can be used to get area preserving maps to display structures which are quite similar to those discussed in Section 4.1, and in particular the Cantor set nature of the underlying phase space.

Working with the Henon-Siegel map, we have seen [see Figures 13,14 and 32], for the value of the parameter α used, that a central region of invariant curves is surrounded by a five fold island chain, which in turn is surrounded by a region of chaos, which eventually leaks out, and moves off to infinity. Initial conditions giving rise to motion on an invariant curve, shown us the curve, but nothing else. However, for the parts of chaotic phase space which eventually dissociate, we can use the "point at infinity" as if it were an attractor in a dissipative map, and ask several questions:

1) What is the geometry of the phase space which eventually dissociates? i.e. what is the Julia Set of the point at infinity?

2) As dissociation takes place where does the flux come from? That is, are points in the phase space which are fated to "dissociate" in a given number of iterations of the map (e.g. vibrational periods) randomly distributed in the chaotic phase space, or is there structure? If structures exist, what are they?

In this Subsection we examine these question empirically; in 4.3 we ask the same sort of question for reactive collisions, and then in 4.4 draw the observations together and relate them to phase space invariants.

In the following series of Figures the Henon-Siegel map is used as a model for a dissociating two degree of freedom system, with a bound degree of freedom sharing energy with a dissociative one. We examine a "collision complex" defined in the phase plane of the dissociative co-ordinate by taking that volume of phase space which 1) does not dissociate in fewer than 5 iterations of the map (5 vibrations of the bound degree of freedom), and; 2) such that for the initial p and q, $(p^2 + q^2) < 2.25$. This latter condition simply implies that when we ask, "what trajectories appear in the exit channel with positive momentum (and thus separate, never to recombine) after a fixed number of vibrational periods?", we do not wish to include those which began far out in the entrance channel, and just happened to show up in the exit channel at the prescribed time, never having really been part of the collision complex. With these conventions, and a choice of $\cos(\alpha) = 0.22205$ (the same as that for Figure 32) we examine the phase space structure of the dissociation process.

Figure 36A. Initial points taken to form the *collision complex* for the Henon-Siegel map. The complex is defined by taking points which are 1) in the neighborhood of the (0,0) fixed point, but dissociate; 2) have a lifetime of at least 5 vibrational periods.

Figure 36 B. Initial conditions which do not dissociate in 10 periods, that is, 5 periods later than in 36A.

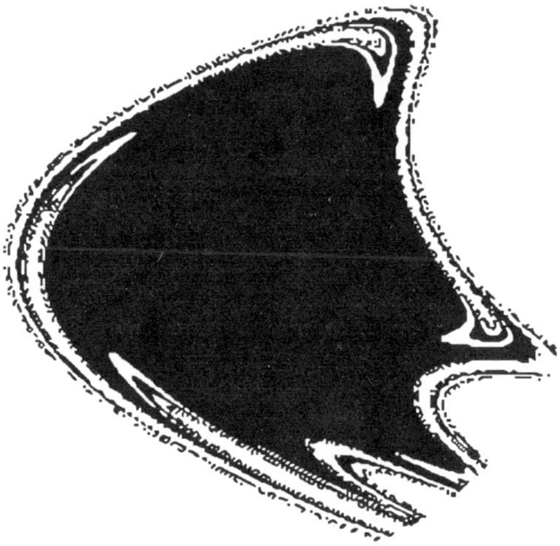

36C
20 periods

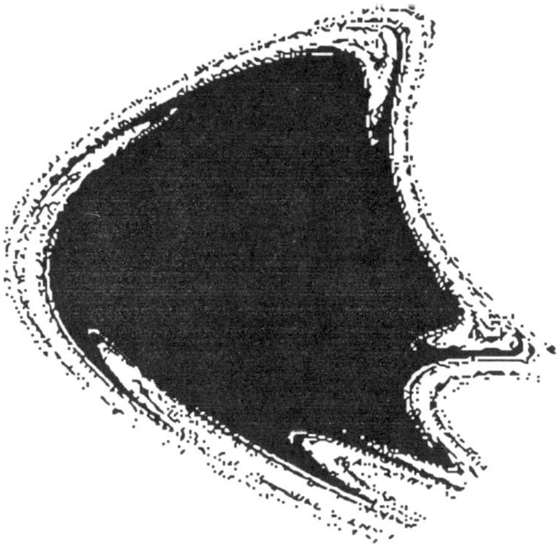

36D
30 periods

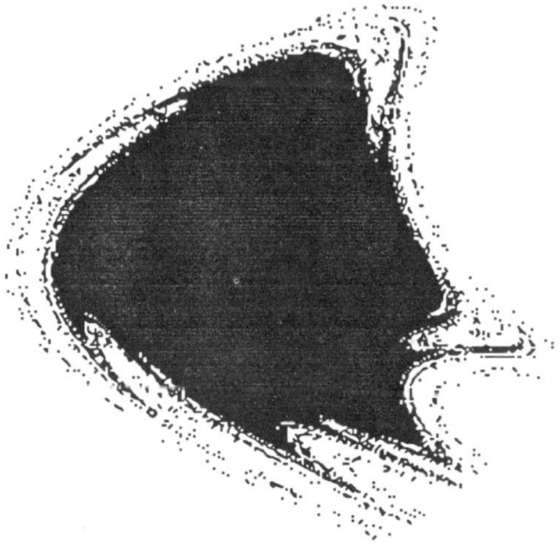

36E
50 periods

Figure 36 A-E. What parts of dissociative phase space dissociate in what times? In A, the black area defines those *initial conditions* in the (p,q) phase plane for the Henon-Siegel map which we will take to be the *collision complex.* These points, which define a contiguous region in the phase space, with a simple rectifiable boundary, are

[continued]

then iterated forward in time, and records kept of what time is taken (measured in iterations of the map, which correspond to vibrational periods of the unseen degree of freedom) for dissociation. In B through E, then are shown those *initial conditions* which are 1) in the collision complex, and 2) which will not dissociate in a time less than the number of periods indicated. What is seen that for this range of times, that while the geometry of the phase spaced is perhaps complicated, it is by no means random.

In each case what is plotted are the regions of initial conditions (within the complex as defined above) which do *not* dissociate within the given number of vibrational periods. Dissociation was defined to be $(p_n^2 + q_n^2) > 150$, with n< (the number of periods of interest). In Figure 36 a global picture is given. It is important to note, again, that the indicated regions of phase space are initial conditions: none have been mapped forward. A brief glance at the series 36 B through E shows a highly regular pattern of dissociation, that is complicated, but not random. A random dissociation would result in a uniform and unstructured depleting in these initial condition plots. Nevertheless, a Monte Carlo sampling of the dissociation process shows that the above sequence displays exponential decay (with a time scale of ca 30 periods). How can this apparently regular process give rise to exponential decay? Additionally, what features of the dynamics give rise to the observed banded structure of the phase space?

In Figure 32, a very long lived dissociating trajectory is shown, and interpreted in terms of slow leakage through a Cantorus. How does this long time dynamics appear in the initial condition plots of the type shown in Figure 36? The sequence of Figure 37A through D explores this, showing those initial conditions which are undissoicated after 100, 500, 1000 and 5000 vibrational periods. A factor of 10 higher resolution is employed.

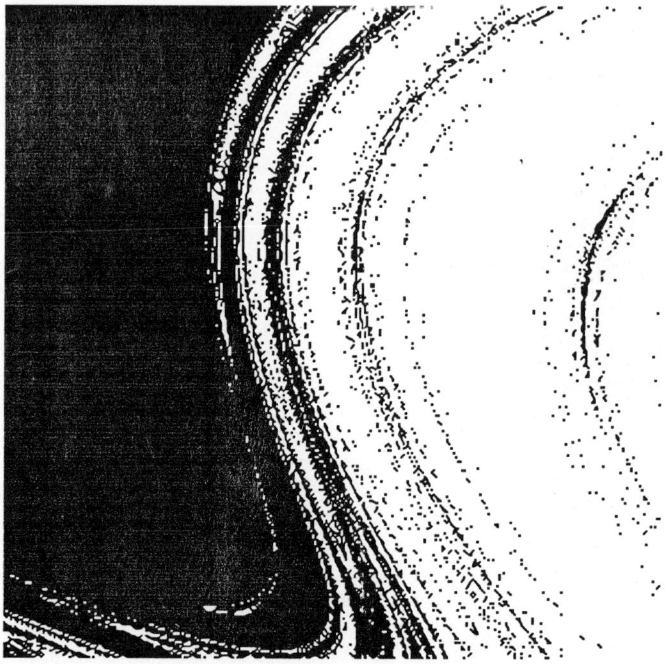

Figure 37A. 100 periods

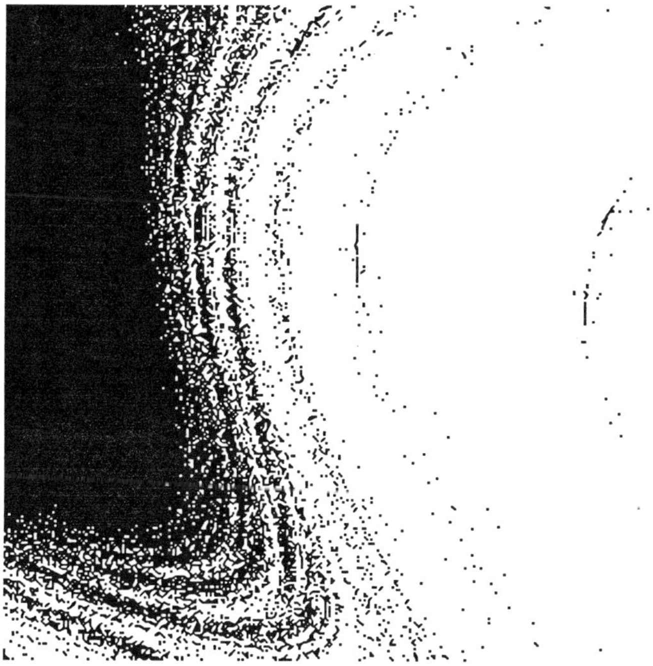

Figure 37B. 500 iterations

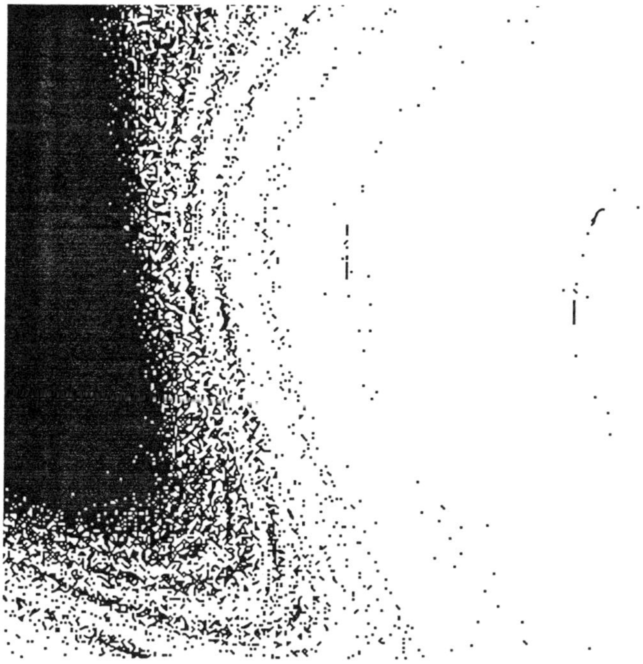

Figure 37C. 1000 iterations

Figure 37D. 5000 iterations. This sequence gives a series of magnifications of the lower right corner of the Henon-Siegel collision complex of Figure 36, at a resolution 10 times that of these earlier figures. What is shown are the initial conditions which do not dissociate in 100, 500, 1000, and 5000 vibrational periods. Comparison of these results with those of Figure 36 shows a second level of decay times with a much longer time scale. Additionally, at these longer times (two orders of magnitude longer than the initial decay of 36A-E) dissociation does seem to take place in a much more random manner: although some structure is apparent, it is less well defined than before and the phase space is at least approximately uniformly depleted.

The long time dissociation dynamics of the type of trapped trajectory of Figure 32 is, indeed, also evident in Figure 36 dissociation continues. This longer time scale dissociation leads to a phase space which is much more uniformly depleted than that resulting from the initial short time dissociation. This more or less random (at least on the resolution scale shown here) depletion of phase space at long times is consistent with the conceptualization of the slow dissociation resulting from penetration of the Cantorus as a Markov process. This if further discussed on Section 4.4.

Further, we note that the Julia Set, related to the point at infinity acting as an attractor, possess fully as complex a Cantor set structure as those of the $z^2 + c$ maps, and as the strange attractors of maps like that of Henon.

4.3 Reactivity Bands in Reactive Dynamics

It has been recognized for at least twenty five years that classical reactive scattering can display surprisingly complex dynamics a function of small continuous variation of a classical initial condition. For example, collinear scattering of A + B-C(v) to give A + B-C(v') *or* A-B(v") + C, variation of the *phase* of the BC oscillator, holding the other variables constant can give a highly structured series of "bands" of alternating reactive and non-reactive behavior.

Such reactivity bands are often associated with the formation of a collision complex, with the chaotic internal dynamics within the long lived complex greatly amplifying dependences on initial conditions. This immediately suggests the possibility of a Cantor Set type classical structure of reactivity as a function of initial conditions. Further there is also the possibility of self-similar structures, giving rise to a renormalization approach to the whole dynamics.

A particularly careful analysis of an example of this latter type has been given by Jaffe and Tiyapan, as shown in Figure 38. These authors considered the scattering of He from I_2 in a region of complex formation. Figures 38A and B show the final I_2 action as a function of initial vibrational phase angle, θ. The sequence of structures near the 1:5; 1:6;... 1:n resonances are shown in 38A, intertwined with chaotic "chattering" regions. Figure 38B shows evidence of an almost exactly self similar set of structures corresponding to the 2:56;2:57;2:58... resonances sandwiched between the 1:50 and 1:50 resonances of 38A. This structure is on a scale five orders of magnitude finer than that of 38A, and once the nature of the self similar structure is identified, scaling laws allow a renormalization analysis of both the classical and semi-classical dynamics.

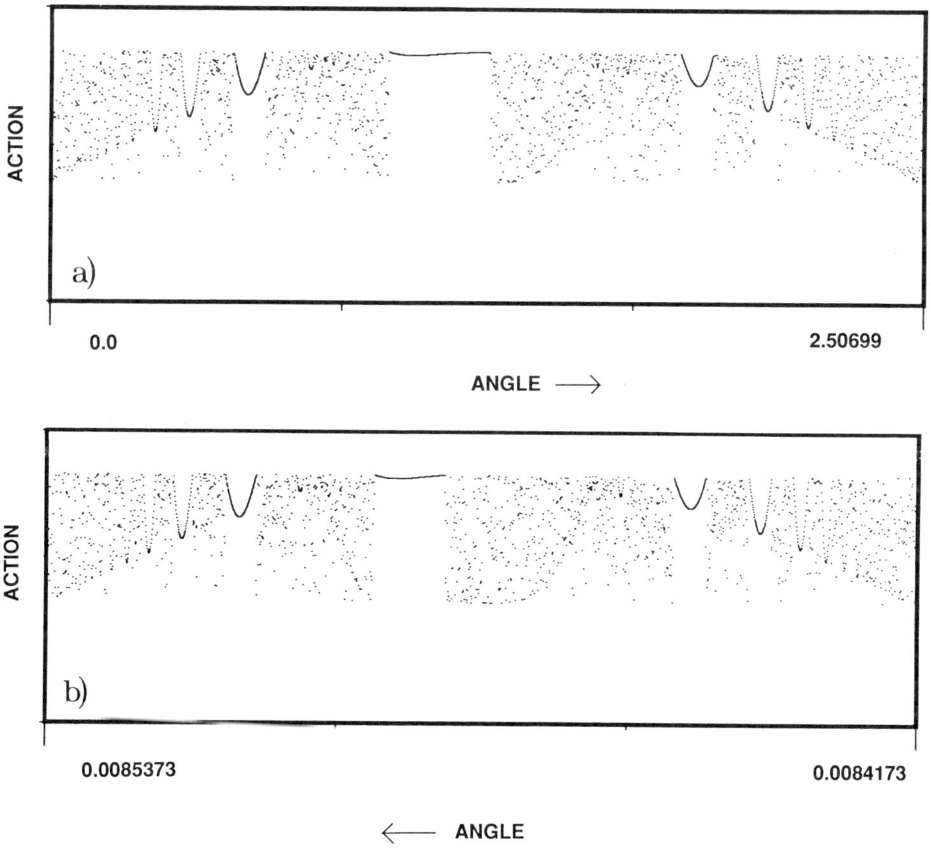

Figure 38A,B. Self similar structure in He-I_2 scattering in a "chattering" region. Final action is plotted as a function of initial angle. B is a five order of magnitude magnification of a small part of A. From the work of Jaffe and Tiyapan. Angle values shown indicate the full range of variation of the angle in each case.

4.4 Exponential Decay: The Rate of Unimolecular Decomposition as Controlled by Cantori & by Homoclinic Oscillations

In this final Section we make a qualitative distinction between exponential decay as controlled by slow leakage through a cantorus, as opposed to empirically observed exponential decay associated with the unstable manifold and its characteristic homo-clinic oscillations. The nature of dissociation of a bounded oscillator coupled to a dissociative mode as modeled by the Henon-Siegel map is used to illustrate this distinction, and to indicate the direction of future theory.

The phase plane of a dissociative oscillator

Consider a one dimensional non-linear oscillator of the form shown in Figure 39A and its associated phase portrait shown in 39B.

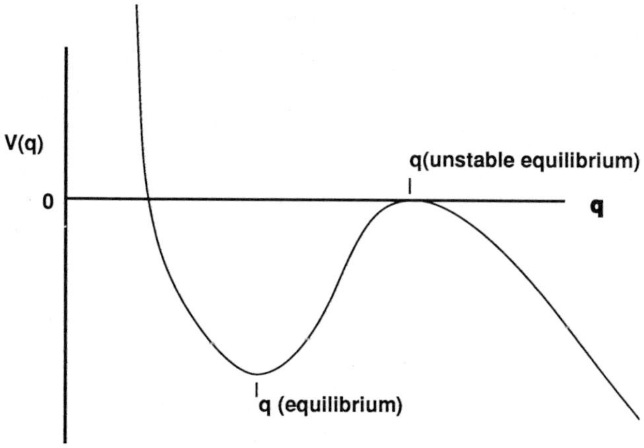

Figure 39A. The potential for a "cubic" oscillator which has one stable, and one unstable fixed point, the latter at the barrier top.

What now happens if a second (assumed bounded) degree of freedom is coupled to this dissociative mode, and we look at the surface of section in the phase plane of the dissociative mode. In particular we ask as to the fate of the invariant curves in the phase plane, especially the separatrix. In the one degree of freedom picture of the dissociation a single trajectory can follow either the inner, bounded, part of the separatrix, or can follow either of the two unbounded branches. We now ask, what happens in the coupled system if we start at the unstable fixed point and move out along the branch of the *unstable manifold* which is the analog of the separatrix.

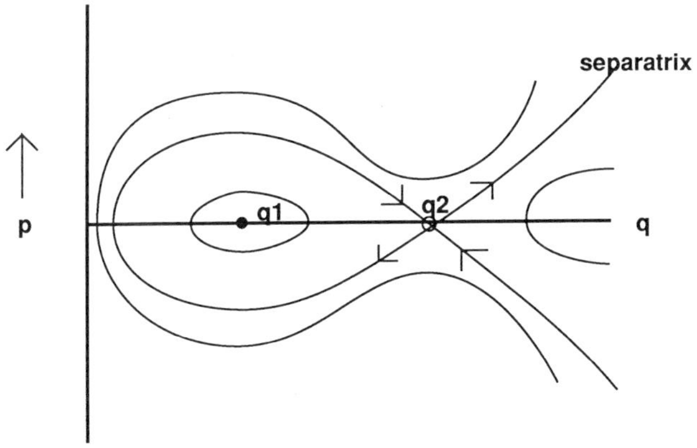

Figure 39B. The phase portrait for the one dimensional oscillator of Figure 39A. q1 is an elliptic fixed point, about which there is stable small amplitude bounded motion. q2 is a hyperbolic (unstable) fixed point, corresponding to a particle at the top of the barrier of Figure 39A, with no kinetic energy. Such a particle is unstable with respect to small perturbations. Motion at energies above and below the barrier top is divided by the indicated separatrix. The arrows leading in and out of the unstable fixed point along the separatrix indicate the direction of motion near that hyperbolic fixed point. The incoming branches represent the contracting of phase space in the direction of the fixed point, while the outgoing branches allow the exponential separation of phase points associated with the onset of chaos in a system with a higher dimensionality.

the answer is well known in non-linear dynamics, Poincaré described the phenomenon in his *New Methods in Celestial Mechanics* by noting that "nothing is more suitable for providing us with an idea of the complex nature of the three body problem, and of all the problems of dynamics in general." The onset of the complexity referred to is illustrated in the sketch of Figure 40. Namely, homoclinic oscillations set in. Comparison of the oscillatons of Figure 40 with the evolution of dissociating phase space in Figures 36B through E shows a high degree of similarity. In fact the whole dissociation of the "outer" part of the phase space associated with the collision complex can be described in terms of the geometry of the oscillating unstable manifold of the Henon-Siegel map. This dissociation is exponential, but the process underlying it is non-Markovian, and in fact no *mixing* of the phase space occurs. This is to be contrasted with the behavior seen in Figure 37, where a much more uniform an seeming more random depletion of phase space has occurred. This *is* in fact a Markov process: the time scale for leakage through the Cantorus is very slow compared to the time for chaotic mixing in the interior, and dissociation is quite properly described by a first order rate process.

These last cursory remarks are ment to stimulate ones thinking, rather than to present a complete theory, although one can be based on them. Let us end by simply noting , that exponential decay of a classical ensemble does not necessarily imply that the dynamics is Markovian: phase space is more complex than that.

5. CONCLUSIONS

We conclude very briefly, and then indicate avenues for further research.

Examination of coupled non-linear systems of two degrees of freedom, and the associated area preserving maps, yields a wealth of new dynamical structures which have no counterpart in the dynamics of one degree of freedom systems: chaotic motion is but a symptom of this complexity. The

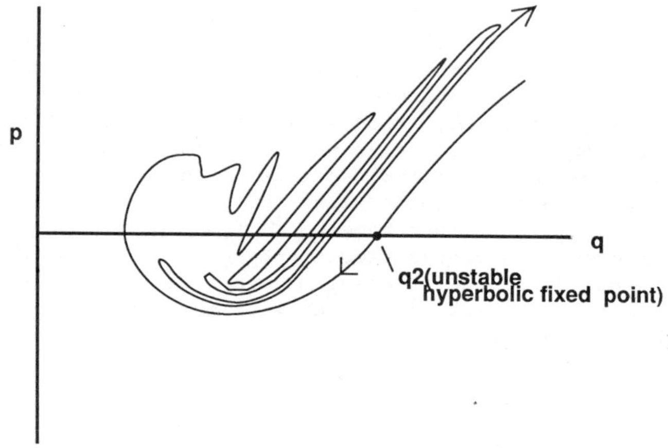

Figure 40. A rough sketch of the manifold obtained by following the outgoing branch of the unstable fixed point. In a one degree of freedom system this curve would rejoin the fixed point, and the trajectory which fell off the fixed point would return in infinite time, see Figures 39A and B. Here then, illustrating the situation for a generic unstable manifold in a non-integrable system, begins an instability, which is then amplified as the manifold comes back to the hyperbolic point, oscillating an infinite number of times on the way. These oscillations, as they involve the same fixed point which spawned the manifold are called *homoclinic,* if a different hyperbolic fixed point were approached the oscillations would be *heteroclinic.* As the oscillations are associated with an area preserving system, as they get geometrically closer, they grow in the transverse direction to preserve area, as this lengthening continues the curves must wrap around one another in a very complex manner, indeed.

underlying dynamics of a chaotic system involves understanding the new types of invariant sets: the Cantori are the remnants of invariant tori, and can form bottlenecks to diffusion; unstable fixed points have a remnant of separatrix structure, the unstable manifolds, which shows a complex oscillatory structure (the homo- and hetero- clinic oscillations) which also control macroscopic rates. All of this leads up to the summary statement that *chaos is not a monolith, simply characterized by a single length or time scale.*

These observation should have a profound affect on our thinking about the foundations of statistical theories of rate processes or collisions leading to complexes, provided that the necessary generalizations can be made. These include the relation of the classical structures to quantum effects, and the relationship of the two degree of freedom results discussed here to what happens in systems of higher dimensionality. Both of these are highly active areas, and some new work in these directions is referenced in the following section, but both are only beginning to shed light on these questions. It will be interesting to return to these issues a decade hence at the third NATO ASI on Atomic and Molecular Collisions!

6. THE LITERATURE

As these notes are from an ASI Lecture Series, I have written them as Lecture Notes and not as a Review Article: although a few names are mentioned here and there in the text, the emphasis was on the pedagogical devolpoment rather than providing a detailed guide to the literature. Such a guide, in abbreviated form is now provided. What is given here is an introduction to the literature, and is not to be thought of as comprehensive. It will suffice, however, to start the interested reader in the direction of the primary and secondary literature.

A first, and general comment is that the Journal of Choice in this area has been *Physica D* for developments in non-linear dynamics. A new section of the Journal of Physics (from the British Physical Society) will focus on non-linear dynamics. When applications to specific physical systems occur these often appear in the Journal of Chemical Physics, Physical Review A, Physical Review Letters, and the Journal of Physics A, and JETP.

The Introduction, and General References

An interesting overview of new trends in mathematics and their relation to the physical sciences has been given by Steen[1], this might well be read in conjunction with Wigner's famous article[2], "The Unreasonable Effectiveness of Mathematics in the Physical Sciences". An introduction to "fractals" with much of the relevant history (including discussions of Weierstrauss, Cantor, Housdorff etc.) is given by Mandelbrot in his two books on fractals[3,4], and further discussed and developed by Peitgen and Richter[5]. A "New York Times Best Seller" introduction, suitable for non-scientists is that by Gleick[6]. Excellent texts on classical and non-linear dynamics have appeared, those by Arnold[7],Abraham and Marsden[8], Lichtenberg and Lieberman[9], and Gukenheimer and Holmes[10] have been useful to the author, as have the more specialized monographs by Cvitanovic[11] (on the Feigenbaum theory, and its possible universality), and Horton, Reichl, and Szebehely[12], which focuses on correlations in the long time chaotic dynamics of Hamiltonian systems. Review articles have appeared in large numbers by now (although few address questions involving collisions) these include those of Helleman[13], Berry[14], Ford[15], on non-linear Hamiltonian systems, and a Special Issue of the Proceedings of the Royal Society[16]. Percival[17], and Reinhardt[18], have reviewed specific issues related to semi-classical quantization on tori, the latter of these emphasizing quantization on ordered structures in chaotic regions. Noid, Koszykowsky and Marcus[19] have discussed semi-classical quantization and molecular dynamics from a modern dynamical perspecitve, focusing on microscopic dynamics and it relation to obeservation; Brumer and Shapiro[20] have taken a more abstract approach. An overview of the role of fractal structures in decision theory has been given by the Maryland Gang[21]. Smale's fundamental articles on the structure of the unstable manifold have been collected[22].

Regular and Irregular Motion

(2.1) The title is taken from M. V. Berry's influential review[14]; the original paper of Henon and Heiles[23] , discussed motion of elliptical galaxies, in the tradition of Contopoulos[24]. A traditional introduction to angle action variables is that if Goldstein[25], which is a valuable historical document, as it offers no geometric interpretations of these variables, and certainly does not disucss non-integrability. Arnold[7] is the more modern treatment.

(2.2) The fact that strongly coupled non-linear systems were not ergodic came as quite a surprise: A classic investigation is the Fermi-Pasta-Ulam[26] paper of 1954, some of Ulam's comments on this work, which Fermi took very seriously, are in his biography[27]. The coupled oscillator systems of interest to Fermi-Pasta-Ulam were of the generic type: mixed chaotic and regular dynamics. The work of KAM...Kolmogorov[28],Arnold[29], and Moser[30]...established in the following decade that this was the generic situation. Fully chaotic and fully integrable systems have been the subject of much interest: the Painleve conjecture has been used to find such unusual systems[31,32]. Fully chaotic systems (for example Refs 32-36) have been of great interst in problems of semi-classical quantiztion: Gutzwiller refers to such systems as displaying *hard chaos* and discusses their quantization[37].

(2.3) Statistical theories of chemical reactions are discussed by Hase[38], Pechukas[39]; a recent review of transition state theory by Truhlar, Hase and Hynes[40] brings these ideas up to date.

(2.4) A important molecular analog of the Fermi-Pasta-Ulam result is that of Wolfe and Hase, who found large volumes of trapped phase space above a classical dissociation limit[41], quantum and semi-classical studies of related systems by Hedges and Reinhardt[42-44] led to the conclusion that stongly non-statistical behavior would be observed quantum mechanically. Quite independently an analog of this was observed by Kennedy and Carrington[45], in their observations of highy excited but long lived states of H_3^+ well above the dissociation limit. A recent theoretical analysis, based on concepts from modern non-linear dyanamics has been given by Taylor, Pollak and co-workers [46]. Examples of chaotic Hamiltonian dynamics with substantial short term order (sometimes refered to as motion on "vague tori") are given by Jaffe and Reinhardt[47], Shirts and Reinhardt[48], Reinhardt[49], and DeLeon and Berne[50] in the context of molecular dynamics and chemcial reaction dynamics. Other examples had appeared earlier in the astronomy literature[51]. See also Ref (12). Discussions of long time correlations and short time trapping of energy in a bond mode in OCS is from the early work of Wagner and Davis[52].Trapping in chemically activated decenes has been discussed by Rabinovitch, Trenwith, Oswald and Flowers[53] in terms of a "sequential relaxation" mechanism introduced by Hutchinson, Hynes and Reinhardt[54], and is further discusssed by Duneczky and Reinhardt[55].

Development of Theory

(3.1) The maps used are the *Cremona maps* of Henon-Siegal[56,57], and the Chirikov[58], or *standard* map.

(3.2) The Poincare-Birkhoff theorem is discussed in Ref (9), pp 168 et.seq.

(3.3) See Refs (28-30) for the KAM papers, and Refs (9 & 14) for more qualitative discussions. The Chirikov "resonance overlap" criterion for the onset of chaos is discussed in these latter refereneces, and in Chirikov's own review[58]. The Fermi-Pasta-Ulam work is in Refs(26-27).

(3.4) A delightful and very readable introduction to the theory of rational approximation has been given by Rademacher[59]; a more traditional introduction is that by Hardy and Wright[60]. John Greene's seminal paper[61] on locating the "last" KAM (or rotational, in the language use here) torus, making full use of the number theoretic concepts outlined here, appeared in 1979.

(3.5) The key references in this area are I. C. Percival[62], Aubry[63], and Mather[64].

(3.6) Use of the ideas of Greene (ref 61), and the new concept of the cantorus(refs 62-64), to give a framework of determining the flux for diffusion through a phase space bottleneck is due to MacKay, Meiss and Percival[65] and Bensimon and Kadanoff[66]. The original ideas for variational determination of flux are implicit in the work of Wigner[67] and Keck[68] on transition state theory, and are generalized by MacKay and Meiss[69]. Earlier work in this area of phase space transition state theory of DeVogelaere[70] and Pollack and Pechukas[71] contains germs of the ideas of refs 65-66. The applications to problems of chemical dynamics originated with the work of Davis[72] on OCS and intramolecular energy trapping; and was then extended to van der Waals complex by Gray and Davis[73], and to collinear reaction dynamics by Davis[74] and Skodje and Davis[75].

The Fractal Structure of Hamiltonian Phase Space

(4.1) Attractors and Julia Sets (Basin Boundaries) are discussed in Refs 3,4,5 and 21. The original "strange attractor" is that of Henon[76].

(4.2) The idea behind figures 36 and 37 was developed by the author[77] at the Telluride Summer Research Center during a tutorial for high school students in the Summer of 1987, subsequently, and independently, others have discovered the same idea as applied to conservative systems: Vasquez, Jefferys and Sivaramakrishnan[78a] had actually looked at the same map(!); Bleher, Grebogi, Ott and Brown[78b], have considered the analog of fractal basin boundaries for a modified version of the Sinai Billiard, which has two "dissociation" channels. Figure 37 suggests that the "edge" of the interior stable region is a bit fuzzy, for other values of the parameter α, this edge is quite sharp and well defined. Such issues are discussed by Greene, MacKay and Stark[79].

(4.3) Observations of "bands" of initial conditions leading to chattering or resonance regions in A + BC collisions have been note by Stein and Marcus[80], Rankin and Miller[81],and Gottdeiner[82];and, a somewhat more thorough analysis given by Noid, Gray and Rice[83]. The first complete analysis is that of Jaffe and Tiyapan[84], who have completely elucidated the self-similar structure and found the appropriate renormaliztion laws, allowg predictive construction of structures such as that shown in Figure 38, reproduced here with permission of these authors.

(4.4) A detailed analysis of the control of exponential decay by the homoclinic oscillations associated with the unstable fixed point, as distinguished from the Markov process of decay through the cantorus, for the Henon-Siegel map has been worked out by the author[85], but, as it was not presented at the NATO meeting, it is omitted here. Smale[22] and more recently Easton[86] have discussed the geometry of the homoclinic "tangle". The quote from Poincare is taken from Easton's paper.

The Future

The most serious omission from these expository lectures is any indication of what happens in more than two dimensions. A new possibility for diffusion in phase space opens up in these higher dimesnions, as the tori are no longer of the appropriate dimension to trap volumes of the energy conserving phase space. This new process is *Arnold diffusion*, which is discussed in Ref 9. Let it be said that at once that the situation in three dimensions is not fully understood. Interesting early results are the identification of the generalization of the Golden mean to the problem of approximation of *three* mutually irrational numbers, see the pedagogical paper of Kim and Ostlund[87]. Martens, Davis and Ezra[88] have carried out a study of the OCS problem in the full 3-dimensional phase space [rather than in a restricted 2-dimensional model as in Ref 72], and have found that transport is quite complex, but that the Spiral mean of Kim and Ostlund does seem to play some role in trapping. Gillilan[89] and Gillilan and Reinhardt[90] have found the multidimensional analogs of the turnstiles of MacKay, Meiss and Percival (ref 65), allowing an understanding of the phase space dynamics of diffusion and trapping of an atom on a three dimensional surface.

ACKNOWLEDGEMENTS

This work has grown out of a number of important interactions with co-workers. In particular discussions over a number of years with I. C. Percival, C. Jaffe and M. Davis have contributed to the author's interest in and understanding of this area. Interactions with many workers at the Telluride Summer Research Center has been highly stimulating. I also wish to acknowledge the enthusiastic support of NATO and the organizers of the present meeting at Cortona, and of the U.S.National Science Foundation.

REFERENCES

1) L. A. Steen, Science, **240**,611(1988)
2) E. P. Wigner, Commun. Pure and Appl. Math. **13**,1(1960). This is reprinted in *Mathematical analysis of physical systems*, R. E. Mickens, Ed. Van Nostrand, New York, 1985, p1.
3) B.B. Mandelbrot, *Fractals: Form, Chance, and Dimension*, W. H. Freeman & Co. San Francisco 1977.
4) B.B. Mandelbrot, *The Fractal Geometry of Nature*, W. H. Freemand & Co. San Francisco, 1983.
5) H.-O. Peitgen and P. H. Richter, The Beauty of Fractals, Springer, New York, Berlin, 1986.
6) J. Gleick, *Chaos, Making a New Science*, Viking, New York, 1987.
7) V.I. Arnold, *Mathematical Methods of Classical Mechanics*, Springer, New York, Berlin, 1980.
8) R. Abraham and J. E. Marsden, *Foundations of Mechanics*, 2nd Edition, Benjamin/Cummings, Reading Mass, 1978.
9) A. J. Lichtenberg and M. A. Lieberman, *Regular and Stochastic Motion*, Springer, New York, Berlin, 1983.
10) J. Guckenheimer and P. Holmes, *Nonlinear Oscillations, Dynamical Systems, and Bifurcations of Vector Fields*, Springer, New York, Berlin, 1983.
11) P. Cvitanovic(Ed.),*Universality in Chaos*,Adam Hilger,Bristol (UK), 1984.
12) C. W. Horton, L. E. Reichl, and V. G. Szebehely (Eds.), *Long Time Prediction in Dynamics*, Wiley-Interscience, New York, 1983.
13) R. H. G. Helleman, in *Fundamental Problems in Statistical Mechanics V*, E. G. D. Cohen, Ed. North-Holland, Amsterdam, 1980, p165. This review contains a very substantial list of references.
14) M. V. Berry, in *Topics in Nonlinear Dynamics a Tribute to Sir Edward Bullard*, S. Jorna, Ed. American Institute of Physics (New York), AIP Conference Proceedings 46 p16, 1978.
15) J. Ford, Adv. Chem. Phys. **24**,155(1973); and in *Fundamental Problems in Statistical Mechanics III*, E. G. D. Cohen, Ed. North-Holland, Amsterdam, 1975, p215.
16) *Dynamical Chaos*, Proc. Roy. Soc.(London) Series A,**413**,3-199(1987).
17) I. C. Percival, Adv. Chem. Phys. **36**,1(1977).
18) W. P. Reinhardt, in Mickens (Ed.) *Mathematical Analysis of Physical Systems*, see Ref. 2, p 169.
19) D. W. Noid, M. L. Koszykowski and R. A. Marcus, Ann. Rev. Phys.Chem.**32**,267(1981).

20) P. Brumer and M. Shapiro, Adv. Chem. Phys. **70**,365(1988).
21) C. Grebogi, E. Ott and J. A. Yorke, Science **238**,632(1987).
22) S. Smale, *The Mathematics of Time, Essays on Dynamical Systems, Economic Processes and Related Topics*, Springer, New York, Berlin, 1980.
23) M. Henon and C. Heiles, Astron. J. **69**,73(1964).
24) G. Contopoulos, Astron J. **68**,1(1963).
25) H. Goldstein, *Classical Mechanics*, Addison-Wesley, Reading, Mass. 1951.
26) E. Fermi, J. Pasta and S. ULam, Los Alamos Technical Report (1954); reprinted in *Nonlinear Wave Motion,* A. C. Newall (Ed.) AMS Lectures in Appl. Math. **15**,143(1974).
27) S. Ulam, *Adventures of a Mathematician*, Scribners, New York, 1976, pp225 et. seq.
28) A. N. Kolmogorov, Address to the 1954 International Congress of Mathematicians, reprinted in English in Ref 8, p741.
29) V. I. Arnold, Russian Mathematical Surveys **18**, 9(1963).
30) J. Moser, Memoirs of the American Manthematical Society, **81**,1(1968).
31) A. Ramnani, B. Dorizzi and B. Grammaticos, Phys. Rev. Letts. **49**,1539(1982).
32) J. Weiss, M. Tabor and G. Carnevale, J. Math. Phys. **24**,522(1983).
33) A. Carnegie and I. C. Percival, J. Phys. A **17**,801(1984).
34) S. G. Matinyan, G. K. Savvidi and N. G. Ter-Arutyunyan-Savvidi, Sov. Phys. JETP **53**,421(1981).
35) S-J. Chang, Phys. Rev. D **29**,259(1984)
36) M. Gutzwiller, Physica D **5**,183(1982).
37) M. Gutzwiller, J. Phys. Chem.**92**,3154(1988).
38) W.L. Hase in *Dynamics of Molecular Collisions*, W. H. Miller, Ed., Modern Theoretical Chemistry Vol **2**, Plenum, New York,1976, p121.
39) P. Pechukas, *ibid.* p269.
40) D. G. Truhlar, W. L. Hase and J. T. Hynes, J. Phys. Chem.**87**,2664(1983).
41) R. Wolfe and W. L. Hase, J. Chem. Phys.**72**,316(1980).
42) R. M. Hedges and W. P. Reinhardt, Chem. Phys. Letts.**91**,241(1982).
43) R. M. Hedges and W. P. Reinhardt, J. Chem. Phys.**78**,3964(1983).
44) R. M. Hedges, R. T. Skodje, F. Borondo, and W. P. Reinhardt in, *Resonances in Electron-Molecule Scattering, Van der Waals Complexes and Reactive Chemical Dynamics*, D. Truhlar, Ed. ACS Symposium **263**,323(1984).
45) A. Carrington and R. A. Kennedy, J. Chem. Phys.**81**,91(1984).
46) J. M. Gomez-Llorente, J. Zakrzewski,H.S. Taylor and K. Kulander, J. Chem. Phys.**89**,5959(1888); J. M. Gomez-Llorente and E. Pollak, J. Chem. Phys. **89**,1195(1988).
47) C. Jaffe and W. P. Reinhardt, J. Chem. Phys.**77**,5191(1982)
48) R. Shirts and W. P. Reinhardt, J. Chem. Phys.**77**,5204(1982).
49) W. P. Reinhardt, J. Phys. Chem.**86**,2158(1982).
50) N. DeLeon and B. Berne, J. Chem. Phys.**75**,3495(1981).
51) G. Contopoulos, Astron. J.**76**,147(1971).
52) M.J. Davis and A. F. Wagner in ACS Symposium **263**,337(1984), see ref 44)
53) A. B. Trenwith and B. S. Rabinovitch, J. Chem. Phys. **85**,1694(1986); A. B. Trenwith, D. A Oswald, B. S. Rabinovitch and M. C. Flowers, J. Phys. Chem. **91**,4398(1987).
54) J. Hutchinson, J. T. Hynes and W. P. Reinhardt, J. Chem. Phys. **79**,4247(1983).
55) C. Duneczky and W. P. Reinhardt, Trans. Faraday. Soc. II, **84**,1511(1988).
56) M. Henon, Quart. Appl. Math. **27**,291(1969).
57) C. L. Siegel, Ann. Math. **43**,607(1942).
58) B. Chirikov, Phys. Reports **52**,263(1979).
59) H. Rademacher, *Higher Mathematics from and Elementary Point of View,* Birkhauser, Boston, Basel, 1983, see especially Chapter 7.
60) G. H. Hardy and E. M. Wright, *An Introduction to the Theory of Numbers,* Oxford, 4th Edition 1971, see Chapters 10,11.

61) J. M. Greene, J. Math. Phys. **20**,1183(1979).
62) I. C. Percival, in Am. Inst. Phys. Conf. Proc. **57**,302(1979).
63) S. Aubry and P. Y. LeDaeron, Physica D **8**,381(1983).
64) J. N. Mather, Topology **21**,457(1982).
65) R. S. MacKay, J. D. Meiss and I. C. Percival, Physical D**13**,55(1984).
66) D. Bensimon and L. P. Kadanoff, *ibid.* D**13**,82(1984).
67) E. P. Wigner,J. Chem. Phys. **5**,720(1937).
68) J. Keck, Adv. Chem. PHys. **13**,85(1967).
69) R. S. MacKay and J. D. Meiss, J. Phys. A **19**,L225(1986).
70) R. De Vogelaere and M. Boudart, J. Chem. Phys. **23**,1236(1955).
71) E. Pollak and P. Pechukas, J. Chem. Phys. **69**,1218(1978).
72) M. J. Davis, J. Chem. Phys. **83**,1016(1985).
73) M. J. Davis and S. K. Gray, J. Chem. Phys. **84**,5389(1986)
74) M. J. Davis, J. Chem. Phys. **86**, 3978(1987).
75) R. T. Skodje and M. J. Davis, J. Chem. Phys. **88**,2429(1988).
76) M. Henon, Commun. Math. Phys. **50**,69(1976).
77) W. P. Reinhardt, unpublished work, presented at the NATO ASI, Summer 1987.
78) a) E. C. Vazquez, W. H. Jefferys and A. Sivaramakrishnan, Physica D **29**,84(1987); b) S. Bleher, C. Grebogi. E. Ott and R. Brown, Phys. Rev. A **39**,930(1988).
79) J. M. Greene, R. S. MacKay and J. Stark, Physica D **21**,267(1986).
80) J. R. Stine and R. A. Marcus, Chem. Phys. Letts **29**,575(1974).
81) C. C. Rankin and W. H. Miller, J. Chem. Phys. **55**,3150(1971).
82) L. Gottdiener, Mol. Phys. **29**, 1585(1975).
83) D. W. Noid, S. K. Gray and S. A. Rice, J. Chem. Phys. **84**,2649(1986).
84) A. Tiyapan and C. Jaffe private communication Summer 1988.
85) W. P. Reinhardt, unpublished work.
86) R. W. Easton, Trans. Am. Math. Soc. **294**,719(1986).
87) S. Kim and S. Ostlund, Phys. Rev. A **34**,3426(1986).
88) C. Martens, M. J. Davis and G. Ezra, Chem. Phys. Letts.**142**,519(1987).
89) R. Gillilan, Thesis, University of Pennsylvania, 1988, unpublished.
90) R. Gillilan and W. P. Reinhardt, Chem. Phys. Letts (submitted).

1. M. Venanzi
2. J.G. Gomez Llorente
3. K.E. Thylwe
4. N. Elander
5. J. Krause
6. O. Roncero
7. C.A. Weatherford
8. P. Brault
9. R.F.M. Lobo
10. B. Lepetit
11. N. Katzenellenbogen
12. A. Peet
13. D. Manolopoulos
14. G. Hose
15. R.A. Stansfield
16. R. Frost
17. S.R. Thareja
18. P. Francken
19. A.N. Brooks
20. M. Cebe
21. M. Monteiro
22. V. Sidis
23. I.S. Curuk
24. M. Gavrila
25. T.T. Scholz
26. F.A. Gianturco
27. O. Ylmaz
28. A. Van der Avoid
29. F. Di Giacomo
30. Filomena R.
31. W.H. Miller
32. R. Maurer
33. D. Grimbert
34. R. Schinke
35. G. Hähner
36. C.J. Gillan
37. N. Walet
38. M. Terao
39. N.C. Carlsund
40. M. Pont
41. F.R. McCourt
42. M.L. Dubernet
43. S.E. Nielsen
44. S. Oss
45. B.M. McLaughlin
46. F. Borondo
47. L. Benmevraïem
48. A. Metropoulos
49. S. Boumsellek
50. M. Thachuk
51. B. Bransden
52. A.J.F. Praxedes
53. A. Martin
54. P.G. Burke
55. P.C. Deshmukh

CONTRIBUTORS

BRIAN. H. BRANSDEN
Dept. of Theoretical Physics
The University, South Road
Durham DH1 3LE
GREAT BRITAIN

PHILIP G. BURKE
Dept. of Applied Mathematics
& Theoretical Physics,
The Queen's University
Belfast BT7 1NN – U.K.

M. GARVILA
FOM Inst. voor Atoom en
Molecuul. Phys.
Kruislaan 407
1098 SJ Amsterdam – HOLLAND

F.A. GIANTURCO
Dip. di Chimica
Città Universitaria
Piaz.le A. Moro, 5
00185 Roma – ITALY

C.J. JOACHAIN
Physique Theorique
Campus Plaine, U.L.B.
Boulvd du Triomphe
1050 Bruxelles – BELGIUM

J. KESSLER
Physikalisches Institut
Wilhelms Universität Münster
Domagkstrasse 75
4400 Münster – F.R.G.

W.H. MILLER
Chem. dept.
Univ. of California
Berkeley, CA 94720
U.S.A.

B. REINHARDT
Chemistry Dept.
Univ. of Pensylvania
Philadelphia, PA 19104
U.S.A.

R. SCHINKE
M.P.I. fur Stömungsforschung
Bunsenstrasse 6-10
D-34 Göttingen
F.R.G.

V. SIDIS
Lab. de Collisions Atomiques
du CNRS
Bat. 351
91405 Orsay Cedex - FRANCE

G. STEFANI
CNR, Ist. Metodologie
Avanzate Inorganiche
Area Ricerca CP10
00016 Monterotondo - ITALY

J. P. TOENNIES
M.P.I. fur Stömungsforschung
Bunsenstrasse 6-10
D-34 Göttingen
F.R.G.

M. TRONC
Lab. de Chimie Physique
11 Rue Pierre et Marie Curie
75231 Paris Cedex
FRANCE

F. Vecchiocattivi
Dip. di Chimica
Università di Perugia
06100 Perugia
ITALY

A. ZECCA
Dip. di Fisica
Università di Trento
38050 Povo (Trento)
ITALY

PARTICIPANTS

L. BENMEURAIEM
Observatoire de Bordeaux
B.P. 21
33270 Floirac
FRANCE

F. BORONDO
Depto de Quimica CXIV
Univ. Autonoma
Canto Blanco
24049 Madrid - SPAIN

S. BOUMSELLEK
LCAM
Univ. Paris Sud
Bat 351
91405 Orsay Cedex - FRANCE

A.N. BROOKS
Theor. Chem. Dept.
Univ. Chem. Lab
Lensfield Rd
Cambridge CB2 1EW - U.K.

N. CARLSUND
Research Institute of Physics
Frescativagen 24
S104 05 Stockholm
SWEDEN

M. CEBE
Kimya Bolumu Baskani
Fen Edebiyat Fak
Uludag Univ.
Bursa - TURKEY

I.S. CURUC
Physics Dept.
Middle East Techn. Univ.
Ankara
TURKEY

P.C. DESHMUKH
Dept. of Physics
Indian Institute of Technol.
Madras
INDIA 600036

J.B.R. DIAZ
Depto Quimica
ETSI Telecomunicacion
Univ. Politec. de Madrid
28040 SPAIN

F. DI GIACOMO
Istituto di Chimica
Facoltà di Ingegneria
V. del Castro Laurenziano, 7
Roma - ITALY

M.L. DUBERNET
Lab de Phys. et Opt. Corp.
P. et M. Curie Univ.
Tour 12, 5 Et.
4, Place Jussieu
75252 Paris Cedex 05 - FRANCE

N. ELANDER
Research Institute of Physics
Frescativagen 24
S-104 05 Stockholm
SWEDEN

A.J. FERROS PRAXEDES
Complexo I-IST
Centro de Fis. Molec.
Av Rovisco Pais
1000 Lisboa - PORTUGAL

A. FILABOZZI
Dip. Di Fisica
Città Universitaria
00185 Roma
ITALY

Ph. FRANCKEN
Phys. Theor.
Fac. des Sciences, CP 227
Univ Libre de Bruxelles
Bruxelles - BELGIUM

M. FRASER MONTEIRO
Physics Dept
Univ. de Lisboa
Av Rovisco Pais Co 1
1000 Lisboa - PORTUGAL

R.J. FROST
Chemistry Dept.
Birmingham University
Edgeboston
Birmingham - U.K.

C.J. GILLAN
Dept. Appl. Maths. and Theor. Phys.
David Bates Building
Queen's University
Belfast BT7 INN - U.K.

J. GOMEZ LLORENTE
Dept. of Chemical Physics
Weizmann Institute of Science
Rehovot
ISRAEL

D. GRIMBERT
LCAM, Univ. Paris Sud
Bat 351
91405 Orsay Cede
FRANCE

G.HAHNER
MPI für Strömungsforschung
Bunsents 10
D-3400 Goettingen
FRG

G. HOSE
Chem. Phys. Dept.
Weizmann Inst of Science
76100 Rehovot
ISRAEL

N. KATZENELLENBOGEN
Chem. Phys. Dept.
Weizmann Inst. of Science
Rehovot 76100
ISRAEL

J. KRAUSE
Chem. Phys. Dept.
Weizmann Inst. of Science
Rehovot 76100
ISRAEL

O. YLMAZ
Physics Dept.
Middle East Tech. Univ.
Ankara
TURKEY

B. LEPETIT
ER 261 CNRS
Observatoire
92195 Meudon
FRANCE

D.E. MANOLOPOULOS
Chem. Labs
University of Cambridge
Lensfield Rd
Cambridge CG2 IEW – U.K.

R.F. Dos Reis MARMONT LOBO
Dpto de Fisica
Fac. de Ciencias e Tecnicas
Univ. nova de Lisboa
2825 Monte de Caparica – PORTUGAL

F. MARTIN
Quimica Dept. CXIV
Univ. Autonoma
Cantoblanco
28049 Madrid – SPAIN

R.J. MAURER
Cyclotron Institute
Texas A and M University
College Station
Texas 77843 – USA

F.R.W. McCOURT
Chem. Dept.
Univ. of Waterloo
Waterloo, Ontario N2L 3G1
CANADA

B. McLAUGHLIN
Chem. Dept.
139 Smith Hall
University of Minnesota
Minneapolis, Minnesota 55455 – USA

D.J. McLAUGHLIN
Physics Dept.
Hartford University
200 Bloomfield ave
West Hartford, CT 06117 - USA

P. McNAUGHTEN
Applied Math. Dept.
Queens University
Belfast BT7 INN
Belfast - U.K.

A. METROPOULOS
Theor. and Phys. Chemistry
Nat. Hellenic Res. Found
48 v. Constantinou ave.
11635 Athens - GREECE

A. MIKLAVC
B. Kidric Inst. of Chemistry
Hajdrihova 19
61000 Ljubljana
YUGOSLAVIA

L.A. MORGAN
Computer Center
Royal Holloway and
Bedford New College
Egham Hill
Egham, Surrey TW20 0EX - U.K.

M. MORALDI
Ist. di El. Quantistica, CNR
Via Panciatichi, 56/30
Firenze
ITALY

S.E. NIELSEN
University of Copenhagen
Chemistry Lab III H.C Orsted Institute
Universitetsparken 5
2100 Copenhagen
DENMARK

L. NENCINI
ENEA, TIB-FIS
P.O. Box 65
00044 Frascati (Roma)
ITALY

L. OPRADOLCE
Inst de Estr. y Fis del Espacio
C.C. 67 - Suc 28
1428 Buenos Aires
ARGENTINA

S. OSS
Physics Dept.
Univ. of Trento
38050 Povo (Trento)
ITALY

A.C. PEET
Chem. Lab
University of Cambridge
Lensfield Rd
Cambridge - U.K.

M. PONT
FOM Ins for At and Mol Phys.
Postbus 41883
1009 DB Amsterdam
HOLLAND

O. RONCERO
Istituto de Estructura de la Materia
C.S.I.C.
Serrano 123
28006 Madrid - SPAIN

P.J. SALAS PERALTA
Depto Quimica
ETSI Telecomunicacion
Univ. Politec de Madrid
28040 Madrid - SPAIN

R.A. STANFIELD
Chem. Lab
University of Cambridge
Lensfield Rd
Cambridge CB2 1EW - U.K.

T. SCHOLZ
Dept. of Applied Mat. and
Theoretical Physics
The Queen's University
Belfast BT 3HY - U.K.

M. THACHUK
Dept. of Chemistry
University of Waterloo
Waterloo, Ontario N2L 3G1
CANADA

S.R. THAREJA
Chemistry Dept.
I.I.T. Kanpur
Kanpur U.P. 208016
INDIA

M. TERAO
Univ. Catholique de Louvain
Institut de Physique
Unité FYAM, Chemin du (Cyclotron 2)
Louvain-la-Neuve B-1348 - BELGIUM

K.-E. V. THYLWE
Inst. of Theoret. Phys.
University of Uppsala
Thunbergsvagen 3
S-752 38 Uppsala - SWEDEN

A. VAN DER AVOIRD
Theor. Chem. Inst.
Nijmegen University
Toornooiveld
6525 ED Nijmegen - HOLLAND

A.G. VELA
Inst. de Estructura de la Materia
CSIC
Serrano 123
28006 Madrid - SPAIN

M. VENANZI
Dept. of Chemistry
University of Rome
Città Universitaria, 00185
Rome - ITALY

N. WALET
FOM Ins for At and Mol Phys.
Postbus 41883
1009 DB Amsterdam
HOLLAND

C.A. WEATHERFORD
Phys. Dept Box 981
Florida A and M University
Tallahassee, Florida 32307
USA

K. WEIDE
MPI für Strömungsforschung
Bunsen st 1
3400 D Goettingen
FRG

INDEX